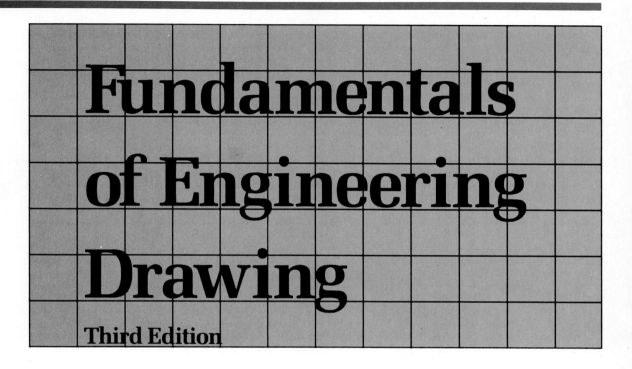

Fundamentals of Engineering Drawing

Third Edition

Cecil Jensen
Former Technical Director
R. S. McLaughlin Collegiate and
Vocational Institute
Oshawa, Ontario, Canada

Jay D. Helsel
Professor and Chairman
Department of Industry
and Technology
California University of Pennsylvania
California, Pennsylvania

GLENCOE

Macmillan/McGraw-Hill

New York, New York
Columbus, Ohio
Mission Hills, California
Peoria, Illinois

Text Designer: Caliber Design Planning, Inc.
Cover Designer: Caliber Design Planning, Inc.

Library of Congress Cataloging-in-Publication Data

Jensen, Cecil Howard, 1925–
    Fundamentals of engineering drawing / Cecil Jensen, Jay D. Helsel.
  —3rd ed.
     p. cm.
    Consists of chapters 1 through 16 of Engineering drawing and
design, fourth ed.
    Includes bibliographical references.
    ISBN 0-07-032560-X
    1. Mechanical drawing. I. Helsel, Jay D. II. Jensen, Cecil
Howard, 1925– Engineering drawing and design. III. Title
T353.J48   1990
604.2—dc20

89-12190
CIP

Fundamentals of Engineering Drawing, Third Edition
Imprint 1993
Copyright © 1990 by the Glencoe Division of Macmillan/McGraw-Hill School Publishing
Company. All rights reserved. Copyright © 1990, 1985, 1979 by McGraw-Hill, Inc. All rights
reserved. This work consists of selected chapters from *Engineering Drawing and Design,* Fourth
Edition. © 1990 by McGraw-Hill, Inc. Printed in the United States of America. Except as
permitted under the United States Copyright Act of 1976, no part of this publication may be
reproduced or distributed in any form or by any means, or stored in a database or retrieval system,
without the prior written permission of the publisher.

Send all inquiries to:
GLENCOE DIVISION
Macmillan/McGraw-Hill
936 Eastwind Drive
Westerville, Ohio 43081

3 4 5 6 7 8 9 10 11 12 13 14 15   VH   00 99 98 97 96 95 94 93

ISBN 0-07-032560-X

C O N T E N T S

P R E F A C E

The third edition of *Fundamentals of Engineering Drawing* builds upon the success of previous editions. Not all students will choose a drafting career. However, an understanding of the language of graphics is necessary for anyone who plans to work in any of the fields of technology.

Drafting, like all technical areas, is constantly changing. The computer has revolutionized the way in which drawings and parts are made. With the advent of the integrated chip, the use of the computer is the most significant development in design and drafting to occur in recent years. Since computer-aided drafting (CADD) is now considered an essential part of any drafting course, CADD has been added to this text.

Because manuals are readily available to give the details of operation of computers and their production of drawing, the emphasis has been placed on the application of CADDs. Chapter 2 covers CADD equipment, linework, and lettering. In other chapters, basic CADD commands pertaining to the particular topic follow the information on manual drafting.

In addition, a complete chapter on Geometric Tolerancing has been added to this text. Whether geometric tolerancing is being applied to a drawing or is being used in interpreting the requirements for the production of a part, a clear understanding of the meaning of the term is essential.

In order to incorporate the important topics of CADD and geometric tolerancing into this text, the chapters on welding, drawing for numerical control, and design concepts were eliminated from this edition of *Fundamentals of Engineering Drawing*.

In this new edition, the authors have made every effort to translate the most current technical information available into the most usable form from the standpoint of both teacher and student. The latest developments and current practices in all areas of graphic communication, functional drafting, materials representation, shop processes, and geometric tolerancing using U.S. Customary and metric drawing practices have been incorporated into this text in a manner that synthesizes, simplifies, and converts complex drafting standards and procedures into understandable instructional units. Extensive author research and visits to drafting rooms throughout the country and the adoption of the latest ANSI drawing standards have resulted in a presentation that combines current drafting practices with practical pedagogical techniques to produce the most efficient learning system yet designed for the instruction of engineering drawing.

Every chapter in *Fundamentals of Engineering Drawing* is divided into a number of single-concept units, each with its own objective, instruction, examples, review, and assignments in both U.S. Customary and metric units of measurements. This organization provides the student with a logical sequence of experiences which can be adjusted to individual needs; it also permits maximum efficiency in learning essential concepts. Development of each unit is from the simple to the complex and from the familiar to the unfamiliar. Checkpoints are included to provide maximum reinforcement at each level.

Both metric and U.S. Customary conventions as utilized on a practical level of American industry have been incorporated into this text. Problems are stated in conventional and metric form. Thus, the text may be used in a completely conventional (Customary) course, in a fully metric-oriented course, or in a course that utilizes both metric and conventional systems. The teacher may also customize the course by selecting appropriate problems or materials to emphasize or deemphasize any degree of metrication.

A solutions manual which provides complete solutions to all graphic problems found in the text is also available.

The authors would like to thank the many users of the previous edition of the textbook for their thoughtful and useful comments.

Cecil Jensen
Jay D. Helsel

CECIL JENSEN is the author or coauthor of many successful technical books, including *Engineering Drawing and Design, Fundamentals of Engineering Drawing, Fundamentals of Engineering Graphics* (formerly called *Drafting Fundamentals), Interpreting Engineering Drawings, Architectural Drawing and Design for Residential Construction, Home Planning and Design,* and *Interior Design.* Some of these books are printed in three languages and are used in many countries.

He has twenty-seven years of teaching experience in mechanical and architectural drafting and was a technical director for a large vocational school in Canada.

Before entering the teaching profession, Mr. Jensen gained several years of design experience in the industry. He has also been responsible for the supervision of the teaching of technical courses for General Motors apprentices in Oshawa, Canada.

He is a member of the Canadian Standards Committee (CSA) on Technical Drawings (which includes both mechanical and architectural drawing) and is chairman of the Committee on Dimensioning and Tolerancing. Mr. Jensen is Canada's representative on the American (ANSI) Standards for Dimensioning and Tolerancing and has represented Canada at two world (ISO) conferences in Oslo (Norway) and Paris on the standardization of technical drawings.

He took an early retirement from the teaching profession in order to devote his full attention to writing.

JAY D. HELSEL is a professor and department chairman of industry and technology at California University of Pennsylvania. He completed his undergraduate work at California State College and was awarded a master's degree from Pennsylvania State University. He has done advanced graduate work at West Virginia and at the University of Pittsburgh, where he completed a doctoral degree in educational communications and technology. In addition, Dr. Helsel holds a certificate in airbrush techniques and technical illustration from the Pittsburgh Art Institute.

He has worked in industry and has taught drafting, metalworking, woodworking, and a variety of laboratory and professional courses at both the secondary and college levels. During the past twenty-five years, he has also worked as a free-lance artist and illustrator. His work appears in many technical publications.

Dr. Helsel is coauthor of *Engineering Drawing and Design, Fundamentals of Engineering Drawing, Programmed Blueprint Reading,* and *Mechanical Drawing.* He is also the author of the series *Mechanical Drawing Film Loops.*

DRAFTING UPDATE (PRESENT DRAWING PRACTICES)		
FEATURE	**SYMBOL**	**ANSI PUBLICATION Y14.5M—1982 (EXCEPT WHERE NOTED, REFER TO CLAUSE NO.)**
LINES (THREE LINE WIDTHS NOW REPLACED BY TWO LINE WIDTHS.)	——— THICK / ——— THIN	Y14.2M—1979
LETTERING (TWO APPROVED STYLES. HEIGHT OF LETTERING DEPENDENT ON DRAWING SIZE.)		Y14.2M—1979
MILLIMETER DIMENSIONING PRACTICES		CLAUSE 1.6.1
METRIC LIMITS AND FITS		B 4.2—1978
DIAMETER SYMBOL (NOW PRECEDES THE DIAMETER VALUE. THE SYMBOL REPLACES THE ABBREVIATION DIA)	Ø	CLAUSE 1.8.1
RADIUS SYMBOL (NOW PRECEDES THE RADIUS VALUE)	R	CLAUSE 1.8.2
REFERENCE DIMENSION	(8.6)	CLAUSE 1.7.6
SURFACE TEXTURE SYMBOL	√ √ √ √	Y14.36—1978
SPECIFYING REPETITIVE FEATURES	X	CLAUSE 1.9.5
COUNTERBORE OR SPOTFACE	⌴	CLAUSE 3.3.10
COUNTERSINK	⌵	CLAUSE 3.3.11
DEPTH	↧	CLAUSE 3.3.12
CONICAL TAPER	—▷ 0.2 : 1	CLAUSE 2.13
FLAT TAPER	—◁ 0.15 : 1	CLAUSE 2.14
SYMMETRICAL OUTLINES		CLAUSE 1.8.8
ALL AROUND	◯	CLAUSE 3.4.2.3
DIMENSIONING CHORDS, ANGLES, AND ARCS	⊢50⊣ ⊢⌒50⌣⊣ 60°	CLAUSE 1.8.3
NOT TO SCALE DIMENSION	◄——1̲2̲0̲——►	CLAUSE 1.7.9
FEATURE CONTROL FRAME (FORMERLY CALLED FEATURE CONTROL SYMBOL. ORDER OF SEQUENCE CHANGED)	⊕ Ø 0.1 A	CLAUSE 3.4.2
DATUM TARGET SYMBOL	Ø10 / A 3	CLAUSE 3.3.3 AND 4.5.1
GEOMETRIC CHARACTERISTIC SYMBOLS	STRAIGHTNESS —	CLAUSE 6.4.1
	FLATNESS ▱	CLAUSE 6.4.2
	PARALLELISM //	CLAUSE 6.6.3
	SYMMETRY ⊕	CLAUSE 5.12
	CIRCULAR RUNOUT ↗	CLAUSE 6.7.2.1
	TOTAL RUNOUT ↗↗	CLAUSE 6.7.2.2

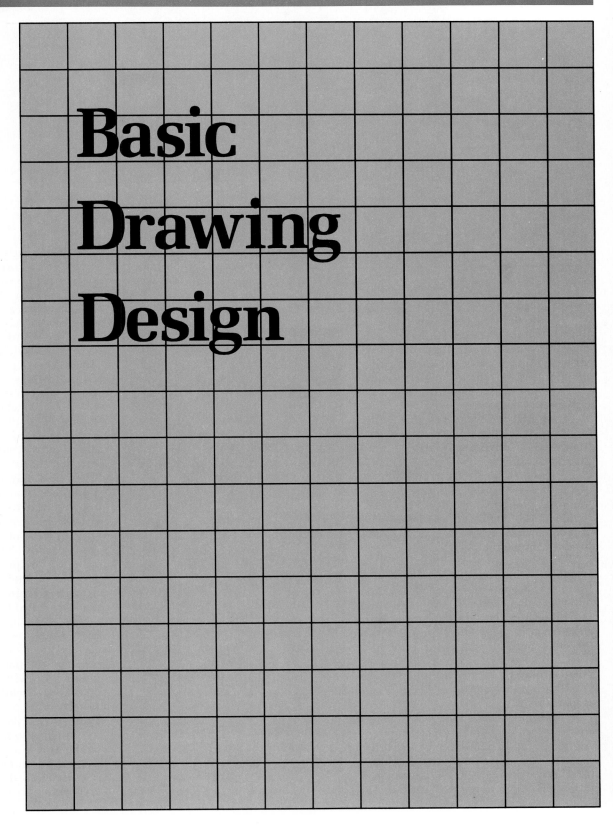

Basic Drawing Design

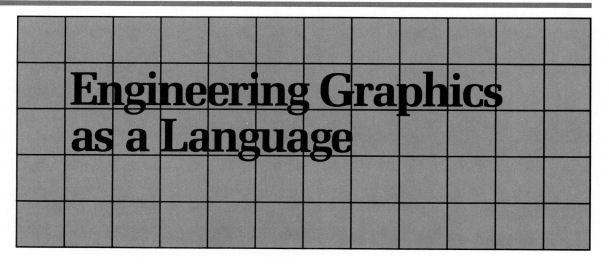

Engineering Graphics as a Language

UNIT 1-1
The Language of Industry

Since earliest times people have used drawings to communicate and record ideas so that they would not be forgotten. Figure 1-1-1 shows builders in an early civilization following technical drawings to construct a building.

The word *graphic* means dealing with the expression of ideas by lines or marks impressed on a surface. A drawing is a graphic representation of a real thing. Drafting, therefore, is a graphic language, because it uses pictures to communicate thoughts and ideas. Because these pictures are understood by people of different nations, drafting is referred to as a "universal language."

Drawing has developed along two distinct lines, with each form having a different purpose. On the one hand, artistic drawing is concerned mainly with the expression of real or imagined ideas of a cultural nature. Technical drawing, on the other hand, is concerned with the expression of technical ideas or ideas of a practical nature, and it is the method used in all branches of technical industry.

Even highly developed word languages are inadequate for describing the size, shape, and relationship of physical objects. For every manufactured object there are drawings that describe its physical shape completely and accurately, communicating engineering concepts to manufacturing. For this reason, drafting is referred to as the "language of industry."

Drafters translate the ideas, rough

Fig. 1-1-1 Early use of drawings in constructing a building.

sketches, specifications, and calculations of engineers, architects, and designers into working plans which are used in making a product. See Fig. 1-1-2. Drafters may calculate the strength, reliability, and cost of materials. In their drawings and specifications, they describe exactly what materials workers are to use on a particular job. To prepare their drawings, drafters use either computer-aided drafting and design (CADD) systems or manual drafting instruments such as compasses, dividers, protractors, templates, and triangles, as well as drafting machines that combine the functions of several devices. They also may use engineering handbooks, tables, and calculators to assist in solving technical problems.

Drafters are often classified according to their type of work or their level of responsibility. Senior drafters (designers) take the preliminary information provided by engineers and architects to prepare design "layouts" (drawings made to scale of the object to be built). Detailers (junior drafters) make drawings of each part shown on the layout, giving dimensions, material, and any other information necessary to make the detailed drawing clear and complete. Checkers carefully examine drawings for errors in computing or recording sizes and specifications.

Drafters also may specialize in a particular area of drafting, such as mechanical, electrical, electronics, aeronautics, structural design, piping, or architecture.

TYPICAL BRANCHES OF ENGINEERING GRAPHICS	ACTIVITIES	PRODUCTS	SPECIALIZED AREAS
MECHANICAL 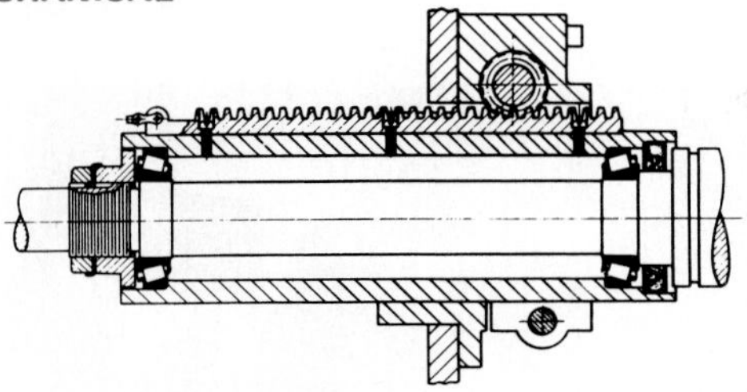	DESIGNING TESTING MANUFACTURING MAINTENANCE CONSTRUCTION	MATERIALS MACHINES DEVICES	POWER GENERATION TRANSPORTATION MANUFACTURING POWER SERVICES ATOMIC ENERGY MARINE VESSELS
ARCHITECTURAL 	PLANNING DESIGNING SUPERVISING	BUILDINGS ENVIRONMENT LANDSCAPE	COMMERCIAL BUILDINGS RESIDENTIAL BUILDINGS INSTITUTIONAL BUILDINGS ENVIRONMENTAL SPACE FORMS
ELECTRICAL 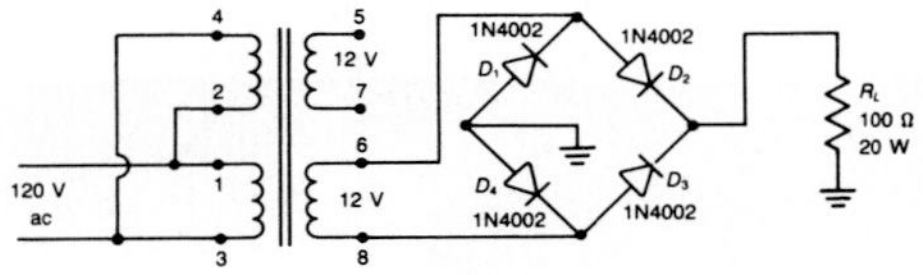	DESIGNING DEVELOPING SUPERVISING PROGRAMMING	COMPUTERS ELECTRONICS POWER ELECTRICAL	POWER GENERATION POWER APPLICATION TRANSPORTATION ILLUMINATION INDUSTRIAL ELECTRONICS COMMUNICATIONS INSTRUMENTATION MILITARY ELECTRONICS
AEROSPACE 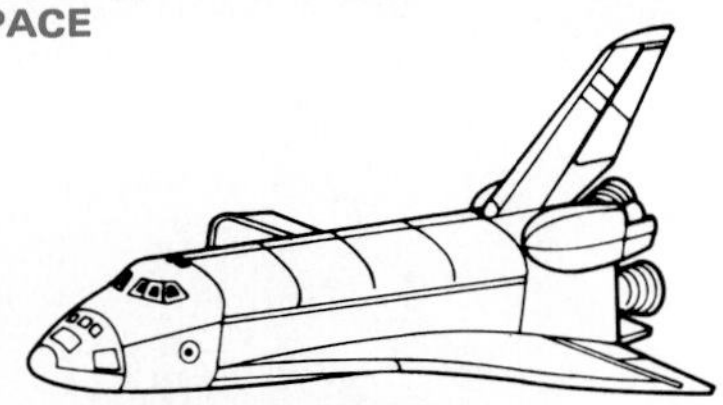	PLANNING DESIGNING TESTING	MISSILES PLANES SATELLITES ROCKETS	AERODYNAMICS STRUCTURAL DESIGN INSTRUMENTATION PROPULSION SYSTEMS MATERIALS RELIABILITY TESTING PRODUCTION METHODS
TECHNICAL ILLUSTRATING 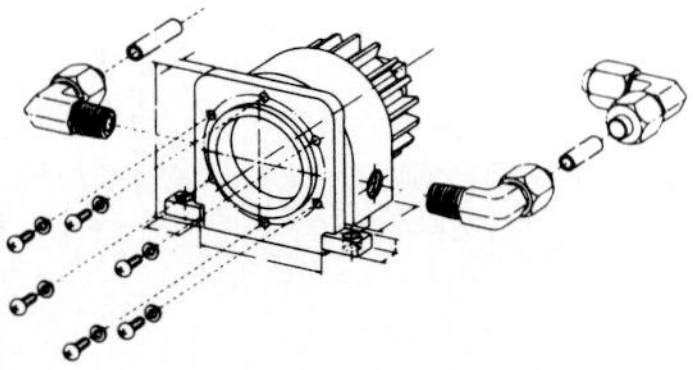	PROMOTION DESIGNING ILLUSTRATING	CATALOGUES MAGAZINES DISPLAYS	NEW PRODUCTS ASSEMBLY INSTRUCTIONS PRESENTATIONS COMMUNITY PROJECTS RENEWAL PROGRAMS

Fig. 1-1-2 Various fields of drafting.

DRAWING STANDARDS

Throughout the long history of drafting, many drawing conventions, terms, abbreviations, and practices have come into common use. It is essential that different drafters use the same practices if drafting is to serve as a reliable means of communicating technical theories and ideas.

In the interest of efficient communication, the American National Standards Institute (ANSI) has adopted a set of drafting standards which are recommended for drawing practice in all fields of engineering and are used and explained throughout this text. These standards apply primarily to end-product drawings. End-product drawings usually consist of detail or part drawings and assembly or subassembly drawings, and are not intended to fully cover other supplementary drawings such as checklists, item lists, schematic diagrams, electrical wiring diagrams, flowcharts, installation drawings, process drawings, architectural drafting, and pictorial drawing.

The information and illustrations shown have been revised to reflect current industrial practices in the preparation and handling of engineering documents. The increased use of reduced-size copies of engineering drawings made from microfilm and the reading of microfilm require the proper preparation of the original engineering document regardless of whether the drawing was made manually or by computer-aided drafting (CAD). All future drawings should be prepared for eventual photographic reduction or reproduction. The observance of the drafting practices described in this text will contribute substantially to the improved quality of photographically reproduced engineering drawings.

THE DRAFTING OFFICE

The drafting office is the focal point for all engineering work. The drawings produced are the main method of communication between all persons concerned with the design, manufacture, and assembly of products.

The last 20 years have brought great changes to the drafting room. Its physical appearance, furnishings, even its drafters and engineers have moved quickly from their battered domain of old into the information age. This era is often referred to as the "technical revolution."

These changes were brought about largely by the integrated circuit chip. It has, in fact, revolutionized the way we work and play. This era has seen dramatic changes in worldwide communications at all levels—personal, professional, industrial—and in every facet of modern-day life. The microchip is on our wrists (quartz digital watches). It is used to help solve math problems (hand-held calculators). It entertains (video games), and helps to run businesses (computers). The technical changes it has launched have affected many careers, and retraining to upgrade job skills has become commonplace. Drafting and design have been at the forefront of the changes. CAD and CADD are familiar acronyms that have swept through the profession.

Drafting room technology has progressed at the same rapid pace as the economy of our country. Many changes have taken place in the modern drafting room as compared to a typical drafting room scene before CAD as shown in Fig. 1-1-3. Not only is there far more equipment, but it is of much higher quality. Noteworthy progress has been and continues to be made.

(A) THE DRAFTING OFFICE AT THE TURN OF THE CENTURY. (Bettman Archive, Inc.)

(B) MANUAL DRAFTING. (Vemco Corp.)

(C) CAD DRAFTING. (Prime/Computervision)
Fig. 1-1-3 Evolution of the drafting office.

UNIT 1-2
Careers in Engineering Graphics

THE STUDENT

While students are learning basic drafting skills (Fig. 1-2-1), they will also be increasing their general technical knowledge, learning about some of the engineering and manufacturing processes involved in production. Not all students will choose a drafting career. However, an understanding of this graphic language is necessary for anyone who works in any of the fields of technology and is essential for those who plan to enter the skilled trades or become a technician, technologist, or engineer.

Because a drawing is a set of instructions that the worker will follow, it must be accurate, clear, correct, and complete. When drawings are made with the use of instruments, they are called *instrumental* (or *manual*) *drawings*. When they are developed with the use of a computer, they are known as *computer-aided drawings*. When made without instruments or the aid of a computer, drawings are referred to as *sketches*. The ability to sketch ideas and designs and to produce accurate drawings is a basic part of drafting skills.

In everyday life, a knowledge of engineering graphics is helpful in understanding house plans and assembly, maintenance, and operating instructions for many manufactured or hobby products.

PLACES OF EMPLOYMENT

There are over 400,000 people working in drafting positions in the United States. A significant number are women. About 9 out of 10 drafters are employed in private industry. Manufacturing industries that employ a large number of drafters are those making machinery, electrical equipment, transportation equipment, and fabricated metal products. Nonmanufacturing industries employing a large number of drafters are engineering and architectural consulting firms, construction companies, and public utilities.

Drafters also work for the government; the majority work for the armed services. Drafters employed by state and local governments work chiefly for highway and public works departments. Several thousand drafters are employed by colleges and universities and by nonprofit organizations.

TRAINING, QUALIFICATIONS, ADVANCEMENT

Any person interested in becoming a drafter can acquire the necessary training from a number of sources, including junior and community colleges, extension divisions of universities, vocational/technical schools, and correspondence schools. Others may qualify for drafting positions through on-the-job training programs combined with part-time schooling.

The prospective drafter's training in post-high school drafting programs should include courses in mathematics and physical sciences, as well as in CAD and CADD. Studying fabrication practices and learning some trade skills also are helpful, since many higher-level drafting jobs require knowledge of manufacturing or construction methods. This is especially true in the mechanical discipline due to the implementation of CAD/CAM (computer-aided drafting–computer-aided manufacturing). Many technical schools offer courses in structural design, strength of materials, physical metallurgy, CAM, and robotics.

As drafters gain skill and experience, they may advance to higher-level positions as checkers, senior drafters, designers, supervisors, and managers.

(A)

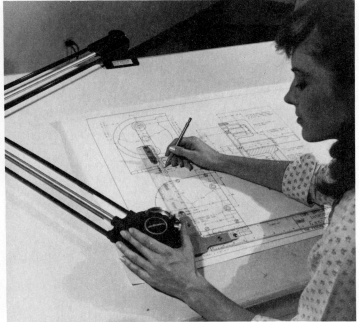
(B)

Fig. 1-2-1 School drafting room. (A) CAD drafting (Courtesy Tektronix). (B) Manual drafting (Bruning).

See Fig. 1-2-2. Drafters who take additional courses in engineering and mathematics are often able to qualify for engineering positions.

Qualifications for success as a drafter include the ability to visualize objects in three dimensions and the development of problem-solving design techniques. Since the drafter is the one who finalizes the details on drawings, attentiveness to detail is a valuable asset.

EMPLOYMENT OUTLOOK

Employment opportunities for drafters are expected to rise rapidly as a result of the increasingly complex design problems of modern products and processes. In addition, computerization is creating many new products and support and design occupations including drafters will continue to grow. On the other hand, photoreproduction of drawings and expanding use of CAD have eliminated many routine tasks done by drafters. This development will probably reduce the need for some less skilled drafters.

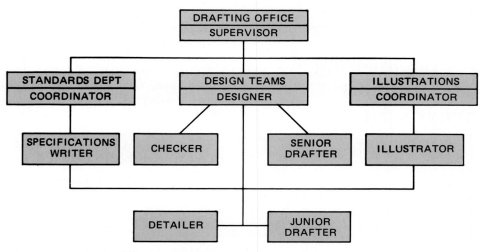

Fig. 1-2-2 Position within the drafting office.

References

1. Charles Bruning Co.
2. *Occupational Outlook Handbook.*

The Drafting Office

U N I T 2 - 1
The Drafting Office

The drafting office is the starting point for all engineering work. Its product, the *engineering drawing*, is the main method of communication between all people concerned with the design and manufacture of parts. Therefore the drafting office must provide accommodations and equipment for the drafters, from designer and checker to detailer or tracer; for the personnel who make copies of the drawings and file the originals; and for the secretarial staff who assist in the preparation of the drawings. Typical drafting workstations are shown in Figs. 2-1-1 and 2-1-2.

Fewer engineering departments rely on manual drafting methods. Computers are replacing drafting boards at a steady pace. In the majority of cases, this is all that is necessary. Equipment for manual and CADD drafting is varied and is steadily being improved. Where a high volume of finished or repetitive work is not necessary, manual drafting does the job adequately and inexpensively. However, for increased productivity CADD is necessary. CADD can serve as a full partner in the design process, enabling the designer to do jobs that are simply not possible or feasible with manual equipment.

Besides increasing the speed with which a job is done, a CAD system can

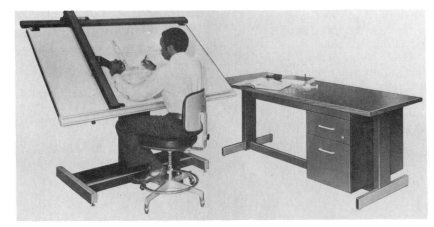

Fig. 2-1-1 Manual drafting workstation.

Fig. 2-1-2 Typical micro-CADD workstation. (Telesis)

perform many of the tedious and repetitive skills ordinarily required of a drafter. This includes such skills as lettering and differentiation of line weights. CAD thus frees the drafter to be more creative while it quickly performs the mundane taks of drafting. It has proved to be, conservatively speaking, at least a 30 percent improvement in production in terms of time spent on drawing.

A CAD system by itself cannot create. A drafter must create the drawing, and thus a strong design and drafting background remains essential.

It may not be practical to handle all the workload in a design or drafting office on a CAD system. Although most design and drafting work most certainly can benefit from it, some functions will continue to be performed by traditional means. Thus some companies will use CAD for only a portion of the workload. Still others use CAD almost exclusively. Whatever the percentage of CAD use, one fact is certain. It has had, and will continue to have, a dramatic effect on design and drafting careers.

Once a CAD system has been installed, the required personnel must be hired or trained. Trained personnel generally originate from one of three popular sources: educational institutions, CAD equipment manufacturer training courses, and individual company programs.

UNIT 2-2
Manual Drafting Equipment

Over the years, the designer's chair and drafting table have evolved into a drafting station which provides a comfortable, integrated work area. Yet much of the equipment and supplies employed years ago are still in use today, although they have been vastly improved.

DRAFTING FURNITURE

Special tables and desks are manufactured for use in single-station or multistation design offices. Typical are desks with attached drafting boards (Fig. 2-2-1). The boards may be used by the occupant of the desk to which it is attached, in which case it may swing out of the way when not in use, or may be reversed for use by the person in the adjoining station.

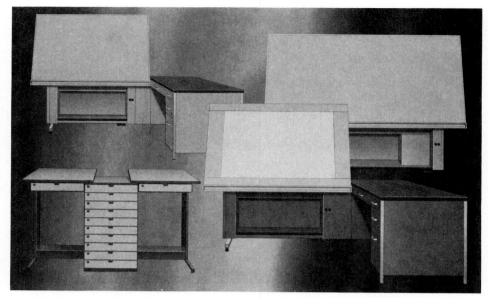

Fig. 2-2-1 Drafting tables are available in a variety of sizes and styles. (A. M. Bruning Co.)

In addition to such special workstations, a variety of individual desks, chairs, tracing tables, filing cabinets, and special storage devices for drawings are available (Fig. 2-2-2).

The simplest manually adjustable tables typically consist of a hinged surface riding on a vertical rod secured by a hand knob screw. The hand knob screw is loosened, the top is set at the desired angle, and the hand knob screw is retightened.

DRAFTING EQUIPMENT

See Fig. 2-2-3 for a variety of drafting equipment.

Fig. 2-2-2 Drafting workstation.

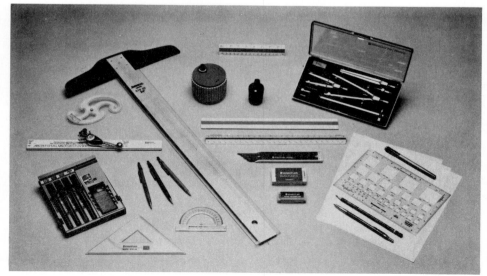

Fig. 2-2-3 Drafting equipment. (Staedtler-Mars)

Drawing Boards

The drawing sheet is attached directly to the surface of a drafting table or a portable drawing board (Fig. 2-2-4). Drafting boards are used in schools and at home and generally have a smaller work surface than what is found on drafting tables. They are designed to stay flat and have straight guiding edges. Most professional drafting tables have a special overlay drawing surface material that "recovers" from minor pinholes and dents. One tradename material is Borco. Some boards also come with magnetic tops and special triangles for automatic parallel lines.

Fig. 2-2-4 Drawing boards. (Teledyne Post.)

Drafting Machines

In the well-equipped engineering department, where the designer is expected to do accurate drafting, the T square has been replaced largely by the drafting machine. This device, which combines the functions of T square, triangles, scale, and protractor, is estimated to save up to 50 percent of the user's time. All positioning is done with one hand, while the other hand is free to draw.

Drafting machines may be attached to any drafting board or table. Two types are currently available. In the track type, a vertical beam carrying the drafting instruments rides along a horizontal beam fastened to the top of the table. In the arm, or elbow type (Fig. 2-2-5), two arms pivot from the top of the machine and are relative to each other.

Track-type drafting machines are especially suitable for long-line work and large drawings.

Fig. 2-2-5 Arm-type drafting machine. (Keuffel & Esser Co.)

Parallel Slide

The parallel slide is used in drawing horizontal lines and for supporting triangles when vertical and sloping lines are being drawn. (See Fig. 2-2-6.) It is fastened on each end to cords, which pass over pulleys. This arrangement permits movement up and down the board while maintaining the parallel slide in a horizontal position.

Fig. 2-2-6 Drafting table with parallel slide.

T Squares

The T square (Fig. 2-2-7) performs the same function as the parallel slide. T squares are made of various materials,

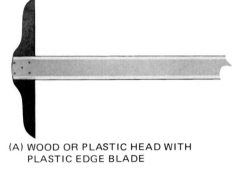

(A) WOOD OR PLASTIC HEAD WITH PLASTIC EDGE BLADE

(B) STEEL BLADE

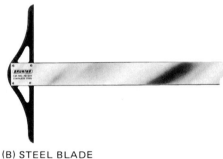

(C) ADJUSTABLE HEAD WITH PLASTIC EDGE BLADE

Fig. 2-2-7 T squares are available in various styles and materials. (A. M. Bruning Co.)

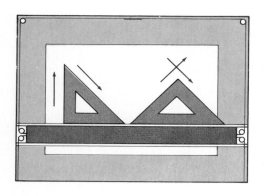

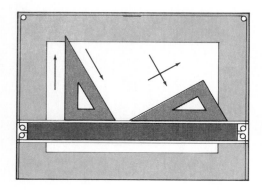

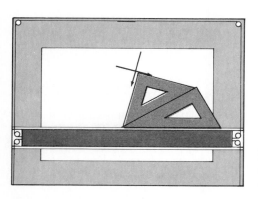

Fig. 2-2-8 The triangles.

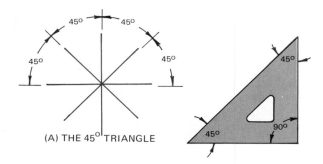

(A) THE 45° TRIANGLE

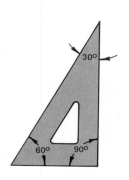

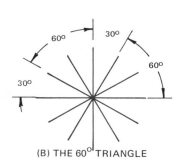

(B) THE 60° TRIANGLE

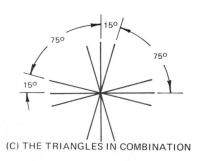

(C) THE TRIANGLES IN COMBINATION

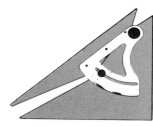

Fig. 2-2-10 Adjustable triangle.

the more popular being plastic-edged wood blades with heads made from wood or plastic.

The head of the T square is placed on the left side of a drawing board for use by right-handed people and on the right side of the drawing board for use by left-handed people.

Triangles

Triangles are used together with the parallel straightedge or T square when you are drawing vertical and sloping lines (Fig. 2-2-8). The triangles most commonly used are the 30/60° and the 45° triangles. Singly or in combination, these triangles can be used to form

angles in multiples of 15°. For other angles, the protractor (Fig. 2-2-9) is used. All angles can be drawn with the adjustable triangle (Fig. 2-2-10); this instrument replaces the two common triangles and the protractor.

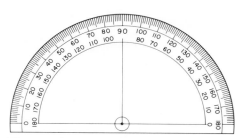

Fig. 2-2-9 A protractor is used to lay out or measure angles.

Scales

"Scale" may refer to the measuring instrument or the size to which a drawing is to be made.

Measuring Instrument Shown in Fig. 2-2-11 are the common shapes of

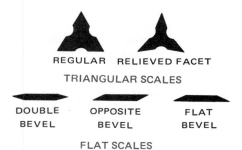

Fig. 2-2-11 End view shapes of scales.

scales used by drafters to make measurements on their drawings. Scales are used only for measuring and are not to be used as a straightedge when drawing lines. It is important that drafters draw accurately to scale. The scale to which the drawing is made must be given in the title block or strip.

Sizes to which Drawings Are Made When objects are drawn at their actual size, the drawing is called *full scale* or *scale 1:1*. Many objects, however, such as buildings, ships, or airplanes, are too large to be drawn full scale, so they must be drawn to a reduced scale. An example would be the drawing of a house to a scale of ¼ in. = 1 ft.

Frequently, objects such as small watch parts are drawn larger than their actual size so that their shape can be seen clearly. Such a drawing has been drawn to an enlarged scale. The minute hand of a wristwatch, for example, could be drawn to a scale of 5:1.

Many mechanical parts are drawn to half scale, 1:2, and quarter scale, 1:4 or nearest metric scale 1:5. Notice that the scale is expressed as an equation. With reference to the 1:5 scale the left side of the equation represents a unit of the size drawn; on the right side, a unit of the actual drawing equals five units of measurement of the actual object.

Scales are made with a variety of combined scales marked on their surfaces. This combination of scales spares the drafter the necessity of calculating the sizes to be drawn when working to a scale other than full size.

Metric Scales The linear unit of measurement for mechanical drawings is the millimeter. Scale multipliers and divisors of 2 and 5 are recommended, which give the scales shown in Fig. 2-2-12.

The numbers shown indicate the difference in size between the drawing and the actual part. For example, the ratio 10:1 shown on the drawing means that the drawing is 10 times the actual size of the part, whereas a ratio of 1:5 on the drawing means the object is 5 times as large as it is shown on the drawing.

The units of measurement for architectural drawings are the meter and millimeter. The same scale multipliers and divisors as used for mechanical drawings are used for architectural drawings.

U.S. Customary Scales

Inch Scales There are three types of scales which show various values that

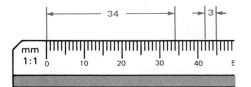
1:1 SCALE (1 mm DIVISIONS)

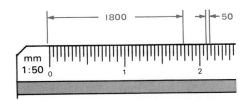
1:2 SCALE (2 mm DIVISIONS)

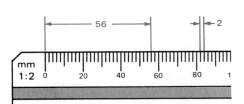
1:5 SCALE (5 mm DIVISIONS)

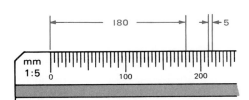
1:50 SCALE (50 mm DIVISIONS)

Fig. 2-2-13 Inch scales.

ENLARGED	SIZE AS	REDUCED
1000 : 1	1 : 1	1 : 2
500 : 1		1 : 5
200 : 1		1 : 10
100 : 1		1 : 20
50 : 1		1 : 50
20 : 1		1 : 100
10 : 1		1 : 200
5 : 1		1 : 500
2 : 1		1 : 1000

Fig. 2-2-12 Metric scales.

are equal to 1 inch (in.) (Fig. 2-2-13). They are the decimal inch scale, the fractional inch scale, and the scale which

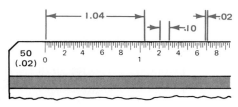
DECIMAL INCH SCALE (FULL SIZE)

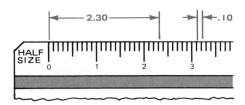
DECIMAL INCH SCALE (HALF SIZE)

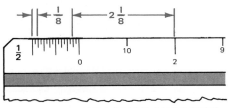
FRACTIONAL INCH SCALE (HALF SIZE)

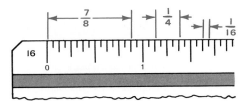
FRACTIONAL INCH SCALE (FULL SIZE)

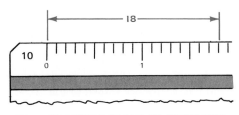
CIVIL ENGINEER SCALE (10 DIVISIONS)

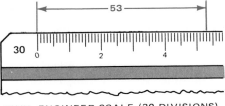
CIVIL ENGINEER SCALE (30 DIVISIONS)

has divisions of 10, 20, 30, 40, 50, 60, and 80 parts to the inch. The last scale is known as the civil engineer's scale. It is used for making maps and charts. The divisions or parts of an inch can be used to represent feet, yards, rods, or miles. This scale is also useful in mechanical drawing when the drafter is dealing with decimal dimensions.

On fractional inch scales, multipliers or divisors of 2, 4, 8, and 16 are used, offering such scales as full size, half size, and quarter size.

Foot Scales These scales are used mostly in architectural work. See Fig. 2-2-14. They differ from the inch scales

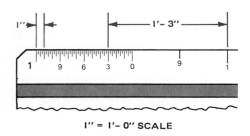

1″ = 1′– 0″ SCALE

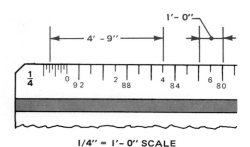

1/4″ = 1′– 0″ SCALE

Fig. 2-2-14 Foot and inch scales.

in that each major division represents a foot, not an inch, and the end units are subdivided into inches or parts of an inch. The more common scales are ⅛ in. = 1 ft, ¼ in. = 1 ft, 1 in. = 1 ft, and 3 in. = 1 ft. The most commonly used inch and foot scales are shown in Fig. 2-2-15.

Compasses

The compass is used for drawing circles and arcs. Several basic types and sizes are available (Fig. 2-2-16).

- *Friction head compass*, standard in most drafting sets.
- *Bow compass*, which operates on the jackscrew or ratchet principle by turning a large knurled nut.
- *Drop bow compass*, mostly used for drawing small circles. The center rod contains the needle point and remains stationary while the pencil or pen leg revolves around it. See Fig. 2-2-17.

DECIMALLY DIMENSIONED DRAWINGS	FRACTIONALLY DIMENSIONED DRAWINGS	DIMENSIONED IN FEET AND INCHES	
		SCALE	EQUIVALENT RATIO
10 : 1	8 : 1	6 IN.= 1 FT	1 : 2
5 : 1	4 : 1	3 IN.= 1 FT	1 : 4
2 : 1	2 : 1	1½ IN.= 1 FT	1 : 8
1 : 1	1 : 1	1 IN.= 1 FT	1 : 12
1 : 2	1 : 2	¾ IN.= 1 FT	1 : 16
1 : 5	1 : 4	½ IN.= 1 FT	1 : 24
1 : 10	1 : 8	⅜ IN.= 1 FT	1 : 32
1 : 20	1 : 16	¼ IN.= 1 FT	1 : 48
ETC.	ETC.	3/16 IN.= 1 FT	1 : 64
		⅛ IN.= 1 FT	1 : 96
		1/16 IN.= 1 FT	1:192

Fig. 2-2-15 Commonly used foot and inch scales.

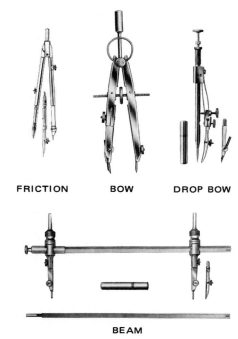

FRICTION   BOW   DROP BOW

BEAM

Fig. 2-2-16 Compasses. (Keuffel & Esser Co.)

- *Beam compass*, a bar with an adjustable needle and pencil-and-pen attachment for drawing large arcs or circles.
- *Circuit scribing instrument*, a modified drop bow compass, used to cut terminal pads and prepare printed-circuit layouts on scribe coat film.

The bow compass is adjusted by turning a screw whose knurled head is

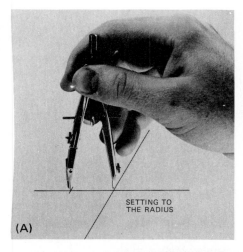

(A)   SETTING TO THE RADIUS

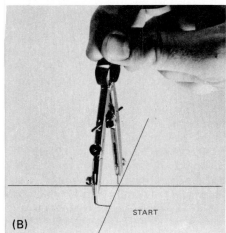

(B)   START

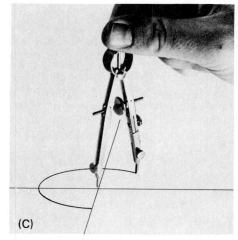

(C)

Fig. 2-2-17 Adjusting the radius for the bow lead compass. (Keuffel & Esser Co.)

located either in the center or to one side. The bow compass can be used and adjusted with one hand as shown in Fig. 2-2-18. The proper technique is:

1. Adjust the compass to the correct radius.

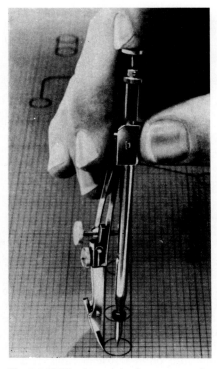

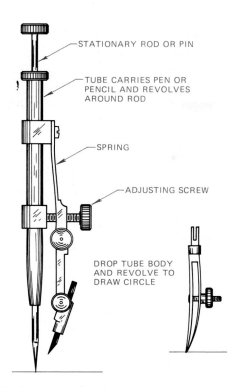

STATIONARY ROD OR PIN

TUBE CARRIES PEN OR PENCIL AND REVOLVES AROUND ROD

SPRING

ADJUSTING SCREW

DROP TUBE BODY AND REVOLVE TO DRAW CIRCLE

Fig. 2-2-18 The drop-spring bow compass is used for drawing very small circles.

2. Hold the compass between the thumb and finger.
3. With greater pressure on the leg with the needle located on the intersection of the center lines, rotate the compass in a clockwise direction. The compass should be slightly tipped in the direction of motion.

Dividers

Lines are divided and distances transferred (moved from one place to another) with dividers. The basic types of dividers are shown in Fig. 2-2-19.

Dividers have a steel pin insert in each leg and come in a variety of sizes and designs, similar to the compasses. A compass can be used as a divider by replacing its lead point with a steel pin.

Drawing Instrument Sets

Many drafters have a complete drawing set, which usually includes several compasses and dividers with extension attachments for making inked drawings (Fig. 2-2-20).

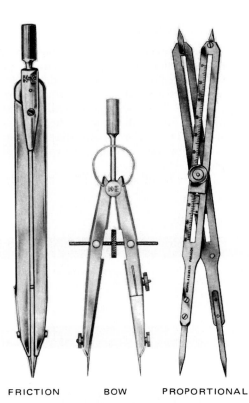

FRICTION BOW PROPORTIONAL

(A) TYPES OF DIVIDERS
Fig. 2-2-19 Dividers. (Keuffel & Esser Co.)

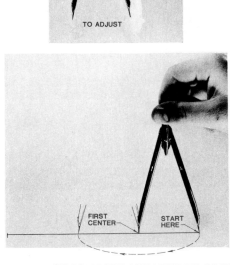

TO ADJUST

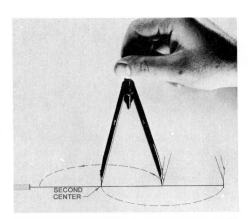

SECOND CENTER

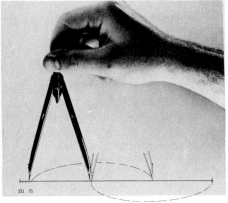

FIRST CENTER   START HERE

m n

(B) DIVIDERS ARE USED TO DIVIDE AND TO TRANSFER DISTANCES

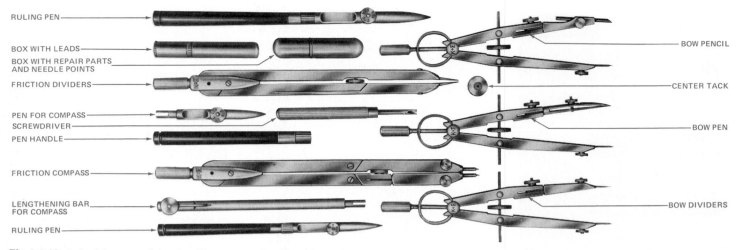

RULING PEN

BOX WITH LEADS

BOX WITH REPAIR PARTS
AND NEEDLE POINTS

FRICTION DIVIDERS

PEN FOR COMPASS

SCREWDRIVER

PEN HANDLE

FRICTION COMPASS

LENGTHENING BAR
FOR COMPASS

RULING PEN

BOW PENCIL

CENTER TACK

BOW PEN

BOW DIVIDERS

Fig. 2-2-20 A three-bow set of drawing instruments. (Keuffel & Esser Co.)

Drafting Leads and Pencils

Leads Because of the drawing media used and the type of reproduction required, pencil manufacturers have marketed three types of lead for the preparation of engineering drawings.

Graphite Lead This is the conventional type of lead which has been used for years. It is made from graphite, clay, and resin. It is available in a variety of grades or hardnesses—9H, 8H, 7H, and 6H (hard); 5H and 4H (medium hard); 3H and 2H (medium); H and F (medium soft); and HB, B, 2B, 3B, 4B, 5B, and 6B (very soft), the latter not being recommended for drafting. The selection of the proper grade of lead is important. A hard lead might penetrate the drawing paper while a soft lead will smear. The next two types of drafting leads were developed as a result of the introduction of film as a drawing medium. A limited number of grades are available in these leads, and they do not correspond to the grades used for graphite lead.

Plastic Lead This type of lead is designed for use on film only. It has good microform reproduction characteristics.

Plastic-Graphite Lead As the name implies, this lead is made of plastic and graphite. There are two basic types: fired and extruded. They are similar in material content to plastic fired lead, but they are processed differently. They are designed for use on film only, erase well, do not readily smear, and produce a good opaque line which is suitable for microform reproduction. The main drawback with this type of lead is that it does not hold a point well.

Drafting Pencils The leads are held either in the conventional wood-bonded cases known as wooden pencils or in metal or plastic cases known as mechanical pencils. See Figs. 2-2-21 and 2-2-22.

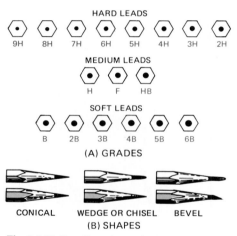

HARD LEADS

9H 8H 7H 6H 5H 4H 3H 2H

MEDIUM LEADS

H F HB

SOFT LEADS

B 2B 3B 4B 5B 6B

(A) GRADES

CONICAL WEDGE OR CHISEL BEVEL
(B) SHAPES

Fig. 2-2-21 Pencil grades and point shapes.

With the latter, the lead is ejected to the desired length of projection from the clamping chuck and then pointed in the same manner as the wood-bonded pencil. The 0.5-mm mechanical pencil has come into popular use because it does not require sharpening and produces a consistent line width. Disposable mechanical pencils now operate just as any mechanical pencil, but they are discarded after the lead has been used.

Lead Pointers

A fast, convenient means of putting a clean drafter's point on mechanical or wood-cased pencils is not only desirable, but necessary. Mechanical sharpeners (Fig. 2-2-23A) are made with special drafter's cutters that remove the wood as shown. The required point shape is then formed by hand sanding or by a special pointer.

When hand sanding (Fig. 2-2-23B),

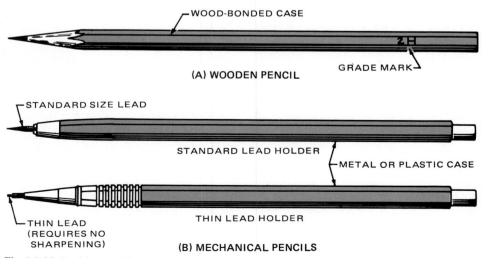

WOOD-BONDED CASE

2H

(A) WOODEN PENCIL

GRADE MARK

STANDARD SIZE LEAD

STANDARD LEAD HOLDER

METAL OR PLASTIC CASE

THIN LEAD
(REQUIRES NO
SHARPENING)

THIN LEAD HOLDER

(B) MECHANICAL PENCILS

Fig. 2-2-22 Drafting pencils.

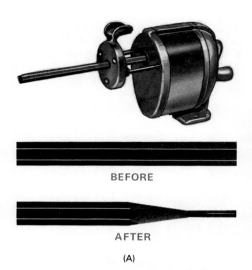

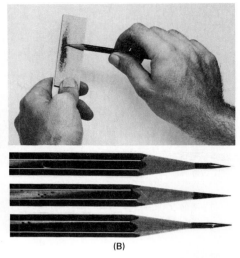

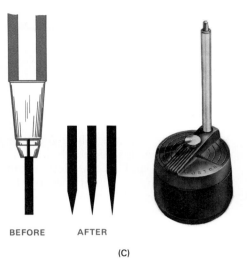

Fig. 2-2-23 Lead pointers. (A) A drafter's pencil sharpener cuts the wood, not the lead. (B) Shaping the lead by hand sanding. (C) Shaping the lead with a lead pointer.

rub the lead back and forth on a sandpaper block or a fine file, while turning it slowly to form the point. Keep the sandpaper block at hand so that you can sharpen the pencil often.

Special pointers are used for shaping the lead, as shown in Fig. 2-2-23C. Such devices may be hand-operated or electrically powered.

Erasers and Cleaners

Erasers A variety of erasers have been designed to do special jobs—remove surface dirt, minimize surface damage on film or vellum, and remove ink or pencil lines. See Fig. 2-2-24.

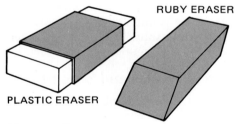

Fig. 2-2-24 Erasers.

Erasing Machines Hardly any pressure is required when using the electrically powered erasing machine because the high-speed rotation of the shaft actually does the clean-up job rapidly and flawlessly. These machines make erasures with pinpoint accuracy. (Fig. 2-2-25).

Cleaners An easy way to clean tracings is to sprinkle them lightly with gum eraser particles while working. Then triangles, scales, etc., stay spotless and clean the surface automatically as they are moved back and forth. The particles contain no grit or abrasive, and will

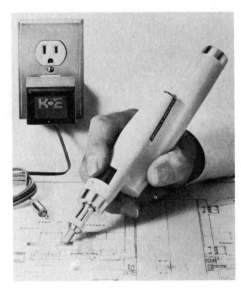

Fig. 2-2-25 Erasing machine. (Keuffel & Esser Co.)

actually improve the ink-taking quality of the drafting surface.

Erasing Shields These thin pieces of metal or plastic have a variety of openings to permit the erasure of fine detail lines or lettering without disturbing nearby work that is to be left on the drawing. See Fig. 2-2-26. Through the use of this device, erasures can be performed quickly and accurately.

Fig. 2-2-26 Erasing shield. (A. M. Bruning Co.)

KEEPING DRAWINGS CLEAN

It goes without saying that the end product, the drawing, must be clean and sharp for reproduction purposes. Dirt from the drafter's hands and instruments, graphite from the pencils, and dirt from the air are the main contributors to dirty drawings. Preventive maintenance goes a long way in keeping drawings clean.

The drafter's hands should always be clean. Dirty, oily, or perspiring hands should never come in contact with drawings.

Drafting instruments, such as triangles, scales, parallel slides, and the blade of the T square should be cleaned at least daily.

The drawing surface of the board must be brushed frequently to remove the dirt particles built up by dirt in the air, from erasing, and drawing lines.

Other recommendations are:

- Keep away from the drawings when sharpening the lead of pencils and compasses. Wipe the lead with a cleaning tissue to remove the loose graphite.
- Always use a brush when cleaning the drawing. Never wipe a drawing with your hands.
- Roll up your shirt sleeves as the cloth and buttons may damage or smudge your drawings.

Brushes

A light brush (Fig. 2-2-27) is used to keep the drawing area clean. By using a brush to remove eraser particles and any accumulated dirt, the drafter avoids smudging the drawing.

Fig. 2-2-27 Drafter's brush.
(A. M. Bruning Co.)

Lettering Aids

Lettering sets or guides (Fig. 2-2-28) are also used when it is desirable to have more uniform and accurate letters and numerals than can be obtained by the freehand method. Lettering sets contain a number of guide templates that give a variety of letter shapes and sizes, as well as different slope angles.

(A)

(B)

Fig. 2-2-28 Lettering aids. (A) Mechanical lettering (A. M. Bruning Co.) (B) Appliques. (Letraset)

Dry transfer lettering is a product which offers a wide variety of lettering of good quality and can be applied speedily. It adheres firmly to paper, wood, glass, and metal and is available in different colors. In case of errors, letters can be removed with cellophane tape or a pencil eraser.

Lettering typewriters have been used in drafting offices for some time for the lettering of bills of material and typing on appliques. But now small, movable typewriters can letter anywhere on the drawing without the need for removing the drawing from the board.

Templates

To save time, drafters use templates for drawing circles and arcs. Standard hole

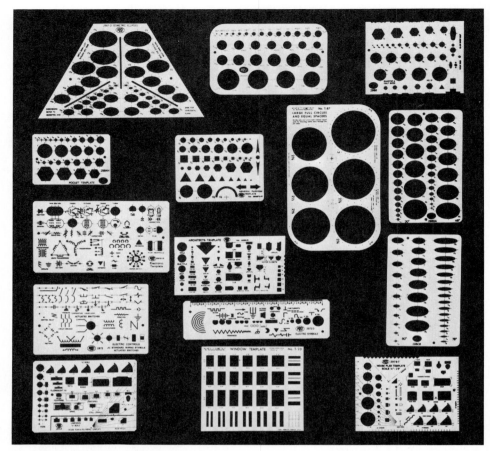

Fig. 2-2-29 Templates. (Teledyne Post.)

size templates ranging from small to 6.00 in. (150 mm) in diameter are available. Templates are also used for drawing standard square, hexagonal, triangular, and elliptical shapes and standard electrical and architectural symbols. See Fig. 2-2-29.

Irregular Curves

For drawing curved lines in which, unlike circular arcs, the radius of curvature is not constant, a tool known as an irregular or French curve (Fig. 2-2-30) is used. The patterns for these curves are based on various combinations of ellipses, spirals, and other math-

Fig. 2-2-30 Irregular curves. (Teledyne Post.)

ematical curves. The curves are available in a variety of shapes and sizes. Generally, the drafter plots a series of points of intersection along the desired path and then uses the French curve to join these points so that a smooth-flowing curve results.

Curved Rules and Splines

Curved rules and splines (Fig. 2-2-31) solve the problem of ruling a smooth curve through a given set of points. They lie flat on the board and are as easy to use as a triangle; yet they can be bent to fit any contour to a 3 in. (75 mm) minimum radius and will hold the position without support. A clear, plastic ruling edge stands away from the board just far enough to prevent ink lines from smearing.

Inking Equipment

Although most production drawings are drawn with pencil, in the last few years the number of ink drawings has been on the increase. The use of this type of

Fig. 2-2-31 Curved rule and spline. (Keuffel & Esser Co.)

drawing for technical illustrations and the demand for good, clear drawings for microform reproduction have brought about the introduction of new and improved inking methods and techniques. Typical inking equipment is shown in Fig. 2-2-32.

Inking Pens Two types of pens are used to produce ink lines. The ruling pen with an adjustable blade for drawing different-width lines is filled by a dropper cap, squeeze bottle, or cartridge tube. Inking compasses are available with permanent or detachable blades, which interchange with the lead attachment part of the compass. Specially designed pens for drawing curved, multiple broken, or hidden lines are also available.

The second type, called a needle-in-tube type pen, has gained wide acceptance with drafters because the line widths are fixed. Also it is suitable for drawing both lines and letters. Different-size needle points are available which produce different-width lines.

The use of a needle-in-tube pen with a circular template has, in many instances, replaced the drawing of circles and arcs by means of a compass. For best results, the pen should be perpendicular to the paper.

Recent developments in plotters have advanced the state of the art in pens and improved inks.

Calculators

Calculators, such as those shown in Fig. 2-2-33, are used by drafters to make fast mathematical calculations using division, multiplication, and extractions of square roots, and to solve problems

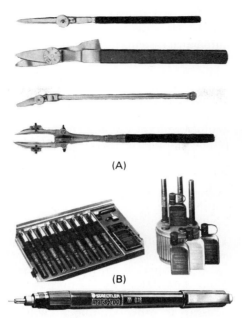

(A)

(B)

Fig. 2-2-32 Inking instruments. (A) Blade type (A. M. Bruning Co.) (B) Needle-in-tube type. (Staedtler-Mars)

involving areas, volumes, masses, strengths of materials, pressures, etc.

BASIC EQUIPMENT

Figure 2-2-34 shows the basic drafting equipment often found in a student drafting kit. This equipment along with other common items is listed below.

Fig. 2-2-33 Calculator.

Drawing board
T square, parallel-ruling straightedge (parallel slide), or drafting machine
Drawing sheets (paper or film)
Drafting tape
Drafting pencils
Pencil sharpener
Lead point and sandpaper
Eraser

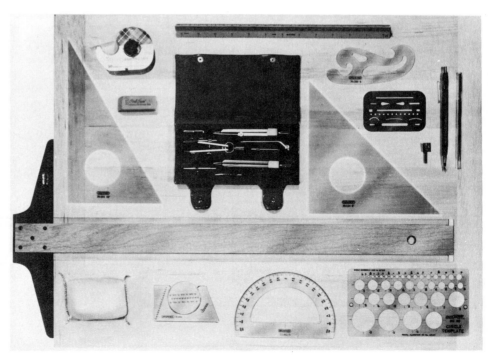

Fig. 2-2-34 Student drafting kit. (A. M. Bruning Co.)

Erasing shield
Triangles, 45° and 30/60° (not
  required with drafting machine)
Scales
Irregular curve
Drawing instrument set
Black drawing ink
Technical fountain pens
Brush
Protractor
Cleaning powder
Calculator

Your drafting instructor can tell you exactly what equipment will be needed for your course.

ASSIGNMENTS

See Assignments 1 through 4 for Unit 2-2 on page 28.

UNIT 2-3
CAD Equipment

COMPONENTS OF A CAD SYSTEM

A CAD system consists of various combinations of equipment. This holds true for small, medium, and large systems. The specific package selected depends largely on the needs of the user. For example, in some cases various types of drawings (check print or finished draw-

ing), referred to as hard copy, may be needed. However, companies using fully automated systems may not require any hard copy whatsoever. This means that while one company will choose a piece of equipment that prepares a drawing in one way, another will select equipment that uses a totally different method. Still another will not use any equipment to produce graphic displays or hard copies.

Generally, CAD equipment is categorized as follows:

- Processing and storage
- Input
- Output

This unit will analyze each of these major pieces of equipment. Figure 2-3-1 shows an operational flowchart for a complete CAD system. The purpose and function of each component and its relationship to other equipment will be described. These are typical components found in any system. It would be unlikely, however, to find all of these items in any one system. The central processing unit (CPU) is considered part of the processing equipment. An alphanumeric (letters and numbers) keyboard is used to manually input data. It may be attached to the graphics display monitor as one unit. In combination, a keyboard and graphics display monitor is referred to as a terminal. A CPU may be reserved for a single purpose. If so, it is normally attached to the terminal; this combination is referred to

as a computer. Thus, a computer is composed of a CPU, alphanumeric keyboard, and graphics display monitor.

A typical system arrangement is interactive. This means that one must cause the interaction between the CPU and the graphics display shown on the monitor screen. An alphanumeric keyboard or other input equipment will aid this process. After the work on the CAD unit has been completed, the information may be transferred to various output devices.

Figure 2-3-2 shows one type of system arrangement. To the left is a monitor, the drives, CPU, and a keyboard. In the center are input devices known as a graphics tablet (digitizer) and a joystick. In the background to the right is a plotter.

PROCESSING AND STORAGE EQUIPMENT

The CAD program, drawings, and symbols are stored on disks. The CPU collects and processes input information to produce the required drawings and data.

Central Processing Unit

Bits and Bytes The CPU is the computing portion of the system. A large number of integrated circuit (IC) chips are combined into a microprocessor.

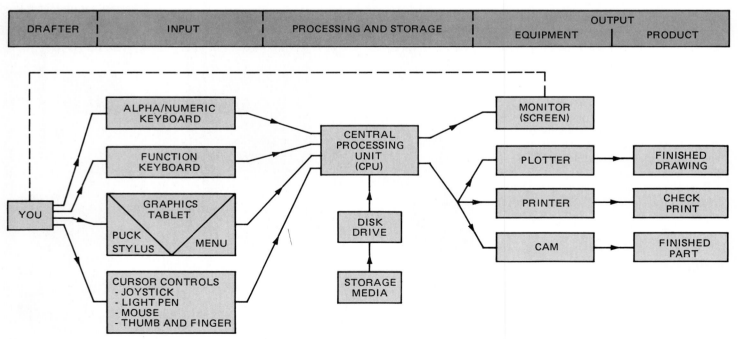

Fig. 2-3-1 Operational flowchart of a CAD system.

This is where the number crunching occurs, i.e., the performance of fundamental computations. The number of computations, or the capacity of the unit, is designated by the number of bytes. *Byte* is a character of memory containing 8 bits. The speed at which number crunching occurs is rapidly increasing as a result of the availability of 16- and 32-bit machines.

The size (capacity) of the CPU will generally determine the type of CAD system. *Micro* is the term used for small systems; *mini*, for medium systems; and *mainframe*, for large systems.

Micro System   The micro unit, used for small applications, has a typical user (or dynamic) memory capacity in the 512K to 640K range. To get a feel for memory size, 128K roughly equals 80 typewritten pages.

A single or dual monitor and alphanumeric keyboard are normally part of a micro unit. See Fig. 2-3-3. These units, known as desktop computers, are considered dedicated. When a system is dedicated, it serves a single user. Micro systems are economically priced, readily available, and are used for general drafting applications. Although its capacity is comparatively small, an operator can accomplish a considerable amount of work with limited training. Thus, large, expensive systems need not be tied up with basic drafting requirements. The data gathered on the micro can later be "translated" to a larger system. As a result, CAD has become "accessible" to many more designers and drafters.

Mini Systems   A mini system is generally a standalone operation. Standalone equipment can process information without having to access a separate or mainframe CPU. It is similar to a micro system, but has added capability. The special feature of a mini is its power. Anything that can be drawn on a traditional drafting board can be created on a mini system. It can also be automatically transformed into three-dimensional (3-D) representation and rotated. Thus, prior to the micro-CAD revolution it constituted the majority of CAD systems. Remember:

Micro systems = accessibility
Mini systems = power

Networking   More than one terminal or workstation of any combination of minis and micros can be connected to a mini system CPU. This is known as *networking*. The network provides the full

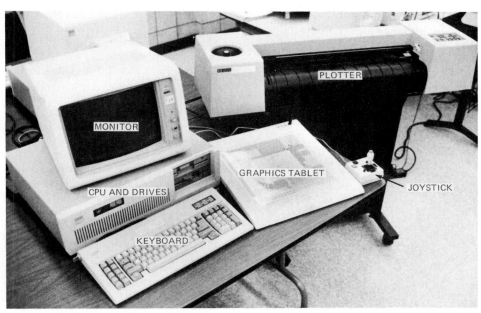

Fig. 2-3-2 CAD system equipment.

power of the system to additional terminals. The result is reduced cost and standardization.

A *queue*, or response time, means a waiting period. This occurs in a network system when simultaneous requests are made by several users. For example, if several operators request a drawing, the response is given in the order requested. The operator whose request was made last will receive the response last. As processing time becomes faster with each generation of computers, the wait is substantially reduced.

Mainframe   A mainframe system has a large-capacity CPU (or host computer). CAD is only one of many functions which it can perform. It offers even more capacity than the micro and

Fig. 2-3-3 Microcomputer. (Telesis)

mini systems. Mainframe terminals are normally found in a remote location of the workplace and are not combined into a single unit as with the micro system. See Fig. 2-3-4.

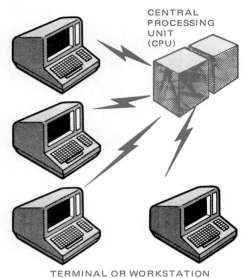

Fig. 2-3-4 Mainframe system.

Software

Program Language Software is a set of instructions called a program and what it does. A program is a group of written instructions logically arranged to perform a task. The instructions tell the computer what, how, and when to do something. They are used to input information into the system. Programs are written in a variety of languages. The most common of these are FORTRAN (FORmula TRANslation) and BASIC (Beginners All-purpose Symbolic Instruction Code). BASIC is popular because it uses English-like and math-like "easy-to-use" language. The notations consist of statements rather than sentences. No matter which language is used, drafters and designers need not necessarily develop a knowledge of them.

A drafter or designer will normally serve as the user of programs rather than the developer. Consequently, the programs will be received complete. Some programs do, however, provide an option to write special features, called *Macros*, into the program. This allows you to "customize" the system for your particular purpose.

Storage CAD systems will magnetically store information on a floppy disk. A disk will economically store drawings and programs plus offer speedy retrieval. Only "seconds" are required to find a program.

Different types of disks are available. There are hard (fixed) types and flexible types, single or dual. This means that data storage may be on one side (single) or both sides (dual) of the disk. A flexible disk is similar in appearance to a 45-rpm phonograph record. Two standard diameters for floppy disks are 5.25 in. and 3.50 in. These are shown in Fig. 2-3-5. Note the protective covering over the larger disk. Disks must be handled gently. Even a small scratch can damage the contents. The cover will protect the disk from dust, dirt, and accidental scratching. It cannot, however, prevent damage caused by mishandling. A disk, for example, cannot be exposed to heat or magnetic fields. The contents will immediately become damaged.

The small tabs shown in Fig. 2-3-5D protect disk contents. It is possible to accidentally write over the data on the disk with new data, thus destroying the original data. The small tab covering the slot prevents this from happening. This is known as *write protect*.

Hard, or fixed, disks are available in various configurations. They handle a larger amount of data than do floppy disks.

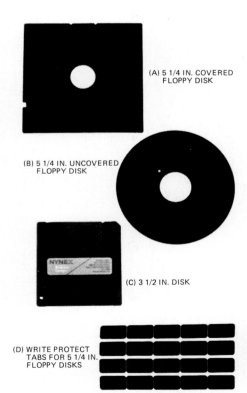

(A) 5 1/4 IN. COVERED FLOPPY DISK

(B) 5 1/4 IN. UNCOVERED FLOPPY DISK

(C) 3 1/2 IN. DISK

(D) WRITE PROTECT TABS FOR 5 1/4 IN. FLOPPY DISKS

Fig. 2-3-5 Floppy disks.

Larger mainframe CAD systems use a different software format. The programs are stored on disks and magnetic tapes. They provide extended capability over the small and medium-sized systems. As in all computers, the language of electrical impulses (the binary system) is used. Each portion of a program will normally represent a particular command. Commands are combined to prepare an engineering drawing. For example, one part will instruct the machine to draw a line, another part a circle, and so on.

Processing Units

Other devices, such as drives and memories, are used to input to or output from the processing equipment. The smaller systems, which use floppy disks, require equipment to drive the software. A disk drive receives the flexible disk directly, as shown in Fig. 2-3-6, and may be used as permanent storage. The information on the disk may be a program or drawings. When loading a disk into a drive, be certain to insert it with the label up. The slotted edge goes in first, with the write-protect slot to the left.

Systems not having a fixed drive use two disk drives. One contains the programs, and the other is used to maintain a permanent record of drawings after they have been developed. The disk drive itself is a piece of hard equipment (hardware). *Hardware* is a term that applies to any and all pieces of equipment having a physical entity.

Fig. 2-3-6 Floppy disk drive.

Computers have permanent or temporary memory systems to store programs and data. The contents in temporary memory are destroyed when power is lost or when a power surge occurs. Thus, power surge protectors should be used on all systems.

Information stored on a disk is not lost as a result of a power interruption. However, a floppy disk must not be removed while inputting or outputting information, i.e., while the drive-busy light on the disk drive is lit. A glitch (discontinuity) may result.

Input Equipment

Skills in the use and operation of input equipment replaces the skills of manual drafting. This equipment instructs the computer to draw lines and circles of various widths and sizes and to apply such graphics as symbols, dimensions, and notes to a drawing.

Alphanumeric Keyboard An alphanumeric keyboard provides for direct communication with the CPU. The alphanumeric keyboard is a separate piece of equipment. It is indispensable and is found on every CAD system. It may be used to manually input data for:

Nongraphic work
Adding text and notes
Exact coordinate input
Command selection

Every CAD system provides a means to select a part of a program that allows the creation of a particular command. On smaller systems, one of the two ways to effect this is by keying in the appropriate letters (word) or numbers. For example, LINE may be the command for line creation. To select that part of the program, key in LINE and press the ENTER key.

The keyboard is an extended version of a standard typewriter keyboard. (See Fig. 2-3-7.) The alphanumeric keys are the same. Usually, though, there are additional keys that allow some specialized commands to be accomplished. *Alpha* refers to the keys that input letters of the alphabet. *Numeric* refers to the other keys, each of which inputs a number. The user may type in an alphanumeric instruction. Input is completed by pressing the carriage return (ENTER key).

Function Keyboard The function keyboard is a piece of input equipment used to retrieve a program or part of a

Fig. 2-3-7 Alphanumeric keyboard. (Epson)

FUNCTION KEYS   ALPHA NUMERIC KEYBOARD   NUMERIC KEYPAD

Fig. 2-3-8 Alphanumeric keyboard, numeric keypad, and function keys. (IBM)

program. It contains several buttons or keys. A part of a program is electronically connected to one of the buttons or keys, which are operated during the execution of a particular function. A function keyboard is referred to differently by various manufacturers. While *program function keyboard* is probably most descriptive, for simplicity, *function keyboard* will be used to describe this input device.

The arrangement of the buttons or keys varies with the hardware manufacturer. Some of the smaller dedicated units combine them directly on the alphanumeric board of the computer. See Fig. 2-3-8. The function keys are located at the left or across the top.

Some larger mini systems have a completely separate function keyboard, as shown in Fig. 2-3-9. Each function associated with the corresponding button may be identified. The functions vary

Fig. 2-3-9 Separate function keyboard. (Cadam Inc.)

depending on what is to be accomplished by a particular system. Figure 2-3-9 represents a typical medium-size function keyboard.

Graphics Tablet The graphics tablet is a flat surface electronically sensitized beneath the surface. When proper contact is made on the surface, electrical impulses are transmitted to the computer. The information that is transmitted provides the programmed instructions to the CPU.

In graphic applications, the graphics tablet is far more important than the keyboard and serves many purposes. One use is quick and accurate graphic conversion. A rough sketch can be converted to a finished drawing by simply transferring point and line locations to the screen. Input is based on the *XY* (horizontal-vertical) line coordinate system and information is entered quickly and efficiently into the computer, using the graphics tablet. The result is graphically displayed on a monitor screen. This process is called *digitizing*. Consequently, a graphics tablet is also referred to as a digitizer. Other functions such as symbol input may be performed with the aid of a tablet.

Any selection on the tablet will retrieve a portion of the program corresponding to the desired title or symbol. For example, if you wish to draw lines, select the menu item labeled LINE. This will call up the part of the program that allows the creation of lines.

Several pieces of equipment are used in conjunction with the graphics tablet. These may include a stylus or pen, a push-button cursor or puck, a power module or console, and a menu. The tablet itself is a flat surface and is available in a wide range of sizes. It may vary from a small surface of 11 x 11 in. (275 x 279 mm) to one which exceeds an E-size drawing, *i.e.*, 36 x 48 in. (910 x 1220 mm). Beneath the surface lies a grid pattern of many horizontal and vertical sensors. These lines are used to detect electrical pulses at desired *XY* coordinate locations using a stylus or puck. Their locations are transferred to the computer. Fig. 2-3-10A shows a tablet and stylus. Fig. 2-3-10B shows a tablet with a push-button puck.

Puck and Stylus The position of each desired point of a drawing or sketch on the tablet is sensed by a puck (often called a "mouse" because of its shape) or stylus (electronic pen). Several styles of pucks and styluses are available. Figure 2-3-11 shows a puck and two styluses.

Each puck has fine black cross hairs (lines) for positioning. Pressing the appropriate button causes horizontal

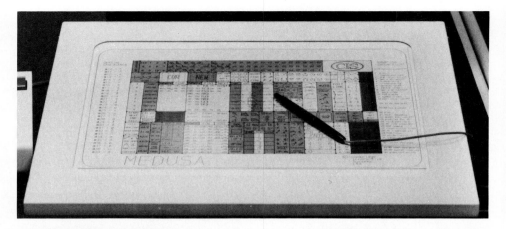

Fig. 2-3-10 (A) Graphics tablet (Prime/Computervision) and (B) digitizer (Intergraph).

(*X*) and vertical (*Y*) data to be accurately sent to the CPU by an electrical signal. The result is displayed on the monitor screen as a bright mark. This method of selecting a position and activating the puck is known as *digitizing*. It is repeated as often as necessary to complete the drawing. The buttons on a puck may be used for various purposes, depending on the manner in which the software has been created.

A stylus (pen) essentially operates the same as a single-button puck. The position is selected by the tip of the pen. As the pen tip is moved across the tablet surface, the position changes correspondingly on the monitor. It will appear as a small bright mark on the monitor. After the desired position has been located, it is digitized by activating the stylus, usually by pressing down on the point.

Tablet Menu CAD systems are menu-driven. This means that a programmed menu selection is made to call up a particular part of a program. Graphics tablets are provided with a menu having a variety of options. A menu may appear on the screen, or it may be placed on the tablet surface. A graphics tablet menu is shown in Fig. 2-3-12A. Each of the small boxes, or cells, is used to select a specific menu item. Many of the terms are abbreviated to fit into the box. To make a selection, place and activate the stylus or puck over the desired item such as the RECTANGLE icon as shown in Fig. 2-3-12A. To draw lines, for example, digitize within the box or cell labeled LINE. Other partial graphics tablet menus are shown in Fig. 2-3-12B. Many tablet overlays have a double coding to identify a command. Notice both a name and an icon within cells shown in Fig.

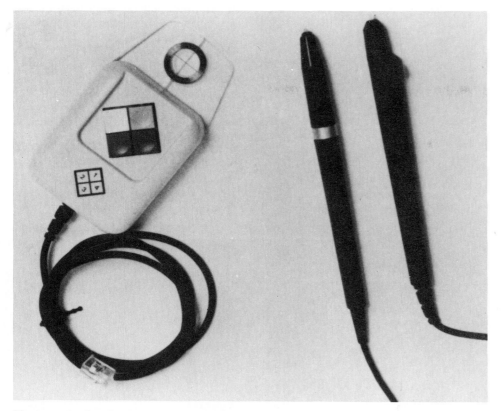

Fig. 2-3-11 Puck and stylus.

2-3-12B. An *icon* is a picture (graphical representation) of what something means. There is an icon glossary (Appendix A) at the back of this text.

A menu will normally occupy only a portion of the tablet surface area. A symbol also may be generated off the menu by the use of the stylus or puck. You can accomplish this by pointing to the menu item with the stylus or puck and touching down. Next, digitize the desired position on the graphics tablet. The result is displayed on the screen. The process is repeated until the drawing is finished.

Cursor Controls

Light Pen A light pen is used as a direct-entry input pointing device. It is also considered a digitizer since it can change displayed points and select menu options on the screen. The pen is electronic and contains a photocell sensory element to detect the presence of light. Hence the term light pen. Attached to one end is a cable through which the signal is transmitted. The other end of the pen may be positioned by hand to a desired screen location. After positioning, touch the screen with the tip of the pen. Depressing it causes the pen to become activated. Figure 2-3-13 shows a drawing being created using a light pen with command selection by a function keyboard (left front). One disadvantage of the light pen is drafting fatigue, which often occurs with prolonged use.

The light pen is moved about the display screen and indicates the current active position under consideration. The position is illuminated by a blinking character (rectangle, arrow, or cross hair) or a very bright spot commonly referred to as a *cursor*. In this text, the term cursor will always refer to the current active position on the screen.

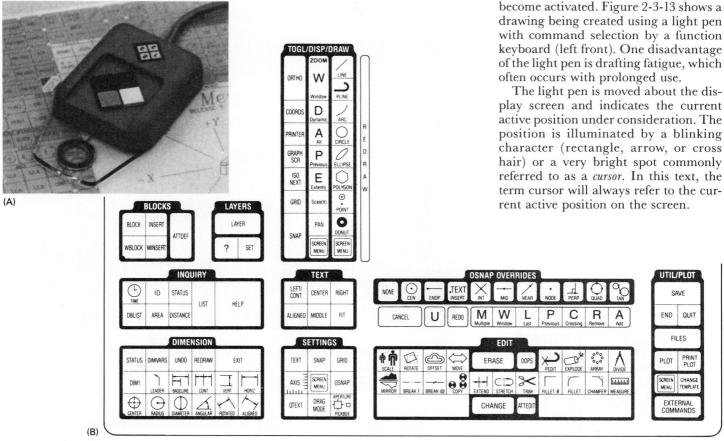

Fig. 2-3-12 Graphics tablet menus. (Autodesk Inc.)

Fig. 2-3-13 Drafting creation using a light pen and function keyboard. (Cadam Inc.)

Joystick A joystick is another type of device used to control the cursor. It can be added to many systems, further enhancing CAD capability. The joysticks shown in Fig. 2-3-14 are simply an extended version of those commonly used with video games. An electrical connection to the computer is made with a cable. The joystick steers a lighted cursor on the monitor screen. Once the cursor is positioned, press the button to set the location.

The rate of speed at which the cursor is moved on the monitor screen is variable. It is proportional to the distance the stick is moved from the vertical position.

Other Input Equipment Several other pieces of input equipment may be used to position the cursor. They include a thumb and finger wheel, touch pad,

Fig. 2-3-15 Thumb and finger wheel. (Tektronix)

mouse, and tracball. While these devices are normally not used, it is possible for one to be found in use. A thumb and finger wheel as shown in Fig. 2-3-15 is used to locate the cursor horizontally (X direction) and vertically (Y direction). Do this by rotating each wheel. The intersection of a horizontal and a vertical line determines the cursor location very accurately.

A tracball or roller ball is seated in a container with a portion of the sphere extended. It is rotated in any direction by the palm of the hand. This rotation causes the cursor to move accordingly.

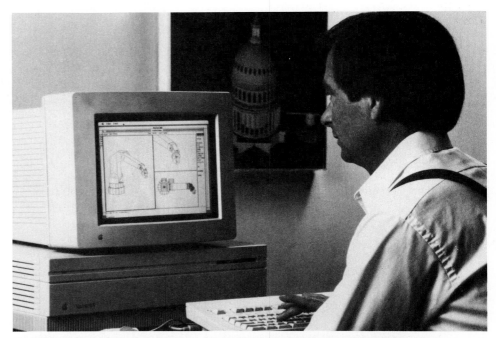

Fig. 2-3-14 Joystick. (Cadam Inc.)

Fig. 2-3-16 Monitors. (top: Autodesk Inc.; bottom: Intergraph Corp.)

When the desired position is located, simply stop the ball.

Output Equipment

The drawings that are being produced are observed on the monitor screen. At any time, a print of what is seen on the screen can readily be obtained from a printer. This print is referred to as a hard copy and is an exact reproduction of what is shown on the screen. Diagonal lines, circles, and arcs are a combination of straight vertical and horizontal lines and appear as zigzag or jagged lines. However, this type of drawing (hard copy) is quite adequate for checking progress while developing a finished design. When the drawing or design is completed, a finished drawing with near-perfect lines and lettering is made on a plotter. This is equivalent to the drawing made manually. Prints or microfilm are then produced from this drawing.

If a company has a fully developed CADD-CAM operation, a drawing may not be required. Instead, a set of instructions in coded form is transmitted directly to the fabricating equipment. The instructions include all location and size information necessary to produce a part or product.

Monitor The monitor (or graphics display station) is used to show the image of the drawing being made. The image displays data in either an alphanumeric (written) or a graphical (pictorial) manner. The user can view a picture of the design as it is being entered into the system. This display can be accomplished in a variety of ways depending on the type of monitor used.

The most popular monitor is the cathode-ray tube (CRT). The display method is similar to a television screen. Figure 2-3-16A illustrates a single-screen display station. Both the drawing and the alphanumerics are superimposed on the screen. Figure 2-3-16B uses a dual-screen setup. The screen to the left is used for the drawing while the screen to the right displays all menus and prompts.

Raster screens have become the dominant type of CRT displays. The raster type uses a grid network to display the image. Each grid is either a dark or a light image that falls within a square area that appears on the screen as a dot (dark) or undot (light). Each dot is known as a pixel (picture element).

The resolution (clearness) of a raster-developed image will depend on the closeness of lines forming the grid pattern. The smaller the pixels, the more resolution the image has. The greater the number of dots per unit area, the greater the resolution. The greater number of dots improves picture quality.

An example of a low-resolution display is shown in Fig. 2-3-17. Notice the unevenness of the top and right-side lines. This is known as jaggies or "stairstepping." Notice also that the circles have a severe case of the jaggies. The top line actually resembles a stair-step, thus the slang terminology. This jagged type of line will always result on a low-resolution raster screen if the line is inclined.

Significant advancement in raster technology has been achieved in recent years. Jaggies appearing on the lines and polygon boundaries have been significantly reduced.

The type of CRT mentioned so far produces the image in one color, similar to black-and-white television. The majority of raster systems are now color-enhanced and are similar to a color television screen. Each of three electron guns emits one of the primary television colors: red, green, or blue. From combinations of these, any color pattern may result. Nearly all of the new graphics display stations contain this option. There are several other types of monitors, but they are less popular and will not likely replace the raster CRT.

Plotters If a CAD system is thought of as an automated drafting machine, the plotter is the part replacing the activity of "laying lead." It produces the finished original drawing that was previously developed and displayed on the monitor. Regardless of screen resolution, a plotter will produce quality lines. There simply are more addressable dots per unit area. Thus, line quality is excellent and virtually jaggie-free.

For economical reasons, the plotter should not be used as a print machine. Once the drawing has been finalized, prints may be produced by a white-printer, photocopier, or microfilm equipment.

There are three types of plotters commonly used by industry to produce a finished drawing. They are:

- Pen plotter
- Electrostatic plotter
- Laser plotter

Pen Plotters A line-type digital pen plotter is an electromechanical graphics output device. Pen plotters (shown in Fig. 2-3-18) are the most popular. They move a pen in two dimensions across paper medium.

Various types of ink pens can be used, such as wet ink, felt-tip, or liquid ball, and are inserted or removed rather quickly. A pen holder is shown in Fig. 2-3-19. They may be a single color or multicolor. More importantly, plotter pens offer a variable line width (weight) option, with modern models now offering a pencil option.

Electrostatic Plotters This type of plotter is replacing the pen plotters in applications requiring high production. Plots

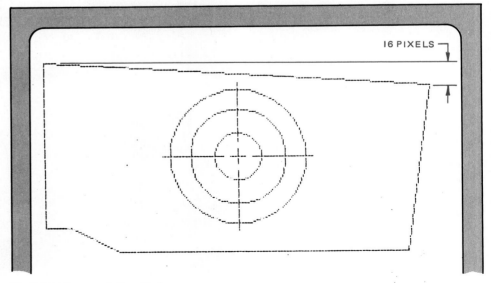

Fig. 2-3-17 Low-resolution display.

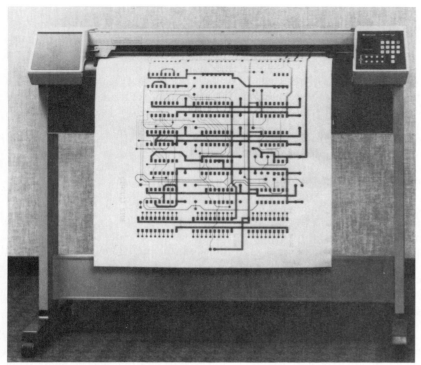

Fig. 2-3-18 Pen plotter. (Hewlett Packard)

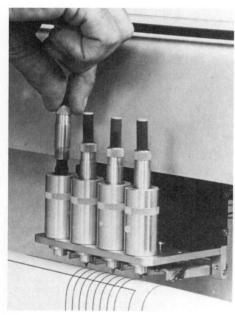

Fig. 2-3-19 Multiple pen holder.

are made on any of the same media as a pen plotter, but at a much faster speed. The main drawback of this plotter is the relatively high cost.

Laser Plotters A laser plotter uses a moving laser beam to alter a point-by-point electrical charge on the surface of a rotating drum. The drum is exposed to dry ink which adheres to the charged areas of the drum. Laser plotters, like

electrostatic plotters, produce drawings at high speed. However, they are expensive, large, and heavy.

Printers A print is a preliminary drawing that is produced by a printer. See Fig. 2-3-20. It produces an image of what is seen on the monitor complete with all jagged lines and circles. All lines have one thickness. The print is used to check preliminary stages of the design. It produces a print more quickly and less expensively than pen plotting. The print, however, does not approach the level of quality produced by the pen plotter. Thus, it is used primarily for preliminary check prints.

Computer-Aided Manufacturing

Numerical Control Computer-aided manufacturing (CAM) uses the result of a computer-aided design. Combining CAD and CAM has dramatically increased productivity and accuracy.

Fig. 2-3-20 Ink jet printer.

After CAD is used to prepare the product design, the instructions for the manufacture of that design are sent directly to the manufacturing location. One method of transmitting the information is known as numerical control (NC) or computer numerical control (CNC). NC tapes and equipment can store the designs that are used with a variety of production-related processes without producing an actual engineering drawing. In other words, the information is transmitted directly from one database to another.

True CAD-CAM is the ultimate goal of industry. CAD-CAM means that an engineering drawing is no longer produced, since it is a direct hard-wired connection. The output of the CAD system is a drawing stored in a geometric database. This drawing is transmitted directly into the CAM equipment. The result is a finished manufactured part.

Robotics The other part of CAM is known as robotics. Robot machinery differs from CNC machinery in that movement is now the prime duty. Automatic manipulators are used to perform a variety of material-handling functions. The robot manipulators are arms and hands. See Fig. 2-3-21. They will grasp, operate,

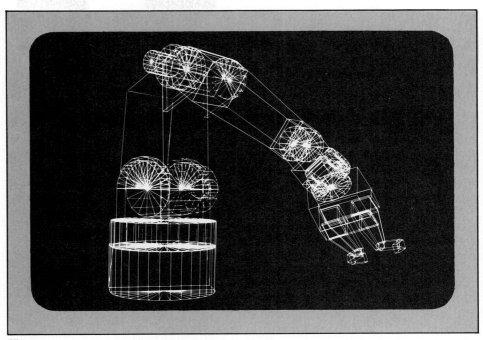

Fig. 2-3-21 Robot display on monitor screen.

assemble, and handle with great consistency and dependability. They are able to perform tasks that are considered too difficult, dangerous, or monotonous for human workers. This is especially true in environments that are intolerable to human beings.

ASSIGNMENTS FOR CHAPTER 2

Assignments for Unit 2-2, Manual Drafting Equipment and Supplies

1. Using the scales shown in Fig. 2-2-A determine lengths A through K.

2. Metric measurement assignment. With reference to Fig. 2-2-B and using the scale:

- 1:1 measure distances A through E
- 1:2 measure distances F through K
- 1:5 measure distances L through P
- 1:10 measure distances Q through U
- 1:50 measure distances V through Z

3. Inch measurement assignment. With reference to Fig. 2-2-B and using the scale:

- 1:1 decimal inch scale measure distances A through F
- 1:1 fractional inch scale measure distances G through M
- 1:2 decimal inch scale measure distances N through T
- 1:2 fractional inch scale measure distances U through Z

4. Foot and inch measurement assignment. With reference to Fig. 2-2-B and using the scale;

- 1″ = 1′ − 0″ measure distances A through F
- 3″ = 1′ − 0″ measure distances G through M
- ¼″ = 1′ − 0″ measure distances N through T
- ⅜″ = 1′ − 0″ measure distances U through Z

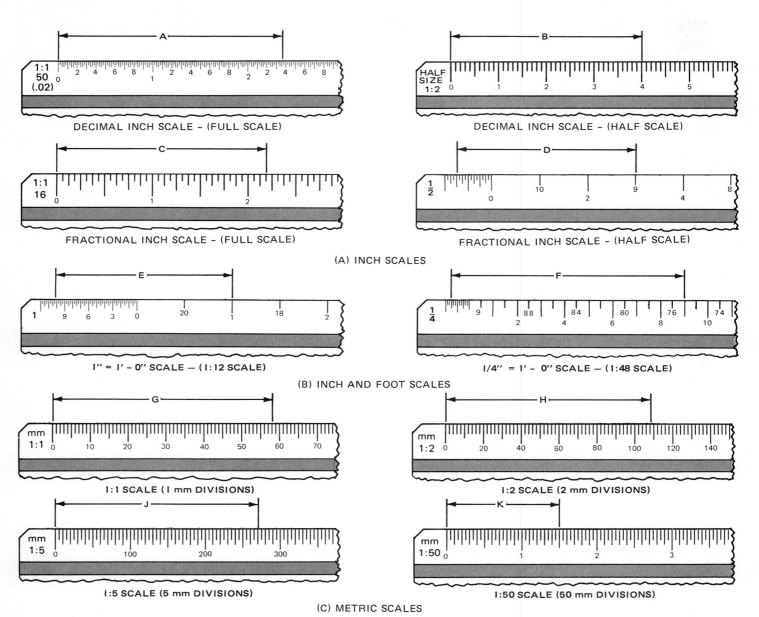

Fig. 2-2-A Reading drafting scales.

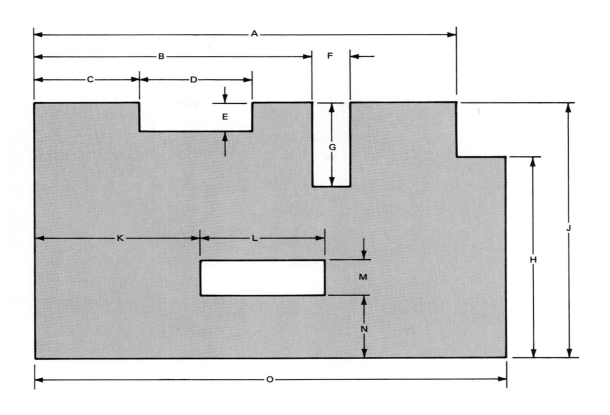

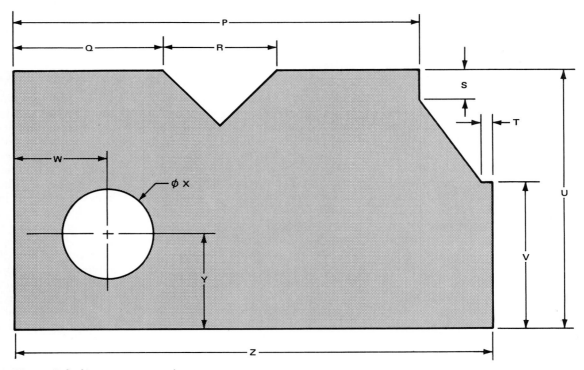

Fig. 2-2-B Scale measurement assignment.

Drawing Media, Filing, Storage, and Reproduction

U N I T 3 - 1
Drawing Media and Format

The families of drafting and tracing materials—*paper and film*—differ sufficiently between and within themselves to provide a wide choice of qualities and characteristics for selection of the perfect material for drawing requirements.

PAPER

Drafting papers come with a wide range of qualities—strength, erasability, permanence, translucency, etc. The distinguishing feature between drawing and tracing paper is translucency.

Opaque papers are used primarily as drawing papers. Because of a change in drafting practices, the need for drawing papers today is limited to the two extremes in the scale of quality—the very highest grade, permanent drawing papers with the best possible erasing quality for maps or master drawings, which are later photographed, and the inexpensive school type of papers for the educational market. Since master drawings often must be revised and corrected, the major consideration in high-grade drawing papers is erasability.

The more translucent papers are used as tracing papers. Formerly, they were used almost exclusively for the ink tracing of pencil drawings made on opaque papers. Translucency was the prime requirement.

Today, the usual drafting practice is to develop the master drawing directly on translucent paper from which reproductions can be made, thereby saving the time, expense, and checking involved in the tracing process. This practice means that, in addition to good translucency, the modern high-grade tracing paper must be able to withstand considerable handling. The paper should retain these qualities for a long time to avoid the eventual necessity of redrawing, either manually or photographically.

FILM

The most recently developed drafting medium is film. The advantages of polyester film as a drawing material are many. Raw polyester has natural dimensional stability, great tearing strength, high transparency, age and heat resistance, nonsolubility, and waterproofness. The outstanding virtue of film over any other drafting medium is that film is almost indestructible. Its amazing permanence safeguards the important investment in engineering drawings and records. It is permanently translucent, waterproof, unaffected by aging, superb for pencil drafting, ink work, and typewriting, and has unequaled erasability.

Polyester materials, however, present some problems. The material must be sufficiently dense to avoid reflection from the copyboard in microforming, but translucent enough for backlighting or contact printing.

Should a diazo (whiteprint) machine be used to make prints of the CAD or manually produced drawing then translucent film or paper is used as the drawing medium. However, reproducible prints (diazo intermediates) can be made from the original drawing from which prints can be made.

If electrostatic reproduction or microfilming is used to produce prints or microfilm of the CAD or manually produced drawing, then either opaque or translucent film or paper can be used as the drawing medium.

PREPRINTED GRID SHEETS

This type of drawing paper makes the job of making manually prepared drawings easier and quicker. Cross-sectional lines printed directly on the paper or used as a liner beneath the drawing paper provide an accurate guide for all drawing work. These cross-sectional lines are available in several grid sizes. The squared and pictorial styles (isometric, perspective, and oblique) are the more common preprinted grid papers used by drafters. These grids, when applied directly on the paper, as shown in Fig. 3-1-1, or on film with a special nonreproducible ink, will not appear on the prints when the drawing is reproduced by diazo and photograph methods.

For freehand work, whether rough sketches in the field or in the drafting room, whether these are preliminary or to be used as the finished drawings, grid-

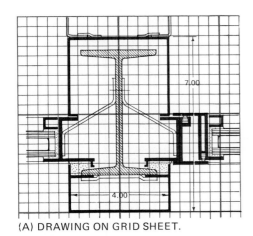

(A) DRAWING ON GRID SHEET.

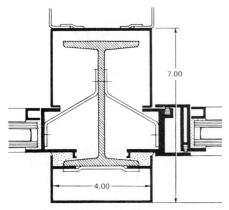

(B) PRINT OF DRAWING SHOWN IN (A).

Fig. 3-1-1 Drawing directly on preprinted grid sheets.

line papers are an invaluable aid. The cross-sectional patterns serve as ready-made guides for base lines, dimensioning, and angles. See Fig. 3-1-2.

STANDARD DRAWING SIZES

Inches Drawing sizes in the inch system are based on dimensions of commercial letterheads, 8.5 × 11 in., and standard rolls of paper or film 36 and 42

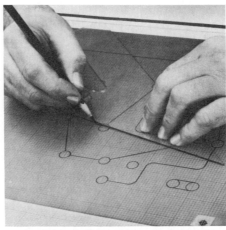

Fig. 3-1-2 Preprinted grid sheet being used as an underlay. (Keuffel & Esser Co.)

in. wide. They can be cut from these standard rolls with a minimum of waste. See Fig. 3-1-3.

Metric Metric drawing sizes are based on the A0 size, having an area of 1 square meter (m^2) and a length-to-width ratio of $1:\sqrt{2}$. Each smaller size has an area half of the preceding size, and the length-to-width ratio remains constant. See Fig. 3-1-4.

CAD After booting the CAD system, the drawing size "limits" must be set prior to starting the drawing. This will be determined by the space that the object to be drawn will require. For example, to create a full-scale single-view drawing of a small object will require only a small drawing size. To prepare a full-scale multiview drawing of a larger object will require a larger drawing size. There are several standard drawing sizes from which to choose.

An alternate to this procedure would be to draw to full scale and then scale down the finished plot and insert it into an appropriate paper size.

A third alternative would be to draw the paper and border at whatever scale will fit the drawing and adjust the scale when plotting to the corrected plotted size.

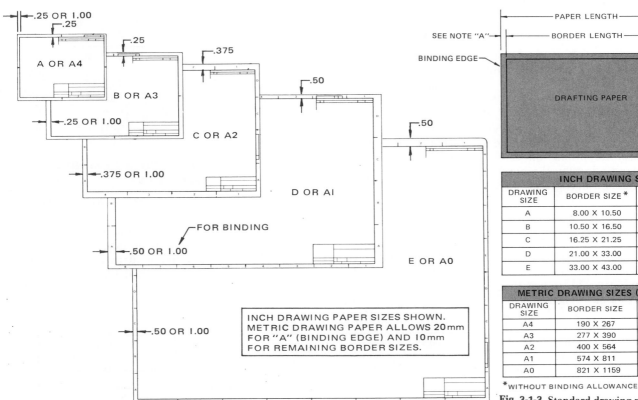

INCH DRAWING SIZES		
DRAWING SIZE	BORDER SIZE *	OVERALL PAPER SIZE
A	8.00 X 10.50	8.50 X 11.00
B	10.50 X 16.50	11.00 X 17.00
C	16.25 X 21.25	17.00 X 22.00
D	21.00 X 33.00	22.00 X 34.00
E	33.00 X 43.00	34.00 X 44.00

METRIC DRAWING SIZES (MILLIMETERS)		
DRAWING SIZE	BORDER SIZE	OVERALL PAPER SIZE
A4	190 X 267	210 X 297
A3	277 X 390	297 X 420
A2	400 X 564	420 X 594
A1	574 X 811	594 X 841
A0	821 X 1159	841 X 1189

*WITHOUT BINDING ALLOWANCE

Fig. 3-1-3 Standard drawing paper sizes.

INCH DRAWING PAPER SIZES SHOWN. METRIC DRAWING PAPER ALLOWS 20 mm FOR "A" (BINDING EDGE) AND 10 mm FOR REMAINING BORDER SIZES.

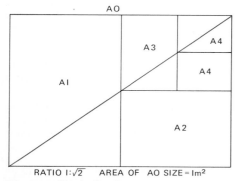
Fig. 3-1-4 Metric drawing paper.

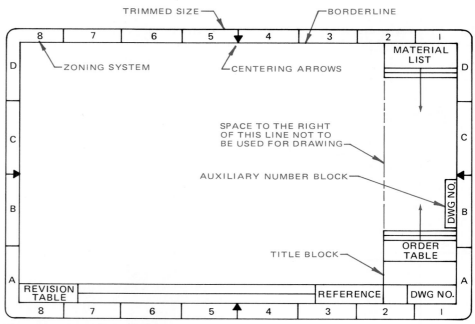
Fig. 3-1-5 Drawing paper format.

DRAWING FORMAT

A general format for drawings is shown in Fig. 3-1-5, which illustrates a drawing trimmed to size. It is recommended that preprinted drawing forms be made to the trimmed size and have rounded corners, as shown, to minimize dog-ears and tears.

Zoning System

Drawings larger than B size may be zoned for easy reference, by dividing the space between the trimmed size and the inside border into zones measuring 4.25 × 5.50 in. These zones are numbered horizontally and lettered vertically, with uppercase letters, from the lower RH (right-hand) corner, as in Fig. 3-1-5, so that any area of the drawing can be identified by a letter and a number, such as B3, similar to reading a road map. Just as with maps, this is useful to locate fine detail on complex drawings.

Marginal Marking

In addition to zone identification, the margin may also carry fold marks to enable folding and a graphical scale to facilitate reproduction to a specific size. In the process of microforming, it is necessary to center the drawing within rather close limits in order to meet standards. To facilitate this operation, it has become common practice to put a centering arrow or mark on at least three sides of the drawing. Most practices include the arrows on each of the four sides. If three sides are used, the arrows should be on the two sides and on the bottom. This helps the camera operator to align the drawing properly since the copyboard usually contains cross hairs through the center of the board at right angles. With any three arrows aligned

on the cross hairs, centering is automatic. The arrows should be on the center of the border which outlines the information area of the drawing, not at the edge of the sheet on which the drawing is made.

TITLE BLOCKS AND TABLES

Title Block

Title blocks vary greatly and are usually preprinted. Drafters rarely are required to make their own.

The title block is located in the lower right-hand corner. The arrangement and size of the title block are optional, but the following information should be included:

1. Drawing number
2. Name of firm or organization
3. Title or description
4. Scale

Provision may also be made within the title block for the date of issue, signatures, approvals, sheet number, draw-

ing size, job, order, or contract number; references to this or other documents; and standard notes such as tolerances or finishes. An example of a typical title block is shown in Fig. 3-1-6. In classrooms, a title strip is often used on A- and B-size drawings, such as shown in Fig. 3-1-7.

NORDALE MACHINE COMPANY		
PITTSBURGH, PENNSYLVANIA		
COVER PLATE		
MATERIAL- M S		NO. REQD- 4
SCALE- 1 : 2	DN BY D Scott	A - 7628
DATE- 3/6/88	CH BY R Jensen	

Fig. 3-1-6 Title block.

Item (Material) List and Order Table

The whole space above the title block, with the exception of the auxiliary number block, should be reserved for tabulating materials, change of order, and revision; drawing in this space should be avoided. On preprinted forms,

DRAFTING TECHNOLOGY CALIFORNIA UNIVERSITY OF PENNSYLVANIA CALIFORNIA, PENNSYLVANIA	NAME:		DWG NAME:	DWG NO.
	COURSE:			
	DATE:	APPD:	SCALE:	

Fig. 3-1-7 Title strip.

the right-hand inner border may be graduated to facilitate ruling for an item list, as shown in Fig. 3-1-8.

AMT	DET	STOCK SIZE	MAT.

NORALE MACHINE CO.
PITTSBURGH, PENNSYLVANIA

MODEL _____
PART NAME _____
PART NO. _____
OPERATION _____
FOR USE ON _____
METAL _____ DIE CLEARANCE _____
TOLERANCE: ±0.5mm UNLESS OTHERWISE SPECIFIED   METRIC
LAYOUT _____ DRAWN _____
CHECKED _____ APPROVED _____
SCALE _____ DATE _____
FROM B.P. _____ DATED _____

SHEETS	SHEET	**NO.**

Fig. 3-1-8 Combined title block, order table, and item list.

Change or Revision Table

All drawings should carry a change or revision table, either down the right-hand side or across the bottom of the drawing. In addition to the description of drawing changes, provision may be made for recording a revision symbol zone location, issue number, date, and approval of the change. Typical revision tables are shown in Fig. 3-1-9.

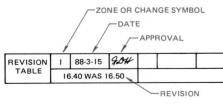

REVISIONS		
SYMBOL	DESCRIPTION	DATE & APPROVAL

(A) VERTICAL REVISION TABLE

ZONE OR CHANGE SYMBOL
DATE
APPROVAL

REVISION TABLE	I	88-3-15	GOH		
	16.40 WAS 16.50				

REVISION

(B) HORIZONTAL REVISION TABLE

Fig. 3-1-9 Revision tables.

Auxiliary Number Blocks

An auxiliary number block, approximately 2 × .25 in. (50 × 10 mm) is placed above the title block so that when prints are folded, the number will appear close to the top RH corner of the print, as in Fig. 3-1-5. This is done to facilitate identification when the folded prints are filed on edge.

Auxiliary number blocks are usually placed within the inside border, but they may be placed in the margin outside the border line if space permits.

References

1. Keuffel and Esser Co.
2. *Machine Design* and National Microfilm
3. Eastman Kodak Company

UNIT 3-2
Filing and Storage

One of the most common and difficult problems facing an engineering department is how to set up and maintain an efficient engineering filing area. Normal office file methods are not considered satisfactory for engineering drawings. To properly serve its function, an engineering filing area must meet two important criteria: accessibility of information and protection of valuable documentation.

For this kind of system to be effective, drawings must be readily accessible. The degree of accessibility is dependent

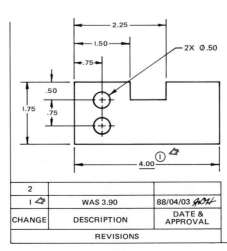

	2		
	I	WAS 3.90	88/04/03 GOH
CHANGE		DESCRIPTION	DATE & APPROVAL
REVISIONS			

(C) APPLICATION

on whether drawings are considered active, semiactive, or inactive.

FILING ORIGINAL DRAWINGS

Unless a company has developed a full microforming system, the original drawings which the drafter produced either manually or by CAD are kept and filed for future use or reference and to make prints as required. Unlike the prints, the originals must *not* be folded to avoid crease lines, which would appear in copies. They are filed in either a flat or rolled position. See Fig. 3-2-1.

In determining what type of equipment to use for engineering files, it should be remembered that different types of drawings require different kinds of files. Also, in planning a filing system, keep in mind that filing requirements are always increasing; unlike normal office files that can be purged each year, the more drawings produced, the more need to be stored. Therefore, any filing system must have the flexibility of being easily expanded, and generally in a minimum of space.

Microform Filing Systems

It seems logical that reducing drawings to tiny images on film would make them

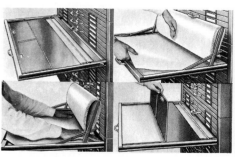

Fig. 3-2-1 Filing systems for original drawing masters and prints. (top: Ulrick Plan File; bottom: A. M. Bruning)

more difficult to locate. However, this is not the case, for while they are reduced in size, they are made more uniform. This results in improved file arrangements.

Roll Film Roll film can be coded in several ways for visual or automatic retrieval. The more common methods used are flash cards, code lines, sequential numbering image control, and binary code patterns.

Aperture Cards In many respects, aperture cards have the same filing and retrieval capabilities as jackets and microfiches. However, there is one important difference. It is possible to use aperture cards in machine records handling systems. They can be printed, punched, and sorted by machine.

Jackets and Microfiches Both these microforms are basically the same with respect to retrieval. See Fig. 3-2-2. Each has a title header for identifying numbers and titles. Each jacket or microfiche contains a group of images arranged in a logical sequence so that the particular images can be readily found.

CAD

Original drawings in CAD are digital information and stored on magnetic media such as tape, floppy disk, and hard disk, or on optical media such as laser disks. (See Fig. 3-2-3.) Since magnetic media can be easily damaged, special procedures are needed to protect original drawings. Paper copies may be stored as permanent records (called hard copy). Optical disks are not easily damaged and make excellent permanent records.

FOLDING OF PRINTS

To facilitate handling, mailing, and filing, prints should be folded to letter size, 8.5 × 11 in. (210 × 297 mm), in such a way that the title block and auxiliary number always appear on the front face and the last fold is always at the top. In filing, this prevents other drawings from being pushed into the folds of filed prints.

Recommended methods of folding standard-size prints are illustrated in Fig. 3-2-4.

On preprinted forms, it is recommended that fold marks be included in

Fig. 3-2-2 Microfilm jackets. (Eastman Kodak Co.)

Fig. 3-2-3 Floppy disk with protective cover. (Tektronix, Inc.)

the margin of the drawings on size B and larger and be identified by number, for example, "fold 1," "fold 2." In zoned prints, the fold lines will coincide with zone boundaries, but they should, nevertheless, be identified.

To avoid loss of clarity by frequent folding, important details should not be placed close to fold areas.

Reference

1. Eastman Kodak Company and "Setting Up and Maintaining an Effective Drafting Filing System," *Reprographics*, March 1975.

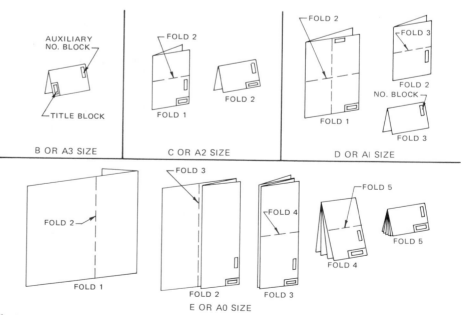

Fig. 3-2-4 Folding prints.

UNIT 3-3
Drawing Reproduction

A revolution in reproduction technologies and methods began in the 1940s and 1950s. It brought with it new equipment and supplies which have made quick copying commonplace. The new technologies make it possible to apply improved systems approaches and new information handling techniques to all types of files ranging from small documents to large engineering drawings. See Fig. 3-3-1.

The pressures on business and government for greater efficiency, space savings, cost reductions, lower investment costs, and equally important factors provided a fertile field for the new reproduction technologies. There is no reason to believe that such pressures will diminish; in fact, as the years go by, it is certain that more and more improvements will occur, newer and better reproduction and information handling equipment and methods will be discovered, and the advantages which they offer will find ever-widening application.

The following reproduction methods apply whether the original drawing was prepared manually or by CAD (plotter).

REPRODUCTION EQUIPMENT

Studies of reproduction facilities, existing or proposed, should first consider the nature of the demand for this service, then the processes which best satisfy the demand, and finally the particular machines which employ the processes. Factors to consider at these stages of study include:

- *Input originals*—sizes, paper mass, color, artwork
- *Quality of output copies*—depending on expected use and degree of legibility required
- *Size of copies*—same size, enlarged, reduced
- *Color*—copy paper and ink
- *Registration*—in multiple-color work
- *Volume*—numbers of orders and copies per order
- *Speed*—machine productivity, convenient start-stop, and load-unload
- *Cost*—direct labor, direct material, overhead
- *Future requirements*

Two general kinds of reproduction are recognized: copying and duplicating.

Copying machines are suitable for both line work and pictorials, often on large sheets. They operate at slow speeds, and so the cost per copy is relatively high.

In contrast, duplicating processes are characterized by high speed, high volume, and low cost per copy. They can be used with a wide variety of papers in many sizes, masses, and finishes. Although duplicators—spirit, stencil, and offset—are used in conjunction with copying machines in the engineering and drafting offices, they will not be covered in this text.

COPIERS

The principal kinds of copiers are diazo, electrostatic, thermographic, and photographic devices.

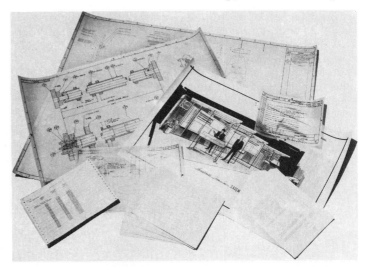

Fig. 3-3-1 Drawing reproduction. (Eastman Kodak Co.)

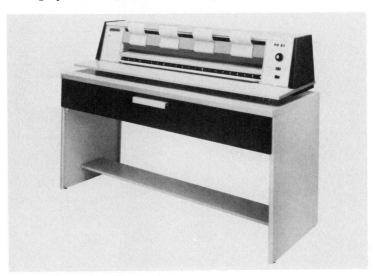

Fig. 3-3-2 A whiteprint machine. (Bruning)

Blueprinting was the classical process for copying engineering drawings. It provided white lines on a blue background. Prints were made on a continuous roll-fed machine and were not intended for reproduction. The process was wet, and the quality of prints was poor. In recent years, blueprinting has been replaced, mainly by diazo.

Diazo (Whiteprint)

In this process (Figs. 3-3-2 and 3-3-3), paper or film coated with a photosensitive diazonium salt is exposed to light passing through an original of translucent paper or film. The exposed coated sheet is then developed by an alkaline agent such as ammonia vapor. Where the light passes through the clear areas of the master, it decomposes the diazonium salt, leaving a clear area on the copy. Where markings on the origi-

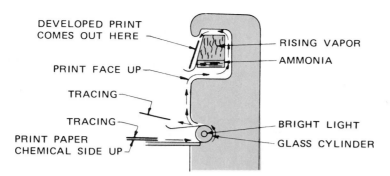

DEVELOPED PRINT
COMES OUT HERE

RISING VAPOR
AMMONIA

PRINT FACE UP

TRACING

TRACING

PRINT PAPER
CHEMICAL SIDE UP

BRIGHT LIGHT
GLASS CYLINDER

ROLLERS MOVE THE TRACING AND PRINT AROUND
THE LIGHT, AND MOVE THE PRINT PAST THE
RISING AMMONIA VAPOR.

Fig. 3-3-3 The diazo printing process.

nal block the light, the ammonia and the unexposed coating produce an opaque dye image of the original markings. A positive original makes a positive copy, and a negative original makes a negative copy. Therefore, the polarity is said to be *nonreversing*. The three diazo processes currently used differ mainly in the way the ammonia is introduced to the diazo coat. These are ammonia vapor, moist developing, and pressure developing.

Perhaps the most significant characteristic of the diazo process is that it allows reproduction of fine detail. Diazo is a high-contrast process and thus ideal for document reproduction.

Photographic

Photography is the process of creating latent images on light-sensitive silver halide material by exposure to light. The images are made visible and permanent by developing and fixing techniques. A camera provides for enlargement or reduction of the image size.

Contact printing and projection printing are the two principal methods of making photographic prints—contact for prints the same size as originals and projection for reduced or enlarged prints.

Electrostatic

Electrostatic reproduction, one form of which is xerography, is a dry copying process which uses electrostatic force to deposit dry powder on copy paper.

Electrostatic transfer enables printing on plain paper, offset paper masters, or transparent materials. Some machines copy only at the same size as the originals, while others can reduce or enlarge. See Fig. 3-3-4.

Another advantage is that the drafter can draw on practically any type of paper because the drawing is photographed and prints or intermediates can be made from the original work.

Such photocopying equipment is excellent for scissors and pasteup drafting.

MICROFORM

Microforming of engineering drawings is now an established practice in many drafting offices. See Fig. 3-3-5. This has come about because of the primary savings in lower transportation, labor, and storage costs of microfilm.

Microform prints are B size, regardless of their original size, and are much easier to handle and store. From the drafting room, drawings are taken to a camera, photographed, and stored in rolls or on cards.

Forms of Film

One way to classify microfilm is according to the physical forms, called *microforms*, in which it is used.

Roll Film   This is the form of the film after it has been removed from the camera and developed. Microfilm comes in four different widths—16, 35, 70, and 105 mm—and is stored in magazines.

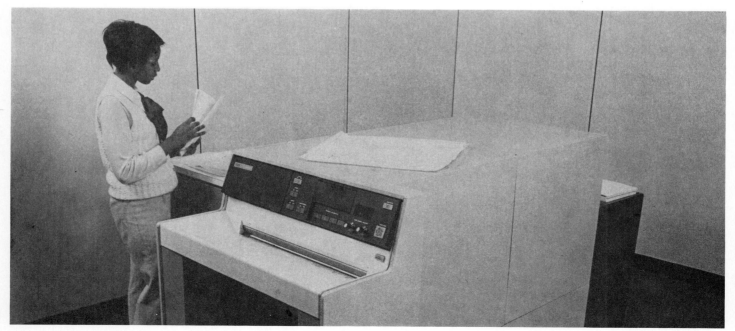

Fig. 3-3-4 Photoreproduction process. (Xerox Corp.)

Fig. 3-3-5 Microforming. (3M Co. and American Motors Co.)

Aperture Cards Perhaps the simplest of the flat microforms is finished roll film cut into separate frames, each mounted on a card having a rectangular hole as shown in Fig. 3-3-6. Aperture cards are available in many sizes.

Fig. 3-3-6 Aperture cards. (Eastman Kodak Co.)

Jackets Jackets are made of thin, clear plastic and have channels into which short strips of microfilm are inserted. They come in a variety of film channel combinations for 16- and/or 35-mm microfilm. Like aperture cards, jackets can be viewed easily.

Microfiche A microfiche is a sheet of clear film containing a number of micro-images arranged in rows. A common size, 100 × 150 mm, frequently is arranged to contain 98 images. Microfiches are especially well suited for quantity distribution of standard information, parts, and service lists.

Readers and Viewers

Microform readers magnify film images large enough to be read and project the images onto a translucent or opaque screen. Some readers accommodate only one microform (rolls, jackets, microfiches, or aperture cards), while others can be used with two or more. Scanning-type readers, having a variable-type magnification, are used when frames containing a large drawing are viewed. Generally, only parts of the drawing can be viewed at one time.

Reader-Printers

Two kinds of equipment are used to make enlarged prints from microform: reader-printers and enlarger-printers. The reader-printer, as illustrated in Fig. 3-3-7, is a reader which incorporates a means of making hard copy from the projected image. The enlarger-printer is designed only for copying and does not include the means for reading.

Fig. 3-3-7 Reader-printer. (Eastman Kodak Co.)

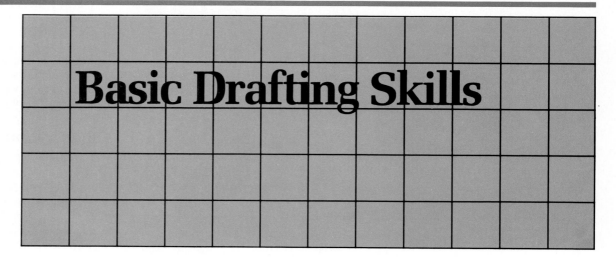

Basic Drafting Skills

UNIT 4-1

Straight Line Work, Lettering, and Erasing— Manual Drafting

LINEWORK

A line is the fundamental, and perhaps the most important, single entity, on a technical drawing. Lines are used to help illustrate and describe the shape of objects which later will become real parts. The various lines used in drawing form the "alphabet" of the drafting language: like letters of the alphabet, they are different in appearance. See Figs. 4-1-1 and 4-1-2. The distinctive features of all lines that form a permanent part of the drawing are the differences in their width and construction. Lines must be clearly visible and stand out in sharp contrast to one another. This line contrast is necessary if the drawing is to be clear and easily understood.

The drafter first draws very light construction lines, setting out the main shape of the object in various views. Since these first lines are very light, they can be erased easily should changes or corrections be necessary. When the drafter is satisfied that the layout is accurate, the construction lines are then changed to their proper type, according to the alphabet of lines. Guidelines, used to ensure uniform lettering, are also drawn very lightly.

Line Widths

Two widths of lines, thick and thin, as shown in Fig. 4-1-3, are recommended for use on drawings. Thick lines are .030 to .038 in. (.05 to 0.8 mm) wide, thin lines between .015 and .022 in. (0.3 to 0.5 mm) wide. The actual width of each line

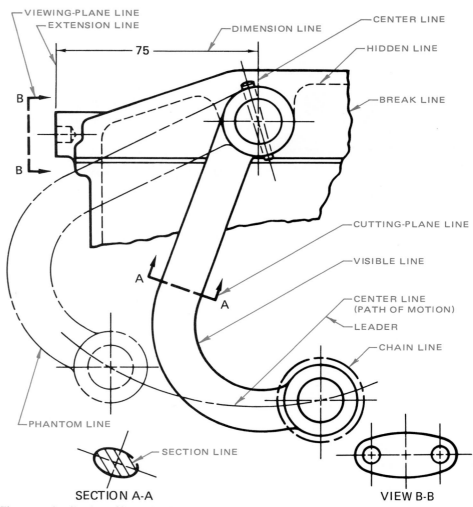

Fig. 4-1-1 Application of lines. (ANSI Y14.2M 1975)

TYPE OF LINE	APPLICATION	DESCRIPTION
HIDDEN LINE — — — — THIN — — — —	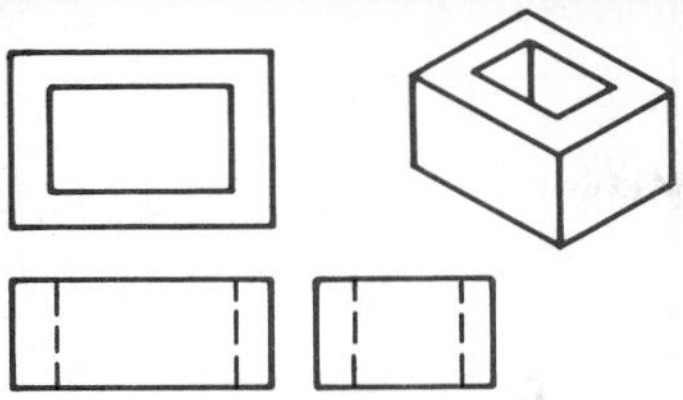	THE HIDDEN OBJECT LINE IS USED TO SHOW SURFACES, EDGES, OR CORNERS OF AN OBJECT THAT ARE HIDDEN FROM VIEW
CENTER LINE THIN ALTERNATE LINE AND SHORT DASHES	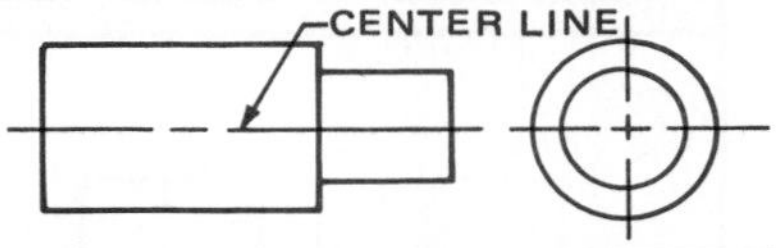	CENTER LINES ARE USED TO SHOW THE CENTER OF HOLES AND SYMMETRICAL FEATURES.
SYMMETRY LINE CENTER LINE THICK SHORT LINES	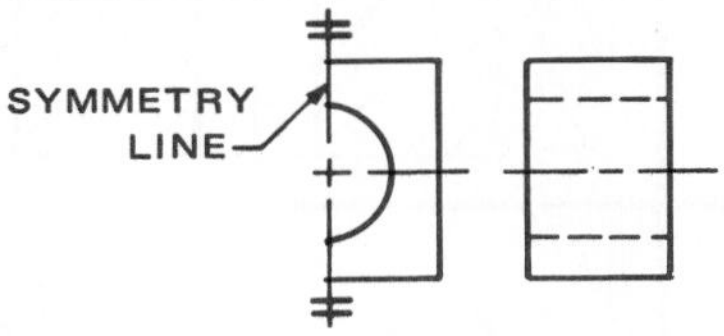	SYMMETRY LINES ARE USED WHEN PARTIAL VIEWS OF SYMMETRICAL PARTS ARE DRAWN. IT IS A CENTER LINE WITH TWO THICK SHORT PARALLEL LINES DRAWN AT RIGHT ANGLES TO IT AT BOTH ENDS.
EXTENSION AND DIMENSION LINES THIN DIMENSION LINE EXTENSION LINE	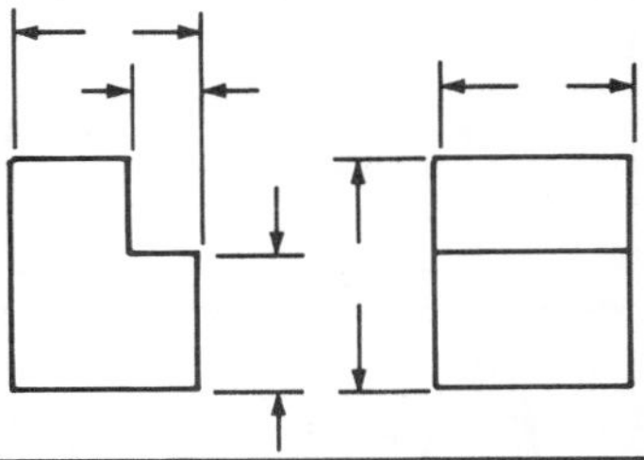	EXTENSION AND DIMENSION LINES ARE USED WHEN DIMENSIONING AN OBJECT.
LEADERS DOT ARROW THIN	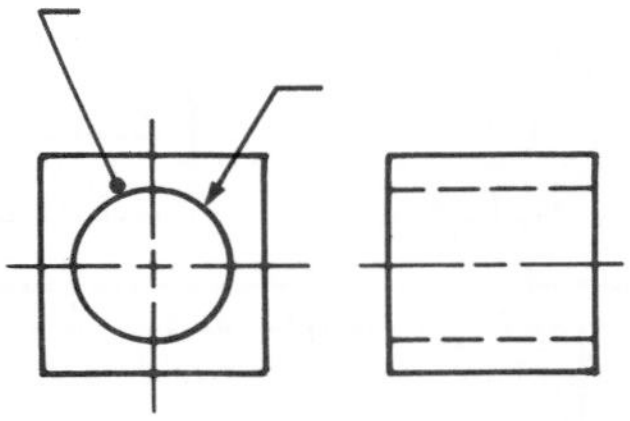	LEADERS ARE USED TO INDICATE THE PART OF THE DRAWING TO WHICH A NOTE REFERS. ARROWHEADS TOUCH THE OBJECT LINES WHILE THE DOT RESTS ON A SURFACE.
BREAK LINES THIN LONG BREAK THICK SHORT BREAK	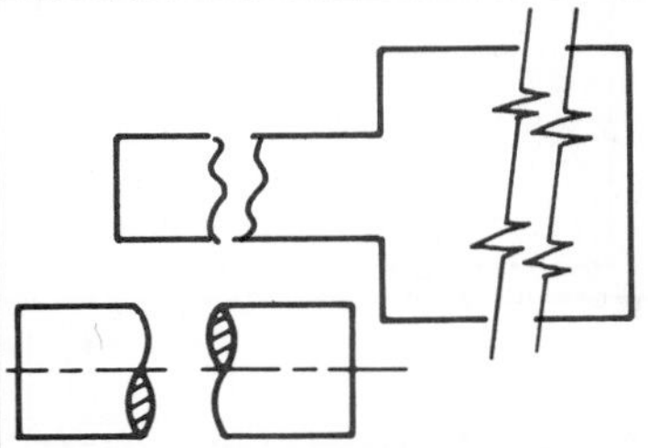	BREAK LINES ARE USED WHEN IT IS DESIRABLE TO SHORTEN THE VIEW OF A LONG PART.
CUTTING-PLANE LINE THICK OR	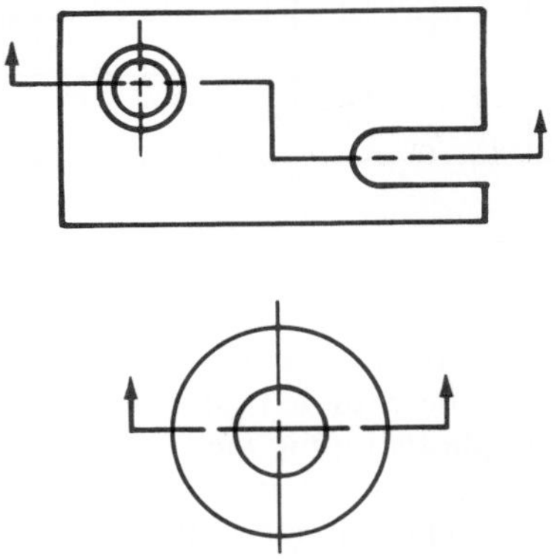	THE CUTTING-PLANE LINE IS USED TO DESIGNATE WHERE AN IMAGINARY CUTTING TOOK PLACE.

Fig. 4-1-2 Types of lines.

TYPE OF LINE	APPLICATION	DESCRIPTION
VISIBLE LINE ———————— THICK ————————	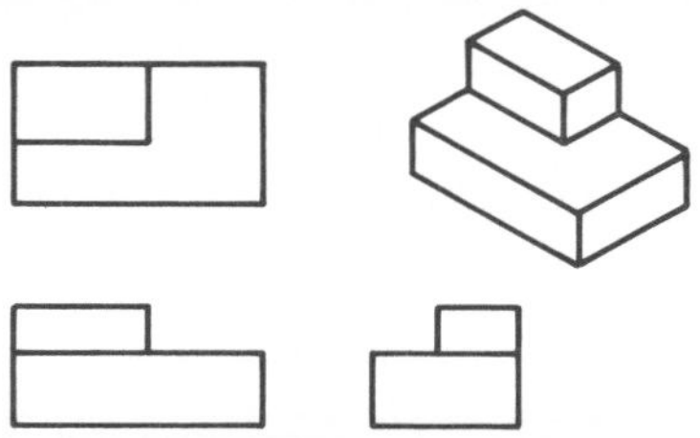	THE VISIBLE LINE IS USED TO INDICATE ALL VISIBLE EDGES OF AN OBJECT. THEY SHOULD STAND OUT CLEARLY IN CONTRAST TO OTHER LINES SO THAT THE SHAPE OF AN OBJECT IS APPARENT TO THE EYE.
SECTION LINES THIN LINES	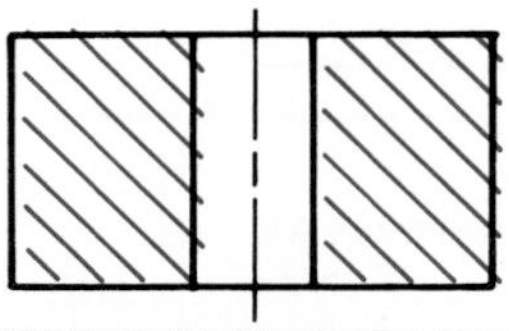	SECTION LINING IS USED TO INDICATE THE SURFACE IN THE SECTION VIEW IMAGINED TO HAVE BEEN CUT ALONG THE CUTTING PLANE LINE.
VIEWING-PLANE LINE THICK OR	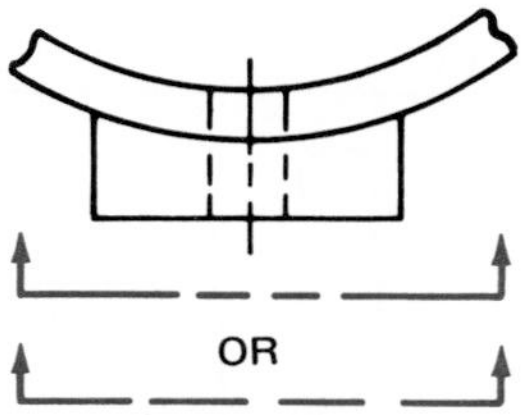	THE VIEWING-PLANE LINE IS USED TO INDICATE DIRECTION OF SIGHT WHEN A PARTIAL VIEW IS USED.
PHANTOM LINE ——— — ——— — ——— THIN	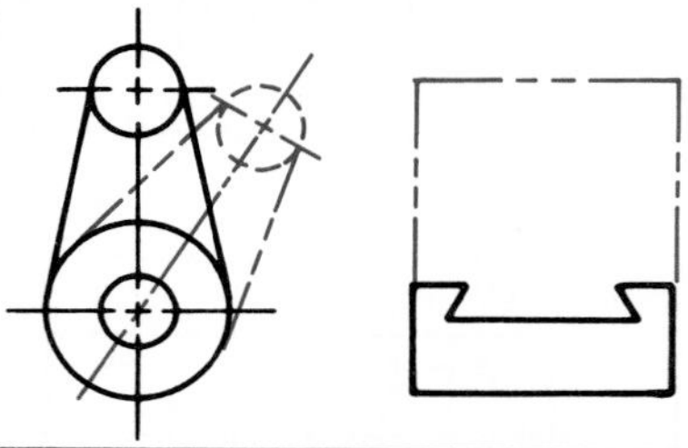	PHANTOM LINES ARE USED TO INDICATE ALTERNATE POSITION OF MOVING PARTS, ADJACENT POSITION OF MOVING PARTS, ADJACENT POSITION OF RELATED PARTS, AND REPETITIVE DETAIL. FOR PHANTOM LINE APPLICATIONS SEE FIGURE 2-6-12.
STITCH LINE — — — — — — — THIN OR SMALL DOTS	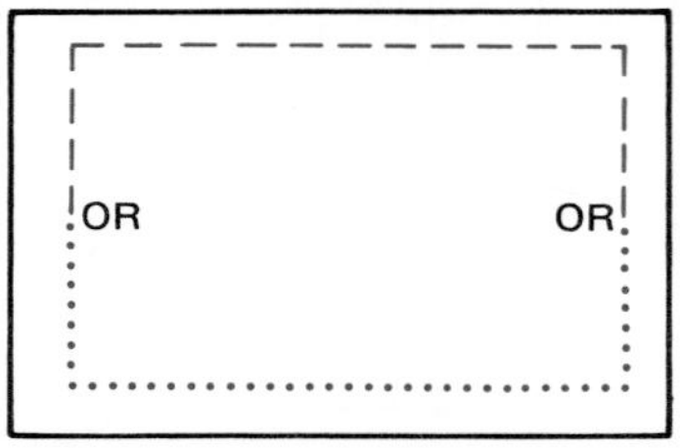	STITCH LINES ARE USED FOR INDICATING A SEWING OR STITCHING PROCESS.
CHAIN LINE ——— — ——— — ——— THICK	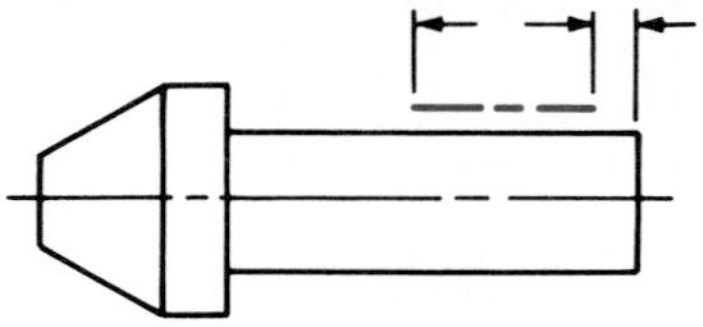	CHAIN LINES ARE USED TO INDICATE THAT A SURFACE OR ZONE IS TO RECEIVE ADDITIONAL TREATMENT OR CONSIDERATIONS.

Fig. 4-1-2 (continued)

is governed by the size and style of the drawing and the smallest size to which it is to be reduced. All lines of the same type should be uniform throughout the drawing. Spacing between parallel lines should be such that there is no fill-in when the copy is reproduced by available photographic methods. Spacing of no less than .12 in. (3 mm) normally meets reproduction requirements.

All lines should be clean-cut, opaque, uniform, and properly spaced for legible reproduction by all commonly used methods, including microforming, in accordance with industry and government requirements. There should be a distinct contrast between the two widths of lines.

Visible Lines

The visible lines should be used for representing visible edges or contours of objects. Visible lines should be drawn so that the views they outline clearly stand out on the drawing with a definite contrast between these lines and secondary lines.

The applications of the other types of lines are explained in detail throughout this text.

Drawing Straight Lines

When using a T square to draw horizontal lines (Figs. 4-1-4 and 4-1-5), hold the head of the T square against the edge of the drawing board and slide the T square either up or down to the desired position. Firmly press down on the blade of the T square to prevent it from moving, then proceed to draw the line. When drawing vertical lines, a triangle, which rests on the top side of the T square, is moved to the desired position and both the blade of the T square and the triangle are held firmly to the drawing board with the hand not holding the pencil.

When a parallel slide is used, as in Fig. 4-1-6, it will always be in a horizontal position as the wire and rollers in the slide move both ends of the slide simultaneously and at the same speed.

A general rule to follow when drawing straight lines is this: Lean the pencil in the direction of the line which you are about to draw. A right-handed person would lean the pencil to the right and draw horizontal lines from left to right. The left-handed person would reverse this procedure. When drawing vertical lines, lean the pencil away from yourself, toward the top of the drafting board, and draw lines from bottom to top. Lines sloping from the bottom to the top right are drawn from bottom to top; lines sloping from the bottom to the top left are drawn from top to bottom. This procedure for sloping lines would be reversed for a left-handed person.

When using a conical-shaped lead, rotate the pencil slowly between your thumb and your forefinger when drawing lines. This keeps the lines uniform in width and the pencil sharp. Do not rotate a pencil having a bevel or wedge-shaped lead.

Many drafters today use the 0.5-mm mechanical pencils. Held perpendicular to the paper, the drafter can produce uniform lines easily. The pencil is not rotated for this procedure. Pencils and leads are available from 0.2 to 0.9 mm in diameter for creating different lines.

THICK

WIDTH .032 IN. (0.7mm)

THIN

WIDTH .016 IN. (0.35mm)

Fig. 4-1-3 Line widths.

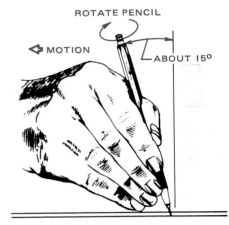

Fig. 4-1-4 Drawing pencil lines.

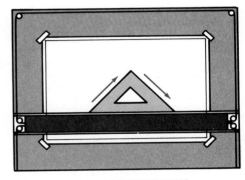

DRAWING SLOPING LINES

Fig. 4-1-6 Drawing sloped lines.

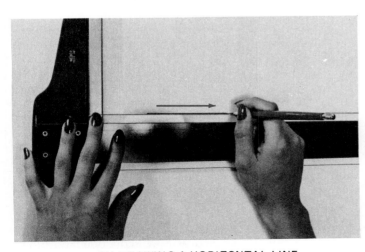

(A) DRAWING A HORIZONTAL LINE

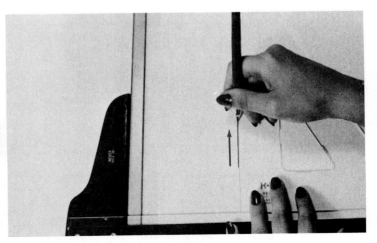

(B) DRAWING A VERTICAL LINE

Fig. 4-1-5 Drawing lines with the aid of a T square.

LETTERING

Single-Stroke Gothic Lettering

The most important requirements for lettering are legibility, reproducibility, and ease of execution. These are particularly important because of the increased use of microforming, which requires optimum clarity and adequate size of all details and lettering. It is recommended that all drawings be made to conform to these requirements and that particular attention be paid to avoid the following common faults:

1. Unnecessarily fine detail
2. Poor spacing of details
3. Carelessly drawn figures and letters
4. Inconsistent delineation
5. Incompleted erasures that leave ghost images
6. Use of differing densities, such as pencil, ink, and typescript on the same drawing

These requirements are met in the recommended single-stroke Gothic characters shown in Fig. 4-1-7 or adaptations thereof, which improve reproduction legibility. One such adaptation by the National Microfilm Association is the vertical Gothic-style Microfont alphabet (Fig. 4-1-8) intended for general usage.

Either inclined or vertical lettering is permissible, but only one style of lettering should be used throughout a drawing. The preferred slope for the inclined characters is 2 in 5, or approximately 68° with the horizontal.

Uppercase letters should be used for all lettering on drawings unless lowercase letters are required to conform with other established standards, equipment nomenclature, or marking.

Lettering for titles, subtitles, drawing numbers, and other uses may be made freehand, by typewriter, or with the aid of mechanical lettering devices such as templates and lettering machines.

Regardless of the method used, all characters are to conform, in general, with the recommended Gothic style and must be legible in full- or reduced-size copy by any accepted method of reproduction.

Letters in words should be spaced so that the background areas between the letters are approximately equal, and words are to be clearly separated by a space equal to the height of the lettering. See Fig. 4-1-9. The vertical space between lines of lettering should be no more than the height of the lettering and no less than half the height of the lettering.

PREFERRED OPEN - TYPE LETTERING

GOOD SPACING OF
CHARACTERS AND EVEN
LINE WEIGHT PRODUCE
CONSISTENTLY GOOD
RESULTS ON MICROFILM

UNDESIRABLE CRAMPED LETTERING

POORLY SPACED AND
FORMED, OR CRAMPED
LETTERING MEANS POOR
RESULTS IN MICROFILMING

Fig. 4-1-9 Spacing of lettering. (National Microfilm Association)

The recommended minimum freehand and mechanical letter heights for various applications are given in Fig. 4-1-10. So that lettering will be uniform and of proper height, light guidelines, properly spaced, are drawn first and then the lettering is drawn between these lines.

Notes should be placed horizontally on drawings and separated vertically by spaces at least equal to double the height of the character size used, to maintain the identity of each note.

Decimal points must be uniform, dense, and large enough to be clearly visible on approved types of reduced copy. Decimal points should be placed in line with the bottom of the associated digits and be given adequate space.

Lettering should not be underlined except when special emphasis is required. The underlining should not be less than .06 in. (1.5 mm) below the lettering.

When drawings are being made for microforming, the size of the lettering is an important consideration. A drawing may be reduced to half size when microformed at 30X reduction and blown back at 15X magnification. (Most microform engineering readers and

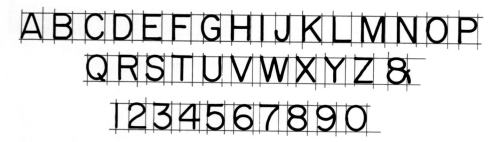

INCLINED LETTERS

VERTICAL LETTERS

Fig. 4-1-7 Approved Gothic lettering for engineering drawings.

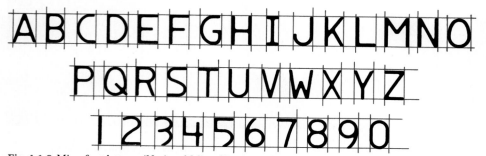

Fig. 4-1-8 Microfont letters. (National Microfilm Assoc.)

USE	INCH		METRIC mm		DRAWING SIZE
	FREEHAND	MECHANICAL	FREEHAND	MECHANICAL	
DRAWING NUMBER IN	0.250	0.240	7	7	UP TO AND INCLUDING 17 x 22 INCHES
TITLE BLOCK	0.312	0.290			LARGER THAN 17 x 22 INCHES
DRAWING TITLE	0.250	0.240	7	7	ALL
SECTION AND TABULATION LETTERS	0.250	0.240	7	7	
ZONE LETTERS AND NUMERALS IN BORDER	0.188	0.175	5	5	
DIMENSION, TOLERANCE, LIMITS, NOTES, SUBTITLES FOR SPECIAL VIEWS, TABLES, REVISIONS, AND ZONE LETTERS FOR THE BODY OF THE DRAWING	0.125	0.120	3.5	3.5	UP TO AND INCLUDING 17 x 22 INCHES
	0.156	0.140	5	5	LARGER THAN 17 x 22 INCHES

THIS IS AN EXAMPLE OF .125 IN. LETTERING

THIS IS AN EXAMPLE OF .188 IN. LETTERING

THIS IS AN EXAMPLE OF .250 IN. LETTERING

Fig. 4-1-10 Recommended lettering heights. (ANSI Y14.2M 1979)

blowback equipment have a magnification of 15X. If a drawing is microformed at 30X reduction, the enlarged blown-back image is 50 percent; at 24X, it is 62 percent of its original size.)

Standards generally do not allow characters smaller than .12 in. (3 mm) for drawings to be reduced 30X, and the trend is toward larger characters. Figure 4-1-11 shows the proportionate size of letters after reduction and enlargement.

The lettering heights, spacing, and proportions in Figs. 4-1-9 and 4-1-10 normally provide acceptable reproduction or camera reduction and blowback. However, manually, mechanically, optimechanically, or electromechanically applied lettering (typewriter, etc.) with heights, spacing, and proportions less than those recommended are acceptable when the reproducibility requirements of the accepted industry or military reproduction specifications are met.

Erasing Techniques

Revision or change practice is inherent in the method of making engineering drawings. It is much more economical to introduce changes or additions on an original drawing than to redraw the entire drawing. Consequently, erasing has become a science all its own. Proper erasing is extremely important since some drawings are revised a great number of times. Consequently, good techniques and materials must be used which permit repeated erasures on the same area. Some recommendations follow.

1. Avoid damaging the surface of the drawing medium by selecting the proper eraser.
2. Lines not thoroughly erased produce ghostlike images on prints, resulting in reduced legibility.
3. A hard, smooth surface, such as a triangle, placed under the lines being removed makes erasing easier.
4. Using an erasing shield protects the adjacent lines and lettering and also eliminates wrinkling. See Fig. 4-1-12.
5. Also erase on the back side of the paper. Lines frequently pick up dirt or graphite on the underside, and if not erased, will still produce lines on the print.

Fig. 4-1-11 Proportionate size of letters after reduction and enlargement. (National Microfilm Association)

Fig. 4-1-12 Using the erasing shield.

6. Be sure to completely remove erasure debris from the drawing surface.
7. When extensive changes are required, it may be more economical to cut and paste or make an intermediate drawing.
8. When erasing, use no more pressure than necessary.
9. The drawing quality of the drawing medium which may have been damaged by erasing may be improved by sprinkling an inking powder on the surface and rubbing it with a cloth.

In addition to these suggestions, it is necessary to match the density of the surrounding background when erasures are made. Often, the erased area is much cleaner than the rest of the drawing. If the change is made on this clean area, the contrast between line and background is different and that area presents a problem in reproduction.

It is usual practice to "smudge" the erased area so that it looks about the same shade as the surrounding area.

Removing Lines on Film

Erasers Lines on photoreproduction film fall into two classifications: photographic lines and pencil-and-ink lines. All these lines can be removed easily so that the erased area can be used for further drafting. Here are some tips for removing lines.

There are three basic types of erasers: rubber, plastic, and liquid. Rubber and plastic erasers may tend to cause a shine on the drafting surface. This is not necessarily detrimental. Good drafting lines

can be drawn easily over areas from which lines have been erased many times. A good general rule to follow is to use a soft, nonabrasive eraser and only enough pressure to remove the line. See Fig. 4-1-13.

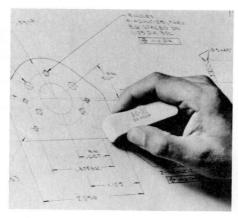

Fig. 4-1-13 Erasing. (Keuffel & Esser Co.)

Erasing machines require special techniques. Many different types of erasing inserts are available. To avoid burning holes in paper or melting the matte surface on film it is necessary to develop a skill of keeping the eraser moving and using a light touch.

Liquid erasers do not put a shine on the drafting surface and can be used to make many erasures in the same place. When plastic erasers are used, the shiny appearance actually may be transparentizing because of the plasticizers used in their manufacture rather than wear on the drafting surface. The transparentizing effect is not detrimental since it does not reduce the ability of the surface to take a pencil line.

When the drafting surface is affected by excessive erasures, it can be repaired by rolling a regular typewriter eraser across the smooth area or by rubbing a small amount of drafting powder into the area with a finger.

Pencil Lines Pencil lines can be removed from all film with a soft, nonabrasive rubber or plastic eraser or with liquid eraser. To keep the drafting surface from becoming too shiny, avoid excessive pressure.

FASTENING PAPER TO THE BOARD

The most common method of holding the drawing paper to the drafting board is with drafting tape.

When fastening the paper to the board, line up the bottom or top edge of the paper with the top horizontal edge of the T square, parallel slide, or horizontal scale of the drafting machine. See Fig. 4-1-14. When refastening a partially completed drawing, use lines on the drawing rather than the edge of the paper for alignment.

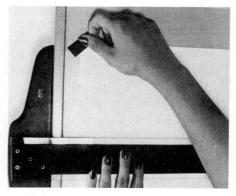

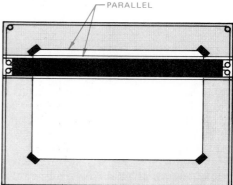

Fig. 4-1-14 Positioning the paper on the board.

Reference and Source Material

1. National Microfilm Association
2. Keuffel and Esser Co.
3. Eastman Kodak Company

ASSIGNMENTS

See Assignments 1 through 4 starting on page 58 for Unit 4-1.

UNIT 4-2
Straight Line Work, Text, and Erasing—CAD

GETTING ACQUAINTED WITH CAD

Skills in drawing lines and lettering are not required with CAD. Rather, skill in operating the CAD equipment becomes essential.

Before learning the fundamentals of drafting, it is recommended that you become familiar with the basics of line and text creation. It will help you to see how easy it is to use a CAD system. Learning these basics will help build confidence and experience needed to master the drafting program that follows.

Prompts

CAD systems guide you through the various aspects of developing a drawing through written instructions that appear on the monitor screen. The instructions, referred to as prompts, cues, options, or message lines, may appear in any location on the screen as shown in Fig. 4-2-1A. Some systems use a dual-screen setup. One screen is for the graphics; the other is for prompting the user. An example of this is shown in Fig. 4-2-1B.

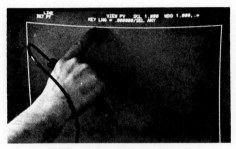

(A) SCREEN MESSAGE OPTIONS

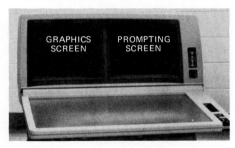

(B) DUAL SCREENS

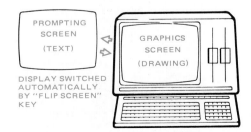

(C) SINGLE-SCREEN CONFIGURATION

Fig. 4-2-1 Prompt locations. (A, CADAM, Inc.; B, Intergraph Corp.)

A third method of displaying the prompts is to alternately "toggle" the same screen as illustrated in Fig. 4-2-1C. The display may be automatically "flipped" by using a simple command. Thus, prompts are found in one of three locations:

- Overlay: prompts on the same screen as the drawing.
- Dual screen: separate screens for the drawing and the prompts.
- Toggling: flip the same screen to alternately display the drawing or the prompts.

Start-Up Procedures

The system must be turned on, or "booted up," before any graphic data can be displayed on the screen. Since each CAD manufacturer has a different start-up procedure, follow the procedures outlined in the CAD manufacturer's handbook.

Menus

CAD systems are "menu-driven." This means you make a menu selection to "call up" a particular part of a program to create graphics. For example, the selection of LINE allows line creation to occur.

There are two categories of menus: ROOT MENU and SUBMENU. Menu selection is made by keyboard input, tablet selection, or moving the cursor to the screen menu item and "setting" it. Menu selection on the keyboard is made by typing in the name of the desired command then pressing ENTER. ENTER is an "end-of-statement" command.

Root Menu A root or master menu for a CAD system is similar in nature to menus found in restaurants. The restaurant master menu lists the foods under seafood, meat, dessert, and so on. The CAD root menu lists major drafting commands such as DRAW, EDIT, DIMENSION, and so on, as shown in Fig. 4-2-1. To create a particular type of graphic, simply select that menu item. For example, automatic sectioning may be created by selecting HATCH. Many of the menu items are abbreviated or shown graphically as icons because of the length of their names. Each, however, is easily identifiable.

When a menu item is selected, there is a time delay. Since the CAD system must search for a command, always remember to wait until the system is ready.

Submenu Each root menu item has subtasks and/or options, also referred to as *submenus*. They allow the selection of various types of graphics. Again using the analogy of a restaurant menu, "Seafood" would have a submenu option list giving the available selections. A CAD root menu item similarly may have various selections as illustrated in Fig. 4-2-2.

ROOT	SUB
SEAFOOD	LOBSTER
MEAT	CRAB LEGS
	SHRIMP
SALADS	SCALLOPS
DESSERTS	HADDOCK

(A) RESTAURANT MENU

ROOT	SUB
DRAW	ARC
	CIRCLE
EDIT	LINE
DIMENSION	TEXT

(B) CAD MENU

Fig. 4-2-2 Similarity between restaurant and CAD menus.

DRAWING STRAIGHT LINES

The lines created using a CAD system provide the shape information used for building and producing the product. The common ways to create horizontal, vertical, and inclined straight lines include:

1. Selection by free input
2. Use of a grid pattern
3. Use of construction lines
4. Keying in distances (coordinate input).

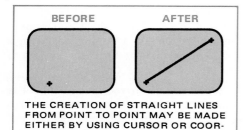

THE CREATION OF STRAIGHT LINES FROM POINT TO POINT MAY BE MADE EITHER BY USING CURSOR OR COORDINATE INPUT.

Fig. 4-2-3 Line (solid) command.

Drawing a Straight Line by Free Input

The first graphics to be placed on the screen will usually be a straight line. (See Fig. 4-2-3.) Two points (or positions) will determine the endpoints of a line while in the LINE command. Depending on the position of the second point relative to the first, the line will be drawn horizontally, vertically, or inclined. The common methods used to place a line on the monitor follows:

1. Select the LINE command by:
   A. Pointing the cursor to the "on screen" menu and selecting it.
   B. Using the appropriate function or alphanumeric keys (e.g., type LINE and press ENTER).
   C. Digitizing by activating a stylus or puck within one of the boxes on the graphics tablet as shown in Fig. 4-2-4.
2. Use the input pointing device to select the first line endpoint as shown in Fig. 4-2-5A. When using a joystick, for example, "set" the cursor by pressing the button on the device.

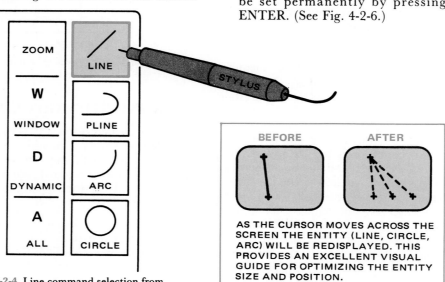

Fig. 4-2-4 Line command selection from graphics tablet.

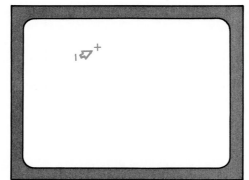

(A) FIRST ENDPOINT

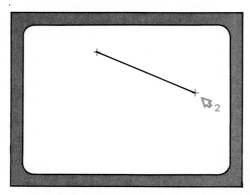

(B) SECOND ENDPOINT AND LINE

Fig. 4-2-5 Drawing a line on the monitor.

With a graphic tablet, activate the stylus (press down) or puck (press a button) at that location. Any dots (grids) on the screen should be disregarded for the time being.

3. The actual line will lengthen as you move the cursor. This is known as *rubberbanding*. Select the second line endpoint. A line will appear as shown in Fig. 4-2-5B. The line may have to be set permanently by pressing ENTER. (See Fig. 4-2-6.)

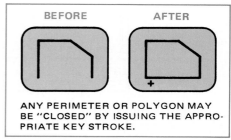

AS THE CURSOR MOVES ACROSS THE SCREEN THE ENTITY (LINE, CIRCLE, ARC) WILL BE REDISPLAYED. THIS PROVIDES AN EXCELLENT VISUAL GUIDE FOR OPTIMIZING THE ENTITY SIZE AND POSITION.

Fig. 4-2-6 Rubberband command.

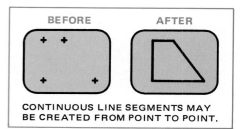

CONTINUOUS LINE SEGMENTS MAY BE CREATED FROM POINT TO POINT.

Fig. 4-2-7 Multi-segment line command.

Drawing Connecting Line Segments

The outline of any object may be created by using the following method:

1. Select the LINE command.
2. Select the first line endpoint at position 1 as shown in Fig. 4-2-8.

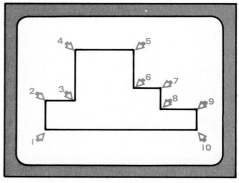

Fig. 4-2-8 Drawing an object.

3. Successively select line endpoints at positions 2, 3, 4, and so on, to position 10.
4. The outline of the object may be completed in one of two ways. Either select position 1 a second time or select the CLOSE option. The object will appear as shown in Fig. 4-2-8. Remember that pressing the ENTER key may be a necessary part of setting the line after reselecting position 1.

ANY PERIMETER OR POLYGON MAY BE "CLOSED" BY ISSUING THE APPROPRIATE KEY STROKE.

Fig. 4-2-9 Close command.

Drawing Separate Lines

You may need to create an additional line, separate from the first. This is accomplished by moving the cursor to a new position. If the enter key was used to place the first line, be certain you are in the LINE command and select the beginning location for the next line. Repositioning the cursor location without drawing a line is known as MOVE NEXT POINT. The concept is the same as in manual drafting where the pencil must be lifted and moved to a new position. In CAD, the cursor is lifted to a new start position by the MOVE NEXT POINT option. See Fig. 4-2-10.

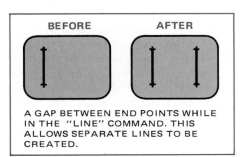

A GAP BETWEEN END POINTS WHILE IN THE "LINE" COMMAND. THIS ALLOWS SEPARATE LINES TO BE CREATED.

Fig. 4-2-10 Move next point command.

You will often draw a line that is not required. If this mistake is immediately realized, it is easy to correct. CAD systems are very forgiving. Every system has an ERASE LAST command option. It may be called ERASE LAST, OOPS, or DELETE LAST ENTITY. The last command executed can easily be erased by simply hitting the appropriate erase command. See Fig. 4-2-11.

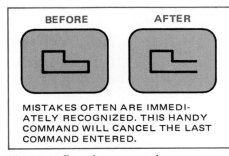

MISTAKES OFTEN ARE IMMEDIATELY RECOGNIZED. THIS HANDY COMMAND WILL CANCEL THE LAST COMMAND ENTERED.

Fig. 4-2-11 Erase last command.

The following procedure is used to create separate lines:

1. Select the LINE command.
2. Locate and set the cursor at position A and then B, as shown in Fig. 4-2-12A. The first line is created as shown.
3. Locate and set the cursor at position C (Fig. 4-2-12A). The second line segment is inadvertently created

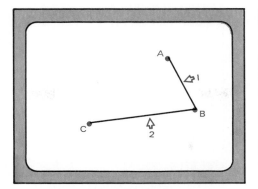

(A) DRAW LINE SEGMENT ABC

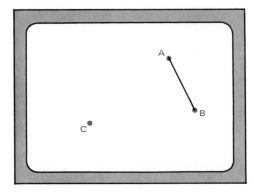

(B) ERASE SEGMENT BC AND MOVE TO NEW POSITION

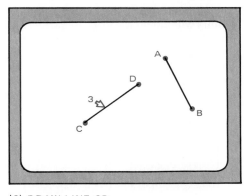

(C) DRAW LINE CD

Fig. 4-2-12 Creating separate lines.

since the MOVE NEXT POINT option was not issued.
4. Immediately select ERASE LAST. The second line will be removed.
5. Select MOVE NEXT POINT. This may be done by selecting the LINE command again or a special key command such as pressing ENTER twice.
6. To begin the second line, locate and set the cursor at position C as shown in Fig. 4-2-12B.
7. Locate and set the cursor at position D as shown in Fig. 4-2-12C. A separate line (3) will be created.
8. Additional separate lines may be created by repeating Steps 5, 6, and 7.

Drawing Horizontal and Vertical Lines

Lines may be created horizontally, vertically, or inclined as previously described. CAD systems have the option to lock onto a horizontal or vertical axis and produce a precise result. This horizontal-vertical constraint, known as ORTHO (short for orthogonalize), will ensure that the lines created will be horizontal or vertical.

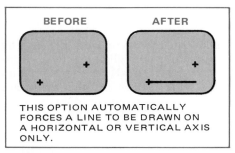

THIS OPTION AUTOMATICALLY FORCES A LINE TO BE DRAWN ON A HORIZONTAL OR VERTICAL AXIS ONLY.

Fig. 4-2-13 Ortho on command.

This is an important feature since technical drawings consist of many such lines. Without ORTHO you would have to estimate the second endpoint position. The likely result is shown in Fig. 4-2-15A. Even though the line is nearly horizontal, it is not exact. This will be shown by jagged rather than smooth lines on the monitor. The result of locking in on an axis is shown in Fig. 4-2-15B. Notice that the second cursor endpoint location (+) would not have produced the desired result. Since the command has been given to lock in on the axis, however, the resulting horizontal line is exact. This option has the additional advantage of obtaining reference positions from adjacent views when constructing multiview drawings.

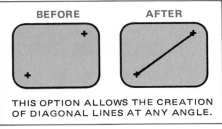

THIS OPTION ALLOWS THE CREATION OF DIAGONAL LINES AT ANY ANGLE.

Fig. 4-2-14 Ortho off command.

The method used to accomplish ORTHO is:

1. Select the LINE command.
2. Select the first line endpoint.
3. Use the appropriate function key or tablet menu to select the ORTHO ON option.

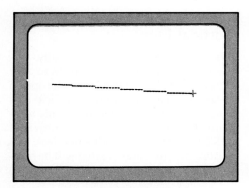

(A) NON-ORTHO LINE

Fig. 4-2-15 Creating a horizontal line.

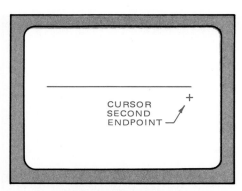

(B) ORTHO LINE

4. Select the second endpoint. The resulting line will be exactly horizontal (or vertical) as shown in Fig. 4-2-15B. Notice also that the length of the line terminates directly above the cursor location.

Drawing Lines Using A Grid Pattern

A grid pattern is useful when learning CAD because it is quick and accurate. An example of a rectangular grid pattern is shown in Fig. 4-2-17. The desired grid pattern size is displayed so that the endpoints of all lines will fall on a grid dot. Grid points are located equidistant

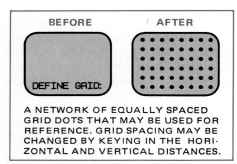

Fig. 4-2-16 Grid command.

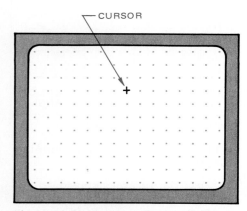

Fig. 4-2-17 Grid pattern and cursor display on screen.

(horizontally and vertically) from each other. When it is desired to position points at the nearest grid dot, use the input device to move the cursor. Set the cursor when it appears near a desired point. Depending on your input device, this is done by digitizing, joystick fire button, or others. When a special command called SNAP is activated, the point snaps to the nearest dot. The location is determined by both the spacing of the grids and the snap increment. The dot spacing is standard with many CAD systems. At times, line endpoints will fall between grid dots. The grid pattern may then be reset to a size where the endpoints will fall on the grid. Regardless of the spacing, the grid is not part of the drawing and is used for reference (snapping to) only. Use the following procedure to use the grid SNAP-ON option:

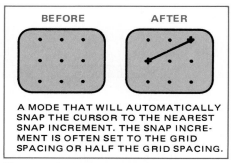

Fig. 4-2-18 Snap-on command.

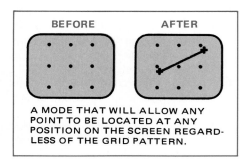

Fig. 4-2-19 Snap-off command.

1. Select the GRID setting command.
2. Follow the prompts by keying in the desired horizontal and vertical grid interval (e.g., .5, and press ENTER once or twice depending on the software). Grid dots will be displayed .50 in. apart in both the *X* (horizontal) and *Y* (vertical) directions as shown in Fig. 4-2-17. *Note:* you will not be able to display a small grid dot spacing of less than .25 in. for B-/(17 in. x 11 in.) and C-/(22 in. x 17 in.) size drawings. This would result in too many grids per unit area. The system will prompt you with a message such as GRID TOO DENSE. A smaller grid will be displayed only by enlarging a portion of the drawing (rescaling).
3. Select the SNAP option (also a sub-menu option).
4. Select ON and a value equal to the grid pattern (e.g., .5, and press ENTER). Use the grid pattern to SNAP to the nearest dot. The SNAP option and GRID pattern can be turned on or off by using a toggle command. A simple word, function, or alphanumeric key input will accomplish this.
5. Select the LINE command.
6. Locate and set the cursor (point 1). It snaps to the closest grid dot (point 2) as shown in Fig. 4-2-20A.
7. Locate and set the cursor (point 3). It snaps to the closest grid dot (point 4) and a line is drawn on the grid pattern (Fig. 4-2-20B).
8. Locate and set the cursor near point 5 and select the CLOSE option. In order to create a triangle, ORTHO must be off. With ORTHO on you cannot make inclined lines, only horizontal and vertical. The triangle shown (Fig. 4-2-20C) is created on the grid pattern. It is exactly 2.00 in. (4 grids) wide and 1.00 in. (2 grids) high. Additional separate lines may be created in a like manner.

The *grid snap* concept is also useful. You are able to create graphics quickly and with accuracy when using a grid system. As the operator becomes more experienced, grids will not be used since most objects are not composed on 1-in. or 10-mm increments. Although a grid pattern may be changed, it should not be used for an entry such as for drawing a line 3.456 in. long. (Creating a line to an accurate length will be covered later in this unit.) All CAD systems have an option to remove the SNAP simply by selecting the SNAP-OFF option. A point

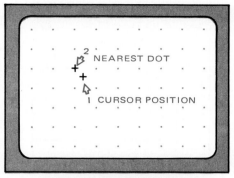

(A) FIRST ENDPOINT GRID SNAP

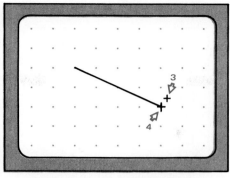

(B) SECOND ENDPOINT GRID SNAP

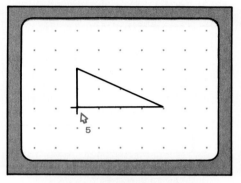

(C) FINISH THE OBJECT USING GRIDS

(D) FIRST LINE DRAWN WITH SNAP OFF

Fig. 4-2-20 Line drawn with and without the snap option.

ent brightness or dashed lines) on the monitor. These lines will allow the transferring of information in both the horizontal and vertical directions. This will assist in the creation of lines of uneven length during the development stage of a drawing. If a construction line option is not available, use any other type of line (e.g., dash or dot) as a guide. After the object is drawn, the construction lines may be removed by using the CONSTRUCTION LINE submenu option or the DELETE command. See Fig. 4-2-22.

Magnification (Zoom)

Any portion of the screen may be temporarily enlarged to any desired magnification. This is a valuable option (not available with manual drafting) to greatly improve drawing accuracy. The command is called ZOOM.

The WINDOW or ZOOM display command is used to change the magnification of an object. You can increase the magnification significantly to more clearly view small details. It will temporarily change the size of the display, but not the scale of the drawing. For

can now be positioned at any location on the screen, regardless of the grid pattern. The selection of points 1 and 3 above, for example, would result in a line drawn exactly from the selected cursor positions (Fig. 4-2-21).

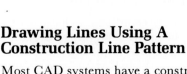

REFERENCE LINES ASSIST MULTI-VIEW DRAWING CONSTRUCTION. THIS IS ESPECIALLY USEFUL WHEN A GRID PATTERN CANNOT BE USED. THE LINES ARE FOR REFERENCE PURPOSES ONLY AND WILL NOT APPEAR ON THE FINISHED PLOT.

Fig. 4-2-21 Construction lines command.

Drawing Lines Using A Construction Line Pattern

Most CAD systems have a construction line option. Construction lines, like a grid pattern, are used as a guide for constructing the drawing. Thus, they will be displayed differently (e.g., differ-

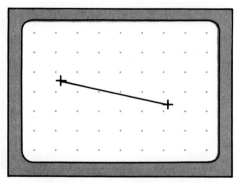

(A) SKETCH OF A PART TO BE DRAWN

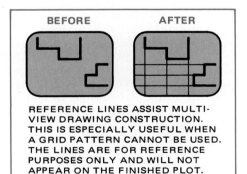

(C) USING CONSTRUCTION LINES AS A GUIDE

Fig. 4-2-22 Use of construction lines.

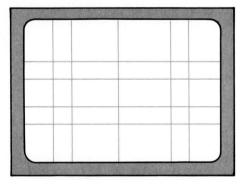

(B) DRAWING CONSTRUCTION LINES

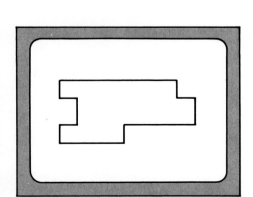

(D) CONSTRUCTION LINES REMOVED

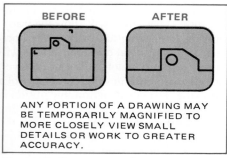

ANY PORTION OF A DRAWING MAY BE TEMPORARILY MAGNIFIED TO MORE CLOSELY VIEW SMALL DETAILS OR WORK TO GREATER ACCURACY.

Fig. 4-2-23 Zoom command.

example, it can be used to double the size of a display so that additional information can be added to the drawing using the original numeric values.

During zooming in it may be desirable to change the area of interest without returning back to the original size. This may be accomplished by a PAN option. PAN enables you to move the zoom window to a different area of concentration without having to return to full window. Thus, it allows access to a part of the drawing outside of the original zoomed-in window area.

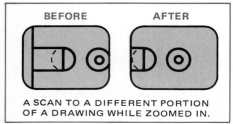

A SCAN TO A DIFFERENT PORTION OF A DRAWING WHILE ZOOMED IN.

Fig. 4-2-24 Pan command.

Application of ZOOM and PAN commands are shown in Fig. 4-2-25.

Coordinate Input

We have seen how a grid pattern is used to great advantage by quickly and accurately creating lines. However, many dimensions do not fall within a standard grid pattern. An example is a line that is 1.356 in. long. Line lengths such as this are created by coordinate input.

The alphanumeric keyboard can be used to generate points in exact locations and lines of exact length. There are three methods of coordinate input:

1. Absolute coordinate
2. Relative coordinate
3. Polar coordinate (line length and angle)

Coordinate input is based on the rectangular (horizontal and vertical) measurement system. All absolute distances are described in terms of their distance from the drawing origin. Relative and polar coordinates may be described with

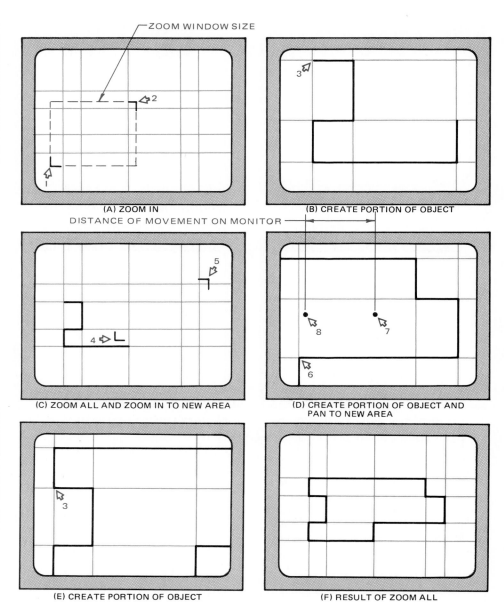

(A) ZOOM IN

(B) CREATE PORTION OF OBJECT

(C) ZOOM ALL AND ZOOM IN TO NEW AREA

(D) CREATE PORTION OF OBJECT AND PAN TO NEW AREA

(E) CREATE PORTION OF OBJECT

(F) RESULT OF ZOOM ALL

Fig. 4-2-25 Use of zoom and pan command.

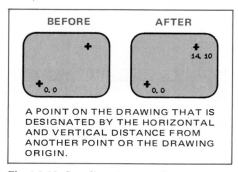

A POINT ON THE DRAWING THAT IS DESIGNATED BY THE HORIZONTAL AND VERTICAL DISTANCE FROM ANOTHER POINT OR THE DRAWING ORIGIN.

Fig. 4-2-26 Coordinate command.

respect to a particular point (position) on the drawing. This position can be at any location on the drawing and is described by two-dimensional coordinates, horizontal (X) and vertical (Y). The X axis is horizontal and is considered the

first and basic reference axis. The Y axis is vertical and is 90° to the X axis. Any distance to the right of the position is considered a positive X value and any distance to the left a negative X value. Distances above the position are considered positive and distances below are negative values.

Absolute Coordinates

An absolute coordinate is located relative to the drawing origin. The origin is normally located at the lower left of the monitor so that all values are positive with respect to the origin. The origin is a base reference point from which all positions on the drawing are measured. Its dimensions are $X = 0$ and $Y = 0$, referred to as 0,0. A point (current

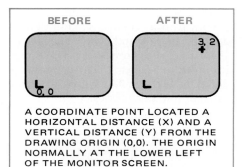

Fig. 4-2-27 Absolute coordinate command.

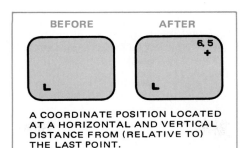

Fig. 4-2-28 Relative coordinate command.

access position) may be accurately located with respect to the origin. Its position is referred to as an absolute coordinate.

The method used to input an absolute coordinate point is:

1. Select the POINT command (and ABSOLUTE COORDINATE option if necessary).
2. Use the alphanumeric keyboard to specify the X (horizontal) and Y (vertical) distance from the origin. For example, if you wish to locate a point 8.8 in. to the right and 6.1 in. above the origin, key in 8.8, 6.1 and press ENTER. Remember to place a comma between the X and Y coordinates. Also, the ENTER key must be used to end the command. You could, for example, key to two decimal places (6.12), three (6.123), and so on. The two coordinates need not be to the same number of decimal places. The absolute coordinate will be located as shown in Fig. 4-2-23A. The coordinate values have been shown for reference purpose only. They will not be displayed on the monitor even though the current access location is identified, as desired, off to the side of the screen.
3. Additional coordinate points may be placed by repeating Step 2. Lines would connect each of the coordinate points if you were in the LINE command.

Relative Coordinates

A relative coordinate is located with respect to the current access location (last cursor position selected) rather than the origin (0,0). In other words, it is located with respect to another point on your drawing. Often, both absolute and relative coordinate input are used during drawing preparation. A line may be created by combining these two methods as follows:

1. Select the LINE command.

2. Select the option for ABSOLUTE COORDINATE input.
3. Key in the desired X and Y values for the first endpoint, e.g., 10.5, 10, and press ENTER. The first endpoint (point 1), which is now the current access location, will be displayed as shown in Fig. 4-2-29A. It is 10.5 in. to the right, and 10 in. above the origin.
4. Select the RELATIVE COORDINATE input option (e.g., use the appropriate keyboard command such as the @ key).
5. Key in the desired X and Y values for the second endpoint, e.g., 5.12, 0. It will be located 5.12 in. to the right "relative" to the last point selected as shown in Fig. 4-2-29B.

6. Press ENTER. The line (2) will appear on the screen at the desired location, as shown in Fig. 4-2-29C. It begins 10.5 in. to the right and 10 in. up from the origin. It will be horizontal and 5.12 in. long extending right (positive direction).
7. Select the RELATIVE COORDINATE input option.
8. Key in −5.12, −5. The new position will be 5 in. below and 5.12 in. to the left of the second position. Notice the result of using the minus sign.
9. Press ENTER. The result is shown in Fig. 4-2-29D.

Polar Coordinates

A polar coordinate is similar to a relative coordinate since it is positioned with

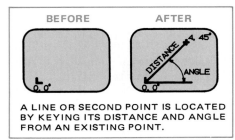

Fig. 4-2-30 Polar coordinate command.

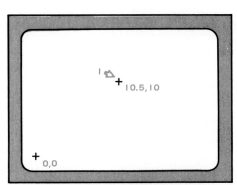

(A) FIRST POSITION (ABSOLUTE COORDINATE)

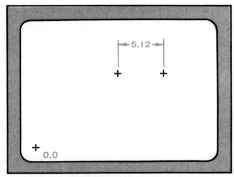

(B) SECOND POSITION (RELATIVE COORDINATE)

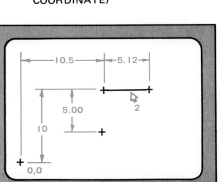

(C) THIRD POSITION (RELATIVE COORDINATE

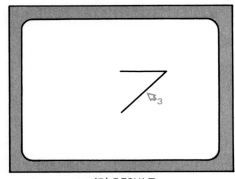

(D) RESULT

Fig. 4-2-29 Line creation using coordinate input.

respect to the current access location. A line, however, will be specified according to its actual length and a direction rather than an *X*, *Y* coordinate distance (Fig. 4-2-31).

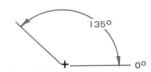

Fig. 4-2-31 Polar coordinate angle.

A line may be created relative to a position using polar coordinate input as follows:

1. Select the LINE command.
2. Locate point 1 as shown in Fig. 4-2-32A.
3. Select the POLAR COORDINATE option and specify the line length and angle (e.g., key in @ 6.5 < 45 and press ENTER). The line, 6.5 in. long, will be created at 45° relative to point 1 as shown in Fig. 4-2-32B.
4. A second line can be created relative to the new access location (point 2, Fig. 4-2-32B) by repeating Step 3. This time, however, by keying in @ 6.5 < 315 and pressing ENTER, it will produce a line having coordi-

nates as shown in Fig. 4-2-32C. The second line is drawn 6.5 in. long and 315° counterclockwise from point 2. The final result is shown in Fig. 4-2-32D.

The identification of the coordinates and angle of these or any lines may easily be checked by using the DISTANCE/ANGLE option.

LINE STYLES

All CAD systems have the option to create different line styles. Refer to Fig. 4-1-1. On larger systems, these options are found on an auxiliary menu shown on the CRT or on the tablet menu.

On some systems, all lines appear the same thickness when developing the drawing on the CRT monitor. Only when the finished drawing is made on the plotter can the variation in line thickness be seen.

Programming the plotter to determine which lines are thin and which are thick is done as the drawing is being developed. The plotter, which produces the finished drawing, is equipped with multiple or interchangeable pens. For example, plotter pen 1 may produce thick lines, plotter pen 2 thin lines. When creating the shape description (solid,

visible lines only), the computer is programmed to use pen 1. To add dimension and hidden and center lines to the drawing, the computer is reprogrammed to use pen 2. The pen task, as shown in Fig. 4-2-33, is used to create different line thicknesses.

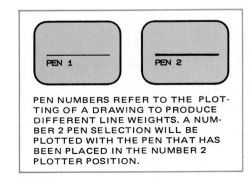

PEN NUMBERS REFER TO THE PLOTTING OF A DRAWING TO PRODUCE DIFFERENT LINE WEIGHTS. A NUMBER 2 PEN SELECTION WILL BE PLOTTED WITH THE PEN THAT HAS BEEN PLACED IN THE NUMBER 2 PLOTTER POSITION.

Fig. 4-2-33 Pen command.

TEXT

The TEXT command allows you to add letters, numbers, words, notes, symbols, and messages to the drawing. Additionally, it can be used for size description (lengths, diameters, etc.) or to provide finishing touches, such as information that cannot otherwise be shown on a drawing. The alphanumeric keyboard is an indispensable input device used to specify the text. Text style and height can be altered to suit the drawing requirements. Each letter, number, or symbol is separately keyed in. Size must be established by the drafter in order to be proportionately correct on the final printed drawing.

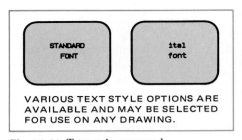

VARIOUS TEXT STYLE OPTIONS ARE AVAILABLE AND MAY BE SELECTED FOR USE ON ANY DRAWING.

Fig. 4-2-34 Text style command.

ERASING

The need for revision or change is inherent in preparing technical drawings. Consequently, a variety of EDIT commands are available. Each is used to facilitate the creation of a drawing. One of the most common EDIT commands is ERASE. Several ways to eliminate various entities from a drawing are:

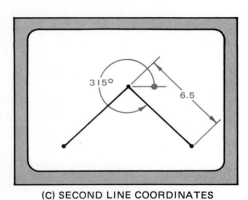

(A) SELECTING A CURRENT ACCESS LOCATION

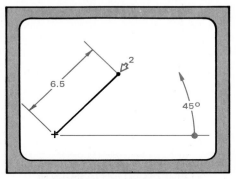

(B) FIRST LINE COORDINATES

(C) SECOND LINE COORDINATES

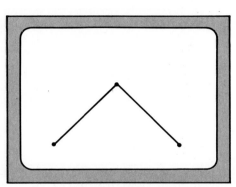

(D) RESULT

Fig. 4-2-32 Polar coordinate input.

1. ERASE LAST entity
2. ERASE ANY entity
3. ERASE BY WINDOW
4. ERASE ALL
5. BREAK

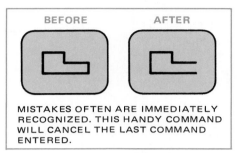

BEFORE   AFTER

MISTAKES OFTEN ARE IMMEDIATELY
RECOGNIZED. THIS HANDY COMMAND
WILL CANCEL THE LAST COMMAND
ENTERED.

Fig. 4-2-35 Erase last command.

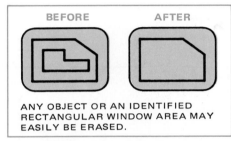

BEFORE   AFTER

ANY OBJECT OR AN IDENTIFIED
RECTANGULAR WINDOW AREA MAY
EASILY BE ERASED.

Fig. 4-2-36 Erase command.

ERASE LAST is used most frequently. You will usually recognize an error immediately after it has been made. When this occurs, select the ERASE LAST command, as previously discussed in this unit under Drawing Separate Lines.

A portion of the drawing can be removed by using the ERASE BY WINDOW command. See Fig. 4-2-37.

ASSIGNMENTS

See Assignments 5 through 19 for Unit 4-2 on pages 59 and 60.

U N I T   4 - 3
CIRCLES AND ARCS

CENTER LINES

Center lines consist of alternating long and short dashes (Fig. 4-3-1). They are used to represent the axis of symmetrical parts and features, bolt circles, and paths of motion. The long dashes of the center lines may vary in length, depending upon the size of the drawing. Center lines should start and end with long

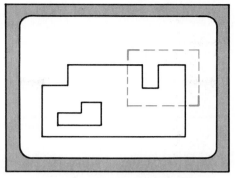

(A) SELECT THE WINDOW

Fig. 4-2-37 Erase by window.

dashes and should not intersect at the spaces between dashes. Center lines should extend uniformly and distinctly a short distance beyond the object or feature of the drawing unless a longer extension is required for dimensioning or for some other purpose. They should not terminate at other lines of the drawing, nor should they extend through the space between views. Very short center lines may be unbroken if no confusion results with other lines.

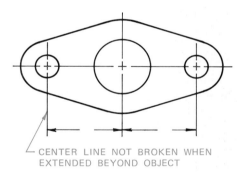

CENTER LINE NOT BROKEN WHEN
EXTENDED BEYOND OBJECT

USE TWO SHORT DASHES
AT POINT OF INTERSECTION

Fig. 4-3-1 Center-line technique.

Center lines are used to locate the center of circles and arcs. They are first drawn as light construction lines, then finished as alternate long and short dashes, with the short dashes intersecting at the center of the circle.

CAD

Center lines may be drawn by following the line commands explained in Unit

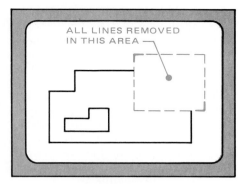

ALL LINES REMOVED
IN THIS AREA

(B) RESULT

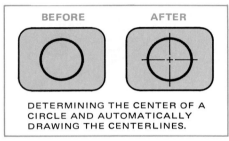

BEFORE   AFTER

DETERMINING THE CENTER OF A
CIRCLE AND AUTOMATICALLY
DRAWING THE CENTERLINES.

Fig. 4-3-2 Center line (automatic) command.

4-2. See Fig. 4-3-2. At the time of this writing many CAD software packages are not programmed to place the short dashes at the center of the circle.

DRAWING CIRCLES AND ARCS

Circles and arcs are drawn with the aid of a compass or template. When drawing circles and arcs with a compass, it is recommended that the compass lead be softer and blacker than the pencil lead being used on the same drawing. For example, if you are drawing with a 2H or 3H pencil, use an H compass lead. This will produce a drawing having similar linework since it is necessary to compensate for the weaker impression left on the drawing medium by the compass lead as compared with the stronger direct pressure of the pencil point. For drawing circles and arcs, see Figs. 4-3-3 and 4-3-4.

It is essential that the compass lead be reasonably sharp at all times in order to ensure proper line width. The compass lead should be sharpened to a bevel point, with the top rounded off as shown in Fig. 4-3-5. The lead is slightly shorter than the needle point.

Drafters find it much easier and faster to use circle templates. There are sets which contain all common sizes and shapes of holes that most drafters are ever called upon to draw.

When using a circle template, choose the correct diameter, line up the marks on the template with the center lines, and trace a dark thick line.

The drawing of arcs should be done before the tangent lines are made heavy. Draw light construction lines to establish the compass point location and check to make certain that the compass lead meets properly with both tangent lines before drawing the arc.

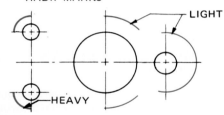

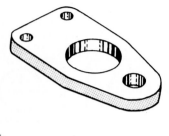

(A) ESTABLISH CENTER LINES AND RADII MARKS

(B) DRAW CIRCLES AND ARCS

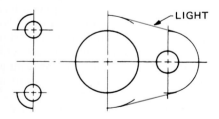

(C) DRAW TANGENT LINES

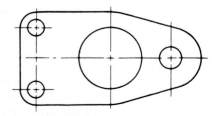

(D) COMPLETE OBJECT LINES

Fig. 4-3-4 Sequence of steps for drawing a view having circles and arcs.

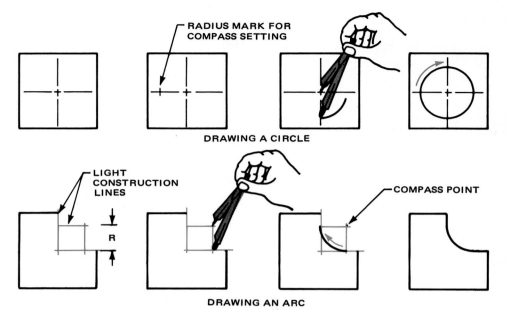

Fig. 4-3-3 Drawing circles and arcs.

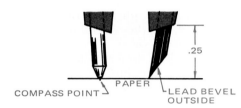

COMPASS POINT   PAPER   LEAD BEVEL OUTSIDE

Fig. 4-3-5 Sharpening and setting the compass lead.

CAD

Circles. There are several common methods used to construct circles. The common methods used to draw circles include: (1) center and radius; (2) center and diameter; (3) three-point circle; and (4) two-point circle. See Fig. 4-3-6. For drawing concentric circles, see Fig. 4-3-7.

Arcs The common methods used to draw arcs include: (1) three points; (2) start, center, and end; (3) start, end, and radius; and (4) start, center, and angle.

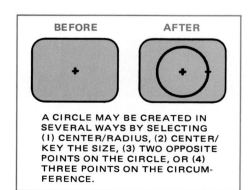

A CIRCLE MAY BE CREATED IN SEVERAL WAYS BY SELECTING (1) CENTER/RADIUS, (2) CENTER/ KEY THE SIZE, (3) TWO OPPOSITE POINTS ON THE CIRCLE, OR (4) THREE POINTS ON THE CIRCUM-FERENCE.

Fig. 4-3-6 Circles command.

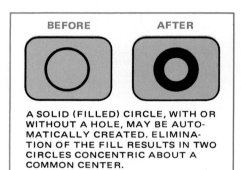

A SOLID (FILLED) CIRCLE, WITH OR WITHOUT A HOLE, MAY BE AUTO-MATICALLY CREATED. ELIMINA-TION OF THE FILL RESULTS IN TWO CIRCLES CONCENTRIC ABOUT A COMMON CENTER.

Fig. 4-3-7 Donut command.

These methods are all covered in the ARC command (Fig. 4-3-8). For draw-ing fillets, arcs, or straight lines tangent to circles, refer to Figs. 4-3-9, 4-3-10, and 4-3-11.

ASSIGNMENTS

See Assignments 20 through 26 for Unit 4-3 on page 62.

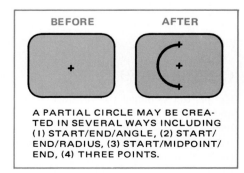

A PARTIAL CIRCLE MAY BE CREA-TED IN SEVERAL WAYS INCLUDING (I) START/END/ANGLE, (2) START/ END/RADIUS, (3) START/MIDPOINT/ END, (4) THREE POINTS.

Fig. 4-3-8 Arc command.

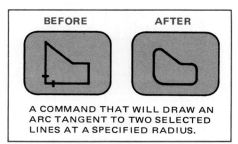

A COMMAND THAT WILL DRAW AN ARC TANGENT TO TWO SELECTED LINES AT A SPECIFIED RADIUS.

Fig. 4-3-9 Fillet command.

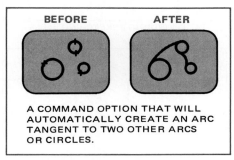

A COMMAND OPTION THAT WILL AUTOMATICALLY CREATE AN ARC TANGENT TO TWO OTHER ARCS OR CIRCLES.

Fig. 4-3-10 Arc tangent command.

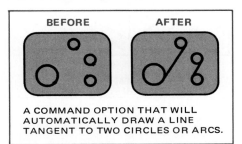

A COMMAND OPTION THAT WILL AUTOMATICALLY DRAW A LINE TANGENT TO TWO CIRCLES OR ARCS.

Fig. 4-3-11 Line tangent command.

U N I T 4 - 4
DRAWING IRREGULAR CURVES

Curved lines may be drawn with the aid of irregular curves, flexible curves, and elliptical templates (Fig. 4-4-1). After you have established the points through which the curved line passes, draw a light freehand line through these points. Next fit the irregular curve or other instrument by trial against a part of the curved line and draw a portion of the line. Move the curve to match the next portion, and so forth. Each new position should fit enough of the part just drawn (overlap) to ensure continuing a smooth line. It is very important to notice whether the radius of the curved line is increasing or decreasing and to place the irregular curve in the same way. If the curved line is symmetrical about the

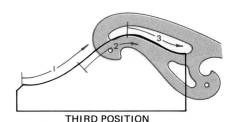

FIRST POSITION

SECOND POSITION

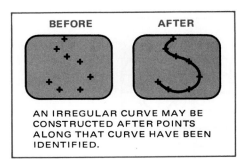

THIRD POSITION

Fig. 4-4-1 Drawing a curved line.

axis, the position of the axis may be marked on the irregular curve with a pencil for one side and then reversed to match and draw the other side.

CAD

An irregular curve is a nonconcentric, nonstraight line drawn smoothly through a series of points. In CAD systems, it is commonly referred to as a spline. See Fig. 4-4-2.

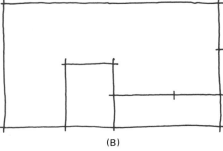

AN IRREGULAR CURVE MAY BE CONSTRUCTED AFTER POINTS ALONG THAT CURVE HAVE BEEN IDENTIFIED.

Fig. 4-4-2 Pline (spline) command.

ASSIGNMENTS

See Assignments 27 and 28 for Unit 4-4 on page 65.

U N I T 4 - 5
SKETCHING

Freehand sketching is a necessary part of drafting because the drafter in industry frequently sketches ideas and designs prior to making instrumental drawings.

The drafter may also use sketches to explain thoughts and ideas to other people in discussions of mechanical parts and mechanisms. Sketching, therefore, is an important method of communication. Practice in sketching helps the student to develop a good sense of proportion and accuracy of observation.

A fairly soft (HB, F, or H) pencil should be used for preliminary practice. Many types of graph or ordinate paper are available and can be used to advantage when close accuracy to scale or proportion is desirable. Freehand sketching of lines, circles, and arcs is illustrated in Fig. 4-5-1.

Since the shapes of objects are made up of flat and curved surfaces, the lines

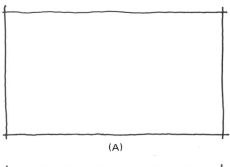

(A)

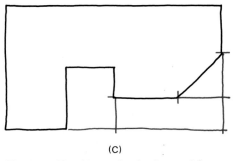

(B)

(C)

Fig. 4-5-1 Sketching a view having straight lines.

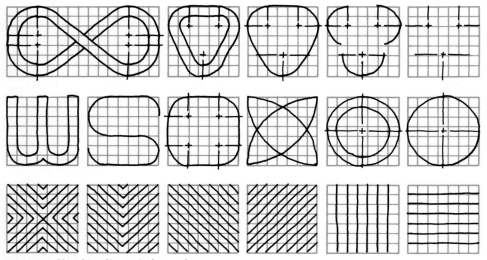

Fig. 4-5-2 Sketching lines, circles, and arcs.

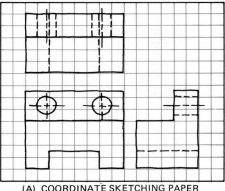

(A) COORDINATE SKETCHING PAPER

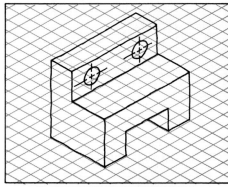

(B) ISOMETRIC SKETCHING PAPER

Fig. 4-5-5 Sketching paper.

forming views of objects will be both straight and curved. Do not attempt to draw long lines with one continuous stroke. First plot points along the desired line path, then connect these points with a series of light strokes.

When you are sketching a view (or views), first lightly sketch the overall size as a rectangular or square shape, estimating its proportions carefully. Then add lines for the details of the shape, and thicken all lines forming the view. See Fig. 4-5-2.

Figure 4-5-3 shows two methods of

sketching circles. Figure 4-5-4 illustrates, both pictorially and orthographically, the use of graph paper for the sketching of a machine part. Coordinate and isometric sketching paper are shown in Fig. 4-5-5.

CAD

Sketching is still a necessary part of drafting because the drafter in industry frequently sketches ideas and designs prior to making a CAD drawing. Sketching may be accomplished manually or on a CAD monitor using the SKETCH command. See Fig. 4-5-6.

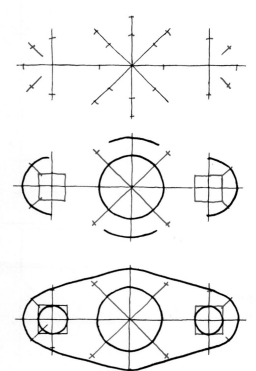

Fig. 4-5-3 Sketching a figure having circles and arcs.

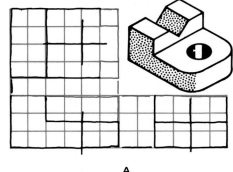

A

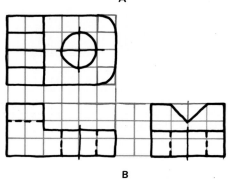

B

Fig. 4-5-4 Usual procedure for sketching three views.

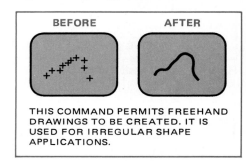

THIS COMMAND PERMITS FREEHAND DRAWINGS TO BE CREATED. IT IS USED FOR IRREGULAR SHAPE APPLICATIONS.

Fig. 4-5-6 Sketch command.

ASSIGNMENTS

See Assignments 29 through 32 for Unit 4-5 on page 65.

UNIT 4-6
INKING AND SCRIBING

INKING

In inking, technique is an important factor. The pen should be moved as it touches the film or paper at the beginning of a line and as it leaves the film or paper at the end of the line. This will minimize belling, or spreading of ink, at the line ends.

There are two inking pens on the market today, the needle-in-tube type and the blade type. Needle-in-tube type pens should be held nearly vertical and moved with a light touch across the drawing medium. See Fig. 4-6-1. Blade-type ruling pens should be inclined at an angle up to 15° from the vertical in the direction of motion.

Normal handling of drawing media is bound to soil them. Ink lines applied over soiled areas do not adhere well and may be chipped off or flake in time. It is always good practice to keep the drawing paper or film clean. Soiled areas can be cleaned effectively with a cleaner.

The use of a needle-in-tube pen with a circular template has, in many instances, replaced the drawing of circles by means of the blade-type pen compass, which must be frequently refilled with ink (Fig. 4-6-2). When you are inking circles by template, the only thing to look out for is to keep the template above the drawing medium. This is accomplished by inserting another template or other thin material between the template and the drawing surface in order to avoid blotting through capillary underflow of ink.

Newer models of pens and improved quick-drying inks have made inking much easier than in previous times.

CAD

Pen plotters, the most commonly used plotters, are used to produce a finished drawing.

SCRIBING

Scribing is a drafting technique which in some areas has replaced pen-and-pencil drafting. It has already done so for some types of close-tolerance work. With regular drafting, using pencil or pen, drawing is made on top of the drawing medium. With scribing, you incise, or cut, lines into a special surface with scribing tools, making lines that are sharp and clean, never vary in width, and can't be smudged. Scribe lines also produce the sharpest prints, direct from the scribed original. See Fig. 4-6-3.

Scribing is used for a variety of applications—as a drafting medium for the tools and templates from which fabricated parts are produced; in aircraft lofting and automobile layouts; in mapping; and for close-tolerance electronic circuitry and microcircuit layout.

In the past, glass, steel, and aluminum were considered the only stable

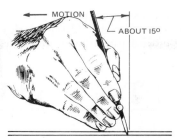
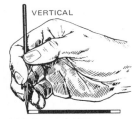
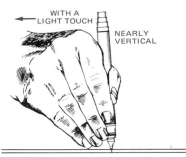
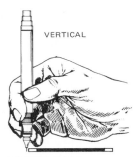

(A) INKING STRAIGHT LINES WITH A BLADE RULING PEN

(B) INKING STRAIGHT LINES WITH A NEEDLE-IN-TUBE PEN

Fig. 4-6-1 Drawing ink lines.

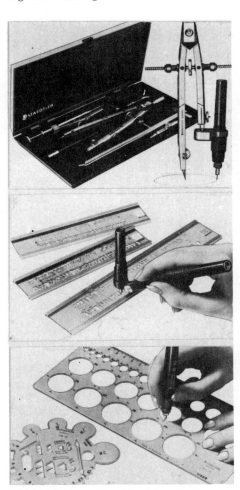

Fig. 4-6-2 Inking. (Staedtler Mars, Inc.)

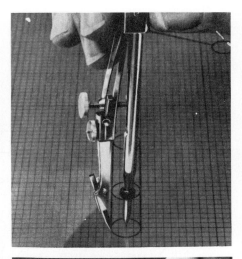
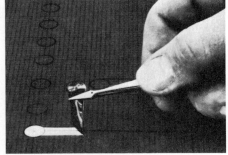

Fig. 4-6-3 Scribing. (Keuffel & Esser Co.)

base materials for scribing because of their extremely high dimensional stability. These materials were adequate but difficult to handle and file. In subsequent years, glass cloth was developed, and it became quite popular as a stable base material. At present, the most useful and stable drafting and reproduction medium is polyester film. It is available in a wide range of surfaces for the precision drafting and reproduction requirements of such industries as mapping and electronics.

There are specially prepared scribe surfaces for the critical line work of mapping and undimensioned drafting—

coatings which can be cut with a blade and then peeled back to the transparent film base.

ASSIGNMENTS FOR CHAPTER 4

Assignments for Unit 4-1, Straight Linework, Lettering, and Erasing—Manual Drafting

A Note About Dual Dimensions

The dual dimensions shown in this book, especially in the assignment sections, are neither hard nor soft conversions. Instead, the sizes are those that would be most commonly used in the particular dimensioning units and so are only approximately equal. Dual dimensioning this way avoids awkward sizes and allows instructor and student to be confident when using either set of dimensions.

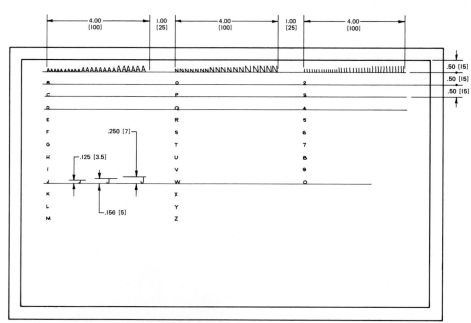

Fig. 4-1-A Lettering assignment.

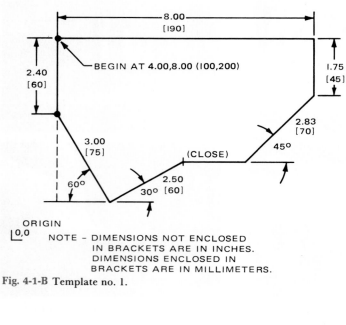

NOTE – DIMENSIONS NOT ENCLOSED IN BRACKETS ARE IN INCHES. DIMENSIONS ENCLOSED IN BRACKETS ARE IN MILLIMETERS.

Fig. 4-1-B Template no. 1.

NOTE - DIMENSIONS NOT ENCLOSED IN BRACKETS ARE IN MILLIMETERS. DIMENSIONS ENCLOSED IN BRACKETS ARE IN INCHES.

1. Lettering assignment. Set up a B (A3) size sheet similar to that shown in Fig. 4-1-A. Using uppercase Gothic lettering shown in Fig. 4-1-7 complete each line. Each letter and number is to be drawn several times to the three recommended lettering heights shown. Very light guidelines must be drawn first. The bracketed dimensions are millimeters.

2. On a B (A3) size sheet draw the template shown in Fig. 4-1-B. Scale 1:1. Do not dimension.

Fig. 4-1-C Template no. 2.

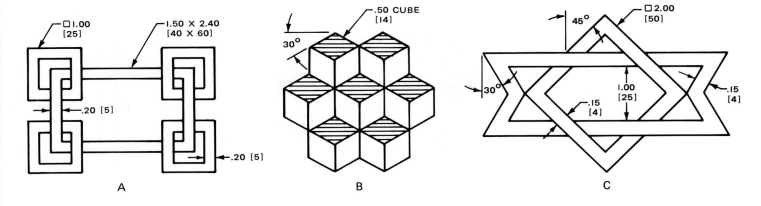

NOTE — DIMENSIONS NOT ENCLOSED IN BRACKETS ARE IN INCHES.
DIMENSIONS IN BRACKETS ARE IN MILLIMETERS.

Fig. 4-1-D Inlay designs.

3. On a B (A3) size sheet draw the template shown in Fig. 4-1-C. Scale 1:1. Do not dimension.

4. On a B (A3) size sheet draw the inlay designs shown in Fig. 4-1-D. Scale 1:1. Do not dimension.

Assignments for Unit 4-2, Straight Linework, Text, and Erasing—CAD

5. Text assignment. Set up a B (A3) size format similar to that shown in Fig. 4-1-A. Using uppercase lettering shown in Fig. 4-1-7, complete each line. Each letter and number is to be keyed several times. Vary each height as shown (bracketed dimensions are in millimeters).

6. Draw the objects shown in Fig. 4-2-A using the FREE INPUT line option. Drawings need not be created to scale but should be about the same proportion as that shown. When a mistake is made, use one of the ERASE options to make the correction.

7. Redraw the objects shown in Fig. 4-2-A but this time use the ORTHO line option. Drawings need not be to scale but should be about the same proportion as that shown. *Note:* For inclined lines, remove the ORTHO option.

8. On a B (A3) size format, draw any two of the objects shown in Fig. 4-2-B. Use either a 1.00-in. or 20-mm grid and snap options to create the views. Scale 1:1.

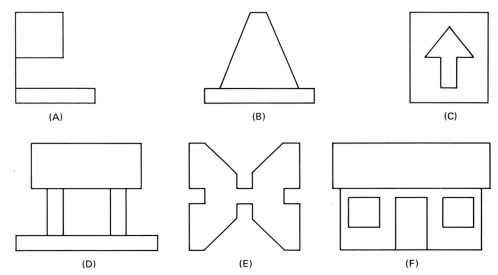

Fig. 4-2-A Drawing straight lines.

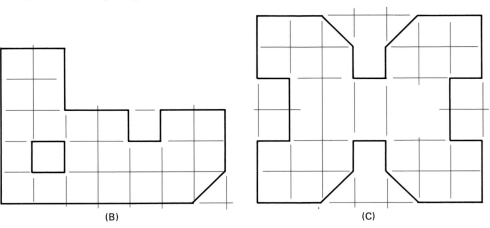

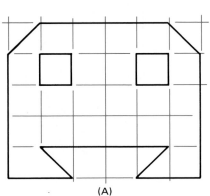

Fig. 4-2-B Assignment.

9. On a B (A3) size format, draw the three objects shown in Fig. 4-2-C. Use either a .50-in. or 10-mm grid and snap options to create the views. Scale 1:1.

10. On a B (A3) size format, draw the three objects shown in Fig. 4-2-D. Use either a .50-in. or 10-mm grid and snap options to create the views. Scale 1:1.

11. On a B (A3) size format, draw the part shown in Fig. 4-2-E. First draw the horizontal and vertical lines as shown in Fig. 4-2-A. Using the BREAK command remove the unwanted portions of the lines as shown by the finished drawing in B. Use either a .50-in. or 10-mm grid. Scale 1:1.

12. Same as Assignment 11 except use any of the problems shown in Fig. 4-2-F. Scale 1:1.

13. Same as Assignment 11 except replace the grid with construction lines for Fig. 4-2-E. Scale 1:1.

14. Same as Assignment 12 except replace the grid with construction lines for Fig. 4-2-F. Scale 1:1.

15. On a B (A3) size format, draw the parts shown in Fig. 4-2-G and plot the relative coordinates of each of the line intersections. In the bottom left corner is the drawing coordinate (absolute) starting point for each part. Move in a clockwise direction. Scale: one grid square equals .50 in. or 10 mm.

16. Same as Assignment 15 except use absolute coordinates.

17. Using absolute coordinates plot Figs. 4-2-H, and J on a B (A3) size format. The bottom left corner of the drawing is the starting point. Scale 1:1.

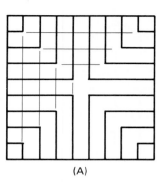
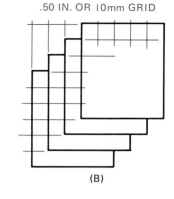
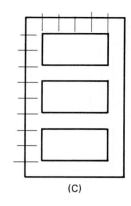

.50 IN. OR 10mm GRID

(A)　　　　(B)　　　　(C)

Fig. 4-2-C Assignment.

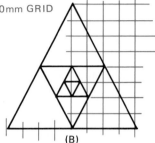
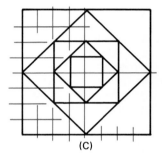
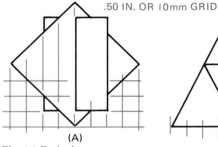

.50 IN. OR 10mm GRID

(A)　　　　(B)　　　　(C)

Fig. 4-2-D Assignment.

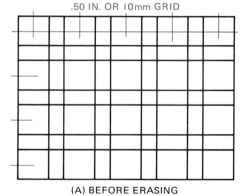

.50 IN. OR 10mm GRID

(A) BEFORE ERASING　　　　(B) AFTER ERASING

Fig. 4-2-E Assignment.

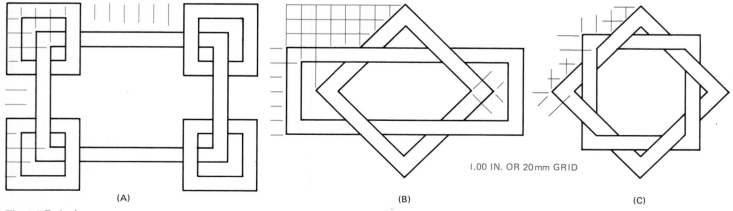

1.00 IN. OR 20mm GRID

(A)　　　　(B)　　　　(C)

Fig. 4-2-F Assignment.

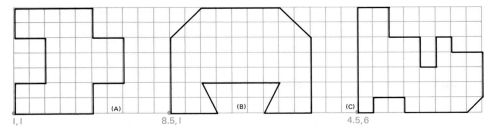

Fig. 4-2-G Assignment.

ABSOLUTE COORDINATES (INCHES)

Point	X Axis	Y Axis	Point	X Axis	Y Axis
1	.25	.25			
2	7.00	.25			
3	8.50	1.00		**New Start**	
4	8.50	2.25	20	.75	1.50
5	7.00	3.00	21	2.25	3.50
6	8.50	3.75	22	.75	3.50
7	8.50	5.25	23	.75	1.50
8	7.00	6.25			
9	5.50	6.25		**New Start**	
10	5.50	4.50	24	3.00	2.25
11	4.75	4.50	25	5.75	2.25
12	4.75	6.25	26	5.75	3.75
13	3.50	6.25	27	3.00	3.75
14	3.50	5.50	28	3.00	2.25
15	1.50	5.50			
16	1.50	6.25		**New Start**	
17	.75	6.25	29	2.75	.75
18	.25	5.5	30	6.00	.75
19	.25	.25	31	5.25	1.50
			32	3.50	1.50
			33	2.75	.75

Fig. 4-2-H Absolute coordinates (inches) assignment.

ABSOLUTE COORDINATES (MILLIMETERS)

Point	X Axis	Y Axis
1	10	10
2	50	10
3	50	20
4	120	20
5	120	10
6	150	10
7	180	30
8	220	30
9	220	100
10	160	100
11	160	130
12	140	130
13	140	160
14	110	160
15	120	140
16	90	140
17	70	100
18	40	120
19	60	160
20	40	160
21	10	140
22	10	80
23	20	40
24	10	40
25	10	10
	New Start	
26	40	50
27	160	50
28	120	90
29	40	70
30	40	50

Fig. 4-2-J Absolute coordinates (metric) assignment.

RELATIVE COORDINATES (INCHES)

Point	X Axis	Y Axis
1	0	0
2	4.50	0
3	0	.75
4	−.75	0
5	0	.75
6	−.75	0
7	0	.75
8	−3.00	0
9	0	−2.25
New Start—Solid		
10	0	.75
11	3.75	0
New Start		
12	−.75	.75
13	−3.00	0
New Start—Solid		
14	0	1.50
15	4.50	0
16	0	2.25
17	−4.50	0
18	0	−2.25
New Start—Solid		
19	0	.75
20	3.75	0
21	0	1.50
New Start—Solid		
22	−.75	0
23	0	−.75
24	−3.00	0
New Start—Solid		
25	5.25	−4.50
26	2.25	0
27	0	2.25
28	−.75	0
29	0	−.75
30	−.75	0
31	0	−.75
32	−.75	0
33	0	−.75
New Start—Solid		
34	.75	.75
35	1.75	0
New Start—Solid		
36	0	.75
37	−.75	0

Fig. 4-2-K Relative coordinates (inches) assignment.

RELATIVE COORDINATES (MILLIMETERS)

Point	X Axis	Y Axis
1	0	0
2	30	0
3	0	10
4	10	0
5	0	−10
6	30	0
7	0	50
8	−10	0
9	0	−15
10	−50	0
11	0	15
12	−10	0
13	0	−50
New Start—Solid		
14	5	10
15	15	0
16	0	20
17	−15	0
18	0	−20
New Start—Solid		
19	−45	0
20	15	0
21	0	20
22	−15	0
23	0	−20

Fig. 4-2-L Relative coordinates (metric) assignment.

18. Using relative coordinates, plot Figs. 4-2-K and L on a B (A3) size format. The bottom left corner of the drawing is the starting (zero) point. Scale 1:1.

19. Using polar coordinates plot Figs. 4-2-M, 4-1-B, and 4-1-C on a B (A3) size format. Scale 1:1.

Assignments for Unit 4-3, Circles and Arcs

20. On a B (A3) size format, draw the four structural steel shapes in Fig. 4-3-A. The fillets and radii are one-half the material thickness. Scale 1:1. Do not dimension.

21. On a B (A3) size format, draw the inlay designs shown in Fig. 4-3-B. Scale 1:1. Do not dimension.

22. On a B (A3) size format, draw the template shown in Fig. 4-3-C. Scale 1:1. Do not dimension.

23. On a B (A3) size format draw the shaft support shown in Fig. 4-3-D. Scale 1:1. Do not dimension.

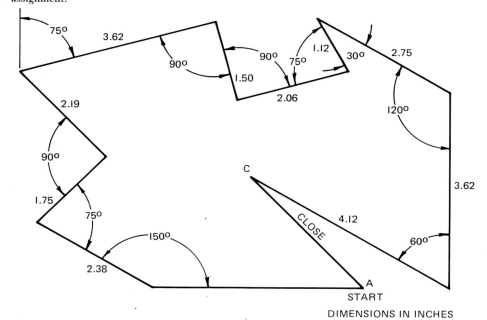

Fig. 4-2-M Template-polar coordinators.

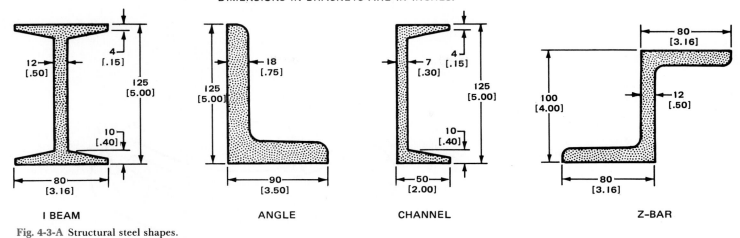

I BEAM · ANGLE · CHANNEL · Z-BAR

Fig. 4-3-A Structural steel shapes.

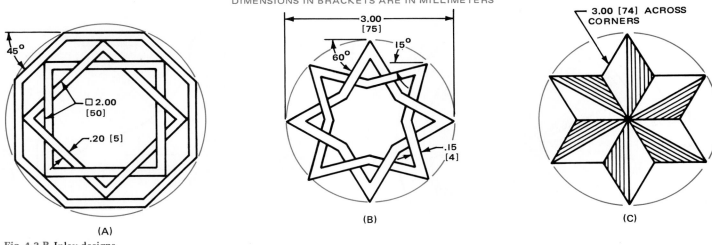

(A) · (B) · (C)

Fig. 4-3-B Inlay designs.

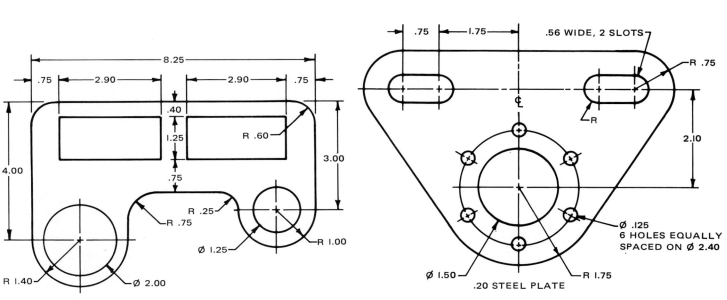

Fig. 4-3-C Template.

Fig. 4-3-D Shaft support.

24. On a B (A3) size format, draw the dial indicator shown in Fig. 4-3-E. Scale 2:1. Do not dimension but add the word DEGREES and the degree numbers shown.

25. On a B (A3) size format, draw the dart board shown in Fig. 4-3-F. Scale 1:2. Use diagonal line shading and add the numbers. Do not dimension.

26. On a B (A3) size format, draw one of the parts shown in Fig. 4-3-G to 4-3-K. Scale 1:1. Do not dimension. Use a .50-in. grid for CAD-constructed drawings.

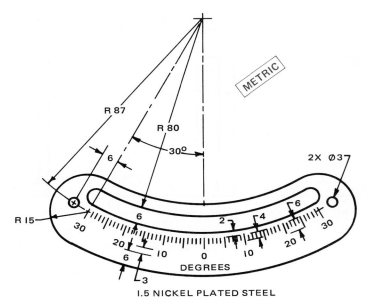

Fig. 4-3-E Dial indicator.

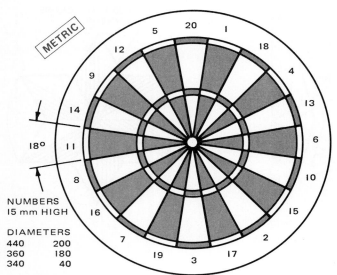

Fig. 4-3-F Dart board.

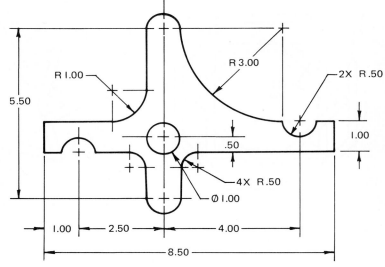

Fig. 4-3-G Pawl.

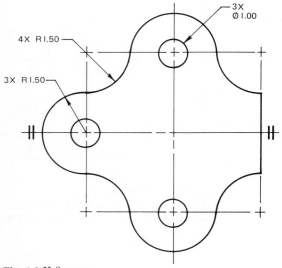

Fig. 4-3-H Spacer.

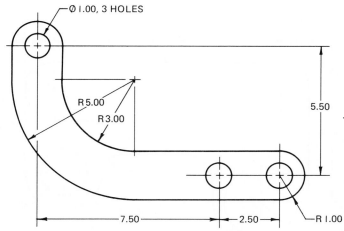

Fig. 4-3-J Rod support.

64 BASIC DRAWING DESIGN

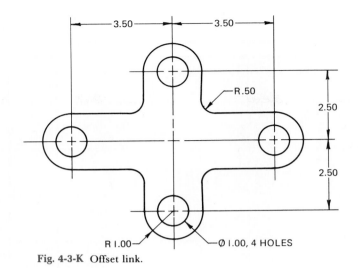

Fig. 4-3-K Offset link.

Assignments for Unit 4-4, Irregular Curves

27. Using grid paper or creating a grid on the monitor, draw the line graph shown in Fig. 4-4-A. Use 25-in. or 5-mm grids.

28. On a B (A3) size format, layout the pattern for the table leg shown in Fig. 4-4-B to the scale 1:2.

Assignments for Unit 4-5, Sketching

29. Sketch the figures shown in Fig. 4-5-2 on grid paper.

30. Sketch the template shown in Fig. 4-3-C on grid paper.

31. Sketch the structural steel shapes shown in Fig. 4-3-A on grid paper.

32. Sketch the shaft support shown in Fig. 4-3-D on grid paper.

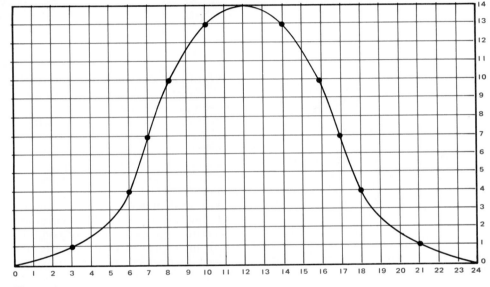

Fig. 4-4-A Line graph.

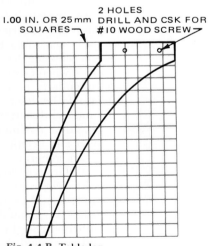

Fig. 4-4-B Table leg.

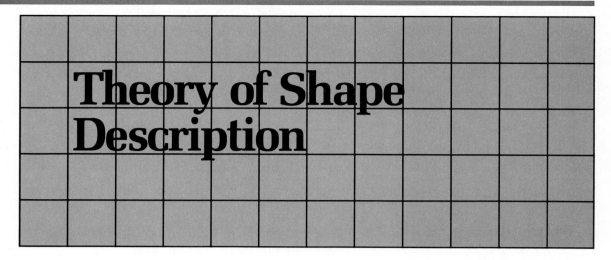

Theory of Shape Description

UNIT 5-1
Theory of Shape Description

Chapter 4 illustrated many simple parts that required only one view to completely describe them. However, in industry, the majority of parts that have to be drawn are more complicated than the ones previously described. More than one view of the object is required to show all the construction features. Except for complex objects of irregular shape, it is seldom necessary to draw more than three views.

Pictorial (three-dimensional) drawings of objects are sometimes used, but the majority of drawings used in mechanical drafting for completely describing an object are multiview drawings as shown in Fig. 5-1-1.

Pictorial projections, such as axonometric, oblique, and perspective projection, are useful for illustrative purposes and are frequently employed in installation and maintenance drawings and design sketches.

As a result of new drawing techniques and equipment, pictorial drawings are becoming a popular form of communication, especially with people not trained to read engineering drawings. Practically all drawings of do-it-yourself projects for the general public or of assembly-line instructions for nontechnical personnel are done in pictorial form.

SHAPE DESCRIPTION BY VIEWS

When looking at objects, we normally see them as three-dimensional, having *width*, *depth*, and *height*, or *length*, *width*, and *height*. The choice of terms used depends on the shape and proportions of the object.

Spherical shapes, such as a basketball, are described as having a certain *diameter* (one term).

Cylindrical shapes, such as a baseball bat, have *diameter* and *length*. However, a hockey puck would have *diameter* and *thickness* (two terms).

Objects which are not spherical or cylindrical require three terms to describe their overall shape. The terms used for a car would probably be length, width, and height; for a filing cabinet, width, height, and depth; for a sheet of drawing paper, length, width, and thickness. These terms are used interchangeably according to the *proportions* of the object being described, and the *position* it is in when being viewed. For example, a telephone pole lying on the ground would be described as having diameter and length, but when placed in a vertical position, its dimensions would be diameter and height.

In general, distances from left to right are referred to as width or length, distances from front to back as depth or width, and vertical distances (except when very small in proportion to the others) as height. On drawings, the multidimensional shape is represented by a

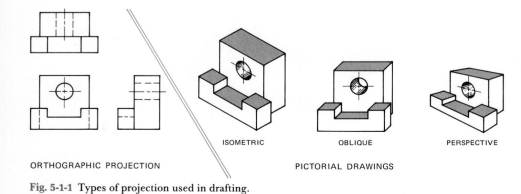

ORTHOGRAPHIC PROJECTION   ISOMETRIC   OBLIQUE   PERSPECTIVE

PICTORIAL DRAWINGS

Fig. 5-1-1 Types of projection used in drafting.

view or views on the flat surface of the drawing paper.

Many mechanical parts do not have a definite "front" or "side" or "top," as do objects, such as refrigerators, desks, or houses, and their shapes vary from the simple to the complex. Decisions have to be made on how many views, and which views, will be drawn. Following are some basic guidelines.

1. Draw those views that are necessary to fully explain the shape.
2. The front view is usually the "key" view; it shows the width or length of the object and gives the most information about its shape. When the longest dimension is drawn in a horizontal position, the object will seem balanced.

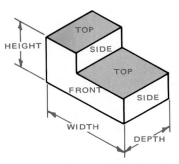

(A) PICTORIAL DRAWING (ISOMETRIC)

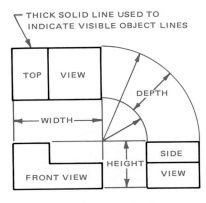

(B) ORTHOGRAPHIC PROJECTION DRAWING (THIRD ANGLE)

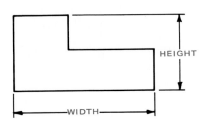

(C) ORTHOGRAPHIC VIEW

Fig. 5-1-2 A simple object shown in pictorial and orthographic projection.

3. Choose those views that will show most of the detailed features of the object as "visible," thus avoiding the extensive use of "hidden" feature lines.

Surface Terms

When describing the shape of an object, reference is often made to the types of surfaces found on the object relative to the three principal viewing planes—horizontal plane, vertical plane, profile plane. These surfaces can be identified as follows:

Parallel flat surfaces that are parallel to the three principal viewing planes;

Hidden surfaces that are hidden in one or more reference planes;

Inclined flat surfaces that are inclined in one plane and parallel to the other two planes;

Oblique flat surfaces that are inclined in all three reference planes;

Circular surfaces that have diameter or radius.

PICTORIAL VIEWS

Pictorial drawings represent the shape with just one view. See Fig. 5-1-2A. How-

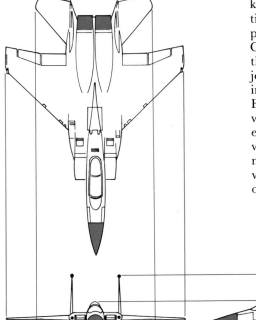

Fig. 5-1-3 Systematic arrangement of views.

ever, the majority of parts manufactured in industry are too complicated in shape and detail to be described successfully by a pictorial view.

ORTHOGRAPHIC PROJECTION

The drafter must represent the part which appears as three-dimensional (width, height, depth) to the eye on the flat plane of the drawing paper. Different views of the object—front, side, and top views—are systematically arranged on the drawing paper to convey the necessary information to the reader (Figs. 5-1-2B and 5-1-3). Features are projected from one view to another. This type of drawing is called an *orthographic projection*. The word *orthographic* is derived from two Greek words: *orthos*, meaning straight, correct, at right angles to; and *graphikos*, meaning to write or describe by drawing lines.

An orthographic view is what you would see looking directly at one side or "face" of the object. See Fig. 5-1-2(C). When looking directly at the front face, you would see width and height (two dimensions) but not the third dimension, depth. Each orthographic view gives two of the three major dimensions.

Orthographic Systems

Two systems of orthographic projection, known as first- and third-angle projection, are used (Fig. 5-1-4). Third-angle projection is used in the United States, Canada, and many other countries throughout the world. First-angle projection, which will be described in detail in Unit 5-8, is used mainly in many European and Asiatic countries. As world trade has brought about the exchange of engineering drawings as well as the end products, drafters are now called upon to communicate in, as well as understand, both types of orthographic projection.

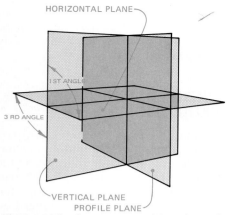

Fig. 5-1-4 The three planes used in orthographic projection.

HORIZONTAL PLANE
1ST ANGLE
3 RD ANGLE
VERTICAL PLANE
PROFILE PLANE

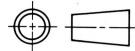

Fig. 5-1-5 ISO projection symbol.

TITLE BLOCK

Fig. 5-1-6 Locating ISO projection symbol on drawing paper.

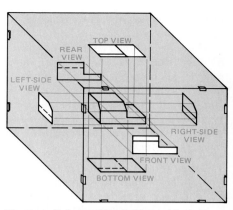

TOP VIEW
REAR VIEW
LEFT-SIDE VIEW
RIGHT-SIDE VIEW
FRONT VIEW
BOTTOM VIEW

Fig. 5-1-7 Relationship of object with viewing planes in third-angle projection.

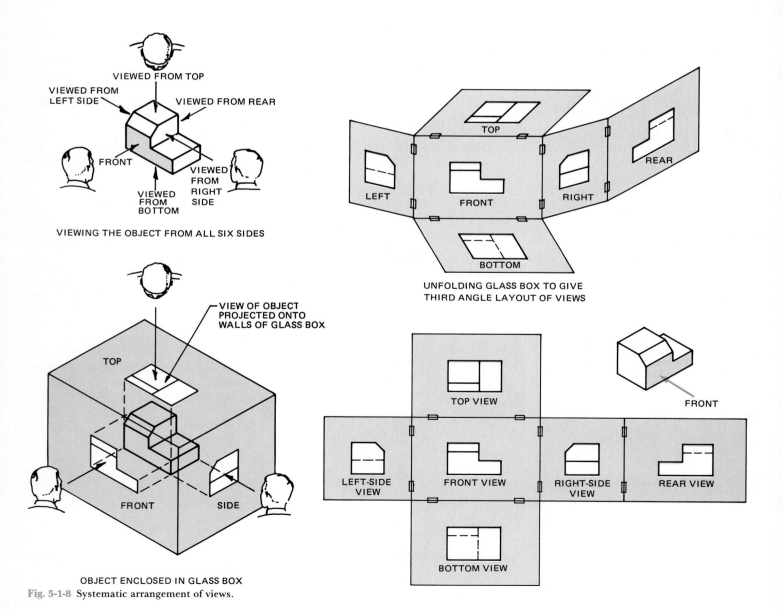

VIEWED FROM TOP
VIEWED FROM LEFT SIDE
VIEWED FROM REAR
FRONT
VIEWED FROM RIGHT SIDE
VIEWED FROM BOTTOM

VIEWING THE OBJECT FROM ALL SIX SIDES

TOP
LEFT
FRONT
RIGHT
REAR
BOTTOM

UNFOLDING GLASS BOX TO GIVE THIRD ANGLE LAYOUT OF VIEWS

VIEW OF OBJECT PROJECTED ONTO WALLS OF GLASS BOX
TOP
FRONT
SIDE

OBJECT ENCLOSED IN GLASS BOX

TOP VIEW
FRONT
LEFT-SIDE VIEW
FRONT VIEW
RIGHT-SIDE VIEW
REAR VIEW
BOTTOM VIEW

Fig. 5-1-8 Systematic arrangement of views.

ISO Projection Symbol

With two types of projection being used on engineering drawings, a method of identifying the type of projection is necessary. The International Standards Organization, known as ISO, has recommended that the symbol shown in Fig. 5-1-5 be shown on all drawings and located preferably in the lower right-hand corner of the drawing, adjacent to the title block (Fig. 5-1-6).

Drawings made in the United States are understood to be shown in third-angle projection if the ISO symbol is not used.

Third-Angle Projection

In third-angle projection, the object is positioned in the third-angle quadrant, as shown in Fig. 5-1-7. The person viewing the object does so from six different positions, namely, from the top, front, right side, left side, rear, and bottom. The views or pictures seen from these positions are then recorded or drawn on the plane located between the viewer and the object. These six viewing planes are then rotated or positioned so that they lie in a single plane, as shown in Fig. 5-1-8. Rarely are all six views used. An exception would be the drawing of a die (one of a pair of dice). See Fig. 5-1-9. Only the views which are necessary to fully describe the object are drawn. Simple objects, such as a gasket, can be described sufficiently by one view alone. However, in mechanical drafting two- or three-view drawings of objects are more common, the rear, bottom, and one of the two side views being rarely used.

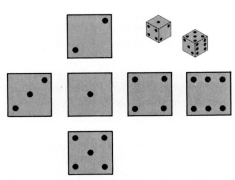

Fig. 5-1-9 Six views required to show a die.

ASSIGNMENTS

See Assignments 1 and 2 for Unit 5-1 on page 78.

UNIT 5-2
Arrangement of Views

SPACING THE VIEWS

It is important for clarity and good appearance that the views be well balanced on the drawing paper, whether the drawing shows one view, two views, three views, or more. The drafter must anticipate the approximate space required. This is determined from the size of the object to be drawn, the number of views, the scale used, and the space between views. Ample space should be provided between views to permit placement of dimensions on the drawing without crowding. Space should also be alotted so that notes can be added without crowding. However, space between views should not be excessive.

Figure 5-2-1 shows how to balance the views for a three-view drawing. For a drawing with two or more views, follow these guidelines:

1. Decide on the views to be drawn and the scale to be used, e.g., 1:1 or 1:2.
2. Make a sketch of the space required for each of the views to be drawn, showing these views in their correct location. (A simple rectangle for each view will be adequate, Fig. 5-2-1B.)
3. Put on the overall drawing sizes for each view. (These sizes are shown as W, D, and H.)
4. Decide upon the space to be left between views. (These spaces should be sufficient for the parallel dimension lines to be placed between views. For most drawing projects, 1.50 in. is sufficient.)
5. Total these dimensions to get the overall horizontal distance (A) and overall vertical distance (B).
6. Select the drawing sheet to best accommodate the overall size of the drawing with suitable open space around the views.
7. Measure the "drawing space" remaining after all border lines, title strip or title block, etc., are in place (Fig. 5-2-1C).
8. Take one-half of the difference between distance A and the horizon-

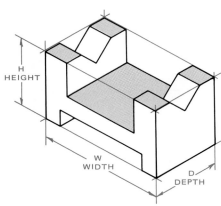

(A) DECIDING THE VIEWS TO BE DRAWN AND THE SCALE TO BE USED

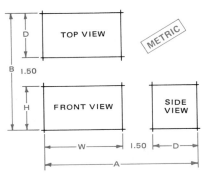

(B) CALCULATING DISTANCES A AND B

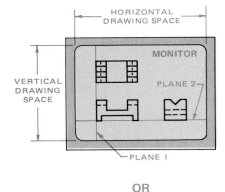

OR

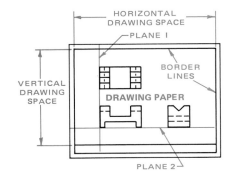

(C) ESTABLISHING LOCATION OF PLANES I AND 2

Fig. 5-2-1 Balancing the drawing on the drawing paper or monitor.

tal "drawing space" to establish Plane 1.

9. Take one-half of the difference between distance B and the vertical "drawing space" to establish Plane 2.

USE OF A MITER LINE

The use of a miter line provides a fast and accurate method of constructing the third view once two views are established (Fig. 5-2-2).

Using a Miter Line to Construct the Right Side View

1. Given the top and front views, project lines to the right of the top view.
2. Establish how far from the front view the side view is to be drawn (distance D).

3. Construct the miter line at 45° to the horizon.
4. Where the horizontal projection lines of the top view intersect the miter line, drop vertical projection lines.
5. Project horizontal lines to the right of the front view and complete the side view.

Using a Miter Line to Construct the Top View

1. Given the front and side views, project vertical lines up from the side view.
2. Establish how far away from the front view the top view is to be drawn (distance D).
3. Construct the miter line at 45° to the horizon.
4. Where the vertical projection lines of

the side view intersect the miter line, project horizontal lines to the left.
5. Project vertical lines up from the front view and complete the top view.

CAD

The working area on the CRT monitor must be established prior to selecting the paper size for the completed drawing.

Construction lines are menu options used in the preparation of multiview drawings on the CRT monitor (Fig. 5-2-3).

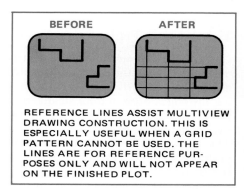

Fig. 5-2-3 CAD construction line command.

ASSIGNMENTS

See Assignments 3 through 6 for Unit 5-2 on pages 78 and 79.

UNIT 5-3
All Surfaces Parallel and All Edges and Lines Visible

To fully appreciate the shape and detail of views drawn in third-angle orthographic projection, the units for this chapter have been designed according to the types of surfaces generally found on objects. These surfaces can be divided into flat surfaces parallel to the viewing planes with and without hidden features; flat surfaces which appear inclined in one plane and parallel to the other two principal reference planes (called *inclined* surfaces); flat surfaces which are inclined in all three reference planes (called *oblique* surfaces); and surfaces which have diameters or radii. These drawings are so designed that

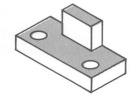

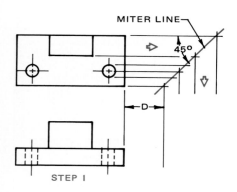

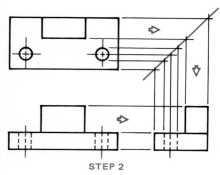

(A) ESTABLISHING WIDTH LINES ON SIDE VIEW

Fig. 5-2-2 Use of a miter line.

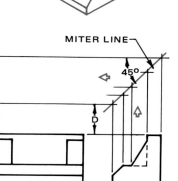

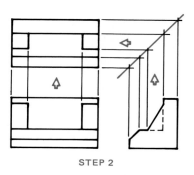

(B) ESTABLISHING WIDTH LINES ON TOP VIEW

only the top, front, and right side views are required.

All Surfaces Parallel to the Viewing Planes and All Edges and Lines Visible When a surface is parallel to the viewing planes, that surface will show as a surface on one view and a line on the other views. The lengths of these lines are the same as the lines shown on the surface view. Figure 5-3-1 shows examples.

ASSIGNMENTS

See Assignments 7 and 8 for Unit 5-3 on pages 79 and 80.

U N I T 5 - 4
Hidden Surfaces and Edges

Most objects drawn in engineering offices are more complicated than the ones shown in Fig. 5-4-1. Many features

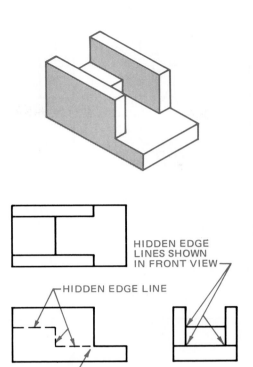

Fig. 5-4-1 Hidden lines.

(lines, holes, etc.), cannot be seen when viewed from outside the piece. These hidden edges are shown with *hidden lines* and are normally required on the drawing to show the true shape of the object.

Hidden lines consist of short, evenly spaced dashes. They should be omitted when not required to preserve the clarity of the drawing. The length of dashes may vary slightly in relation to the size of the drawing.

Lines depicting hidden features and phantom details should always begin and end with a dash in contact with the line at which they start and end, except when such a dash would form a continuation of a visible detail line. Dashes should join at corners. Arcs should start with dashes at the tangent points (Fig. 5-4-2). Figure 5-4-3 shows additional examples of objects requiring hidden lines.

CAD

All CAD systems have the option to create different line styles. On large systems, these options are found on the aux-

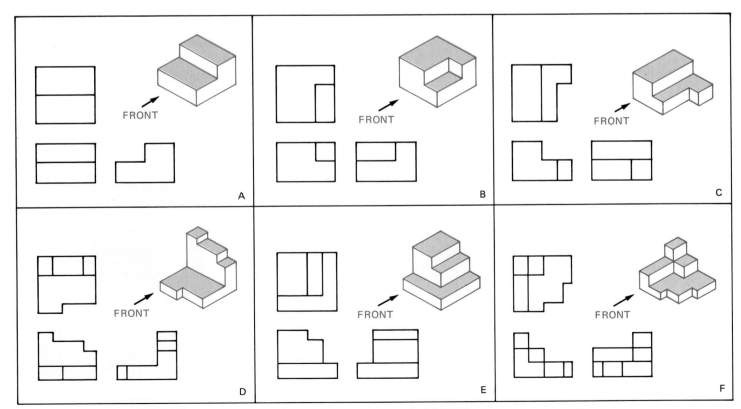

NOTE: ARROWS INDICATE DIRECTION OF SIGHT WHEN LOOKING AT THE FRONT VIEW.

Fig. 5-3-1 Illustrations of objects drawn in third-angle orthographic projection.

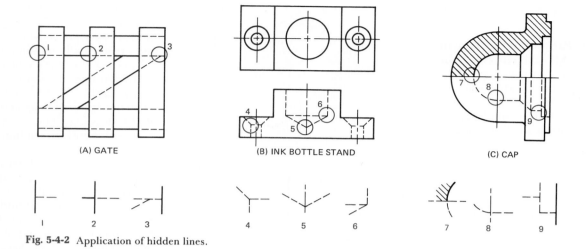

(A) GATE   (B) INK BOTTLE STAND   (C) CAP

Fig. 5-4-2 Application of hidden lines.

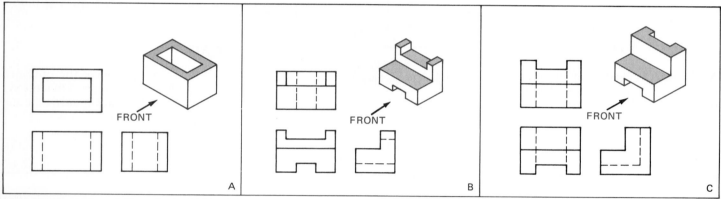

A   B   C

Fig. 5-4-3 Illustration of objects having hidden features.

iliary menu. On smaller systems, the line style selection is made directly from the tablet menu. Any style of line may be drawn by following the line commands explained in Unit 4-2.

ASSIGNMENTS

See Assignments 9 through 12 for Unit 5-4 on pages 80 and 81.

UNIT 5-5
Inclined Surfaces

If the surfaces of an object lie in either a horizontal or a vertical position, the surfaces appear in their true shapes in one of the three views, and these surfaces appear as a line in the other two views.

When a surface is inclined or sloped in only one direction, then that surface is

not seen in its true shape in the top, front, or side view. It is, however, seen in two views as a distorted surface. On the third view it appears as a line.

The true length of surfaces A and B in Fig. 5-5-1 is seen in the front view only. In the top and side views, only the

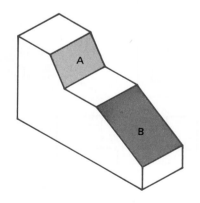

Fig. 5-5-1 Sloping surfaces.

width of surfaces A and B appears in its true size. The length of these surfaces is foreshortened. Figure 5-5-2 shows additional examples.

Where an inclined surface has important features that must be shown clearly and without distortion, an *auxiliary*

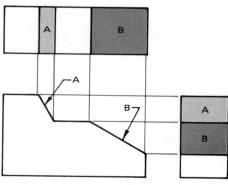

NOTE: THE TRUE SHAPE OF SURFACES A AND B DO NOT APPEAR ON THE TOP OR SIDE VIEWS.

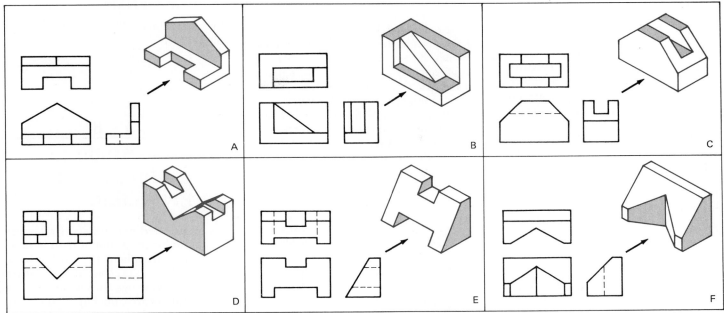

Fig. 5-5-2 Illustrations of objects having sloping surfaces.

NOTE: ARROW INDICATES DIRECTION OF FRONT VIEW

or helper view must be used. This type of view will be discussed in detail in Chap. 15.

ASSIGNMENTS

See Assignments 13 through 16 for Unit 5-5 on page 83.

U N I T 5 - 6
Circular Features

Typical parts with circular features are illustrated in Fig. 5-6-1. Note that the circular feature appears circular in one view only and that no line is used to show where a curved surface joins a flat surface. Hidden circles, like hidden flat surfaces, are represented on drawings by a hidden line.

The intersection of unfinished surfaces, such as found on cast parts, that are rounded or filleted at the point of theoretical intersection, may be indicated conventionally by a line. See Unit 8-18.

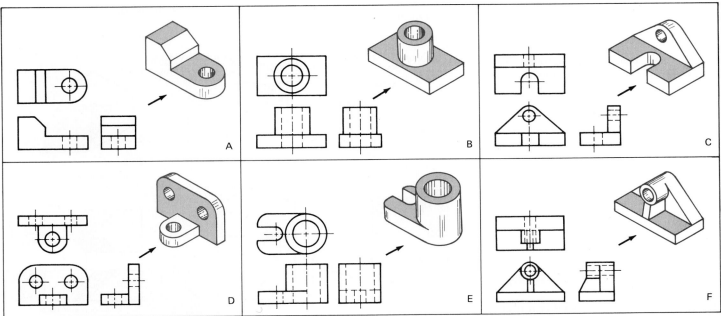

NOTE: ARROWS INDICATE DIRECTION OF FRONT VIEW

Fig. 5-6-1 Illustrations of objects having circular features.

Center Lines

A center line is drawn as a thin, broken line of long and short dashes, spaced alternately. Such lines may be used to locate center points, axes of cylindrical parts, and axes of symmetry, as shown in Fig. 5-6-2. Solid center lines are often used when the circular features are small. Center lines should project for a short distance beyond the outline of the part or feature to which they refer. They must be extended for use as extension lines for dimensioning purposes, but in this case the extended portion is not broken.

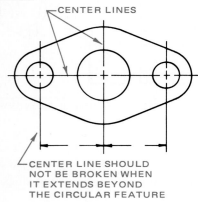

CENTER LINES

CENTER LINE SHOULD NOT BE BROKEN WHEN IT EXTENDS BEYOND THE CIRCULAR FEATURE

NOTE: AT TIME OF WRITING, CAD SYSTEMS DO NOT CONFORM TO THIS STANDARD.

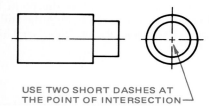

USE TWO SHORT DASHES AT THE POINT OF INTERSECTION

Fig. 5-6-2 Center line applications.

On views showing the circular features, the point of intersection of the two center lines is shown by the two intersecting short dashes.

ASSIGNMENTS

See Assignments 17 through 21 for Unit 5-6 on page 87 and 88.

U N I T 5 - 7
Oblique Surfaces

When a surface is sloped so that it is not perpendicular to any of the three viewing planes, it will appear as a surface in all three views but never in its true shape. This is referred to as an *oblique surface* (Fig. 5-7-1). Since the oblique surface is not perpendicular to the viewing planes, it cannot be parallel to them and consequently appears foreshortened. If a true view is required for this surface, two auxiliary views—a primary and a secondary view—need to be drawn. This is discussed in detail under Secondary Auxiliary Views in Unit 15-4. Figure 5-7-2 shows additional examples of objects having oblique surfaces.

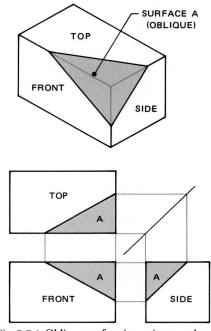

SURFACE A (OBLIQUE)

TOP

FRONT

SIDE

TOP

A

FRONT

A

A

SIDE

Fig. 5-7-1 Oblique surface is not its true shape in any of the three views.

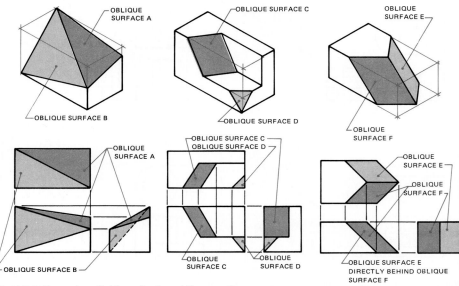

Fig. 5-7-2 Examples of objects having oblique surfaces.

ASSIGNMENTS

See Assignments 22 and 23 for Unit 5-7 on page 91.

U N I T 5 - 8
First-Angle Orthographic Projection

As mentioned previously, first-angle orthographic projection is used in many countries throughout the world. Today with global marketing and the interchange of drawings with different countries, drafters are called upon to prepare and interpret drawings in both first- and third-angle projection. In first-angle projection, all the views are projected onto the planes located behind the objects rather than onto the planes lying between the objects and the viewer, as in third-angle projection. This is shown in Fig. 5-8-2. The unfolding and positioning of the views in one plane are shown in Fig. 5-8-3. Note that the views are on opposite sides of the front view with the exception of the rear view. A comparison between the views of first- and third-angle projections is shown in Figs. 5-8-1 and 5-8-4. Remember that the views are identical in shape and detail, and only their location in reference to the front view has changed.

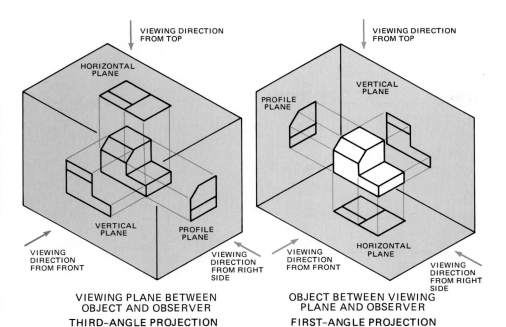

VIEWING PLANE BETWEEN
OBJECT AND OBSERVER

THIRD–ANGLE PROJECTION

OBJECT BETWEEN VIEWING
PLANE AND OBSERVER

FIRST–ANGLE PROJECTION

Fig. 5-8-1 A comparison between third- and first-angle projection.

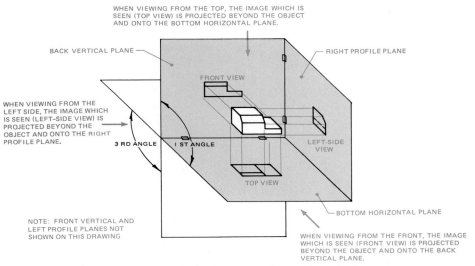

WHEN VIEWING FROM THE TOP, THE IMAGE WHICH IS
SEEN (TOP VIEW) IS PROJECTED BEYOND THE OBJECT
AND ONTO THE BOTTOM HORIZONTAL PLANE.

BACK VERTICAL PLANE

RIGHT PROFILE PLANE

FRONT VIEW

WHEN VIEWING FROM THE
LEFT SIDE, THE IMAGE WHICH
IS SEEN (LEFT-SIDE VIEW) IS
PROJECTED BEYOND THE
OBJECT AND ONTO THE RIGHT
PROFILE PLANE.

3 RD ANGLE   I ST ANGLE

LEFT-SIDE
VIEW

TOP VIEW

NOTE: FRONT VERTICAL AND
LEFT PROFILE PLANES NOT
SHOWN ON THIS DRAWING

BOTTOM HORIZONTAL PLANE

WHEN VIEWING FROM THE FRONT, THE IMAGE
WHICH IS SEEN (FRONT VIEW) IS PROJECTED
BEYOND THE OBJECT AND ONTO THE BACK
VERTICAL PLANE.

Fig. 5-8-2 Relationship of object with viewing planes in first-angle projection.

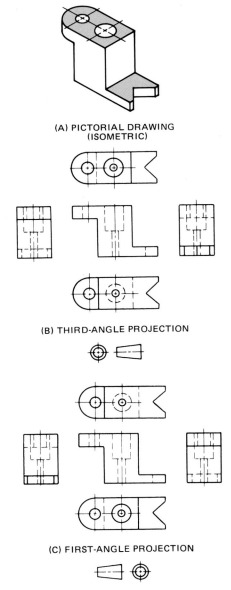

(A) PICTORIAL DRAWING
(ISOMETRIC)

(B) THIRD-ANGLE PROJECTION

(C) FIRST-ANGLE PROJECTION

Fig. 5-8-4 A simple object shown in pictorial and orthographic.

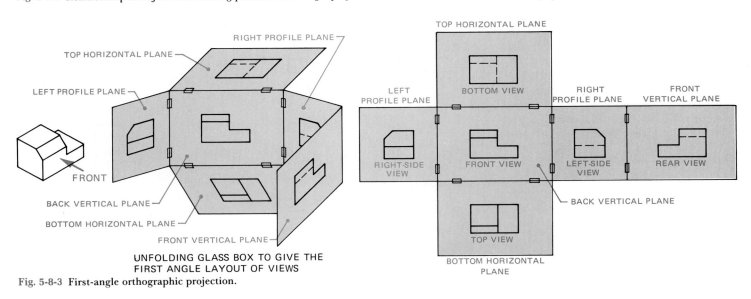

UNFOLDING GLASS BOX TO GIVE THE
FIRST ANGLE LAYOUT OF VIEWS

Fig. 5-8-3 First-angle orthographic projection.

See Assignments 24 and 25 for Unit 5-8 on page 92.

UNIT 5-9
One- and Two-View Drawings

VIEW SELECTION

Views should be chosen that will best describe the object to be shown. Only the minimum number of views that will completely portray the size and shape of the part should be used. They should also be chosen to avoid hidden feature lines whenever possible, as shown in Fig. 5-9-1.

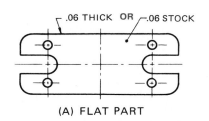
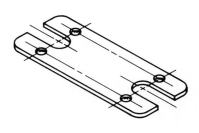

(A) FLAT PART

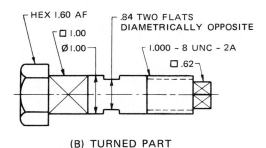
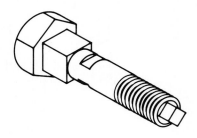

(B) TURNED PART

Fig. 5-9-2 One-view drawings.

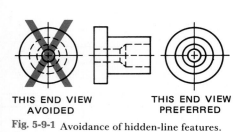

THIS END VIEW AVOIDED   THIS END VIEW PREFERRED

Fig. 5-9-1 Avoidance of hidden-line features.

Except for complex objects of irregular shape, it is seldom necessary to draw more than three views. For representing simple parts, one- or two-view drawings will often be adequate.

ONE-VIEW DRAWINGS

In one-view drawings, the third dimension, such as thickness, may be expressed by a note or by descriptive words or abbreviations, such as DIA, Ø, or HEXAGON ACROSS FLATS. Square sections may be indicated by light crossed diagonal lines. This applies whether the face is parallel or inclined to the drawing plane. These are illustrated in Fig. 5-9-2.

When cylindrically shaped surfaces include special features such as a keyseat, a side view (often called an *end view*) is required.

TWO-VIEW DRAWINGS

Frequently the drafter will decide that only two views are necessary to explain

fully the shape of an object (Fig. 5-9-3). For this reason, some drawings consist of two adjacent views, such as the top and front views only, or front and right side views only. Two views are usually sufficient to explain fully the shape of cylindrical objects; if three views were used, two of them would be identical, depending on the detail structure of the part.

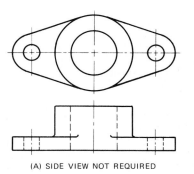

(A) SIDE VIEW NOT REQUIRED

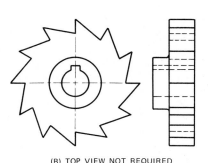

(B) TOP VIEW NOT REQUIRED

Fig. 5-9-3 Two-view drawings.

ASSIGNMENT

See Assignment 26 for Unit 5-9 on page 93.

UNIT 5-10
Special Views

PARTIAL VIEWS

Symmetrical objects may often be adequately portrayed by half views (Fig. 5-10-1A). A center line is used to show the axis of symmetry. Two short thick lines, above and below the view of the object, are drawn at right angles to and on the center line to indicate the line of symmetry.

Partial views, which show only a limited portion of the object with remote details omitted, should be used, when necessary, to clarify the meaning of the drawing (Fig. 5-10-1B). Such views are used to avoid the necessity of drawing many hidden features.

On drawings of objects where two side views can be used to better advantage than one, each need not be complete if together they depict the shape. Show only the hidden lines of features immediately behind the view (Fig. 5-10-1C).

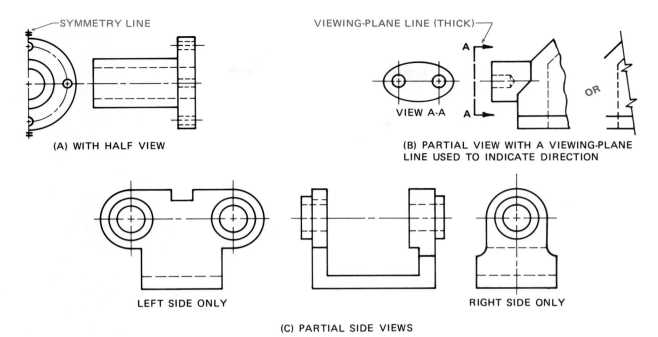

(A) WITH HALF VIEW

(B) PARTIAL VIEW WITH A VIEWING-PLANE LINE USED TO INDICATE DIRECTION

LEFT SIDE ONLY

RIGHT SIDE ONLY

(C) PARTIAL SIDE VIEWS

Fig. 5-10-1 Partial views.

REAR VIEWS AND ENLARGED VIEWS

Placement of Views

When views are placed in the relative positions shown in Fig. 5-1-8, it is rarely necessary to identify them. When they are placed in other than the regular projected position, the removed view must be clearly identified.

Whenever appropriate, the orientation of the main view on a detail drawing should be the same as on the assembly drawing. To avoid the crowding of dimensions and notes, ample space must be provided between views.

Rear Views

Rear views are normally projected to the right or left. When this projection is not practical, because of the length of the part, particularly for panels and mounting plates, the rear view must not be projected up or down. Doing so would result in the part being shown upside down. Instead, the view should be drawn as if it were projected sideways but located in some other position, and it should be clearly labeled REAR VIEW REMOVED (Fig. 5-10-2).

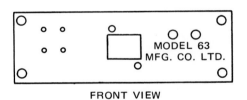

FRONT VIEW

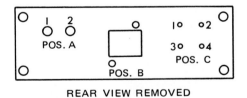

REAR VIEW REMOVED

Fig. 5-10-2 Removed rear views.

Enlarged Views

Enlarged views are used when it is desirable to show a feature in greater detail or to eliminate the crowding of details or dimensions (Fig. 5-10-3). The enlarged view should be oriented in the same manner as the main view. However, if an enlarged view is rotated, state the direction and the amount of rotation of the detail. The scale of enlargement must be shown, and both views should be identified by one of the three methods shown.

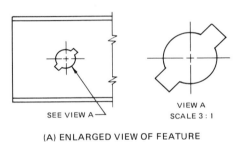

SEE VIEW A

VIEW A
SCALE 3 : 1

(A) ENLARGED VIEW OF FEATURE

DETAIL A
SCALE 5 : 1

METRIC

SEE DETAIL A

SCALE SHOWN ON DRAWING

(B) ENLARGED VIEW OF ASSEMBLY

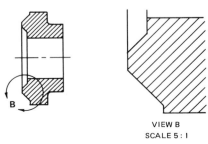

VIEW B
SCALE 5 : 1

(C) ENLARGED REMOVED VIEW

Fig. 5-10-3 Enlarged views.

Key Plans

A method particularly applicable to structural work is to include a small key plan using bold lines on each sheet of a drawing series that shows the relationship of the detail on that sheet to the whole work, as in Fig. 5-10-4.

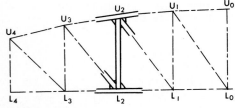

Fig. 5-10-4 Key plan.

Opposite-Hand Views

Where parts are symmetrically opposite, such as for right- and left-hand usage, one part is drawn in detail and the other is described by a note such as PART B SAME EXCEPT OPPOSITE HAND. It is preferable to show both part numbers on the same drawing (Fig. 5-10-5).

ASSIGNMENT

See Assignments 27 through 29 for Unit 5-10 on pages 94 and 95.

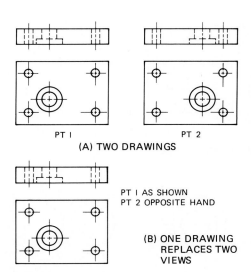

PT I

PT 2

(A) TWO DRAWINGS

PT I AS SHOWN
PT 2 OPPOSITE HAND

(B) ONE DRAWING
REPLACES TWO
VIEWS

Fig. 5-10-5 Opposite-hand views.

ASSIGNMENTS FOR CHAPTER 5

Note: CAD may be substituted for manual drafting for any assignments in this chapter.

Assignments for Unit 5-1, Theory of Shape Description

1. Make a two-view sketch, at a suitable scale, of one of the following: a coffee mug, a dinner plate, a drinking glass or mug. Identify the terms that would be used to best describe the object's overall size.

2. Make a three-view sketch, at a suitable scale, of one of the following: a kitchen table, a filing cabinet, a writing desk, a car, a house, a chest of drawers. Identify the terms that would best describe the object's overall size.

Assignments for Unit 5-2, Arrangement of Views

3. Make a sketch similar to Fig. 5-2-1B and C and establish the distance between Plane 1 and the left border line and between Plane 2 and the bottom border line, given the following: top, front, and right side views; scale 1:1; drawing space 8.00 × 10.50 in., part size: W = 4.10, H = 1.40, D = 2.10; space between views (X and Y), 1.50 in.

4. Make a sketch similar to Fig. 5-2-1B and C and establish the distance between Plane 1 and the left border line

and between Plane 2 and the bottom border line, given the following: top, front, and right side views; scale 1:2; drawing space 8.00 × 10.50 in.; part size: W = 8.50, H = 4.90, D = 4.50; space between views (X and Y), 1.50 in.

5. Step block, Fig. 5-2-A, sheet size A (A4), scale 1:1. Make a three-view drawing using a miter line to complete the right side view. Space between views to be 1.50 in.

6. Stop block, Fig. 5-2-B, sheet size A (A4), scale 1:1. Make a three-view drawing using a miter line to complete the top view. Space between views to be 1.50 in.

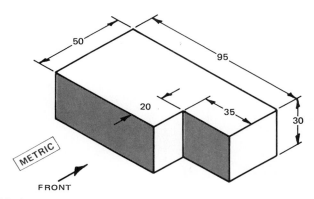

Fig. 5-2-A Step block.

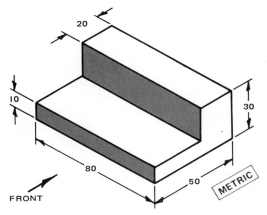

Fig. 5-2-B Stop block.

Assignments for Unit 5-3, All Surfaces Parallel and All Edges and Lines Visible

7. On an A (A4) size sheet of preprinted grid paper (.25 in. or 10 mm grids) sketch three views of each of the objects shown in Figs. 5-3-A and 5-3-B. Each square shown on the objects represents one square on the grid paper. Allow one grid space between views and a minimum of two grid spaces between objects. Identify the type of projection used by placing the appropriate ISO projection symbol at the bottom of the drawing.

8. On an A (A4) size sheet draw three views of one of the parts shown in Figs. 5-3-C to 5-3-F. Allow 1 in. or 25 mm between views. Scale full or 1:1. Do not dimension.

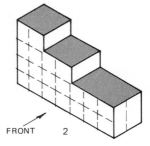
FRONT 1

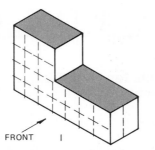
FRONT 2

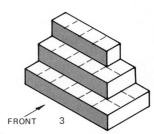
FRONT 3

Fig. 5-3-A Sketching assignment.

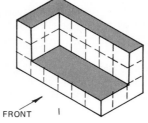
FRONT 1

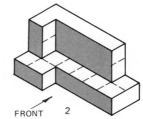
FRONT 2

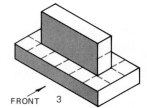
FRONT 3

Fig. 5-3-B Sketching assignment.

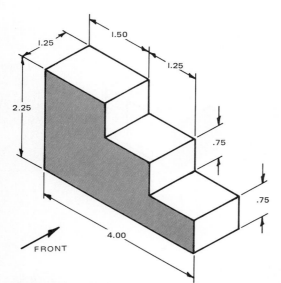

Fig. 5-3-C Step support.

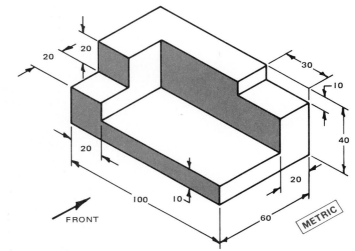
FRONT

METRIC

Fig. 5-3-D Corner block.

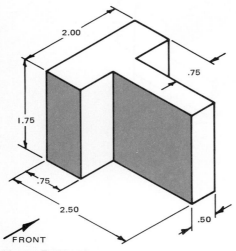
FRONT

Fig. 5-3-E T bracket.

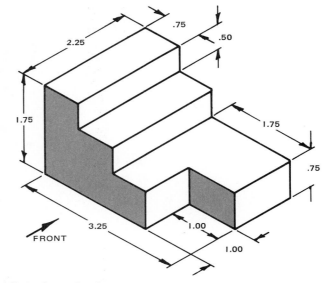
FRONT

Fig. 5-3-F Angle step bracket.

Assignments for Units 5-4, All Surfaces Parallel to the Viewing Plane with Some Edges and Surfaces Hidden

9. On an A (A4) size sheet of preprinted grid paper (.25 in. or 10 mm grids) sketch three views of each of the objects shown in Figs. 5-4-A and 5-4-B. Each square shown on the objects represents one square on the grid paper. Allow one grid space between views and a minimum of two spaces between objects. Identify the type of projection by placing the ISO projection symbol at the bottom of the drawing.

10. Same as Assignment 9 except sketch the objects shown in Figs. 5-4-C and 5-4-D.

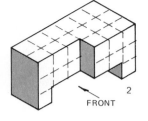
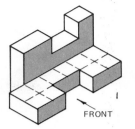
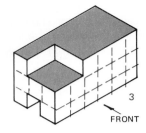

Fig. 5-4-A Sketching assignment.

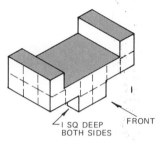
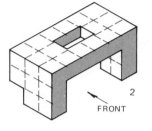
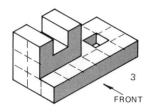

Fig. 5-4-B Sketching assignment.

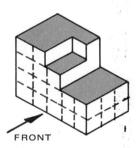

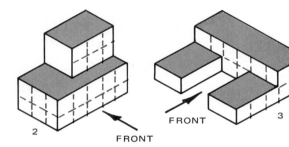

Fig. 5-4-C Sketching assignment.

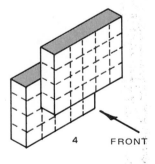
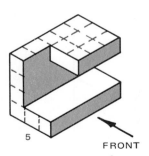
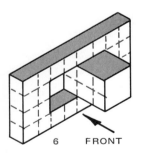

Fig. 5-4-D Sketching assignment.

11. On an A (A4) size sheet make a three-view drawing of one of the parts shown in Figs. 5-4-E to 5-4-H. Scale 1:1. Allow 1.20 in. (30 mm) between views. Do not dimension.

12. Matching test. Match the pictorial drawings to the orthographic drawings shown in Fig. 5-4-J.

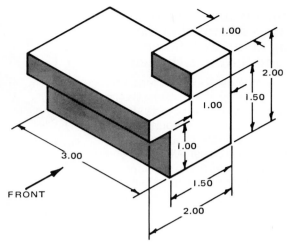

Fig. 5-4-G Bracket.

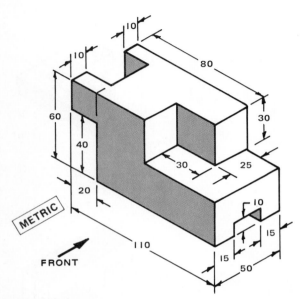

Fig. 5-4-E Adapter.

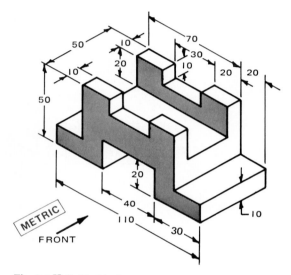

Fig. 5-4-H Guide block.

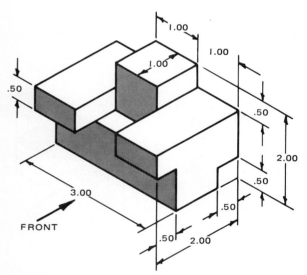

Fig. 5-4-F Link.

Fig. 5-4-J Match pictorial drawings A through M with orthographic drawings.

Assignments for Unit 5-5, Inclined Surfaces

13. On an A (A4) size sheet of preprinted grid paper (.25 in. or 10 mm grids) sketch three views of each of the objects shown in Figs. 5-5-A and 5-5-B. Each square shown on the objects represents one square on the grid paper. Allow one grid space between views and a minimum of two grid spaces between objects. The sloped (inclined) surfaces on each of the three objects are identified by a letter. Identify the sloped surfaces on each of the three views with a corresponding letter. Also identify the type of projection used by placing the appropriate ISO symbol at the bottom of the drawing.

14. On a B (A3) size sheet, make a three-view drawing of one of the parts shown in Figs. 5-5-C to 5-5-F. Allow 1.20 in. (30 mm) between views. Do not dimension. Scale 1:1.

15. Sketching assignment. Make three-view sketches of the parts shown in Figs. 5-5-G through 5-5-K.

16. Matching test. Match the pictorial drawings to the orthographic drawings shown in Fig. 5-5-L.

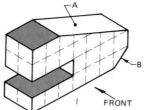
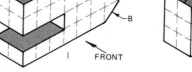
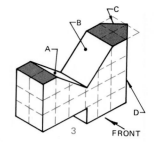

Fig. 5-5-A Sketching assignment.

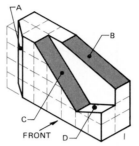
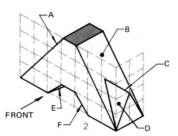
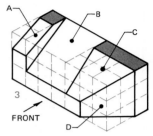

Fig. 5-5-B Sketching assignment.

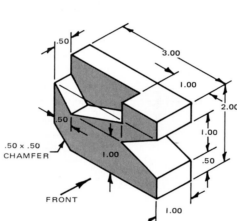
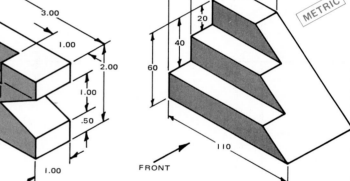

Fig. 5-5-C Slide bar.

Fig. 5-5-E Adjusting guide.

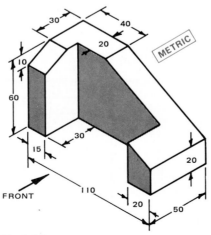
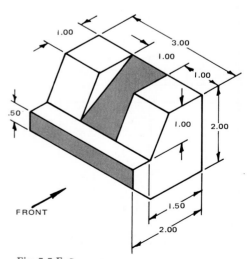

Fig. 5-5-D Flanged support.

Fig. 5-5-F Separator.

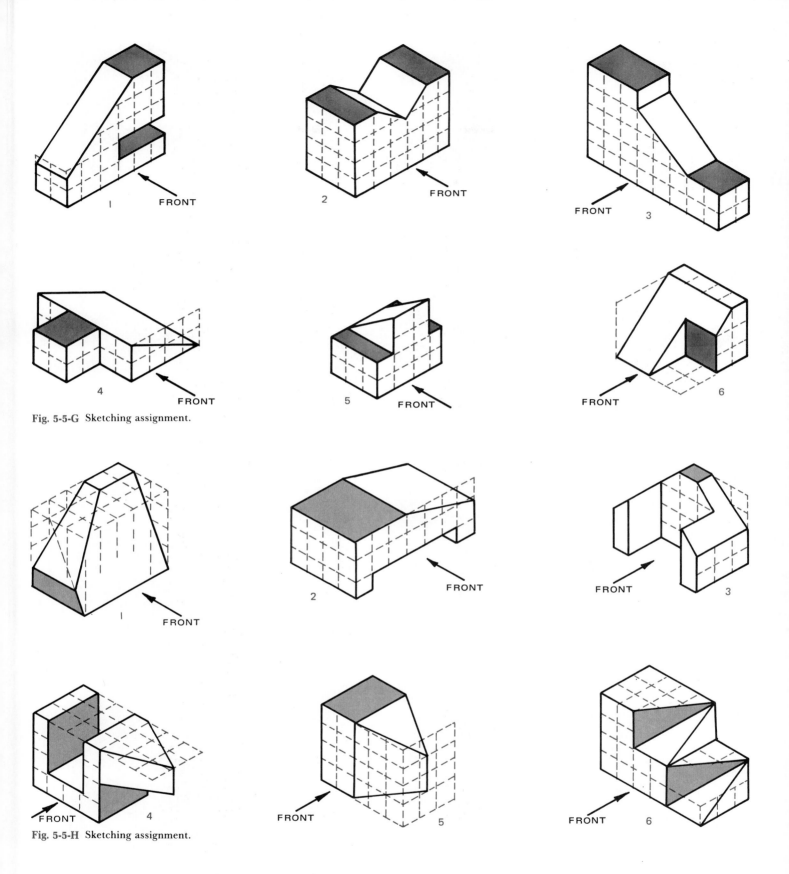

Fig. 5-5-G Sketching assignment.

Fig. 5-5-H Sketching assignment.

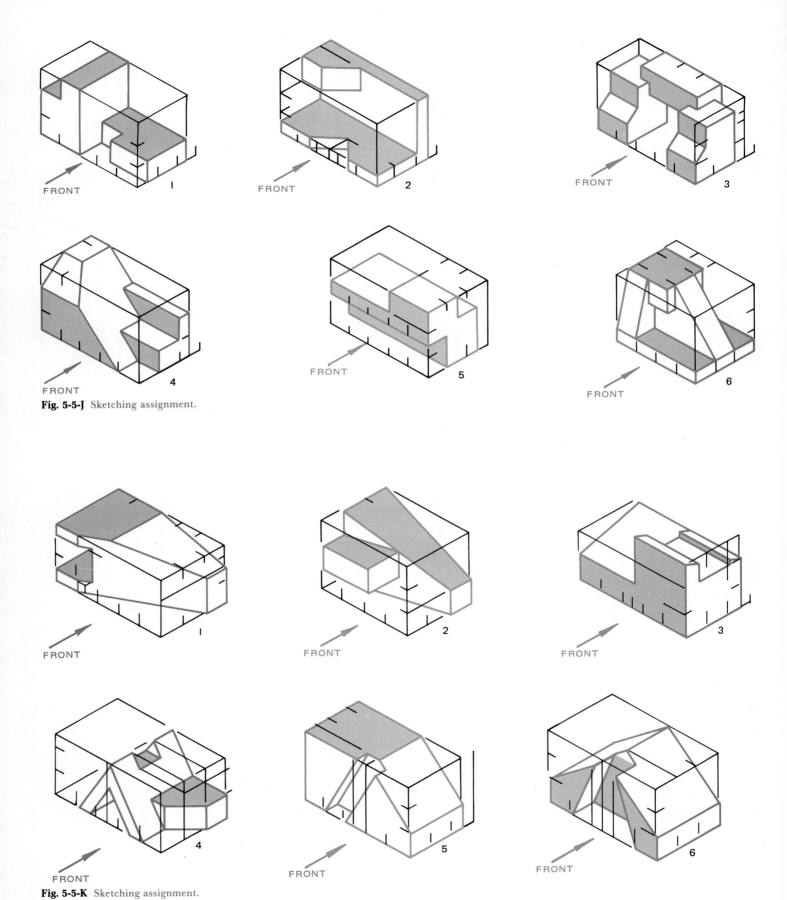

FRONT

FRONT 2

FRONT 3

FRONT 4

Fig. 5-5-J Sketching assignment.

FRONT 5

FRONT 6

FRONT

FRONT 2

FRONT 3

FRONT 4

Fig. 5-5-K Sketching assignment.

FRONT 5

FRONT 6

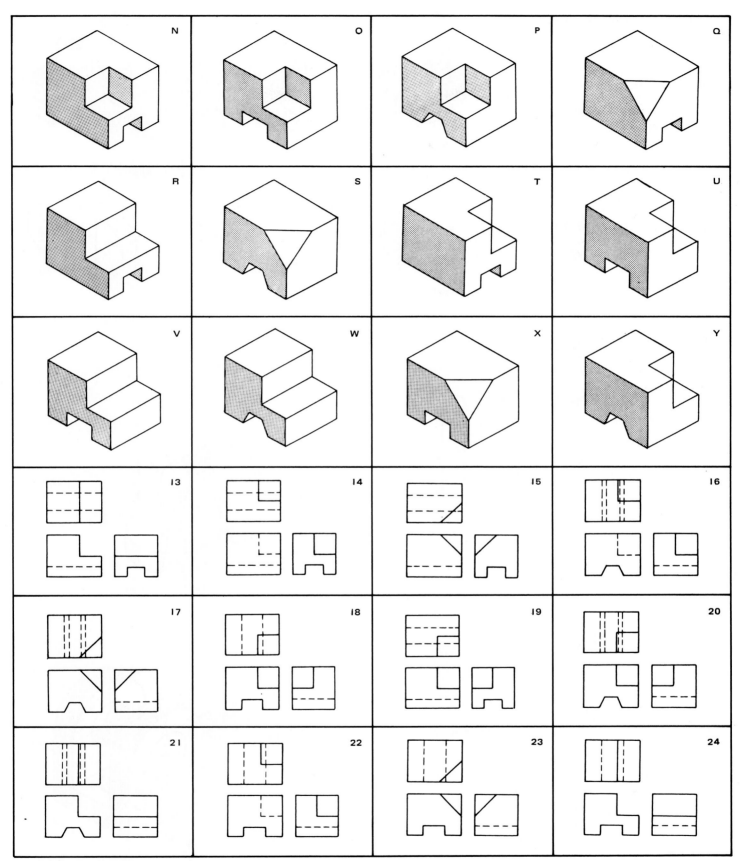

Fig. 5-5-L Matching test.

Assignments for Unit 5-6, Circular Features

17. On an A (A4) size sheet of preprinted grid paper (.25 in. or 10 mm grids) sketch three views of each of the objects shown in Figs. 5-6-A and 5-6-B. Each square shown on the objects represents one square on the grid paper. Allow one grid space between views and a minimum of two grid spaces between objects. Identify the type of projection used by placing the appropriate ISO projection symbol at the bottom of the drawing.

18. One a B (A3) size sheet, make a three-view drawing of one of the parts shown in Figs. 5-6-C to 5-6-F. Allow 1.20 in. (30 mm) between views. Do not dimension. Scale 1:1.

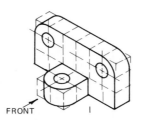
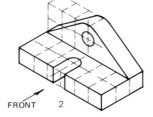
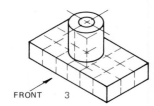

Fig. 5-6-A Sketching assignment.

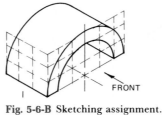
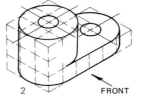
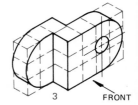

Fig. 5-6-B Sketching assignment.

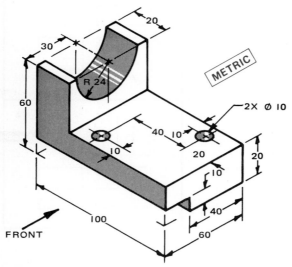

Fig. 5-6-C Rod support.

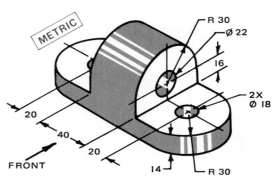

Fig. 5-6-D Pillow block.

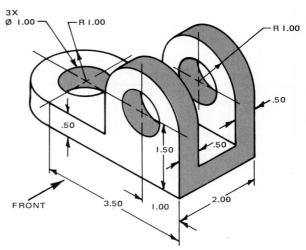

Fig. 5-6-E Cradle support.

Fig. 5-6-F Rocker arm.

19. Sketching assignment. Make three-view sketches of the parts shown in Fig. 5-6-G.

20. Sketching assignment. Sketch the views needed for a multiview drawing for the parts shown in Fig. 5-6-H. Choose your own sizes and estimate proportions.

21. Completion test. See Fig. 5-6-J.

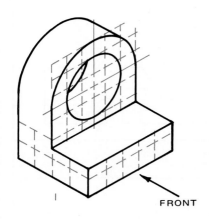

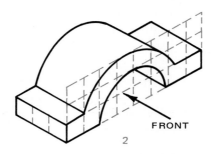

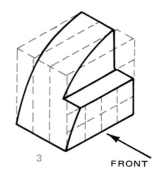

FRONT

FRONT

2

3

FRONT

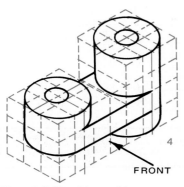

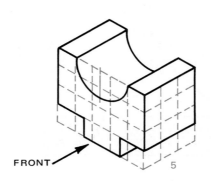

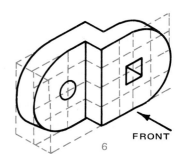

FRONT

4

FRONT

5

6

FRONT

Fig. 5-6-G Sketching assignment.

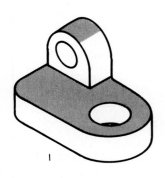

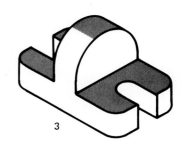

I

2

3

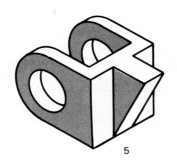

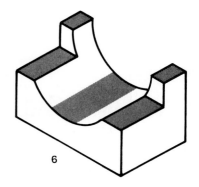

4

5

6

Fig. 5-6-H Sketching assignment.

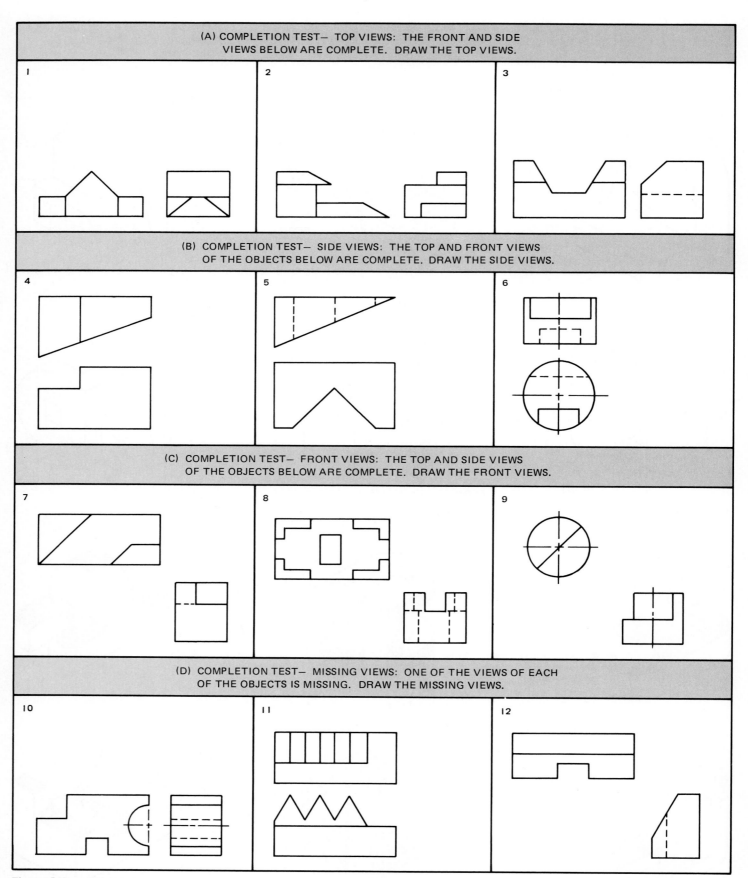

(A) COMPLETION TEST— TOP VIEWS: THE FRONT AND SIDE VIEWS BELOW ARE COMPLETE. DRAW THE TOP VIEWS.

1 2 3

(B) COMPLETION TEST— SIDE VIEWS: THE TOP AND FRONT VIEWS OF THE OBJECTS BELOW ARE COMPLETE. DRAW THE SIDE VIEWS.

4 5 6

(C) COMPLETION TEST— FRONT VIEWS: THE TOP AND SIDE VIEWS OF THE OBJECTS BELOW ARE COMPLETE. DRAW THE FRONT VIEWS.

7 8 9

(D) COMPLETION TEST— MISSING VIEWS: ONE OF THE VIEWS OF EACH OF THE OBJECTS IS MISSING. DRAW THE MISSING VIEWS.

10 11 12

Fig. 5-6-J Completion tests.

Assignments for Unit 5-7, Oblique Surfaces

22. On an A (A4) size sheet of preprinted grid paper (.25 in. or 10 mm grids) sketch three views of each of the objects shown in Figs. 5-7-A to 5-7-C. Draw three objects on each sheet. Each square on the objects represents one square on the grid paper. Allow one grid space between views and a minimum of two grid spaces between objects. The oblique surfaces on the objects are identified by a letter. Identify the oblique surfaces on each of the three views with a corresponding letter. Also identify the type of projection used by placing the appropriate ISO symbol at the bottom of the drawing.

23. On a B (A3) size sheet, make a three-view drawing of one of the parts shown in Figs. 5-7-D to 5-7-G. Allow 1.20 in. (30 mm) between views. Do not dimension.

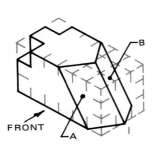
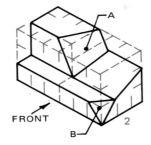
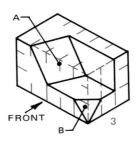

Fig. 5-7-A Sketching assignment.

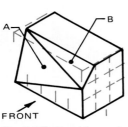
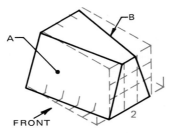
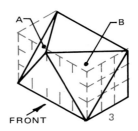

Fig. 5-7-B Sketching assignment.

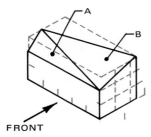
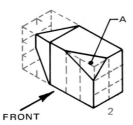
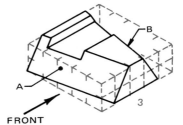

Fig. 5-7-C Sketching assignment.

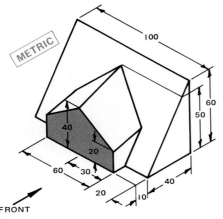

Fig. 5-7-D Base plate.

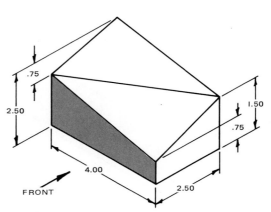

Fig. 5-7-E Angle brace.

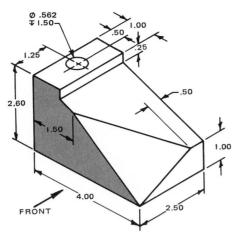

Fig. 5-7-F Support.

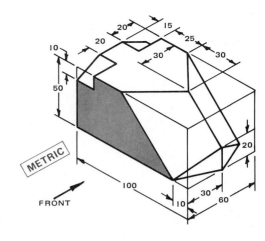

Fig. 5-7-G Locking base.

Assignments for Unit 5-8, First-Angle Orthographic Projection

24. On two A (A4) size sheets of pre-printed grid paper (.25 in. or 10 mm grids) sketch three views in first-angle orthographic projection of each of the objects shown in Figs. 5-8-A and 5-8-B. Draw three objects on each sheet. Each square on the objects represents one square on the grid paper. Allow one grid space between views and a minimum of two grid spaces between objects. Identify the type or projection used by placing the ISO projection symbol at the bottom of the drawing.

25. On a B (A3) size sheet, make a three-view drawing in first-angle projection of one of the parts shown in Figs. 5-8-C to 5-8-F. Allow 1.20 in. (30 mm) between views. Do not dimension.

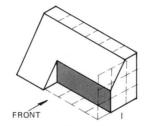
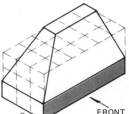
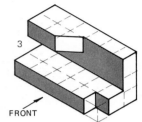

Fig. 5-8-A Sketching assignment.

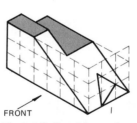
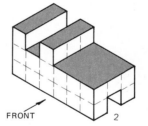

Fig. 5-8-B Sketching assignment.

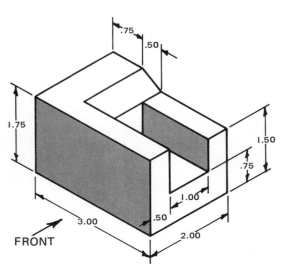

Fig. 5-8-C Spacer.

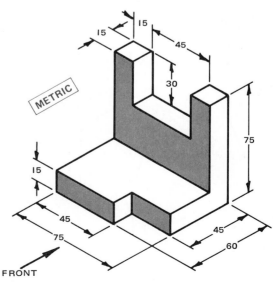

Fig. 5-8-D Spacer slide.

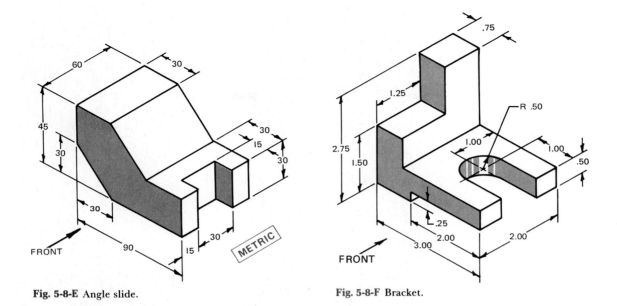

Fig. 5-8-E Angle slide.

Fig. 5-8-F Bracket.

Assignments for Unit 5-9, One- and Two-View Drawings

26. On a B (A3) size sheet, select any four of the objects shown in Fig. 5-9-A or 5-9-B and draw only the necessary views in orthographic third-angle projection which will completely describe each part. Use symbols or abbreviations where possible. The drawings need not be to scale but should be drawn in proportion to the illustrations shown.

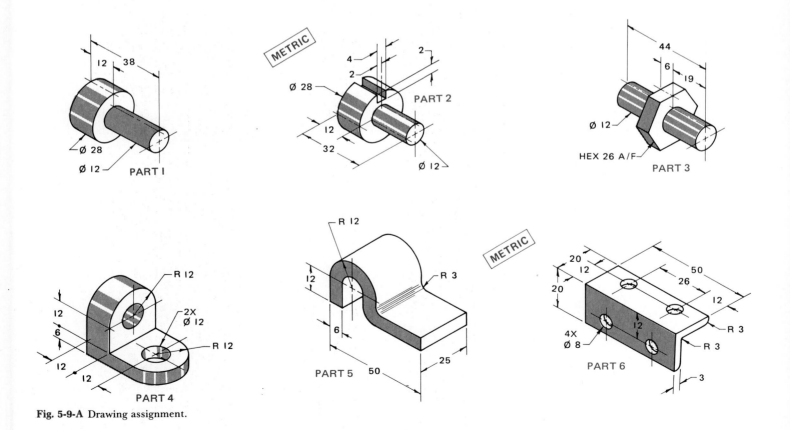

Fig. 5-9-A Drawing assignment.

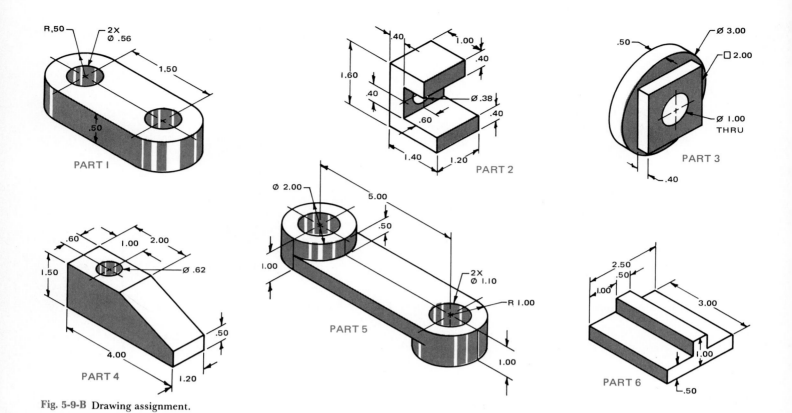

Fig. 5-9-B Drawing assignment.

Assignments for Unit 5-10, Special Views

27. On a B (A3) size sheet, select any one of the objects shown in Figs. 5-10-A to 5-10-D and draw only the necessary views (full and partial) which will completely describe each part. Add dimensions and machining symbols where required. Scale 1:1.

28. On a B (A3) size sheet, select one of the panels shown in Fig. 5-10-E or 5-10-F and make a detail drawing of the part. Enlarged views are recommended. Panels such as these, where labeling is used to identify the terminals, are used extensively in the electrical and electronics industry.

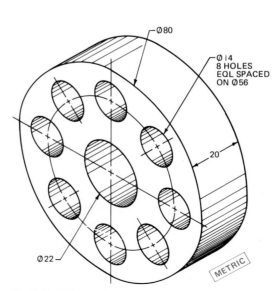

Fig. 5-10-A Round flange.

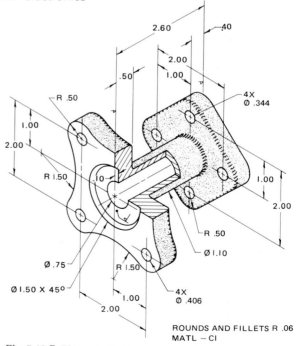

Fig. 5-10-B Flanged adapter.

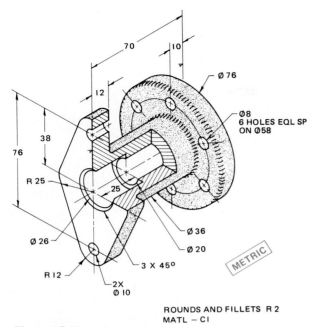

Fig. 5-10-C Flanged coupling.

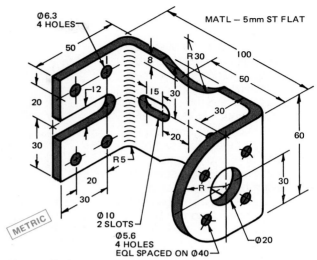

Fig. 5-10-D Connector.

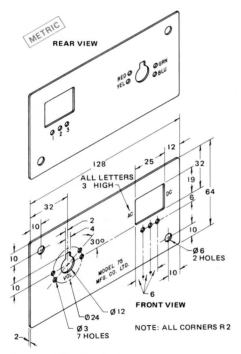

Fig. 5-10-E Radio cover plate.

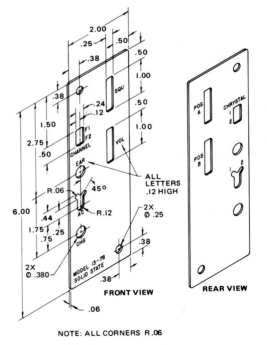

Fig. 5-10-F Transceiver cover plate.

29. With the most truss drawings, the scale used on the overall assembly is such that intricate detail cannot be clearly shown. As a result, enlarged detail views are added. With this type of assembly, many parts are opposite-hand to their counterparts.

The primary problem in the assembly of large-span multimember space structures has been to find some simple, inexpensive, and repetitive way to connect many members into and through a typical joint. The approach shown in this assignment is to shop-fabricate as much as possible, to ship the subassemblies to the site, and finally to complete the assembly by bolting the subassemblies together. The number, size, and strength of bolts are calculated by using the loading requirements at each connection.

Benefits of shop prefabrication of large units include minimization of field erection time and less possible error in the field resulting from the greater tolerance control in the shop.

On a B-size sheet, draw the enlarged views of the gusset assemblies shown in Fig. 5-10-G to a scale of 1 in. = 1 ft.

Shop Bolting Data. All structural members will be bolted to the gusset plate with five .375 in. high-strength bolts. Spacing is 1.50 in. from end and 3.00 in. center to center.

Field Bolting Data. All connections are to be made with five .375 in. high-strength bolts. Spacing is 1.50 in. from end and 3.00 in. center to center.

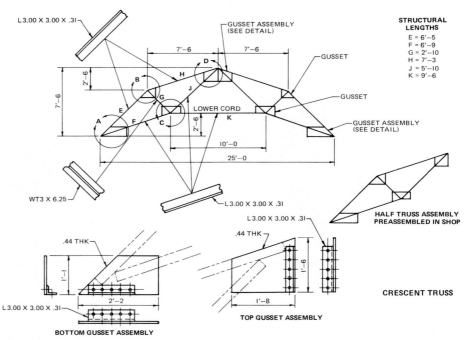

Fig. 5-10-G Crescent truss.

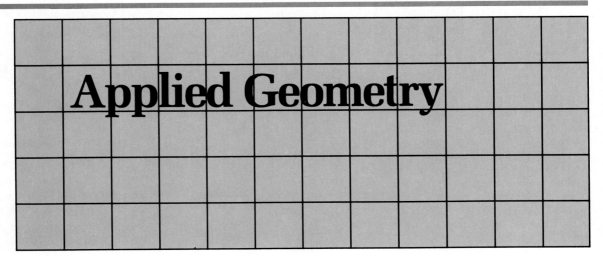

Applied Geometry

UNIT 6-1
Straight Lines

All of the lines forming the views on engineering drawings can be drawn using CAD commands or manual drafting with the instruments and equipment described in Chap. 2. However, geometric constructions have important uses, both in making drawings and in solving problems with graphs and diagrams. Sometimes it is necessary to use geometric constructions, particularly if when doing manual drafting the drafter does not have the advantages afforded by a drafting machine, an adjustable triangle, or templates for drawing hexagonal and elliptical shapes.

Definitions

Bisect Cut or divide into two.
Circumscribe Placing a figure around another touching it at points but not cutting it.
Inscribe Placing a figure within another so that all angular points of it lie on the boundary (circumference).
Parallel Continuously equidistant of lines or surfaces.
Perpendicular At right angles (90°) to a line or surface.
Tangent Meeting a line or surface at a point but not intersecting it.

To Draw a Line or Lines Parallel to and at a Given Distance from an Oblique Line

1. Given line *AB* (Fig. 6-1-1), erect a perpendicular *CD* to *AB*.
2. Space the given distance from the line *AB* by scale measurement or by an arc along line *CD*.
3. Position a triangle, using a second triangle or a T square as base, so that one side of the triangle is parallel with the given line.

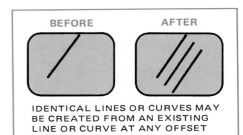

Fig. 6-1-2 CAD parallel (offset) command.

4. Slide this triangle along the base to the point at the desired distance from the given line, and draw the required line. Refer to Figs. 6-1-2 and 6-1-3 for the CAD methods of drawing parallel and perpendicular lines.

To Draw a Straight Line Tangent to Two Circles

Place a T square or straightedge so that the top edge just touches the edges of the

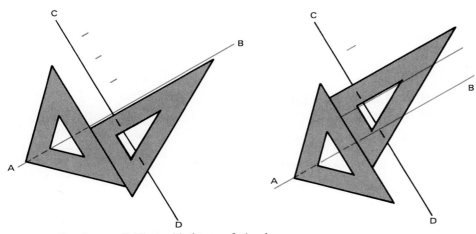

Fig. 6-1-1 Drawing parallel lines with the use of triangles.

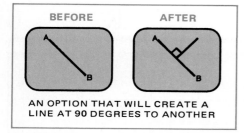

Fig. 6-1-3 CAD perpendicular line command.

circles, and draw the tangent line (Fig. 6-1-4). Perpendiculars to this line from the centers of the circles give the tangent points T_1 and T_2. Refer to Fig. 6-1-5 for the CAD method of drawing a line tangent to a circle.

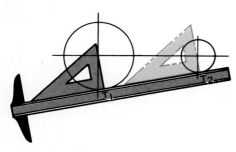

Fig. 6-1-4 Drawing a straight line tangent to two circles.

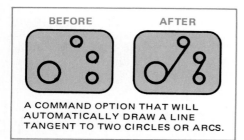

Fig. 6-1-5 CAD line tangent command.

To Bisect a Straight Line

1. Given line AB (Fig. 6-1-6), set the compass to a radius greater than ½ AB.

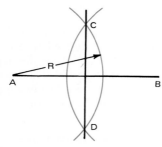

Fig. 6-1-6 Bisecting a line.

2. Using centers at A and B, draw intersecting arcs above and below line AB.

A line CD drawn through the intersections will divide AB into two equal parts and will be perpendicular to line AB.

To Bisect an Arc

1. Given arc AB (Fig. 6-1-7), set the compass to a radius greater than ½ AB.

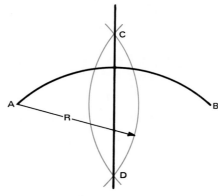

Fig. 6-1-7 Bisecting an arc.

2. Using points A and B as centers, draw intersecting arcs above and below arc AB. A line drawn through the intersections C and D will divide the arc AB into two equal parts.

To Bisect an Angle

1. Given angle ABC, with center B and a suitable radius (Fig. 6-1-8) draw an arc to cut BC at D and BA at E.
2. With centers D and E and equal radii, draw arcs to intersect at F.
3. Join B and F and extend to G. Line BG is the required bisector.

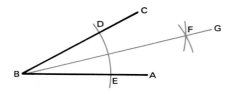

Fig. 6-1-8 Bisecting an angle.

To Divide a Line into a Given Number of Equal Parts

1. Given line AB and the number of equal divisions desired (12, for example), draw a perpendicular from A.
2. Place the scale so that the desired number of equal divisions is conveniently included between B and the perpendicular. Then mark these divisions, using short vertical marks from the scale divisions as in Fig. 6-1-9.

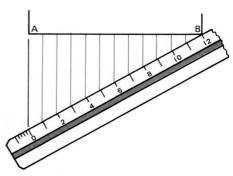

Fig. 6-1-9 Dividing a straight line into equal parts.

3. Draw perpendiculars to line AB through the points marked, dividing the line AB as required.

Refer to Fig. 6-1-10 for the CAD method of drawing a series of equally spaced lines.

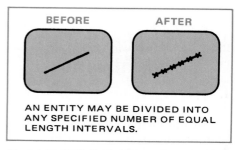

Fig. 6-1-10 CAD divide (measure) command.

ASSIGNMENT

See Assignment 1 for Unit 6-1 on page 104.

UNIT 6-2
Arcs and Circles

To Draw an Arc Tangent to Two Lines at Right Angles to Each Other

Given radius R of the arc (Fig. 6-2-1).

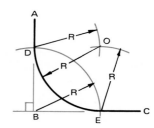

Fig. 6-2-1 Arc tangent to two lines at right angles to each other.

1. Draw an arc having radius R with center at B, cutting the lines AB and BC at D and E, respectively.

2. With D and E as centers and with the same radius R, draw arcs intersecting at O.
3. With center O, draw the required arc. The tangent points are D and E.

To Draw an Arc Tangent to the Sides of an Acute Angle

Given radius R of the arc (Fig. 6-2-2).

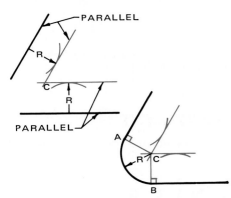

Fig. 6-2-2 Drawing an arc tangent to the sides of an acute angle.

1. Draw lines inside the angle, parallel to the given lines, at distance R away from the given lines. The center of the arc will be at C.
2. Set the compass to radius R, and with center C draw the arc tangent to the given sides. The tangent points A and B are found by drawing perpendiculars through point C to the given lines.

To Draw an Arc Tangent to Two Sides of an Obtuse Angle

Follow the same procedure as for an acute angle (Fig. 6-2-3).

Refer to Fig. 6-2-4 for the CAD method of drawing fillets.

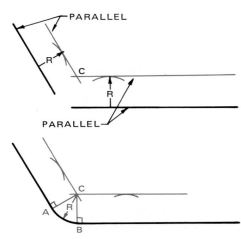

Fig. 6-2-3 Drawing an arc tangent to the side of an obtuse angle.

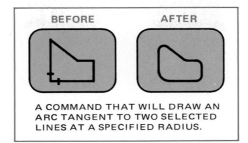

BEFORE   AFTER

A COMMAND THAT WILL DRAW AN ARC TANGENT TO TWO SELECTED LINES AT A SPECIFIED RADIUS.

Fig. 6-2-4 CAD fillet command.

To Draw a Circle on a Regular Polygon

1. Given the size of the polygon (Fig. 6-2-5), bisect any two sides; for example, BC and DE. The center of the polygon is where bisectors FO and GO intersect at point O.
2. The inner circle radius is OH, and the outer circle radius is OA.

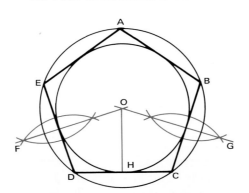

Fig. 6-2-5 Drawing a circle on a regular polygon.

To Draw a Reverse, or Ogee, Curve Connecting Two Parallel Lines

1. Given two parallel lines AB and CD and distances X and Y (Fig. 6-2-6),

join points B and C with a line.
2. Erect a perpendicular to AB and CD from points B and C, respectively.
3. Select point E on line BC where the curves are to meet.
4. Bisect BE and EC.
5. Points F and G where the perpendiculars and bisectors meet are the centers for the arcs forming the ogee curve.

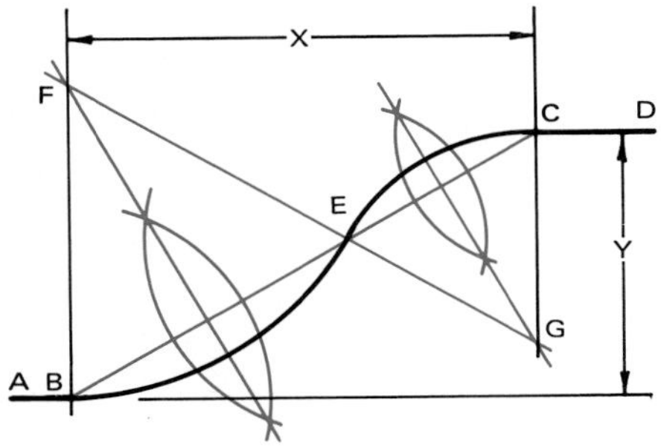

To Draw an Arc Tangent to a Given Circle and Straight Line

1. Given R, the radius of the arc (Fig. 6-2-7), draw a line parallel to the given straight line between the circle and the line at distance R away from the given line.
2. With the center of the circle as center and radius R_1 (radius of the circle plus R), draw an arc to cut the parallel straight line at C.
3. With center C and radius R, draw the required arc tangent to the circle and the straight line.

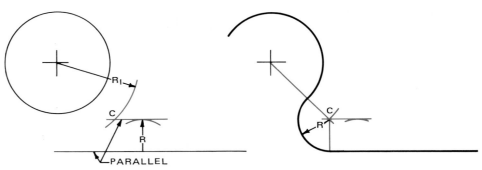

Fig. 6-2-7 Drawing an arc tangent to a circle and a straight line.

Fig. 6-2-6 Drawing a reverse (ogee) curve connecting two parallel planes.

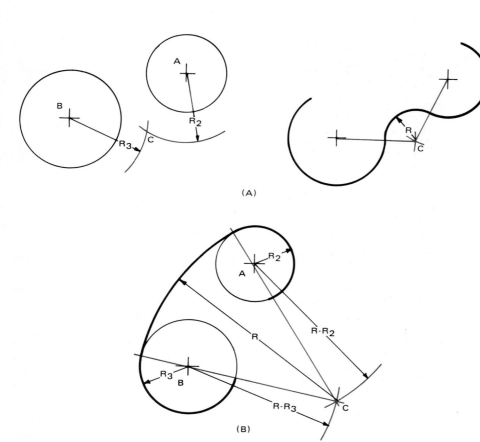

(A)

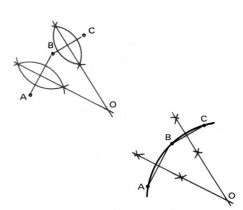

Fig. 6-2-10 Drawing an arc or circle through three points not in a straight line.

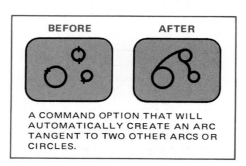

Fig. 6-2-8 Drawing an arc tangent to two circles.

(B)

To Draw an Arc Tangent to Two Circles

Figure 6-2-8A

1. Given the radius of arc R, with the center of circle A as center and radius R_2 (radius of circle A plus R), draw an arc in the area between the circles.
2. With the center of circle B as center and radius R_3 (radius of circle B plus R), draw an arc to cut the other arc at C.
3. With center C and radius R, draw the required arc tangent to the given circles.

Figure 6-2-8B

1. Given radius of arc R, with the center of circle A as center and radius $R - R_2$, draw an arc in the area between the circles.
2. With the center of circle B as center and radius $R - R_3$, draw an arc to cut the other arc at C.
3. With center C and radius R, draw the required arc tangent to the given circles.

See Fig. 6-2-9 to see a CAD method for drawing an arc tangent.

BEFORE	AFTER

A COMMAND OPTION THAT WILL AUTOMATICALLY CREATE AN ARC TANGENT TO TWO OTHER ARCS OR CIRCLES.

Fig. 6-2-9 CAD Arc tangent command.

To Draw an Arc or Circle Through Three Points Not in a Straight Line

1. Given points A, B, and C (Fig. 6-2-10), join points A, B, and C as shown.
2. Bisect lines AB and BC and extend bisecting lines to intersect at O. Point O is the center of the required circle or arc.
3. With center O and radius OA draw an arc.

ASSIGNMENT

See Assignment 2 for Unit 6-2 on page 105.

U N I T   6 - 3
Polygons

A polygon is a plane figure bounded by five or more straight lines not necessarily of equal length. A regular polygon is a plane figure bounded by five or more straight lines of equal length and contains angles of equal size.

To Draw a Hexagon (Fig. 6-3-1), Given the Distance Across the Flats

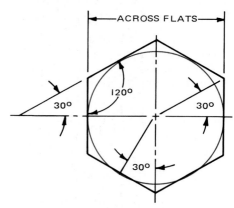

Fig. 6-3-1 Constructing a hexagon, given distance across flats.

1. Establish horizontal and vertical center lines for the hexagon.
2. Using the intersection of these lines as center, with radius one-half the distance across the flats, draw a light construction circle.

3. Using the 60° triangle, draw six straight lines, equally spaced, passing through the center of the circle.

4. Draw tangents to these lines at their intersection with the circle.

To Draw a Hexagon, Given the Distance Across the Corners (Fig. 6-3-2)

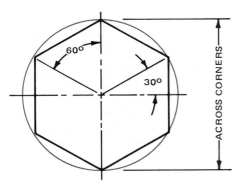

Fig. 6-3-2 Constructing a hexagon, given distance across corners.

1. Establish horizontal and vertical center lines, and draw a light construction circle with radius one-half the distance across the corners.

2. With a 60° triangle, establish points on the circumference 60° apart.

3. Draw straight lines connecting these points.

To Draw an Octagon, Given the Distance Across the Flats (Fig. 6-3-3)

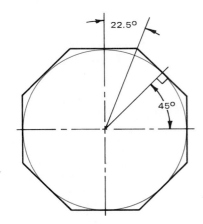

Fig. 6-3-3 Constructing an octagon, given distance across flats.

1. Establish horizontal and vertical center lines and draw a light construction circle with radius one-half the distance across the flats.

2. Draw horizontal and vertical lines tangent to the circle.

3. Using the 45° triangle, draw lines tangent to the circle at a 45° angle from the horizontal.

To Draw an Octagon, Given the Distance Across the Corners (Fig. 6-3-4)

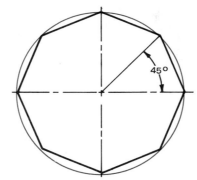

Fig. 6-3-4 Constructing an octagon, given distance across corners.

1. Establish horizontal and vertical center lines and draw a light construction circle with radius one-half the distance across the corners.

2. With the 45° triangle, establish points on the circumference between the horizontal and vertical center lines.

3. Draw straight lines connecting these points to the points where the center lines cross the circumference.

To Draw a Regular Polygon, Given the Length of the Sides

As an example, let a polygon have seven sides.

1. Given the length of side *AB* (Fig. 6-3-5), with radius *AB* and *A* as center, draw a semicircle and divide it into seven equal parts using a protractor.

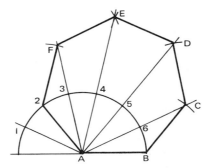

Fig. 6-3-5 Constructing a regular polygon, given length of one side.

2. Through the second division from the left, draw radial line *A2*.

3. Through points 3, 4, 5, and 6 extend radial lines as shown.

4. With *AB* as radius and *B* as center, cut line *A6* at *C*. With the same radius and *C* as center, cut line *A5* at *D*. Repeat at *E* and *F*.

5. Connect these points with straight lines.

These steps can be followed in drawing a regular polygon with any number of sides.

To Inscribe a Regular Pentagon in a Given Circle

1. Given circle with center *O* (Fig. 6-3-6), draw the circle with diameter *AB*.

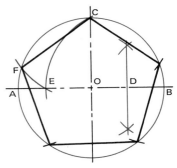

Fig. 6-3-6 Inscribing a regular pentagon in a given circle.

2. Bisect line *OB* at *D*.

3. With center *D* and radius *DC*, draw arc *CE* to cut the diameter at *E*.

4. With *C* as center and radius *CE*, draw arc *CF* to cut the circumference at *F*. Distance *CF* is one side of the pentagon.

5. With radius *CF* as a chord, mark off the remaining points on the circle. Connect the points with straight lines.

Refer to Fig. 6-3-7 for the CAD method of drawing polygons.

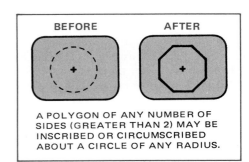

BEFORE   AFTER

A POLYGON OF ANY NUMBER OF SIDES (GREATER THAN 2) MAY BE INSCRIBED OR CIRCUMSCRIBED ABOUT A CIRCLE OF ANY RADIUS.

Fig. 6-3-7 CAD polygon command.

ASSIGNMENT

See Assignment 3 for Unit 6-3 on page 105.

UNIT 6-4
Ellipse

The *ellipse* is the plane curve generated by a point moving so that the sum of the distances from any point on the curve to two fixed points, called *foci*, is a constant.

Often a drafter is called upon to draw oblique and inclined holes and surfaces which take the form of an ellipse. Several methods, true and approximate, are used for its construction. The terms *major diameter* and *minor diameter* will be used in place of *major axis* and *minor axis* so the reader won't become confused with the mathematical X and Y axes.

To Draw an Ellipse—Two-Circle Method

1. Given the major and minor diameters (Fig. 6-4-1), construct two concentric circles with diameters equal to AB and CD.

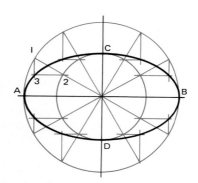

Fig. 6-4-1 Drawing an ellipse—two-circle method.

2. Divide the circles into a convenient number of equal parts. Figure 6-4-1 shows 12.
3. Where the radial lines intersect the outer circle, as at 1, draw lines parallel to line CD inside the outer circle.
4. Where the same radial line intersects the inner circle, as at 2, draw a line parallel to axis AB away from the inner circle. The intersection of these lines, as at 3, gives points on the ellipse.
5. Draw a smooth curve through these points.

To Draw an Ellipse—Four-Center Method

1. Given the major diameter CD and the minor diameter AB (Fig. 6-4-2), join points A and C with a line.

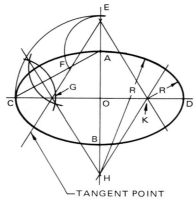

Fig. 6-4-2 Drawing an ellipse—four-center method.

2. Draw an arc with point O as the center and radius OC and extend line OA to locate point E.
3. Draw an arc with point A as the center and radius AE to locate point F.
4. Draw the perpendicular bisector of line CF to locate points G and H.
5. Locate points J and K where $OJ=OH$ and $OK=OG$.
6. Draw arcs with G and K as centers and radii HA and JB to complete the ellipse.

To Draw an Ellipse—Parallelogram Method

1. Given the major diameter CD and minor diameter AB (Fig. 6-4-3), construct a parallelogram.

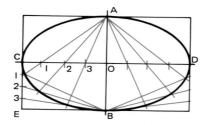

Fig. 6-4-3 Drawing an ellipse—parallelogram method.

2. Divide CO into a number of equal parts. Divide CE into the same number of equal parts. Number the points from C.
3. Draw a line from B to point 1 on line CE. Draw a line from A through

point 1 on CO, intersecting the previous line. The point of intersection will be one point on the ellipse.
4. Proceed in the same manner to find other points on the ellipse.
5. Draw a smooth curve through these points.

Refer to Fig. 6-4-4 for the CAD method of drawing an ellipse.

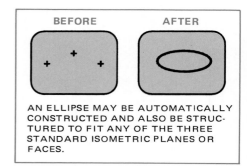

BEFORE	AFTER

AN ELLIPSE MAY BE AUTOMATICALLY CONSTRUCTED AND ALSO BE STRUCTURED TO FIT ANY OF THE THREE STANDARD ISOMETRIC PLANES OR FACES.

Fig. 6-4-4 CAD ellipse command.

ASSIGNMENT

See Assignment 4 for Unit 6-4 on page 106.

UNIT 6-5
Helix and Parabola

HELIX

The *helix* is the curve generated by a point that revolves uniformly around and up or down the surface of a cylinder. The *lead* is the vertical distance that the point rises or drops in one complete revolution.

To Draw a Helix

1. Given the diameter of the cylinder and the lead (Fig. 6-5-1), draw the top and front views.
2. Divide the circumference (top view) into a convenient number of parts (use 12) and label them.
3. Project lines down to the front view.
4. Divide the lead into the same number of equal parts and label them as shown in Fig. 6-5-1.
5. The points of intersection of lines with corresponding numbers lie on the helix. *Note:* Since points 8 to 12 lie on the back portion of the cylinder, the helix curve starting at point 7 and

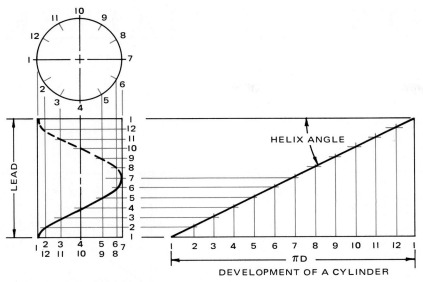

Fig. 6-5-1 Drawing a cylindrical helix.

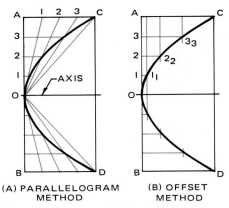

passing through points 8, 9, 10, 11, 12 to point 1 will appear as a hidden line.

6. If the development of the cylinder is drawn, the helix will appear as a straight line on the development.

PARABOLA

The *parabola* is a plane curve generated by a point that moves along a path equidistant from a fixed line (directrix) and a fixed point (focus).

To Construct a Parabola—Parallelogram Method

1. Given the sizes of the enclosing rectangle, distances AB and AC (Fig. 6-5-2A), construct a parallelogram.
2. Divide AC into a number of equal parts. Divide AO into the same number of equal parts. Number the points as shown.
3. Draw a line from O to point 1 on line AC. Draw a line parallel to the axis through point 1 on line AO, intersecting the previous line O-1. The point of intersection will be one point on the parabola.

4. Proceed in the same manner to find other points on the parabola.
5. Connect the points using an irregular curve.

To Construct a Parabola—Offset Method

1. Given the sizes of the enclosing rectangle, distances AB and AC (Fig. 6-5-2B), construct a parallelogram.
2. Divide OA into four equal parts.

(A) PARALLELOGRAM METHOD   (B) OFFSET METHOD

Fig. 6-5-2 Common methods used to construct a parabola.

3. The offsets vary in length as the square of their distances from O. Since OA is divided into four equal parts, distance AC will be divided into 4^2, or 16, equal divisions. Thus since $O1$ is one-fourth the length of OA, the length of line 1-1_1 will be $(\frac{1}{4})^2$, or $\frac{1}{16}$, the length of AC.
4. Since distance $O2$ is one-half the length of OA, the length of line 2-2_1 will be $(\frac{1}{2})^2$, or $\frac{1}{4}$, the length of AC.
5. Since distance $O3$ is three-fourths the length of OA, the length of line 3-3_1 will be $(\frac{3}{4})^2$, or $\frac{9}{16}$, the length of AC.
6. Complete the parabola by joining the points with an irregular curve.

CAD

The helix and parabola can readily be drawn by using the CAD commands shown in Figs. 6-5-3 and 6-5-4.

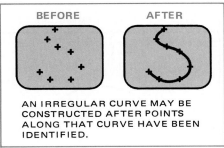

AN IRREGULAR CURVE MAY BE CONSTRUCTED AFTER POINTS ALONG THAT CURVE HAVE BEEN IDENTIFIED.

Fig. 6-5-3 CAD pline (spline) command.

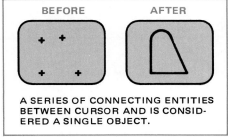

A SERIES OF CONNECTING ENTITIES BETWEEN CURSOR AND IS CONSIDERED A SINGLE OBJECT.

Fig. 6-5-4 CAD polyline command.

ASSIGNMENT

See Assignment 5 for Unit 6-5 on page 106.

ASSIGNMENTS FOR CHAPTER 6

Assignment for Unit 6-1,
Straight Lines

1. Divide a B (A3) size sheet as shown in Fig. 6-1-A. In the designated areas draw the geometric constructions.

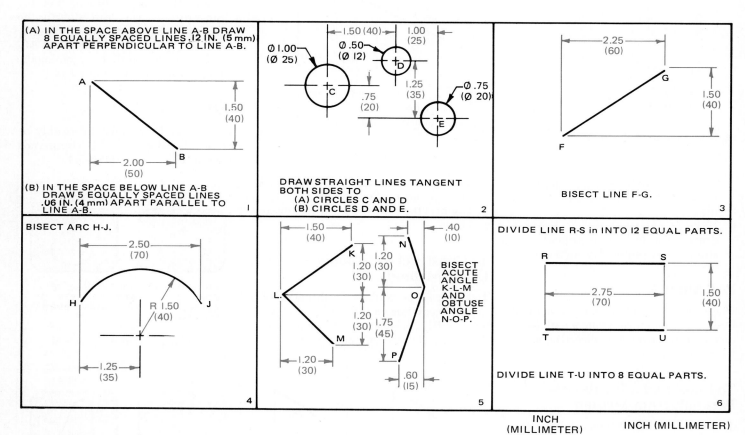

Fig. 6-1-A Straight-line construction.

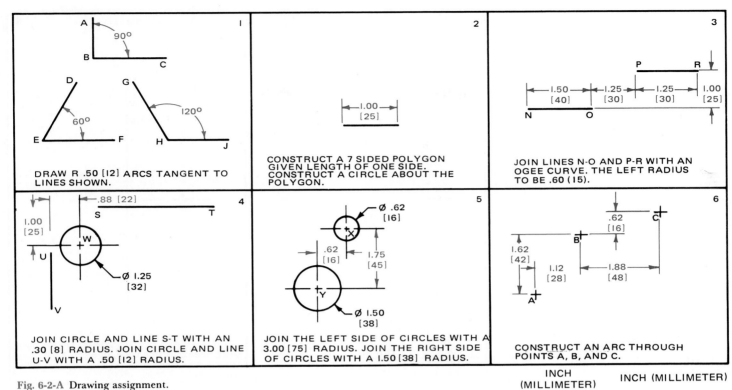

Fig. 6-2-A Drawing assignment.

Assignment for Unit 6-2, Arcs and Circles

2. Divide a B (A3) size sheet as shown in Fig. 6-2-A. In the designated areas draw the geometric constructions.

Assignment for Unit 6-3, Polygons

3. Divide a B (A3) size sheet as shown in Fig. 6-3-A. In the designated areas, draw the geometric constructions.

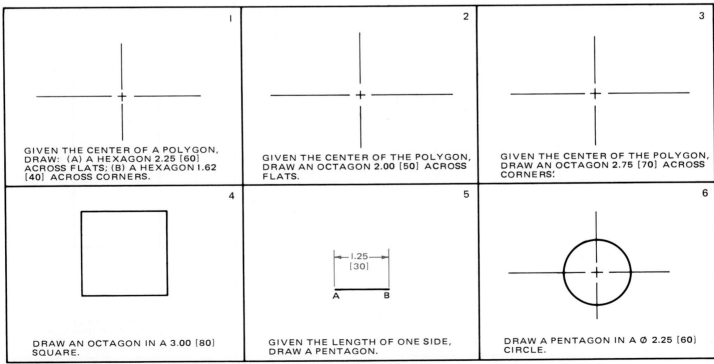

Fig. 6-3-A Drawing assignment.

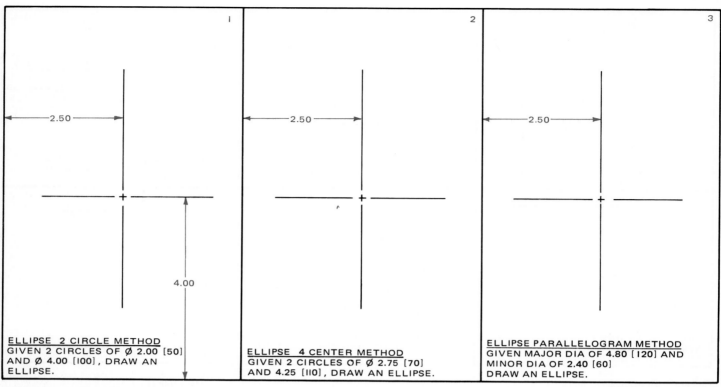

1
2
3
2.50
2.50
2.50
4.00
ELLIPSE 2 CIRCLE METHOD
GIVEN 2 CIRCLES OF Ø 2.00 [50] AND Ø 4.00 [100], DRAW AN ELLIPSE.
ELLIPSE 4 CENTER METHOD
GIVEN 2 CIRCLES OF Ø 2.75 [70] AND 4.25 [110], DRAW AN ELLIPSE.
ELLIPSE PARALLELOGRAM METHOD
GIVEN MAJOR DIA OF 4.80 [120] AND MINOR DIA OF 2.40 [60] DRAW AN ELLIPSE.

Fig. 6-4-A Drawing assignment.

INCH (MILLIMETER)

Assignment for Unit 6-4, The Ellipse

4. Divide a B (A3) size sheet as shown in Fig. 6-4-A. In the designated areas draw the geometric constructions.

Assignment for Unit 6-5, Helix and Parabola

5. Divide a B (A3) size sheet, as shown in Fig. 6-5-A. In the designated areas, draw the geometric constructions.

Review Assignments

6. On a B (A3) size sheet draw one of the parts shown in Figs. 6-6-A to 6-6-D. Do not erase construction lines. Scale 1:1. Do not dimension.

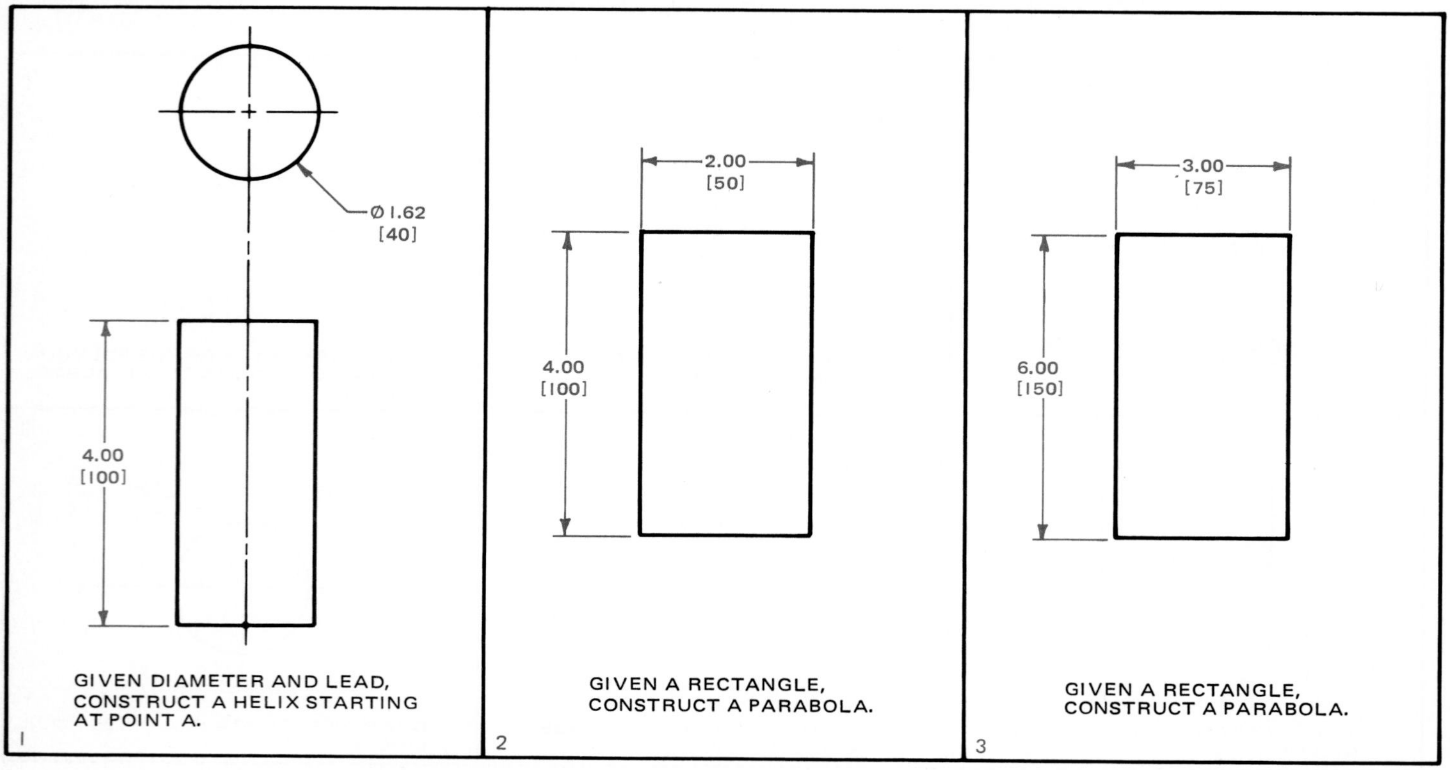

Ø 1.62
[40]
4.00
[100]
GIVEN DIAMETER AND LEAD, CONSTRUCT A HELIX STARTING AT POINT A.
1
2.00
[50]
4.00
[100]
GIVEN A RECTANGLE, CONSTRUCT A PARABOLA.
2
3.00
[75]
6.00
[150]
GIVEN A RECTANGLE, CONSTRUCT A PARABOLA.
3

Fig. 6-5-A Drawing assignment.

INCH
[MILLIMETER]

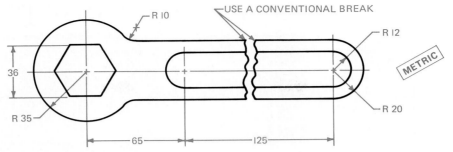

Fig. 6-6-A **Hex wrench.**

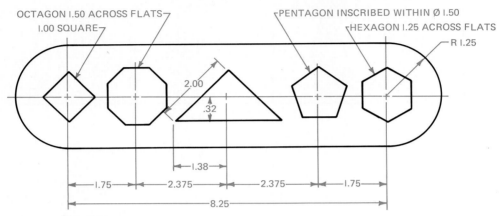

Fig. 6-6-B **Template.**

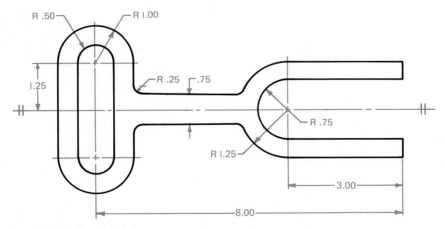

Fig. 6-6-C **Adjustable fork.**

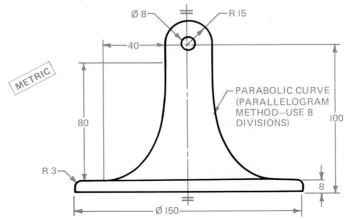

Fig. 6-6-D **Fan base.**

Basic Dimensioning

UNIT 7-1
Basic Dimensioning

A working drawing is one from which a tradesperson can produce a part. The drawing must be a complete set of instructions, so that it will not be necessary to give further information to the people fabricating the object. A working drawing, then, consists of the views necessary to explain the shape, the dimensions needed by the tradesperson, and required specifications, such as material and quantity needed. The latter information may be found in the notes on the drawing, or it may be located in the title block.

DIMENSIONING

Dimensions are indicated on drawings by extension lines, dimension lines, leaders, arrowheads, figures, notes, and symbols. They define geometric characteristics such as lengths, diameters, angles, and locations. See Fig. 7-1-1. The lines used in dimensioning are thin in contrast to the outline of the object. The dimensions must be clear and concise and permit only one interpretation. In general, each surface, line, or point is located by only one set of dimensions. Deviations from the approved rules for dimensioning should be made only in exceptional cases, when they will

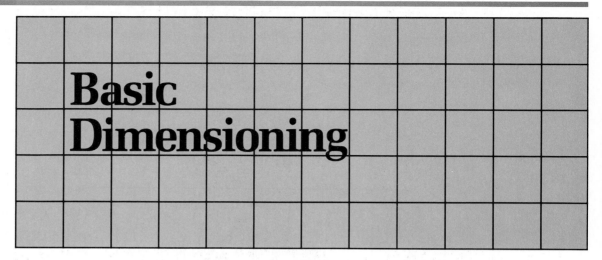

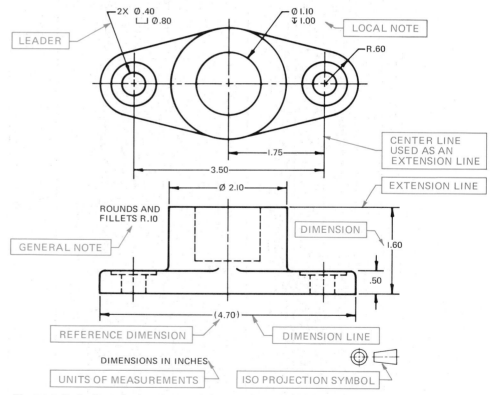

Fig. 7-1-1 Basic dimensioning elements.

improve the clarity of the dimensions. An exception to these rules is for arrowless and tabular dimensioning, which is discussed in Unit 7-4.

In general, each surface, line, or point is located by only one set of dimensions. These dimensions are not duplicated in other views.

Drawings for industry require some form of tolerancing on dimensions so that components can be properly assembled and manufacturing and production requirements can be met. This chapter deals only with basic dimensioning and tolerancing techniques. Geometric tolerancing, such as true positioning and tolerance of form, is covered in detail in Chap. 14.

Dimension and Extension Lines

Dimension lines are used to determine the extent and direction of dimensions, and they are normally terminated by uniform arrowheads, as shown in Fig. 7-1-2. Using an oblique line in lieu of an arrowhead is a common method used in architectural drafting. The recommended length and width of arrowheads should be in a ratio of 3:1 (Fig. 7-1-3B). The length of the arrowhead should be equal to the height of the dimension numerals.

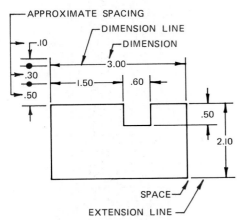

Fig. 7-1-2 Dimension and extension lines.

A single style of arrowhead should be used throughout the drawing. An industrial practice is, where space is limited, to use a small filled-in circle in lieu of an arrowhead (Fig. 7-1-3D).

Preferably, dimension lines should be broken for insertion of the dimension which indicates the distance between the extension lines. Where dimension lines are not broken the dimension is placed above the dimension line.

When several dimension lines are directly above or next to one another, it is good practice to stagger the dimensions in order to improve the clarity of the drawing. The spacing suitable for most drawings between parallel dimension lines is .38 in. (8 mm), and the spacing between the outline of the object and the nearest dimension line should be approximately .50 in. (10 mm). When the space between the extension lines is too small to permit the placing of the dimension line complete with arrowheads and dimension, then the

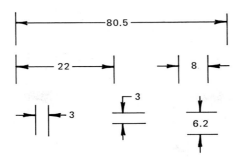

(A) PLACEMENT OF DIMENSIONS

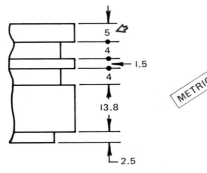

A SMALL CIRCULAR DOT MAY BE USED IN LIEU OF ARROWHEADS WHERE SPACE IS RESTRICTED.

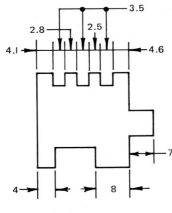

(D) DIMENSIONING IN RESTRICTED AREAS

Fig. 7-1-3 Dimensioning linear features.

alternate method of placing the dimension line, dimension, or both outside the extension lines are used. See Fig. 7-1-3D. Center lines should never be used for dimension lines. Every effort should be made to avoid crossing dimension lines by placing the shortest dimension closest to the outline (Fig. 7-1-3E).

Avoid dimensioning to hidden lines. In order to do so, it may be necessary to use a sectional view or a broken-out section. When the termination for a dimension is not included, as when used on partial or sectional views, the dimension line should extend beyond the center of the feature being dimensioned and

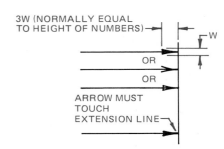

(B) ARROWHEAD SIZE AND STYLES

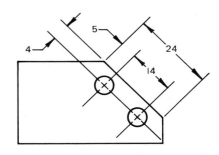

(C) OBLIQUE DIMENSIONING

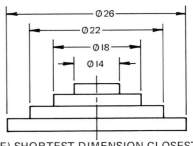

(E) SHORTEST DIMENSION CLOSEST TO OUTLINE

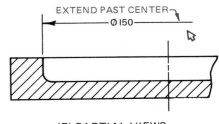

(F) PARTIAL VIEWS

shown with only one arrowhead (Fig. 7-1-3F). Refer to Figs. 7-1-4 and 7-1-5 for the CAD commands for basic dimensioning.

Dimension lines should be placed outside the view where possible and should extend to extension lines rather than visible lines. However, when readability is improved by avoiding either extra long extension lines (Fig. 7-1-6) or the crowding of dimensions, placing of dimensions on views is permissible.

Extension (*projection*) lines are used to indicate the point or line on the drawing to which the dimension applies. See Fig. 7-1-7. A small gap is left between the

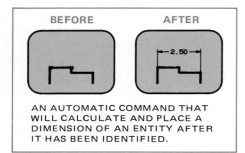

BEFORE   **AFTER**

AN AUTOMATIC COMMAND THAT WILL CALCULATE AND PLACE A DIMENSION OF AN ENTITY AFTER IT HAS BEEN IDENTIFIED.

Fig. 7-1-4 CAD linear dimensioning command.

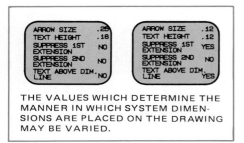

ARROW SIZE	.25
TEXT HEIGHT	.18
SUPPRESS 1ST EXTENSION	NO
SUPPRESS 2ND EXTENSION	NO
TEXT ABOVE DIM. LINE	NO

ARROW SIZE	.12
TEXT HEIGHT	.12
SUPPRESS 1ST EXTENSION	YES
SUPPRESS 2ND EXTENSION	NO
TEXT ABOVE DIM. LINE	YES

THE VALUES WHICH DETERMINE THE MANNER IN WHICH SYSTEM DIMENSIONS ARE PLACED ON THE DRAWING MAY BE VARIED.

Fig. 7-1-5 CAD dimension variables command.

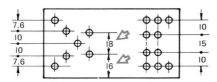

(A) IMPROVING READABILITY OF DRAWING

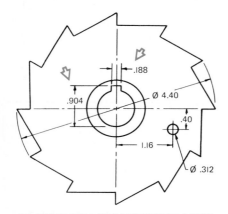

(B) AVOIDING LONG EXTENSION LINES

Fig. 7-1-6 Placing dimensions on view.

extension line and the outline to which it refers, and the extension line extends about .12 in. (3 mm) beyond the outermost dimension line. However, when extension lines refer to points, as in Fig. 7-1-7E, they should extend through the points. Extension lines are usually drawn perpendicular to dimension lines. However, to improve clarity or when there is overcrowding, extension lines may be drawn at an oblique angle as long as clarity is maintained.

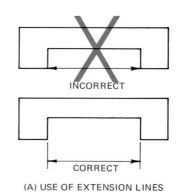

INCORRECT

CORRECT

(A) USE OF EXTENSION LINES

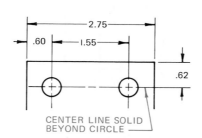

CENTER LINE SOLID BEYOND CIRCLE

(B) CENTER LINE USED AS EXTENSION LINE

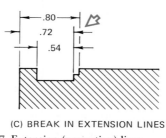

(C) BREAK IN EXTENSION LINES

Fig. 7-1-7 Extension (projection) lines.

Center lines may be used as extension lines in dimensioning. The portion of the center line extending past the circle is not broken, as in Fig. 7-1-7B

Where extension lines cross other extension lines, dimension lines, or visible lines, they are not broken. However, if extension lines cross arrowheads or dimension lines close to arrowheads, a break in the extension line is recommended. See Fig. 7-1-7C.

Leaders

Leaders are used to direct notes, dimensions, symbols, item numbers, or part numbers to features on the drawing. See Fig. 7-1-8. A leader should generally be a

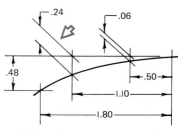

(D) OBLIQUE EXTENSION LINES

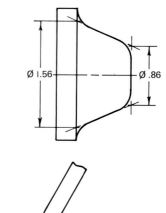

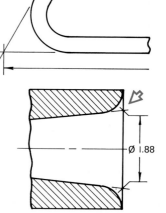

(E) EXTENSION LINE FROM POINTS

single straight inclined line (not vertical or horizontal) except for a short horizontal portion extending to the center of the height of the first or last letter or digit of

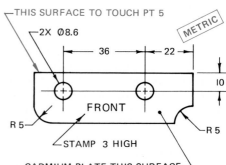

THIS SURFACE TO TOUCH PT 5
2X Ø8.6
METRIC
36
22
10
FRONT
R 5
R 5
STAMP 3 HIGH
CADMIUM PLATE THIS SURFACE

Fig. 7-1-8 Leaders.

the note. The leader is terminated by an arrowhead or a dot of at least .06 in. (1.5 mm) in diameter. Arrowheads should always terminate on a line; dots should be used within the outline of the object and rest on a surface. Leaders should not be bent in any way unless it is unavoidable. Leaders should not cross one another, and two or more leaders adjacent to one another should be drawn parallel if practicable. It is better to repeat dimensions or references than to use long leaders.

Where a leader is directed to a circle or circular arc, its direction should point to the center of the arc or circle. Regardless of the reading direction used, aligned or unidirectional, all notes and dimensions used with leaders are placed in a horizontal position.

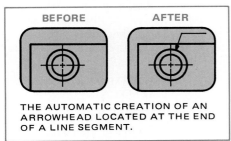

THE AUTOMATIC CREATION OF AN ARROWHEAD LOCATED AT THE END OF A LINE SEGMENT.

Fig. 7-1-9 CAD leader command.

Notes

Notes are used to simplify or complement dimensioning by giving information on a drawing in a condensed and systematic manner. They may be *general* or *local* notes, and should be in the present or future tense.

General Notes These refer to the part or the drawing as a whole. They should be shown in a central position below the view to which they apply or placed in a general note column. Typical examples of this type of note are

- FINISH ALL OVER
- ROUNDS AND FILLETS R .06
- REMOVE ALL SHARP EDGES

Local Notes These apply to local requirements only and are connected by a leader to the point to which the note applies.

Repetitive features and dimensions may be specified in the local note by the use of a X in conjunction with the numeral to indicate the "number of times" or "places" they are required. See Figs. 7-1-1 and 7-1-8. A full space is left between the X and the feature dimension. For additional information refer to Unit 7-3.

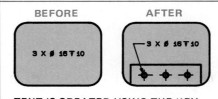

TEXT IS CREATED USING THE KEYBOARD AND TRANSFERRED TO THE DRAWING BY CURSOR POSITION SELECT.

Fig. 7-1-10 CAD text (note) command.

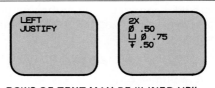

ROWS OF TEXT MAY BE "LINED UP" IN ANY OF SEVERAL ARRANGEMENTS. LEFT JUSTIFY, FOR EXAMPLE, WILL LINE THE STARTING POINT FOR EACH ROW OF TEXT WITH RESPECT TO A VERTICAL LINE.

Fig. 7-1-11 CAD justify command.

Typical examples are:

- 4 × Ø6
- 2 × 45°
- Ø3
  ∨ Ø 11.5 × 86°
- M12 × 1.25

Refer to Figs. 7-1-10 and 7-1-11 for the CAD commands used for adding notes to drawings.

UNITS OF MEASUREMENTS

Although the metric system of dimensioning is becoming the official international standard of measurement, most drawings in the United States are still dimensioned in inches or feet and inches. For this reason, drafters should be familiar with all the dimensioning systems which they may encounter. The dimensions used in this book are primarily decimal inch. However, metric and dual dimensions shown in the problems in this text are also used very frequently.

On drawings where all dimensions are either in inches or millimeters, individual identification of linear units is not required. However, the drawing should contain a note stating the units of measurement.

Where some inch dimensions, such as nominal pipe sizes, are shown on a millimeter-dimensioned drawing, the abbreviation IN must follow the inch values.

Inch Units of Measurement

Decimal-Inch System (U.S. Customary linear units) Parts are designed in basic decimal increments, preferably .02 in., and are expressed with a minimum of two figures to the right of the decimal point. See Fig. 7-1-12. Using the .02 in. module, the second decimal place (hundredths) is an even number or zero. By using the design modules having an even number for the last digit, dimensions can be halved for center distances without increasing the number of decimal places. Decimal dimensions which are not multiples of .02, such as .01, .03, and .15, should be used only when it is essential to meet design requirements such as to provide clearance, strength, smooth curves, etc. When greater accuracy is required, sizes are expressed as three- or four-place decimal numbers, for example, 1.875.

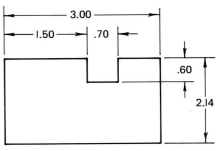

(A) DECIMAL INCH

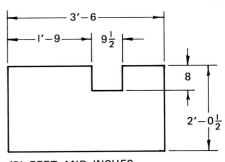

(B) FEET AND INCHES

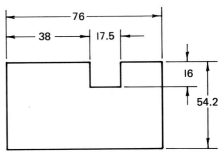

(C) MILLIMETERS

Fig. 7-1-12 Dimensioning units.

Whole dimensions will show a minimum of two zeros to the right of the decimal point.

24.00 not 24

An inch value of less than 1 is shown without a zero to the left of the decimal point.

.44 not 0.44

In cases where parts have to be aligned with existing parts or commercial products, which are dimensioned in fractions, it may be necessary to use decimal equivalents of fractional dimensions.

Fractional-Inch System In this system, parts are designed in basic units of common fractions down to 1/64 in. Decimal dimensions are used when finer divisions than 1/64 in. must be made. Common fractions may be used for specifying the size of holes that are produced by drills ordinarily stocked in fraction sizes and for the sizes of standard screw threads.

When common fractions are used on drawings, the fraction bar must not be omitted and should be horizontal except when applied with a typewriting machine which does not have a horizontal fraction bar.

When a dimension intermediate between 1/64 increments is necessary, it is expressed in decimals, such as .30, .257, or .2575 in.

The inch marks (″) are not shown with the dimensions. A note such as

DIMENSIONS ARE IN INCHES

should be clearly shown on the drawing. The exception is when the dimension "1 in." is shown on the drawing. The 1 should then be followed by the inch marks—1″, <i>not</i> 1.

Foot and Inch System Feet and inches are often used for installation drawings, drawings of large objects, and floor plans associated with architectural work. In this system, all dimensions 12 in. or greater are specified in feet and inches. For example, 24 in. is expressed as 2′–0, and 27 in. is expressed as 2′–3. Parts of an inch are usually expressed as common fractions, rather than as decimals.

The inch marks (″) are not shown on drawings. The drawing should carry a note such as

DIMENSIONS ARE IN FEET
AND INCHES UNLESS
OTHERWISE SPECIFIED

A dash should be placed between the foot and inch values. For example, 1′–3, not 1′3.

SI Metric Units of Measurement

The standard metric units on engineering drawings are the millimeter (mm) for linear measure and micrometer (μm) for surface roughness. See Fig. 7-1-11. For architectural drawings, meter and millimeter units are used.

Whole numbers from 1 to 9 are shown without a zero to the left of the number or a zero to the right of the decimal point.

2 not 02 or 2.0

A millimeter value of less than 1 is shown with a zero to the left of the decimal point.

0.2 not .2 or .20

0.26 not .26

Decimal points should be uniform and large enough to be clearly visible on reduced-size prints. They should be placed in line with the bottom of the associated digits and be given adequate space.

Neither commas nor spaces are used to separate digits into groups in specifying millimeter dimensions on drawings.

32545 not 32 545

Identification A metric drawing should include a general note, such as

UNLESS OTHERWISE
SPECIFIED DIMENSIONS
ARE IN MILLIMETERS

and be identified by the word METRIC prominently displayed near the title block.

Units Common to Either System

Some measurements can be stated so that the callout will satisfy the units of both systems. For example, tapers such as .006 in. per inch and 0.006 mm per millimeter can both be expressed simply as the ratio 0.006:1 or in a note such as TAPER 0.006:1. Angular dimensions are also specified the same in both inch and metric systems.

Standard Items

Fasteners and Threads Either inch or metric fasteners and threads may be used. Refer to the Appendix and Chap. 9 for additional information.

Hole Sizes Tables showing standard inch and metric drill sizes are shown in the Appendix.

DUAL DIMENSIONING

With the great exchange of drawings taking place between the United States and the rest of the world, at one time it became advantageous to show drawings in both inches and millimeters. As a result, many international-operated companies adopted a dual system of dimensioning. Today, however, this type of dimensioning should be avoided if possible. When it may be necessary or desirable to give dimensions in both inches and millimeters on the same drawing as shown in Figs. 7-1-13 and 7-1-14, then a note or illustration should be located near the title block or strip to identify the inch and millimeter dimensions.

MILLIMETER
─────────
INCH

and/or

MILLIMETER (INCH)

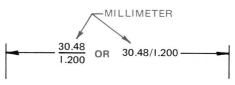

(A) POSITION METHOD

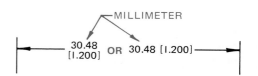

(B) BRACKET METHOD

Fig. 7-1-13 Dual dimensioning.

ANGULAR UNITS

Angles are measured in degrees. The decimal degree is now preferred over the use of degrees, minutes, and seconds. For example, the use of 60.5° is preferred to the use of 60°30′. Where only minutes or seconds are specified, the number of minutes or seconds is preceded by 0°, or 0°0′, as applicable. Some examples follow.

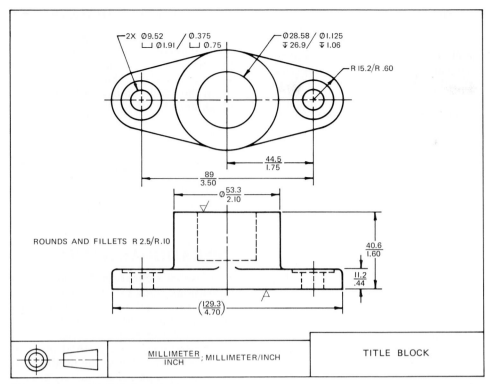

ROUNDS AND FILLETS R 2.5/R.10

MILLIMETER / INCH ; MILLIMETER/INCH   TITLE BLOCK

Fig. 7-1-14 Dual dimensioned drawing.

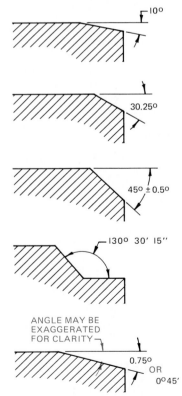

ANGLE MAY BE
EXAGGERATED
FOR CLARITY

Fig. 7-1-15 Angular units.

Decimal Degree	Degrees, Minutes, and Seconds
10° ± 0.5°	10° ± 0°30′
0.75°	0°45′
0.004°	0°0′15″
90° ± 1.0°	90° ± 1°
25.6° ± 0.2°	25°36′ ± 0°12′
25.51°	25°30′36″

The dimension line of an angle is an arc drawn with the apex of the angle as the center point for the arc, wherever practicable. The position of the dimension varies according to the size of the angle and appears in a horizontal position. Recommended arrangements are shown in Fig. 7-1-15.

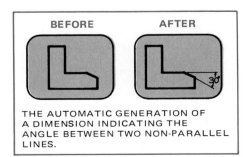

BEFORE   AFTER

THE AUTOMATIC GENERATION OF
A DIMENSION INDICATING THE
ANGLE BETWEEN TWO NON-PARALLEL
LINES.

Fig. 7-1-16 CAD angular dimensioning command.

READING DIRECTION

Dimensions and notes should be placed to be read from the bottom of the drawing for engineering drawings (unidirectional system). For architectural and structural drawings the aligned system of dimensioning is used. See Fig. 7-1-17.

In both methods, angular dimensions and dimensions and notes shown with leaders should be aligned with the bottom of the drawing.

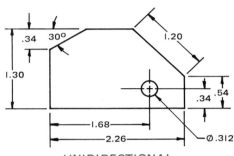

UNIDIRECTIONAL
USED ON ENGINEERING DRAWINGS

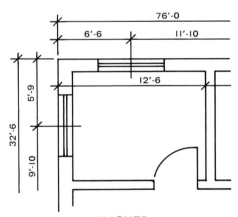

ALIGNED
USED ON ARCHITECTURAL AND
STRUCTURAL DRAWINGS

Fig. 7-1-17 Reading direction of dimensions.

BASIC RULES FOR DIMENSIONING

Refer to Fig. 7-1-18.

- Place dimensions between the views when possible.
- Place the dimension line for the shortest length, width, or height nearest the outline of the object. Parallel dimension lines are placed in order of their size, making the longest dimension line the outermost.
- Place dimensions with the view that best shows the characteristic contour or shape of the object. In following

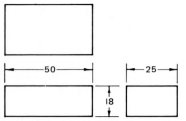

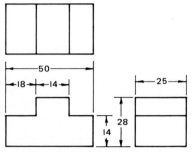

(A) PLACE DIMENSIONS BETWEEN VIEWS

(B) PLACE SMALLEST DIMENSION NEAREST THE VIEW BEING DIMENSIONED

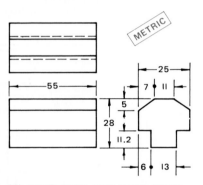

(C) DIMENSION THE VIEW THAT BEST SHOWS THE SHAPE

Fig. 7-1-18 Basic dimensioning rules.

this rule, dimensions will not always be between views.

- On large views, dimensions can be placed on the view to improve clarity.
- Use only one system of dimensions, either the unidirectional or the aligned, on any one drawing.
- Dimensions should not be duplicated in other views.
- Dimensions should be selected so that it will not be necessary to add or subtract dimensions in order to define or locate a feature.

SYMMETRICAL OUTLINES

A part is said to be symmetrical when the features on each side of the center line or median plane are identical in size, shape, and location. Partial views are often drawn for the sake of economy

or space. With CAD duplicating, the other half of the view requires little effort. However, space constraints may preclude this possibility. When only one-half the outline of a symmetrically shaped part is drawn, symmetry is indicated by applying the symmetry symbol to the center line on both sides of the part. See Fig. 7-1-19.

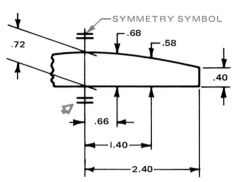

Fig. 7-1-19 Dimensioning symmetrical outlines or features.

REFERENCE DIMENSIONS

A reference dimension is shown for information only, and it is not required for manufacturing or inspection purposes. It is enclosed in parentheses, as shown in Fig. 7-1-20. Formerly the abbreviation REF was used to indicate a reference dimension.

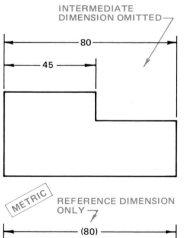

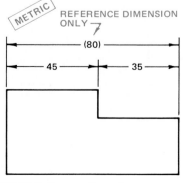

Fig. 7-1-20 Reference dimensions.

NOT-TO-SCALE DIMENSIONS

When a dimension on a drawing is altered, making it not to scale, it should be underlined (underscored) with a straight, thick line (Fig. 7-1-21), except when the condition is clearly shown by break lines.

Fig. 7-1-21 NOT-TO-SCALE dimensions.

OPERATIONAL NAMES

The use of operational names with dimensions, such as turn, bore, grind, ream, tap, and thread, should be avoided. While the drafter should be aware of the methods by which a part can be produced, the method of manufacture is better left to the producer. If the completed part is adequately dimensioned and has surface texture symbols showing finish quality desired, it remains a shop problem to meet the drawing specifications.

ABBREVIATIONS

Abbreviations and symbols are used on drawings to conserve space and time, but only where their meanings are quite clear. Therefore, only commonly accepted abbreviations such as those shown in the Appendix should be used on drawings.

References

1. ANSI Y14.5M, *Dimensioning and Tolerancing*.

ASSIGNMENTS

See Assignments 1 through 3 for Unit 7-1 on pages 141 and 142.

UNIT 7-2
Dimensioning Circular Features

DIAMETERS

Where the diameter of a single feature or the diameters of a number of concentric cylindrical features are to be specified, it is recommended that they be shown on the longitudinal view. See Fig. 7-2-1.

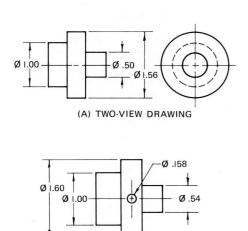

(A) TWO-VIEW DRAWING

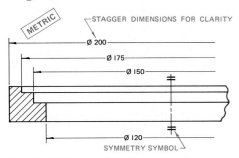

(B) ONE-VIEW DRAWING

Fig. 7-2-1 Diameters.

Where space is restricted or when only a partial view is used, diameters may be dimensioned as illustrated in Fig. 7-2-2. Regardless of where the diameter

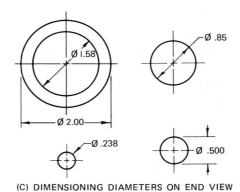

(C) DIMENSIONING DIAMETERS ON END VIEW

dimension is shown, the numerical value is preceded by the diameter symbol Ø for both customary and metric dimensions.

RADII

The general method of dimensioning a circular arc is by giving its radius. A radius dimension line passes through, or is in line with, the radius center and terminates with an arrowhead touching the arc. See Fig. 7-2-3. An arrowhead is never used at the radius center. The size of the dimension is preceded by the abbreviation R for both customary and metric dimensioning. Where space is limited, as for a small radius, the radial dimension line may extend through the radius center. Where it is inconvenient to place the arrowhead between the radius center and the arc, it may be placed outside the arc, or a leader may be used (Fig. 7-2-3A).

Where a dimension is given to the center of the radius, a small cross should be drawn at the center (Fig. 7-2-3B). Extension lines and dimension lines are used to locate the center. Where the location of the center is unimportant, a radial arc may be located by tangent lines (Fig. 7-2-3E).

Where the center of a radius is outside the drawing or interferes with another view, the radius dimension line may be foreshortened (Fig. 7-2-3D). The portion of the dimension line next to the arrowhead should be radial relative to the curved line. Where the radius dimension line is foreshortened and the center is located by coordinate dimensions, the dimensions locating the center should be shown as foreshortened or the dimension shown as not to scale.

Simple fillet and corner radii may also be dimensioned by a general note, such as

ALL ROUNDS AND FILLETS
UNLESS OTHERWISE
SPECIFIED R.20
or
ALL RADII R5

Where a radius is dimensioned in a view that does not show the true shape of the radius, TRUE R is added before the radius dimension, as illustrated in Fig. 7-2-5.

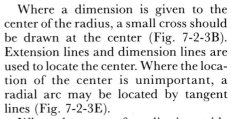

Fig. 7-2-2 Dimensioning diameters where space is restricted.

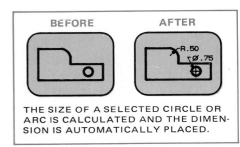

Fig. 7-2-4 CAD radial dimensioning command.

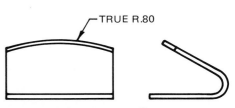

Fig. 7-2-5 Indicating true radius.

Rounded Ends

Overall dimensions should be used for parts or features having rounded ends. For fully rounded ends, the radius (R) is shown but not dimensioned (Fig. 7-2-6A). For parts with partially rounded ends, the radius is dimensioned (Fig. 7-2-6B). Where a hole and radius have the same center and the hole location is

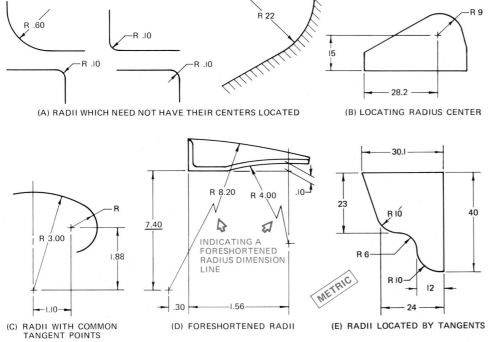

(A) RADII WHICH NEED NOT HAVE THEIR CENTERS LOCATED

(B) LOCATING RADIUS CENTER

(C) RADII WITH COMMON TANGENT POINTS

(D) FORESHORTENED RADII

(E) RADII LOCATED BY TANGENTS

Fig. 7-2-3 Radii.

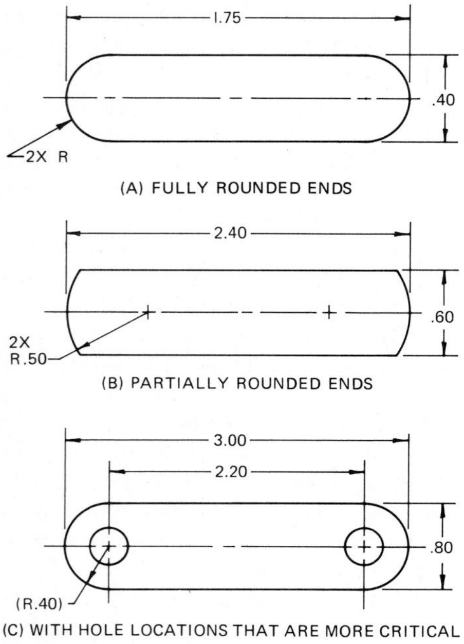

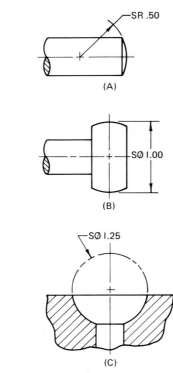

(A) FULLY ROUNDED ENDS
- 1.75
- .40
- 2X R

(B) PARTIALLY ROUNDED ENDS
- 2.40
- .60
- 2X R .50

(C) WITH HOLE LOCATIONS THAT ARE MORE CRITICAL
- 3.00
- 2.20
- .80
- (R.40)

Fig. 7-2-6 External surfaces with rounded ends.

more critical than the location of a radius, then either the radius or the overall length should be shown as a reference dimension (Fig. 7-2-6C).

Dimensioning Chords, Arcs, and Angles

The difference in dimensioning chords, arcs, and angles is shown in Fig. 7-2-7.

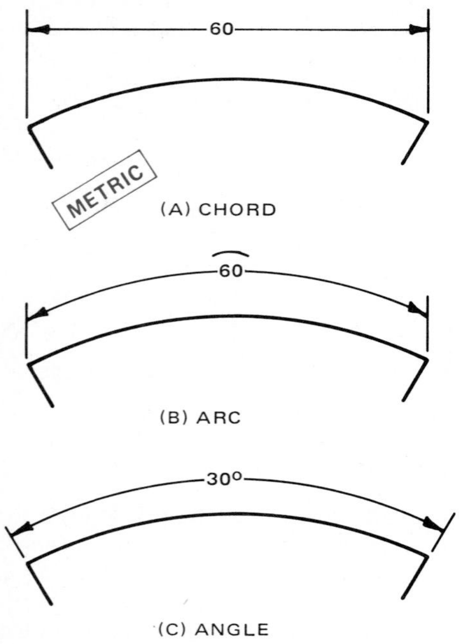

Spherical Features

Spherical surfaces may be dimensioned as diameters or radii, but the dimension should be used with the abbreviations SR or SØ. See Fig. 7-2-8.

- SR .50 (A)
- SØ 1.00 (B)
- SØ 1.25 (C)

Fig. 7-2-8 Spherical surfaces.

Cylindrical Holes

Plain, round holes are dimensioned in various ways, depending upon design and manufacturing requirements (Fig. 7-2-9). However, the leader is the method most commonly used. When a leader is used to specify diameter sizes, as with small holes, the dimension is identified as a diameter by preceding the numerical value with the diameter symbol Ø.

The size, quantity, and depth may be shown on a single line, or on several lines if preferable. For through holes, the abbreviation THRU should follow the dimension if the drawing does not make this clear. The depth dimension of a blind hole is the depth of the full diameter and is normally included as part of the dimensioning note.

When more than one hole of a size is required, the number of holes should be specified. However, care must be taken to avoid placing the hole size and quantity values together without adequate spacing. It may be better to show the note on two or more lines than to use a line note which might be misread (Fig. 7-2-9D).

Minimizing Leaders If too many leaders would impair the legibility of a drawing, letters or symbols as shown in

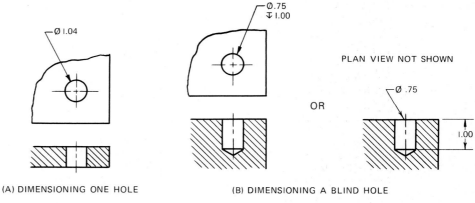

(A) CHORD — 60 — METRIC

(B) ARC — 60

(C) ANGLE — 30°

Fig. 7-2-7 Dimensioning chords, arcs, and angles.

(A) DIMENSIONING ONE HOLE — Ø 1.04

(C) ADDING THE WORD THRU WHEN IT IS NOT CLEAR THE HOLE GOES THROUGH — Ø .62 THRU

Fig. 7-2-9 Cylindrical holes.

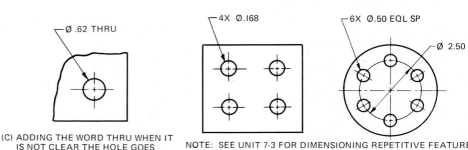

(B) DIMENSIONING A BLIND HOLE — Ø .75 ⊽ 1.00

PLAN VIEW NOT SHOWN — OR — Ø .75 — 1.00

(D) DIMENSIONING A GROUP OF HOLES — 4X Ø.168 — 6X Ø.50 EQL SP — Ø 2.50

NOTE: SEE UNIT 7-3 FOR DIMENSIONING REPETITIVE FEATURES

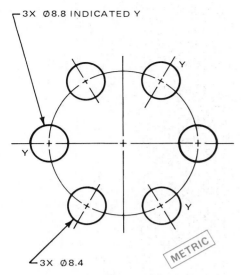

Fig. 7-2-10 Minimizing leaders.

Fig. 7-2-10 should be used to identify the features.

Slotted Holes

Elongated holes and slots are used to compensate for inaccuracies in manufacturing and to provide for adjustment. See Fig. 7-2-11. The method selected to locate the slot would depend on how the slot was made. The method shown in Fig. 7-2-11B is used when the slot is punched out and the location of the punch is given. Figure 7-2-11A shows the dimensioning method used when the slot is machined out.

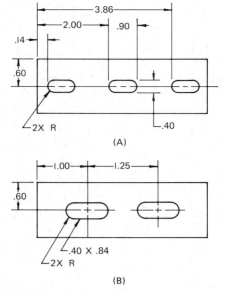

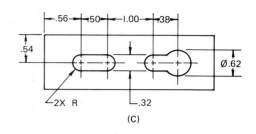

Fig. 7-2-11 Slotted holes.

Countersinks, Counterbores, and Spotfaces

Counterbores, spotfaces, and countersinks are specified on drawings by means of dimension symbols or abbreviations, the symbols being preferred. The symbols or abbreviations indicate the form of the surface only and do not restrict the methods used to produce that form. The dimensions for them are usually given as a note, preceded by the size of the through hole. See Figs. 7-2-12 and 7-2-13.

A **countersink** is an angular-sided recess to accommodate the head of flathead screws, rivets, and similar items. The diameter at the surface and the included angle are given. When the depth of the tapered section of the countersink is critical, it is specified in the note or by a dimension. For counterdrilled holes, the diameter, depth, and included angle of the counterdrill are given.

A **counterbore** is a flat-bottomed, cylindrical recess which permits the head of a fastening device, such as a bolt, to lie recessed into the part. The diame-

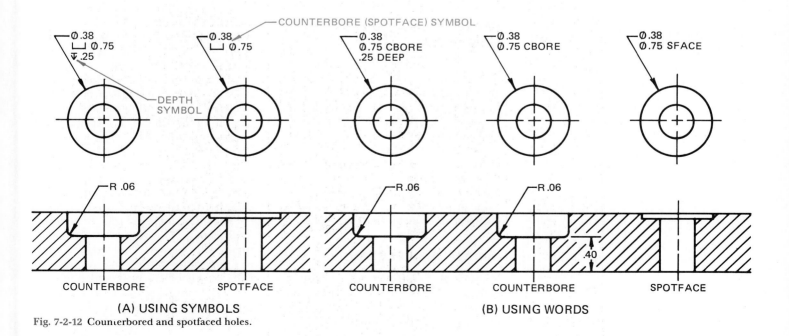

Fig. 7-2-12 Counterbored and spotfaced holes.

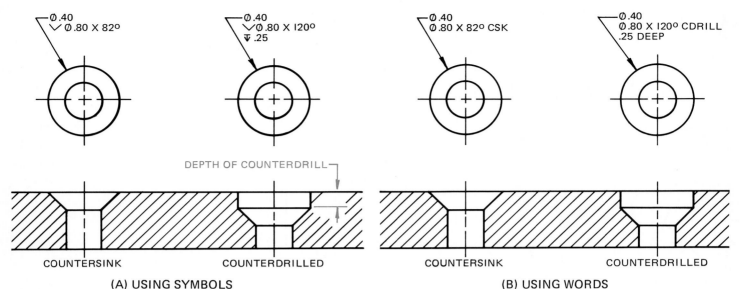

(A) USING SYMBOLS　　　　　　　　　　　　**(B) USING WORDS**

Fig. 7-2-13 Countersunk and counterdrilled holes.

ter, depth, and corner radius are specified in a note. In some cases, the thickness of the remaining stock may be dimensioned rather than the depth of the counterbore.

A **spotface** is an area where the surface is machined just enough to provide smooth, level seating for a bolt head, nut, or washer.

The diameter of the faced area and either the depth or the remaining thickness are given. A spotface may be specified by a general note and not delineated on the drawing. If no depth or remaining thickness is specified, it is implied that the spot-facing is the minimum depth necessary to clean up the surface to the specified diameter.

The symbols for counterbore or spotface, countersink, and depth are shown in Figs. 7-2-12 and 7-2-13. In each case the symbol precedes the dimension.

If CAD is used and the dimensional symbols covered so far in this chapter are not part of the existing CAD library, then this would be an excellent opportunity to develop your own library by using the LIBRARY command. See Fig. 7-2-14.

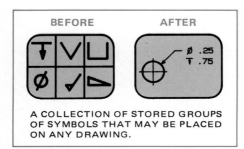

A COLLECTION OF STORED GROUPS OF SYMBOLS THAT MAY BE PLACED ON ANY DRAWING.

Fig. 7-2-14 CAD library command.

References and Source Materials

1. ANSI Y14.5M, *Dimensioning and Tolerancing*.

ASSIGNMENTS

See Assignments 4 through 6 for Unit 7-2 starting on page 143.

UNIT 7-3
Dimensioning Common Features

Repetitive Features and Dimensions

Repetitive features and dimensions may be specified on a drawing by the use of an X in conjunction with the numeral to indicate the "number of times" or "places" they are required. A space is shown between the X and the dimension.

An X which means "by" is often used between coordinate dimensions specified in note form. Where both are used on a drawing, care must be taken to ensure each is clear. See Fig. 7-3-1.

Symmetrical Outlines

Symmetrical outlines may be dimensioned on one side of the axis of symmetry only. See Fig. 7-3-2. Where only part of the outline is shown,

because of functional drafting procedures, the size of the part, or space limitations, symmetrical shapes may be shown by only one-half of the outline, and the symmetry is indicated by applying the symbol for part symmetry to the center line. In such cases, the outline of the part should extend slightly beyond the center line and terminate with a break line. Note the dimensioning method of extending the dimension lines to act as extension lines for the perpendicular dimensions.

Chamfers

The process of chamfering, that is, cutting away the inside or outside piece, is done to facilitate assembly. Chamfers are normally dimensioned by giving their angle and linear length. See Fig. 7-3-3. When the chamfer is 45°, it may be specified as a note.

When a very small chamfer is permissible, primarily to break a sharp corner, it may be dimensioned but not drawn, as in Fig. 7-3-3C. If not otherwise specified, an angle of 45° is understood.

Internal chamfers may be dimensioned in the same manner, but it is often desirable to give the diameter over the chamfer. The angle may also be given as the included angle if this is a design requirement. This type of dimensioning is generally necessary for larger diameters, especially those over 2 in. (50 mm), whereas chamfers on small holes are usually expressed as countersinks. Chamfers are never measured along the angular surface.

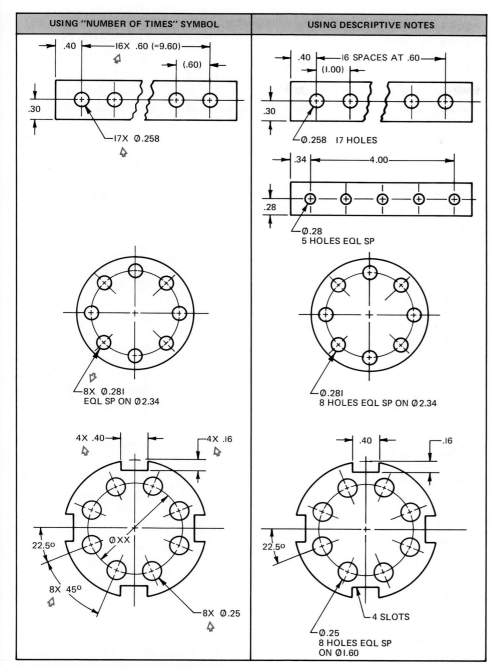

USING "NUMBER OF TIMES" SYMBOL	USING DESCRIPTIVE NOTES

Fig. 7-3-1 Dimensioning repetitive detail.

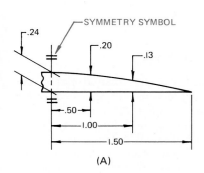

Fig. 7-3-2 Dimensioning symmetrical features.

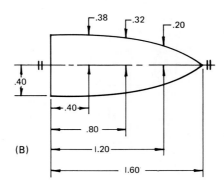

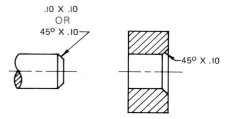

(A) FOR 45° CHAMFERS ONLY

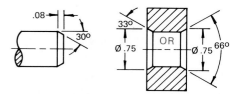

(B) FOR ALL CHAMFERS

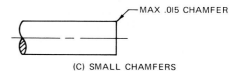

(C) SMALL CHAMFERS

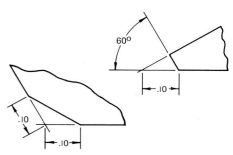

(D) CHAMFERS BETWEEN SURFACES AT OTHER THAN 90°

Fig. 7-3-3 Dimensioning chamfers.

Slopes and Tapers

Slope A slope is the inclination of a line representing an inclined surface. It is expressed as a ratio of the difference in the heights at right angles to the base line, at a specified distance apart. See Fig. 7-3-4. Fig. 7-3-4D is the preferred method of dimensioning slopes on architectural and structural drawings.

The following dimensions and symbol may be used, in different combinations, to define the slope of a line or flat surface:

- The slope specified as a ratio combined with the slope symbol (Fig. 7-3-4A).
- The slope specified by an angle (Fig. 7-3-4B).
- The dimensions showing the difference in the heights of two points from the base line and the distance between them (Fig. 7-3-4C).

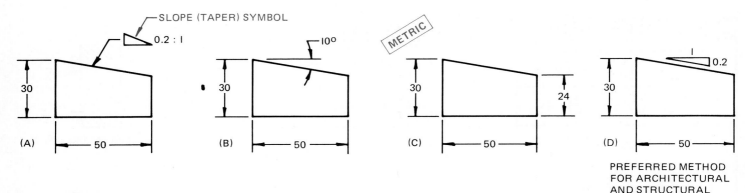

Fig. 7-3-4 Dimensioning slopes.

PREFERRED METHOD
FOR ARCHITECTURAL
AND STRUCTURAL
DRAWINGS

Taper

Taper A taper is the ratio of the difference in the diameters of two sections (perpendicular to the axis of a cone to the distance between these two sections). See Fig. 7-3-5. When the taper symbol is used, the vertical leg is always shown to the left and precedes the ratio figures. The following dimensions may be used, in suitable combinations, to define the size and form of tapered features:

- The diameter (or width) at one end of the tapered feature
- The length of the tapered feature
- The rate of taper
- The included angle
- The taper ratio
- The diameter at a selected cross section
- The dimension locating the cross section

Knurls

Knurling is specified in terms of type, pitch, and diameter before and after knurling. See Fig. 7-3-6. The letter P precedes the pitch number. Where control is not required, the diameter after knurling is omitted. Where only portions of a feature require knurling, axial dimensions must be provided. Where required to provide a press fit between parts, knurling is specified by a note on the drawing which includes the type of knurl required, the pitch, the toleranced diameter of the feature prior to knurling, and the minimum acceptable diameter after knurling. Commonly used types are straight, diagonal, spiral, convex, raised diamond, depressed diamond, and radial. The pitch is usually

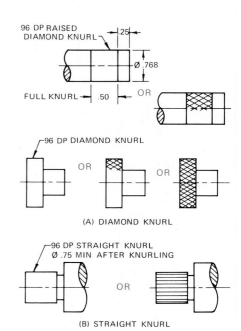

Fig. 7-3-6 Dimensioning knurls.

expressed in terms of teeth per inch or millimeter and may be the straight pitch, circular pitch, or diametral pitch. For cylindrical surfaces, the latter is preferred.

The knurling symbol is optional and is used only to improve clarity on working drawings.

Formed Parts

In dimensioning formed parts, the inside radius is usually specified, rather than the outside radius, but all forming dimensions should be shown on the same side if possible. Dimensions apply to the side on which the dimensions are shown unless otherwise specified. See Fig. 7-3-7.

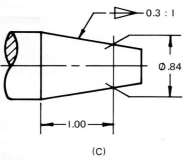

Fig. 7-3-5 Dimensioning tapers.

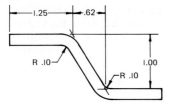

Fig. 7-3-7 Dimensioning theoretical points of intersection.

Undercuts

The operation of undercutting or necking, that is, cutting a recess in a diameter, is done to permit two parts to come together, as illustrated in Fig. 7-3-8A. It is indicated on the drawing

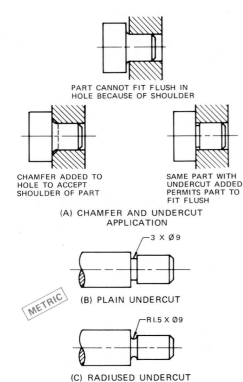

PART CANNOT FIT FLUSH IN HOLE BECAUSE OF SHOULDER

CHAMFER ADDED TO HOLE TO ACCEPT SHOULDER OF PART

SAME PART WITH UNDERCUT ADDED PERMITS PART TO FIT FLUSH

(A) CHAMFER AND UNDERCUT APPLICATION

METRIC

(B) PLAIN UNDERCUT

(C) RADIUSED UNDERCUT

Fig. 7-3-8 Dimensioning undercuts.

by a note listing the width first and then the diameter. If the radius is shown at the bottom of the undercut, it will be assumed that the radius is equal to one-half the width unless otherwise specified, and the diameter will apply to the center of the undercut. When the size of the undercut is unimportant, the dimension may be left off the drawing.

Limited Lengths and Areas

Sometimes it is necessary to dimension a limited length or area of a surface to indicate a special condition. In such instances, the area or length is indicated by a chain line (Fig. 7-3-9A). When indicating a length of surface, the chain line is drawn parallel and adjacent to the surface. When indicating an area of surface, the area is crosshatched within the chain line boundary. See Fig. 7-3-9B.

Wire, Sheet Metal, and Drill Rod

Wire, sheet metal, and drill rod, which are manufactured to gage or code sizes, should be shown by their decimal dimensions; but gage numbers, drill letters, etc., may be shown in parentheses following those dimensions.

EXAMPLES
Sheet —.141 (No. 10 USS GA)
—.081 (No. 12 B & S GA)

Source Materials

1. ANSI Y14.5M, *Dimensioning and Tolerancing*.

ASSIGNMENTS

See Assignments 7 through 11 for Unit 7-3 on pages 146 and 147.

Dimensioning Methods

The choice of the most suitable dimensions and dimensioning methods will depend, to some extent, on how the part will be produced and whether the drawings are intended for unit or mass production.

Unit production refers to cases where each part is to be made separately, using general-purpose tools and machines.

Mass production refers to parts produced in quantity, where special tools and gages are usually provided.

Either linear or angular dimensions may locate features with respect to one another (point-to-point) or from a datum. Point-to-point dimensions may be adequate for describing simple parts. Dimensions from a datum may be necessary if a part with more than one critical dimension must mate with another part.

The following systems of dimensioning are used more commonly for engineering drawings.

RECTANGULAR COORDINATE DIMENSIONING

This is a method for indicating distance, location, and size by means of linear dimensions measured parallel or perpendicular to reference axes or datum planes that are perpendicular to one another.

Coordinate dimensioning with dimension lines must clearly identify the datum features from which the dimensions originate. See Fig. 7-4-1.

CAD systems provide the DATUM (BASELINE) technique as a dimension submenu option. See Fig. 7-4-2.

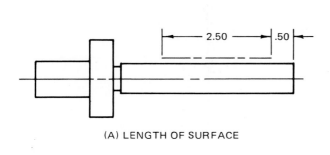

(A) LENGTH OF SURFACE

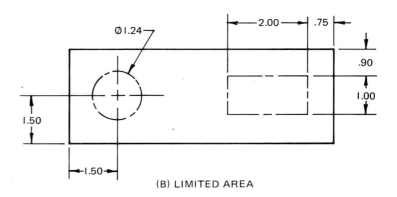

(B) LIMITED AREA

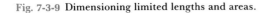

Fig. 7-3-9 Dimensioning limited lengths and areas.

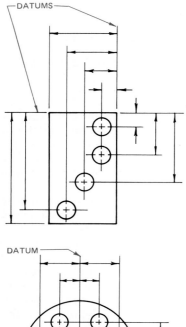

Fig. 7-4-1 Rectangular coordinate dimensioning.

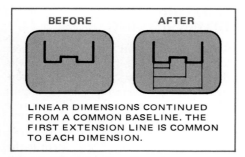

Fig. 7-4-2 CAD baseline (datum) dimensioning.

Rectangular Coordinates for Arbitrary Points Coordinates for arbitrary points of reference without a grid appear adjacent to each point (see Fig. 7-4-3) or

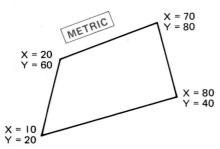

Fig. 7-4-3 Coordinates for arbitrary points.

in tabular form (see Fig. 7-4-4). CAD systems will automatically display any point coordinate when it is picked.

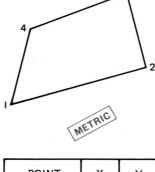

POINT	X	Y
1	10	20
2	80	40
3	70	80
4	20	60

Fig. 7-4-4 Coordinates for arbitrary points in tabular form.

HOLE SYMBOL	HOLE SIZE
A	.246
B	.189
C	.154
D	.125

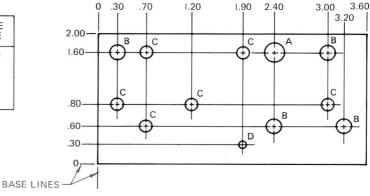

Fig. 7-4-5 Rectangular coordinate dimensioning without dimension lines (arrowless dimensioning).

HOLE DIA	HOLE SYMBOL	LOCATION		
		X	Y	Z
5.6	A₁	60	40	18
4.8	B₁	10	40	THRU
	B₂	75	40	THRU
	B₃	60	16	THRU
	B₄	80	16	THRU
4	C₁	18	40	THRU
	C₂	55	40	THRU
	C₃	10	20	THRU
	C₄	30	20	THRU
	C₅	75	20	THRU
	C₆	18	16	THRU
3.2	D₁	55	8	12
8.1	E₁	42	20	12

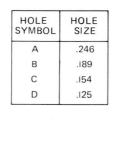

Fig. 7-4-6 Tabular dimensioning.

Rectangular Coordinate Dimensioning Without Dimension Lines
Dimensions may be shown on extension lines without the use of dimension lines or arrowheads. The base lines may be zero coordinates or they may be labeled as X, Y, and Z. See Fig. 7-4-5.

Tabular Dimensioning Tabular dimensioning is a type of coordinate dimensioning in which dimensions from mutually perpendicular planes are listed in a table on the drawing rather than on the pictorial delineation. This method is used on drawings which require the location of a large number of similarly shaped features when dimensioning parts for numerical control. See Fig. 7-4-6.

It may be advantageous to have a part dimensioned symmetrically about its center, as shown in Fig. 7-4-7. When the center lines are designated as the base (zero) lines, then positive and negative values will occur. These values are shown with the dimensions locating the holes. For more information on this type of dimensions, refer to Chap. 15.

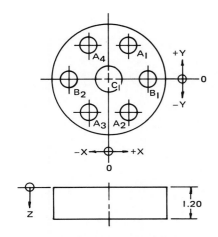

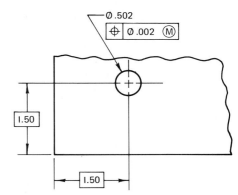

Φ .502

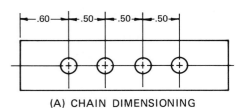

Fig. 7-4-10 True position dimensioning.

Chordal Dimensioning

The chordal dimensioning system may also be used for the spacing of points on the circumference of a circle relative to a datum, where manufacturing methods indicate that this will be more convenient. See Fig. 7-4-9.

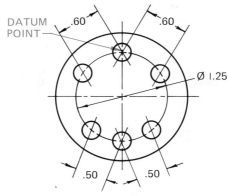

Fig. 7-4-9 Chordal dimensioning.

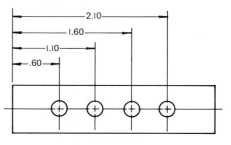

(A) CHAIN DIMENSIONING

Fig. 7-4-11 Chain dimensioning.

HOLE DIA	HOLE SYMBOL	LOCATION		
		X	Y	Z
.375	A₁	.50	.75	THRU
	A₂	.50	−.75	THRU
	A₃	−.50	−.75	THRU
	A₄	−.50	.75	THRU
.250	B₁	1.00	0	.60
	B₂	−1.00	0	.60
.500	C₁	0	0	THRU

Fig. 7-4-7 Rectangular coordinate dimensioning in tabular form with base (zero) line located at the center of the part.

Polar Coordinate Dimensioning

Polar coordinate dimensioning is commonly used in circular planes or circular configurations of features. It is a method of indicating the position of a point, line, or surface by means of a linear dimension and an angle, other than 90°, that is implied by the vertical and horizontal center lines. See Fig. 7-4-8.

True Position Dimensioning

True position dimensioning has many advantages over the coordinate dimensioning system. See Fig. 7-4-10. Because of its scope, it is covered as a complete topic in Chap. 14.

Chain Dimensioning

When a series of dimensions is applied on a point-to-point basis, it is called *chain dimensioning*. See Fig. 7-4-11. A possible disadvantage of this system is that it may result in an undesirable accumulation of tolerances between individual features. See Unit 7-5.

Datum or Common-Point Dimensioning

When several dimensions emanate from a common reference point or line, the method is called *common-point* or *datum dimensioning*. Dimensioning from reference lines may be executed as parallel dimensioning or as a superimposed running dimensioning. See Fig. 7-4-12.

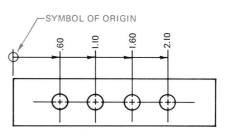

(A) PARALLEL METHOD

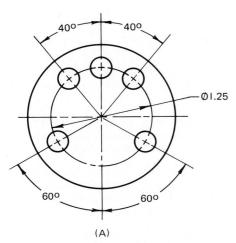

(A)

Fig. 7-4-8 Polar coordinate dimensioning.

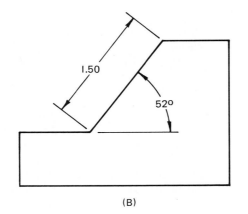

(B)

(B) SUPERIMPOSED METHOD

Fig. 7-4-12 Common-point (baseline) dimensioning.

Superimposed running dimensioning is simplified parallel dimensioning and may be used where there are space problems. Dimensions should be placed near the arrowhead, in line with the corresponding extension line as shown in Fig. 7-4-12B.

The origin is indicated by a circle and the opposite end of each dimension is terminated with an arrowhead.

It may be advantageous to use superimposed running dimensions in two directions. In such cases, the origins may be shown as in Fig. 7-4-13, or at the center of a hole, or other feature.

References

1. ANSI Y14.5M, *Dimensioning and Tolerancing*.

ASSIGNMENTS

See Assignments 12 to 15 for Unit 7-4 starting on page 147.

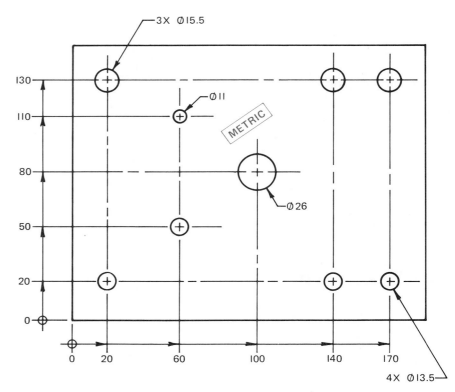

Fig. 7-4-13 Superimposed running dimensions in two directions.

UNIT 7-5
Limits and Tolerances

In the 6000 years of the history of technical drawing as a means for the communication of engineering information, it seems inconceivable that such an elementary practice as the tolerancing of dimensions, which we take so much for granted today, was introduced for the first time about 70 years ago.

Apparently, engineers and fabricators came in a very gradual manner to the realization that exact dimensions and shapes could not be attained in the manufacture of materials and products.

The skilled tradespeople of old prided themselves on being able to work to exact dimensions. What they really meant was that they dimensioned objects with a degree of accuracy greater than that with which they could measure. The use of modern measuring instruments would have shown the deviations from the sizes which they called exact.

As soon as it was realized that variations in the sizes of parts had always been present, that such variations could be restricted but not avoided, and also that slight variations in the size which a part was originally intended to have could be tolerated without its correct

functioning being impaired, it was evident that interchangeable parts need not be identical parts, but that it would be sufficient if the significant sizes which controlled their fits lay between definite limits. Accordingly, the problem of interchangeable manufacture evolved from the making of parts to a would-be exact size, to the holding of parts between two limiting sizes lying so closely together that any intermediate size would be acceptable.

Tolerances are the permissible variations in the specified form, size, or location of individual features of a part from that shown on the drawing. The finished form and size into which material is to be fabricated are defined on a drawing by various geometric shapes and dimensions.

As mentioned previously, the manufacturer cannot be expected to produce the exact size of parts as indicated by the dimensions on a drawing; so a certain amount of variation on each dimension must be tolerated. For example, a dimension given as 1.500 ± .004 in. means that the manufactured part can be anywhere between 1.496 and 1.504 in. and that the tolerance permitted on this dimension is .008 in. The largest and smallest permissible sizes (1.504 and 1.496 in., respectively) are known as the *limits*.

Greater accuracy costs more money, and since economy in manufacturing would not permit all dimensions to be held to the same accuracy, a system for dimensioning must be used. See Fig. 7-5-1. Generally, most parts require only a few features to be held to high accuracy.

In order that assembled parts may function properly and to allow for interchangeable manufacturing, it is necessary to permit only a certain amount of tolerance on each of the mating parts and a certain amount of allowance between them.

In order to calculate limit dimensions, the following definitions should be clearly understood (refer to Fig. 7-5-2).

Basic Size The *basic size* of a dimension is the theoretical size from which the limits for that dimension are derived, by the application of the allowance and tolerance.

Limits of Size These limits are the maximum and minimum sizes permissible for a specific dimension.

Tolerance The *tolerance* on a dimension is the total permissible variation in its size. The tolerance is the difference between the limits of size.

Maximum Material Size The *maximum material size* is that limit of size of a

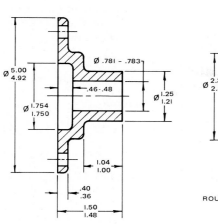

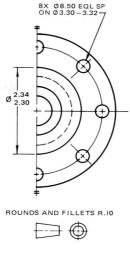

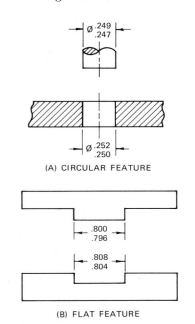

8X Ø8.50 EQL SP
ON Ø3.30 – 3.32

Ø 2.34
2.30

ROUNDS AND FILLETS R.10

Fig. 7-5-1 A working drawing.

BASIC SIZE		1.500
BASIC SIZE WITH TOLERANCE ADDED		1.500 ± .004
HALF OF TOTAL TOLERANCE		
LIMITS—LARGEST AND SMALLEST SIZES PERMITTED		1.504 1.496
TOLERANCE— DIFFERENCE BETWEEN MIN AND MAX LIMITS		.008

Fig. 7-5-2 Limit and tolerance terminology.

feature which results in the part containing the maximum amount of material. Thus it is the maximum limit of size for a shaft or an external feature, or the minimum limit of size for a hole or internal feature.

TOLERANCING

All dimensions required in the manufacture of a product have a tolerance, except those identified as reference, maximum, minimum, or stock. Tolerances may be expressed in one of the following ways:

- As specified limits of tolerances shown directly on the drawing for a specified dimension. See Fig. 7-5-3.
- As plus-and-minus tolerancing
- Combining a dimension with a tolerance symbol. See Unit 7-6.
- In a general tolerance note, referring to all dimensions on the drawing for which tolerances are not otherwise specified
- In the form of a note referring to specific dimensions
- Tolerances on dimensions that locate features may be applied directly to the locating dimensions or by the posi-

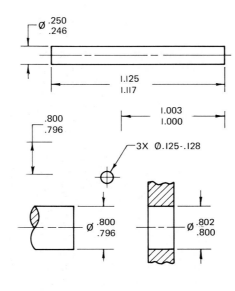

3X Ø.125-.128

(A) TWO LIMITS

25.2°
25.1°

25°30′45″
25°30′15″

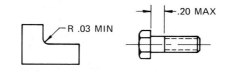

R .03 MIN

.20 MAX

(B) SINGLE LIMITS

Fig. 7-5-3 Methods of indicating tolerances on drawings.

tional tolerancing method described in Chap. 16.

- Tolerancing applicable to the control of form and runout, referred to as *geometric tolerancing*, is also covered in detail in Chap. 16.

Direct Tolerancing Methods

A tolerance applied directly to a dimension may be expressed in two ways.

Limit Dimensioning For this method, the high limit (maximum value) is placed above the low limit (minimum value). When it is expressed in a single line, the low limit precedes the high limit and they are separated by a dash. See Figs. 7-5-3 and 7-5-4.

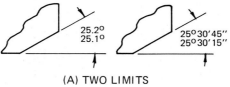

Ø .249
.247

Ø .252
.250

(A) CIRCULAR FEATURE

.800
.796

.808
.804

(B) FLAT FEATURE

Fig. 7-5-4 Limit dimensioning application.

Where limit dimensions are used and where either the maximum or minimum dimension has digits to the right of the decimal point, the other value should have the zeros added so that both the limits of size are expressed to the same number of decimal places. This applies to both U.S. Customary and metric drawings. For example:

30.75 30.00	*not*	30.75 30
.750 .748	*not*	.75 .748

Plus-and-Minus Tolerancing (Refer to Fig. 7-5-5.) For this method the dimension of the specified size is given first and is followed by a plus or minus expression of tolerancing. The plus value should be placed above the minus value. This type of tolerancing can be broken down into bilateral and unilateral tolerancing. In a bilateral tolerance, the plus-and-minus tolerances should generally be equal, but special design considerations may sometimes

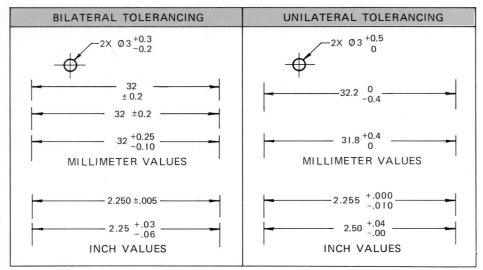

BILATERAL TOLERANCING	UNILATERAL TOLERANCING

2X Ø3 $^{+0.3}_{-0.2}$

32 ±0.2

32 ±0.2

32 $^{+0.25}_{-0.10}$

MILLIMETER VALUES

2.250 ±.005

2.25 $^{+.03}_{-.06}$

INCH VALUES

2X Ø3 $^{+0.5}_{0}$

32.2 $^{0}_{-0.4}$

31.8 $^{+0.4}_{0}$

MILLIMETER VALUES

2.255 $^{+.000}_{-.010}$

2.50 $^{+.04}_{-.00}$

INCH VALUES

Fig. 7-5-5 Plus-and-minus tolerancing.

dictate unequal values. See Fig. 7-5-6. The specified size is the design size, and the tolerance represents the desired control of quality and appearance.

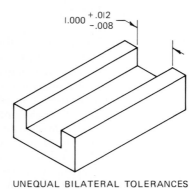

2.000 ± .008

2.000 ± .008

EQUAL BILATERAL TOLERANCES

1.000 $^{+.012}_{-.008}$

UNEQUAL BILATERAL TOLERANCES

Fig. 7-5-6 Applications of bilateral tolerances.

Metric Tolerancing In the metric system the dimension need not be shown to the same number of decimal places as **its tolerance.** For example:

$$1.5 \pm 0.04 \quad not \quad 1.50 \pm 0.04$$
$$10 \pm 0.1 \quad not \quad 10.0 \pm 0.1$$

Where bilateral tolerancing is used both the plus and minus values have the same number of decimal places, using zeros where necessary. For example,

$$30 ^{+0.15}_{-0.10} \quad not \quad 30 ^{+0.15}_{-0.1}$$

Where unilateral tolerancing is used and either the plus or minus value is nil, a single zero is shown without a plus or minus sign. For example,

$$40 ^{0}_{-0.15} \quad or \quad 40 ^{+0.15}_{0}$$

An application of unilateral tolerancing is shown in Fig. 7-5-7.

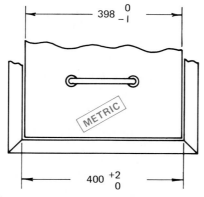

398 $^{0}_{-1}$

METRIC

400 $^{+2}_{0}$

Fig. 7-5-7 Applications of unilateral tolerances.

Inch Tolerancing In the inch system the dimension is given to the same number or decimal places as its tolerance. For example,

Bilateral:

$$.500 \pm .004 \quad not \quad .50 \pm .004$$

Unilateral:

$$.750 ^{+.005}_{-.000} \quad not \quad .750 ^{+.005}_{-0}$$

$$30.0° \pm .2° \quad not \quad 30° \pm .2°$$

Conversion charts for tolerances are shown in Fig. 7-5-8.

TOTAL TOLERANCE IN INCHES		MILLIMETER CONVERSION ROUNDED TO
AT LEAST	LESS THAN	
.00004	.0004	4 DECIMAL PLACES
.0004	.004	3 DECIMAL PLACES
.004	.04	2 DECIMAL PLACES
.04	.4	1 DECIMAL PLACE
.4 AND OVER		WHOLE mm

TOTAL TOLERANCE IN MILLIMETERS		INCH CONVERSION ROUNDED TO
AT LEAST	LESS THAN	
0.002	0.02	5 DECIMAL PLACES
0.02	0.2	4 DECIMAL PLACES
0.2	2	3 DECIMAL PLACES
2 AND OVER		2 DECIMAL PLACES

Fig. 7-5-8 Conversion charts for tolerances.

Either bilateral or unilateral tolerances may be readily added to any dimension using CAD. See Fig. 7-5-9.

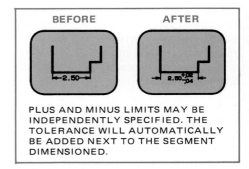

BEFORE	AFTER

2.50

2.50 $^{+.02}_{-.04}$

PLUS AND MINUS LIMITS MAY BE INDEPENDENTLY SPECIFIED. THE TOLERANCE WILL AUTOMATICALLY BE ADDED NEXT TO THE SEGMENT DIMENSIONED.

Fig. 7-5-9 CAD tolerance command.

General Tolerance Notes The use of general tolerance notes greatly simplifies the drawing and saves much layout in its preparation. The following examples illustrate the wide field of application of this system. The values given in the examples are typical.

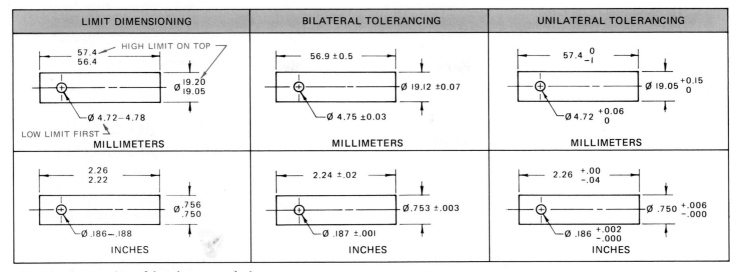

LIMIT DIMENSIONING	BILATERAL TOLERANCING	UNILATERAL TOLERANCING

Fig. 7-5-10 A comparison of the tolerance methods.

EXAMPLE 1
EXCEPT WHERE STATED
OTHERWISE, TOLERANCES
ON FINISHED DECIMAL
DIMENSIONS ±0.1.

EXAMPLE 2
EXCEPT WHERE STATED
OTHERWISE, TOLERANCES
ON FINISHED DIMENSIONS
TO BE AS FOLLOWS:

Dimension (in.)	Tolerance
Up to 4.00	± .004
From 4.01 to 12.00	± .003
From 12.01 to 24.00	± .02
Over 24.00	± .04

A comparison between the tolerancing methods described is shown in Fig. 7-5-10.

Tolerance Accumulation

It is necessary also to consider the effect of each tolerance with respect to other tolerances, and not to permit a chain of tolerances to build up a cumulative tolerance between surfaces or points that have an important relation to one another. Where the position of a surface in any one direction is controlled by more than one tolerance, then the tolerances are cumulative. Figure 7-5-11 compares the tolerance accumulation resulting from three different methods of dimensioning.

Chain Dimensioning The maximum variation between any two features is equal to the sum of the tolerances on the intermediate dis-

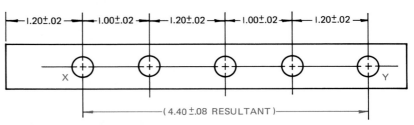

(A) CHAIN DIMENSIONING (GREATEST TOLERANCE ACCUMULATION)

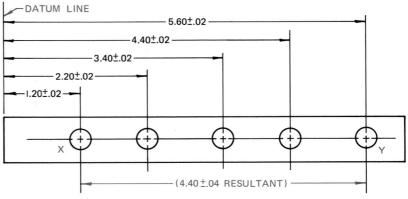

(B) DATUM DIMENSIONING (LESSER TOLERANCE ACCUMULATION)

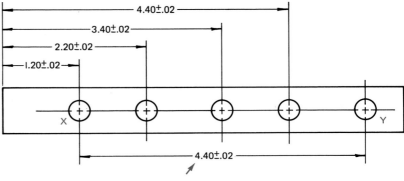

(C) DIRECT DIMENSIONING (LEAST TOLERANCE ACCUMULATION)

Fig. 7-5-11 Dimensioning method comparison.

tances. This results in the greatest tolerance accumulation, as illustrated by the ± .08 variation between holes X and Y, as shown in Fig. 7-5-11A.

Datum Dimensioning The maximum variation between any two features is equal to the sum of the tolerances on the two dimensions from the datum to the feature. This reduces the tolerance accumulation, as illustrated by the ± .04 variation between holes X and Y. See Fig. 7-5-11B.

Direct Dimensioning The maximum variation between any two features is controlled by the tolerance on the dimension between the features. This results in the least tolerance accumulation, as illustrated by the ± .02 variation between holes X and Y. See Fig. 7-5-11C.

References

1. ANSI Y14.5M, *Dimensioning and Tolerancing*.

ASSIGNMENT

See Assignment 16 for Unit 7-5 on page 149.

UNIT 7-6
Fits and Allowances

In order that assembled parts may function properly and to allow for interchangeable manufacturing, it is necessary to permit only a certain amount of tolerance on each of the mating parts and a certain amount of allowance between them.

DEFINITIONS
Fits

The fit between two mating parts is the relationship between them with respect to the amount of clearance or interference present when they are assembled.

There are three basic types of fits: clearance, interference, and transition.

Clearance Fit A fit between mating parts having limits of size so prescribed that a clearance always results in assembly.

Interference Fit A fit between mating parts having limits of size so

prescribed that an interference always results in assembly.

Transition Fit A fit between mating parts having limits of size so prescribed as to partially or wholly overlap, so that either a clearance or interference may result in assembly.

Allowance

An allowance is an intentional difference in correlated dimensions of mating parts. It is the minimum clearance (positive allowance) or maximum interference (negative allowance) between such parts.

The most important terms relating to limits and fits are shown in Fig. 7-6-1. The terms are defined as follows:

Basic Size The size to which limits or deviations are assigned. The basic size is the same for both members of a fit.

Deviation The algebraic difference between a size and the corresponding basic size.

Upper Deviation The algebraic difference between the maximum limit of size and the corresponding basic size.

Lower Deviation The algebraic difference between the minimum limit of size and the corresponding basic size.

Tolerance The difference between the maximum and minimum size limits on a part.

Tolerance Zone A zone representing the tolerance and its position in relation to the basic size.

Fundamental Deviation The deviation closest to the basic size.

DESCRIPTION OF FITS
Running and Sliding Fits

These fits, for which tolerances and clearances are given in the Appendix, represent a special type of clearance fit. These are intended to provide a similar running performance, with suitable lubrication allowance, throughout the range of sizes.

Locational Fits

Locational fits are intended to determine only the location of the mating parts; they may provide rigid or accurate location, as with interference fits, or some freedom of location, as with clearance fits. Accordingly, they are divided into three groups: clearance fits, transition fits, and interference fits.

Locational clearance fits are intended for parts which are normally stationary but which can be freely assembled or disassembled. They run from snug fits for parts requiring accuracy of location, through the medium clearance fits for parts such as ball, race, and housing, to

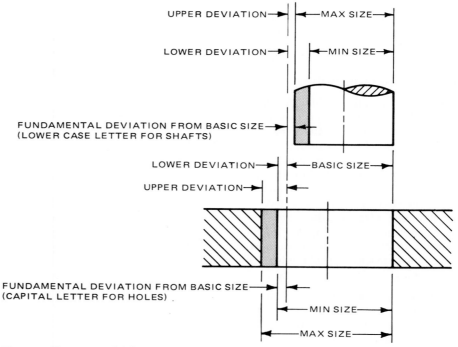

Fig. 7-6-1 Illustration of definitions.

the looser fastener fits where freedom of assembly is of prime importance.

Locational transition fits are a compromise between clearance and interference fits, for application where accuracy of location is important but a small amount of either clearance or interference is permissible.

Locational interference fits are used where accuracy of location is of prime importance and for parts requiring rigidity and alignment with no special requirements for bore pressure. Such fits are not intended for parts designed to transmit frictional loads from one part to another by virtue of the tightness of fit; these conditions are covered by force fits.

Drive and Force Fits

Drive and force fits constitute a special type of interference fit, normally characterized by maintenance of constant bore pressures throughout the range of sizes. The interference therefore varies almost directly with diameter, and the difference between its minimum and maximum values is small, to maintain the resulting pressures within reasonable limits.

Interchangeability of Parts

Increased demand for manufactured products led to the development of new production techniques. Interchangeability of parts became the basis for mass-production, low-cost manufacturing, and it brought about the refinement of machinery, machine tools, and measuring devices. Today it is possible and generally practical to design for 100 percent interchangeability.

No part can be manufactured to exact dimensions. Tool wear, machine variations, and the human factor all contribute to some degree of deviation from perfection. It is therefore necessary to determine the deviation and permissible clearance, or interference, to produce the desired fit between parts.

Modern industry has adopted three basic approaches to manufacturing.

1. *The completely interchangeable assembly.* Any and all mating parts of a design are toleranced to permit them to assemble and function properly without the need for machining or fitting at assembly.
2. *The fitted assembly.* Mating features of a design are fabricated either simultaneously or with respect to one another. Individual members of

mating features are not interchangeable.
3. *The selected assembly.* All parts are mass-produced, but members of mating features are individually selected to provide the required relationship with one another.

STANDARD INCH FITS

Standard fits are designated for design purposes in specifications and on design sketches by means of the symbols as shown in Fig. 7-6-2. These symbols,

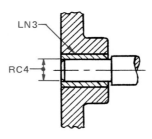

(A) SHAFT IN BUSHED HOLE

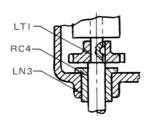

(B) GEAR AND SHAFT IN BUSHED BEARING

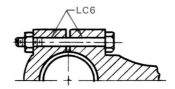

(C) CONNECTING-ROD BOLT

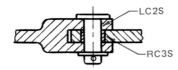

(D) LINK PIN (SHAFT BASIS FITS)

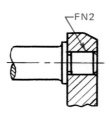

(E) CRANK PIN IN CAST IRON

Fig. 7-6-2 Typical design sketches showing classes of fits.

however, are not intended to be shown directly on shop drawings; instead the actual limits of size are determined and specified on the drawings.

The letter symbols used are as follows:

RC Running and sliding fit
LC Locational clearance fit
LT Locational transition fit
LN Locational interference fit
FN Force or shrink fit

These letter symbols are used in conjunction with numbers representing the class of fit; for example, FN4 represents a class 4, force fit.

Each of these symbols (two letters and a number) represents a complete fit, for which the minimum and maximum clearance or interference, and the limits of size for the mating parts, are given directly in Appendix tables 43 through 47. See Fig. 7-6-3.

Running and Sliding Fits

RC1 Precision Sliding Fit This fit is intended for the accurate location of parts which must assemble without perceptible play, for high-precision work such as gages.

RC2 Sliding Fit This fit is intended for accurate location, but with greater maximum clearance than class RC1. Parts made to this fit move and turn easily but are not intended to run freely, and in the larger sizes may seize with small temperature changes.

Note: LC1 and LC2 locational clearance fits may also be used as sliding fits with greater tolerances.

RC3 Precision Running Fit This fit is about the closest fit which can be expected to run freely, and is intended for precision work for oil-lubricated bearings at slow speeds and light journal pressures, but is not suitable where appreciable temperature differences are likely to be encountered.

RC4 Close Running Fit This fit is intended chiefly as a running fit for grease- or oil-lubricated bearings on accurate machinery with moderate surface speeds and journal pressures, where accurate location and minimum play are desired.

RC5 and RC6 Medium Running Fits These fits are intended for higher running speeds and/or where temperature variations are likely to be encountered.

RC7 Free Running Fit This fit is intended for use where accuracy is not

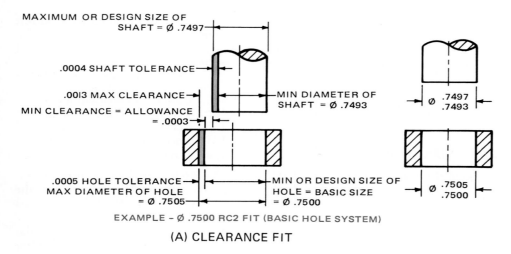

EXAMPLE – Ø .7500 RC2 FIT (BASIC HOLE SYSTEM)

(A) CLEARANCE FIT

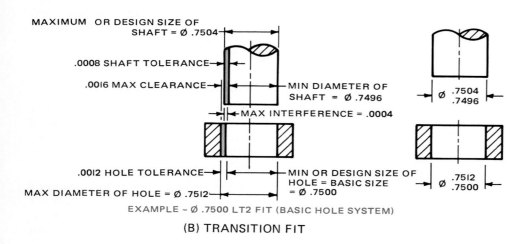

EXAMPLE – Ø .7500 LT2 FIT (BASIC HOLE SYSTEM)

(B) TRANSITION FIT

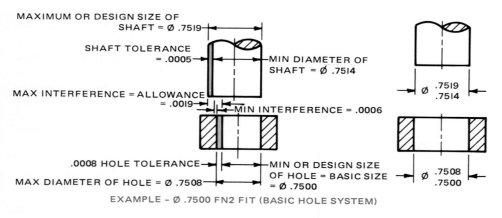

EXAMPLE – Ø .7500 FN2 FIT (BASIC HOLE SYSTEM)

(C) INTERFERENCE FIT

Fig. 7-6-3 Types of inch fits.

essential, and/or where large temperature variations are likely to be encountered.

RC8 and RC9 Loose Running Fits
These fits are intended for use where materials made to commercial toler-

ances are involved such as cold-rolled shafting, tubing, etc.

Locational Clearance Fits

Locational clearance fits are intended

for parts which are normally stationary, but which can be freely assembled or disassembled. They run from snug fits for parts requiring accuracy of location, through the medium clearance fits for parts such as spigots, etc., to the looser fastener fits where freedom of assembly is of prime importance.

These are classified as follows:

LC1 to LC4 These fits have a minimum zero clearance, but in practice the probability is that the fit will always have a clearance. These fits are suitable for location of nonrunning parts and spigots, although classes LC1 and LC2 may also be used for sliding fits.

LC5 and LC6 These fits have a small minimum clearance, intended for close location fits for nonrunning parts. LC5 can also be used in place of RC2 as a free-slide fit, and LC6 may be used as a medium running fit having greater tolerances than RC5 and RC6.

LC7 to LC11 These fits have progressively larger clearances and tolerances, and are useful for various loose clearances for assembly of bolts and similar parts.

Locational Transition Fits

Locational transition fits are a compromise between clearance and interference fits, for application where accuracy of location is important, but either a small amount of clearance or interference is permissible.

These are classified as follows:

LT1 and LT2 These fits average a slight clearance, giving a light push fit, and are intended for use where the maximum clearance must be less than for the LC1 to LC3 fits, and where slight interference can be tolerated for assembly by pressure or light hammer blows.

LT3 and LT4 These fits average virtually no clearance, and are for use where some interference can be tolerated, for example: to eliminate vibration. These are sometimes referred to as an easy keying fit, and are used for shaft keys and ball race fits. Assembly is generally by pressure or hammer blows.

LT5 and LT6 These fits average a slight interference, although appreciable assembly force will be required when extreme limits are encountered, and selective assembly may be desirable. These fits are useful for heavy keying, for ball race fits subject to heavy duty and vibration, and as light press fits for steel parts.

Locational Interference Fits

Locational interference fits are used where accuracy of location is of prime importance, and for parts requiring rigidity and alignment with no special requirements for bore pressure. Such fits are not intended for parts designed to transmit frictional loads from one part to another by virtue of the tightness of fit, as these conditions are covered by force fits.

These are classified as follows:

LN1 and LN2 These are light press fits, with very small minimum interference, suitable for parts such as dowel pins, which are assembled with an arbor press in steel, cast iron, or brass. Parts can normally be dismantled and reassembled, as the interference is not likely to overstrain the parts, but the interference is too small for satisfactory fits in elastic materials or light alloys.

LN3 This is suitable as a heavy press fit in steel and brass, or a light press fit in more elastic materials and light alloys.

LN4 to LN6 While LN4 can be used for permanent assembly of steel parts, these fits are primarily intended as press fits for more elastic or soft materials, such as light alloys and the more rigid plastics.

Force or Shrink Fits

Force or shrink fits constitute a special type of interference fit, normally characterized by maintenance of constant bore pressures throughout the range of sizes. The interference therefore varies almost directly with diameter, and the difference between its minimum and maximum values is small to maintain the resulting pressures within reasonable limits.

These fits may be described briefly as follows:

FN1 Light Drive Fit Requires light assembly pressure, and produces more or less permanent assemblies. It is suitable for thin sections or long fits, or in cast iron external members.

FN2 Medium Drive Fit Suitable for ordinary steel parts, or as a shrink fit on light sections. It is about the tightest fit that can be used with high-grade cast iron external members.

FN3 Heavy Drive Fit Suitable for heavier steel parts or as a shrink fit in medium sections.

FN4 and FN5 Force Fits Suitable for parts which can be highly stressed, and/or for shrink fits where the heavy pressing forces required are impractical.

Basic Hole System

In the basic hole system, which is recommended for general use, the basic size will be the design size for the hole, and the tolerance will be plus. The design size for the shaft will be the basic size minus the minimum clearance, or plus the maximum interference, and the tolerance will be minus, as given in the tables in the Appendix. For example (see Table 43, Appendix), for a 1 in. RC7 fit, values of $+ .0020$, $.0025$, and $- .0012$ are given; hence, limits will be

$$\text{Hole} \, \emptyset \, 1.0000 \quad {+ .0020 \atop - .0000}$$

$$\text{Shaft} \, \emptyset \, .9975 \quad {+ .0000 \atop - .0012}$$

Basic Shaft System

Fits are sometimes required on a basic shaft system, especially in cases where two or more fits are required on the same shaft. This is designated for design purposes by a letter S following the fit symbol; for example, RC7S.

Tolerances for holes and shaft are identical with those for a basic hole system, but the basic size becomes the design size for the shaft and the design size for the hole is found by adding the minimum clearance or subtracting the maximum interference from the basic size.

For example, for a 1 in. RC7S fit, values of $+ .0020$, $.0025$, and $- .0012$ are given; therefore, limits will be

$$\text{Hole} \, \emptyset \, 1.0025 \quad {+ .0020 \atop - .0000}$$

$$\text{Shaft} \, \emptyset \, 1.000 \quad {+ .0000 \atop - .0012}$$

PREFERRED METRIC LIMITS AND FITS

The ISO system of limits and fits for mating parts is approved and adopted for general use in the United States. It establishes the designation symbols used to define specific dimensional limits on drawings.

The general terms "hole" and "shaft" can also be taken as referring to the space containing or contained by two parallel faces of any part, such as the width of slot, or the thickness of a key.

An "International Tolerance grade" establishes the magnitude of the tolerance zone or the amount of part size variation allowed for internal and external dimensions alike. See Table 48, Appendix. There are eighteen tolerance grades which are identified by the prefix IT, such as IT6, IT11, etc. The smaller the grade number the smaller the tolerance zone. For general applications of IT grades, see Fig. 7-6-4.

Grades 1 to 4 are very precise grades intended primarily for gage making and similar precision work, although grade 4 can also be used for very precise production work.

Grades 5 to 16 represent a progressive series suitable for cutting operations, such as turning, boring, grinding, milling, and sawing. Grade 5 is the most precise grade, obtainable by fine grinding and lapping, while 16 is the coarsest grade for rough sawing and machining.

Grades 12 to 16 are intended for manufacturing operations such as cold heading, pressing, rolling, and other forming operations.

As a guide to the selection of tolerances, Fig. 7-6-5 has been prepared to show grades which may be expected to be held by various manufacturing processes for work in metals. For work in other materials, such as plastics, it may be necessary to use coarser tolerance grades for the same process.

A fundamental deviation establishes the position of the tolerance zone with respect to the basic size. Fundamental

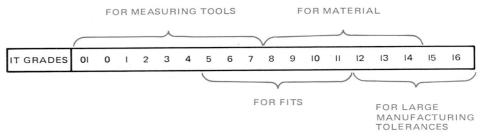

Fig. 7-6-4 Applications of international tolerance (IT) grades.

MACHINING PROCESSES	TOLERANCE GRADES									
	4	5	6	7	8	9	10	11	12	13
LAPPING & HONING										
CYLINDRICAL GRINDING										
SURFACE GRINDING										
DIAMOND TURNING										
DIAMOND BORING										
BROACHING										
REAMING										
TURNING										
BORING										
MILLING										
PLANING & SHAPING										
DRILLING										

Fig. 7-6-5 Tolerance grades for machining processes.

deviations are expressed by "tolerance position letters." Capital letters are used for internal dimensions, and lower case letters for external dimensions.

Symbols

Tolerance Symbol For metric application of limits and fits, the tolerance may be indicated by a basic size and tolerance symbol. By combining the IT grade number and the tolerance position letter, the tolerance symbol is established which identifies the actual maximum and minimum limits of the part. The toleranced sizes are thus defined by the basic size of the part followed by the symbol composed of a letter and a number. See Fig. 7-6-6.

Preferred Tolerance Zones

The preferred tolerance zones are shown in Table 48 of the Appendix. The encircled tolerance zones (13 each) are first choice, the framed tolerance zones are second choice, and the open tolerance zones are third choice.

Hole Basis Fits System

In the hole basis fits system (see Tables 49 and 51 of the Appendix) the basic size will be the minimum size of the hole. For example, for a Ø25H8/f7 fit, which is a Preferred Hole Basis Clearance Fit, the limits for the hole and shaft will be as follows:

Refer to Tables 49 and 51 of the Appendix.
Hole limits = Ø 25.000 − Ø 25.033
Shaft limits = Ø 24.959 − Ø 24.980
Minimum clearance = 0.020
Maximum clearance = 0.074

If a Ø25H7/s6 Preferred Hole Basis Interference Fit is required, the limits for the hole and shaft will be as follows:

Refer to Tables 49 and 51 of the Appendix.

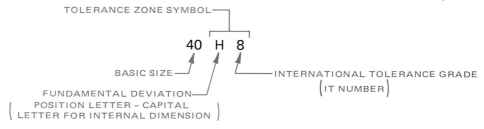

(A) INTERNAL DIMENSION (HOLES)

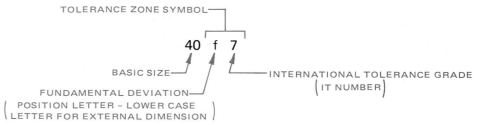

(B) EXTERNAL DIMENSION (SHAFTS)

Fig. 7-6-6 Metric tolerance symbol.

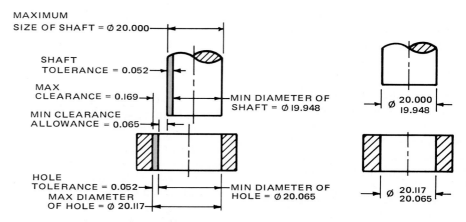

EXAMPLE — D9/h9 PREFERRED SHAFT BASIS FIR FOR A Ø20 SHAFT

(A) CLEARANCE FIT

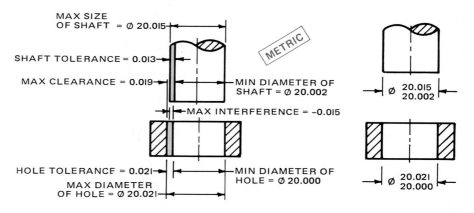

EXAMPLE — H7/k6 PREFERRED HOLE BASIS FIT FOR A Ø20 HOLE

(B) TRANSITION FIT

Fig. 7-6-7 Types of metric fits.

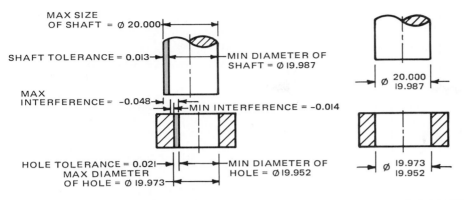

MAX SIZE OF SHAFT = Ø 20.000

SHAFT TOLERANCE = 0.013 → MIN DIAMETER OF SHAFT = Ø 19.987

MAX INTERFERENCE = −0.048 → MIN INTERFERENCE = −0.014

Ø 20.000 / 19.987

HOLE TOLERANCE = 0.021 → MIN DIAMETER OF HOLE = Ø 19.952
MAX DIAMETER OF HOLE = Ø 19.973

Ø 19.973 / 19.952

EXAMPLE — S7/h6 PREFERRED SHAFT BASIS FIT FOR A Ø20 SHAFT

(C) INTERFERENCE FIT

Fig. 7-6-7 (Continued)

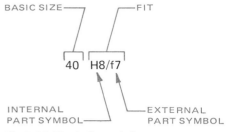

BASIC SIZE — FIT

40 H8/f7

INTERNAL PART SYMBOL — EXTERNAL PART SYMBOL

Fig. 7-6-8 Metric fit symbol.

Hole limits = Ø 25.000 − Ø 25.021
Shaft limits = Ø 25.035 − Ø 25.048
Minimum interference = −0.014
Maximum interference = −0.048

The limits of size for the shaft having a tolerance symbol 40f7 (see Table 51) is

Ø 39.975 MAXIMUM LIMIT
Ø 39.950 MINIMUM LIMIT

The method shown in Fig. 7-6-9A is recommended when the system is first introduced. In this case limit dimensions are specified, and the basic size and tolerance symbol are identified as reference.

As experience is gained, the method shown in Fig. 7-6-9B can be used. When the system is established and standard

Shaft Basis Fits System

Where more than two fits are required on the same shaft, the shaft basis fits system is recommended. Tolerances for holes and shaft are identical with those for a basic hole system. However, the basic size becomes the maximum shaft size. For example, for a Ø16 C11/h11 fit, which is a Preferred Shaft Basis Clearance Fit, the limits for the hole and shaft will be as follows:

Refer to Tables 50 and 52 of the Appendix.
Hole limits = Ø 16.095 − Ø 16.205
Shaft limits = Ø 15.890 − Ø 16.000
Minimum clearance = 0.095
Maximum clearance = 0.315

Preferred Fits

First choice tolerance zones are shown to relative scale in Table 48 of the Appendix. Hole basis fits have a fundamental deviation of "H" on the hole, and shaft basis fits have a fundamental deviation of "h" on the shaft. Normally, the hole basis system is preferred. Figure 7-6-7 shows examples of three common fits.

Fit Symbol A fit is indicated by the basic size common to both components, followed by a symbol corresponding to each component, with the internal part symbol preceding the external part symbol. See Fig. 7-6-8.

The limits of size for a hole having a tolerance symbol 40H8 (see Table 51) is

Ø 40.039 MAXIMUM LIMIT
Ø 40.000 MINIMUM LIMIT

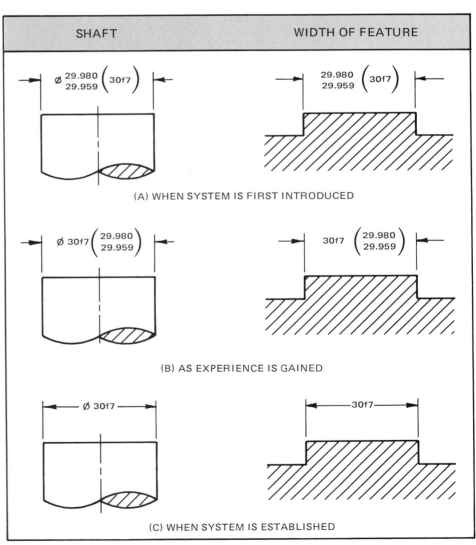

SHAFT	WIDTH OF FEATURE
Ø 29.980 / 29.959 (30f7)	29.980 / 29.959 (30f7)
(A) WHEN SYSTEM IS FIRST INTRODUCED	
Ø 30f7 (29.980 / 29.959)	30f7 (29.980 / 29.959)
(B) AS EXPERIENCE IS GAINED	
Ø 30f7	30f7
(C) WHEN SYSTEM IS ESTABLISHED	

Fig. 7-6-9 Metric tolerance symbol applications.

tools, gages, and stock materials are available with size and symbol identification, the method shown in Fig. 7-6-9C may be used.

This would result in a clearance fit of 0.025 to 0.089 mm. A description of the preferred metric fits is shown in Fig. 7-6-10.

References

1. ANSI B4.2, *Preferred Metric Limits and Fits.*

ASSIGNMENTS

See Assignments 17 through 20 for Unit 7-6 starting on page 150.

UNIT 7-7
Surface Texture

Modern development of high-speed machines has resulted in higher loadings and increased speeds of moving parts. To withstand these more severe operating conditions with minimum friction and wear, a particular surface finish is often essential, making it necessary for the designer to accurately describe the required finish to the persons who are actually making the parts.

For accurate machines it is no longer sufficient to indicate the surface finish by various grind marks, such as "g," "f," or "fg." It becomes necessary to define surface finish and take it out of the opinion or guesswork class.

All surface finish control starts in the drafting room. The designer has the responsibility of specifying the right surface to give maximum performance and service life at the lowest cost. In selecting the required surface finish for any particular part, their designers base decisions on past experience with similar parts, on field service data, or on engineering tests. Such factors as size and function of the parts, type of loading, speed and direction of movement, operating conditions, physical characteristics of both materials on contact, whether they are subjected to stress reversals, type and amount of lubricant, contaminants, temperature, etc., influence the choice.

There are two principal reasons for surface finish control:

1. To reduce friction
2. To control wear

ISO SYMBOL		DESCRIPTION
HOLE BASIS	SHAFT BASIS	
H11/c11	C11/h11	LOOSE RUNNING FIT FOR WIDE COMMERCIAL TOLERANCES OR ALLOWANCES ON EXTERNAL MEMBERS.
H9/d9	D9/h9	FREE RUNNING FIT NOT FOR USE WHERE ACCURACY IS ESSENTIAL, BUT GOOD FOR LARGE TEMPERATURE VARIATIONS, HIGH RUNNING SPEEDS, OR HEAVY JOURNAL PRESSURES.
H8/f7	F8/h7	CLOSE RUNNING FIT FOR RUNNING ON ACCURATE MACHINES AND FOR ACCURATE LOCATION AT MODERATE SPEEDS AND JOURNAL PRESSURES.
H7/g6	G7/h6	SLIDING FIT NOT INTENDED TO RUN FREELY, BUT TO MOVE AND TURN FREELY AND LOCATE AC-CURATELY.
H7/h6	H7/h6	LOCATIONAL CLEARANCE FIT PROVIDES SNUG FIT FOR LOCATING STATIONARY PARTS; BUT CAN BE FREELY ASSEMBLED AND DISASSEMBLED.
H7/k6	K7/h6	LOCATIONAL TRANSITION FIT FOR ACCURATE LOCATION, A COMPROMISE BETWEEN CLEARANCE AND INTERFERENCE.
H7/n6	N7/h6	LOCATIONAL TRANSITION FIT FOR MORE AC-CURATE LOCATION WHERE GREATER INTER-FERENCE IS PERMISSIBLE.
H7/p6	P7/h6	LOCATIONAL INTERFERENCE FIT FOR PARTS REQUIRING RIGIDITY AND ALIGNMENT WITH PRIME ACCURACY OF LOCATION BUT WITHOUT SPECIAL BORE PRESSURE REQUIREMENTS.
H7/s6	S7/h6	MEDIUM DRIVE FIT FOR ORDINARY STEEL PARTS OR SHRINK FITS ON LIGHT SECTIONS, THE TIGHTEST FIT USABLE WITH CAST IRON.
H7/u6	U7/h6	FORCE FIT SUITABLE FOR PARTS WHICH CAN BE HIGHLY STRESSED OR FOR SHRINK FITS WHERE THE HEAVY PRESSING FORCES REQUIRED ARE IMPRACTICAL.

The left margin of the table is labeled: CLEARANCE FITS, TRANSITION FITS, INTERFERENCE FITS. The right margin is labeled: MORE CLEARANCE, MORE INTERFERENCE.

Fig. 7-6-10 Description of preferred metric fits.

Whenever a film of lubricant must be maintained between two moving parts, the surface irregularities must be small enough so they will not penetrate the oil film under the most severe operating conditions. Bearings, journals, cylinder bores, piston pins, bushings, pad bearings, helical and worm gears, seal surfaces, machine ways, and so forth, are examples where this condition must be fulfilled.

Surface finish is also important to the wear of certain pieces which are subject to dry friction, such as machine tool bits, threading dies, stamping dies, rolls, clutch plates, brake drums, etc.

Smooth finishes are essential on certain high-precision pieces. In mechanisms such as injectors and high-pressure cylinders, smoothness and lack of waviness are essential to accuracy and pressure-retaining ability.

Surfaces, in general, are very complex in character. Only the height, width, and direction of surface irregularities will be covered in this section since these are of practical importance in specific applications.

Surface Texture Characteristics

Refer to Fig. 7-7-1.

Microinch A microinch is one-millionth of an inch (.000 001 in.). For written specifications or reference to surface roughness requirements, microinch may be abbreviated as μin.

Micrometer A micrometer is one-millionth of a meter (0.000 001 m). For written specifications or reference to surface roughness requirements, micrometer may be abbreviated as μm.

Roughness Roughness consists of the finer irregularities in the surface texture, usually including those which result from the inherent action of the production process. These are considered to include traverse feed marks and other irregularities within the limits of the roughness-width cutoff.

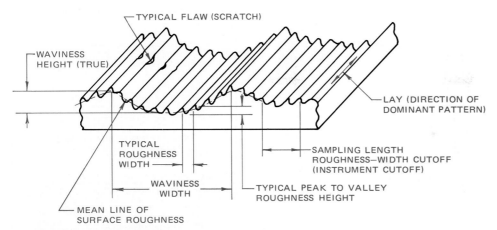

Fig. 7-7-1 Surface texture characteristics.

Flaws Flaws are irregularities which occur at one place or at relatively infrequent or widely varying intervals in a surface. Flaws include such defects as cracks, blow holes, checks, ridges, and scratches. Unless otherwise specified, the effect of flaws is not included in the roughness-height measurements.

Surface Texture Symbol

Surface characteristics of roughness, waviness, and lay may be controlled by applying the desired values to the surface texture symbol, shown in Figs. 7-7-2 and 7-7-3, in a general note, or both.

Roughness-Height Value Roughness-height value is rated as the arithmetic average (AA) deviation expressed in microinches or micrometers measured normal to the center line. ISO and many European countries use the term CLA (center line average) in lieu of AA. Both have the same meaning.

Roughness Spacing Roughness spacing is the distance parallel to the nominal surface between successive peaks or ridges which constitute the predominant pattern of the roughness. Roughness spacing is rated in inches or millimeters.

Roughness-Width Cutoff The greatest spacing of repetitive surface irregularities is included in the measurement of average roughness height. Roughness-width cutoff is rated in inches or millimeters and must always be greater than the roughness width in order to obtain the total roughness height rating.

Waviness Waviness is the usually widely-spaced component of surface texture and is generally of wider spacing than the roughness-width cutoff. Waviness may result from such factors as machine or work deflections, vibration, chatter, heat treatment, or warping strains. Roughness may be considered as superimposed on a "wavy" surface. Although waviness is not currently in ISO Standards, it is included as part of the surface texture symbol to follow present industrial practices in the United States.

Lay The direction of the predominant surface pattern, ordinarily determined by the production method used, is the *lay*. Lay symbols are specified as shown in Fig. 7-7-10.

X = FIGURE HEIGHT OF VALUES.
HORIZONTAL EXTENSION BAR REQUIRED WHEN WAVINESS RATINGS ARE SHOWN.

Fig. 7-7-2 Basic surface texture symbol.

PRESENT SYMBOLS VALUES SHOWN IN CUSTOMARY OR METRIC		FORMER SYMBOLS VALUES SHOWN IN MICROINCHES AND INCHES	
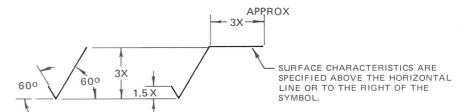			
BASIC SURFACE TEXTURE SYMBOL	√	√	BASIC SURFACE TEXTURE SYMBOL
ROUGHNESS HEIGHT RATING IN MICROINCHES OR MICROMETERS AND N SERIES ROUGHNESS NUMBERS	36 √ N8 √	63 √	ROUGHNESS HEIGHT RATING IN MICROINCHES.
MAXIMUM AND MINIMUM ROUGHNESS HEIGHT IN MICROINCHES OR MICROMETERS	63/32 √	63/32 √	MAXIMUM AND MINIMUM ROUGHNESS HEIGHT RATINGS IN MICROINCHES
WAVINESS HEIGHT IN INCHES OR MILLIMETERS (F)	63/32 F √	63/32 .002 √	WAVINESS HEIGHT IN INCHES
WAVINESS SPACING IN INCHES OR MILLIMETERS (G)	63/32 F–G √	63/32 .002 –1 √	WAVINESS WIDTH IN INCHES
LAY SYMBOL (D)	63/32 √⊥	63/32 √⊥	LAY SYMBOL
MAXIMUM ROUGHNESS SPACING IN INCHES OR MILLIMETERS (B)	63/32 √⊥ B	√⊥ .008	SURFACE ROUGHNESS WIDTH IN INCHES
ROUGHNESS SAMPLING LENGTH OR CUT OFF RATING IN INCHES OR MILLIMETERS (C)	63/32 √⊥ C	.002 –1 .030 √⊥ .008	ROUGHNESS WIDTH CUTOFF IN INCHES

Fig. 7-7-3 Location of notes and symbols on surface texture symbol.

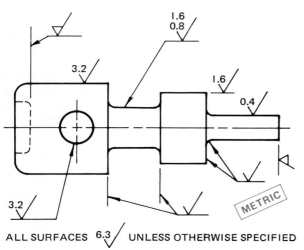

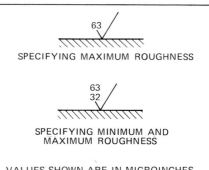

SPECIFYING MAXIMUM ROUGHNESS

SPECIFYING MINIMUM AND
MAXIMUM ROUGHNESS

VALUES SHOWN ARE IN MICROINCHES

RECOMMENDED ROUGHNESS HEIGHT VALUES		N SERIES OF ROUGHNESS GRADE NUMBERS
MICROINCHES μ in.	MICROMETERS μm	
2000	50	N 12
1000	25	N 11
500	12.5	N 10
250	6.3	N 9
125	3.2	N 8
63	1.6	N 7
32	0.8	N 6
16	0.4	N 5
8	0.2	N 4
4	0.1	N 3
2	0.05	N 2
1	0.025	N 1

Fig. 7-7-6 Roughness height ratings.

Fig. 7-7-4 Application of surface texture symbols and notes.

ALL SURFACES 6.3 UNLESS OTHERWISE SPECIFIED

NOTE: VALUES SHOWN ARE IN MICROMETERS

Where only the roughness value is indicated, the horizontal extension line on the symbol may be omitted. The horizontal bar is used whenever any surface characteristics are placed above the bar or to the right of the symbol. The point of the symbol should be located on the line indicating the surface, on an extension line from the surface, or on a leader pointing to the surface or extension line. See Fig. 7-7-4. When numerical values accompany the symbol, the symbol should be in an upright position in order to be readable from the bottom. This means that the long leg and extension line are always on the right. When no numerical values are shown on the symbol, the symbol may also be positioned to be readable from the right side. If necessary, the symbol may be connected to the surface by a leader line terminating in an arrow. The symbol applies to the entire surface, unless otherwise specified. The symbol for the same surface should not be duplicated on other views.

If CAD is used, draw each of the required symbols once and block it. Copy/rotate as many additional symbols as required. See Fig. 7-7-5.

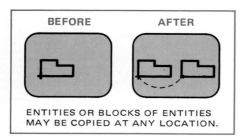

Fig. 7-7-5 CAD copy command.

An alternative to this would be to add these symbols to the CAD library.

Application

Plain (Unplated or Uncoated) Surfaces Surface texture values specified on plain surfaces apply to the completed surface unless otherwise noted.

Plated or Coated Surfaces Drawings or specifications for plated or coated parts must indicate whether the surface texture value applies before, after, or both before and after plating or coating.

Surface Texture Ratings The roughness value rating is indicated at the left of the long leg of the symbol. See Fig. 7-7-4. The specification of only one rating indicates the maximum value, and any lesser value is acceptable. The specification of two ratings indicates the minimum and maximum values, and anything lying within that range is acceptable. See Fig. 7-7-6. The maximum value is placed over the minimum.

Typical surface roughness-height applications are shown in Fig. 7-7-7.

The surface roughness range for common production methods is shown in Fig. 7-7-8.

Waviness-height rating is specified in inches or millimeters and is located above the horizontal extension of the symbol (Fig. 7-7-3). Any lesser value is acceptable.

Waviness spacing is indicated in inches or millimeters and is located above the horizontal extension and to the right, separated from the waviness-height rating by a dash (Fig. 7-7-3). Any

lesser value is acceptable. If the waviness value is a minimum, the abbreviation MIN should be placed after the value.

Lay symbols which indicate the directional pattern of the surface texture are shown in Fig. 7-7-10. The symbol is located to the right of the long leg of the symbol. On surfaces having parallel or perpendicular lay designated, the lead resulting from machine feeds may be objectionable. In these cases, the symbol should be supplemented by the words NO LEAD.

Roughness sampling length or cutoff rating is in inches or millimeters and is located below the horizontal extension (Fig. 7-7-3). Unless otherwise specified, roughness sampling length is .03 in. (0.8 mm). See Fig. 7-7-11.

Notes

Notes relating to surface roughness can be local or general.

General Note Normally, a general note is used where a given roughness requirement applies to the whole part or the major portion. Any exceptions to the general note are given in a local note. See Fig. 7-7-9.

MICROINCHES AA RATING	MICROMETERS AA RATING	APPLICATION
1000	25.2	ROUGH, LOW GRADE SURFACE RESULTING FROM SAND CASTING, TORCH OR SAW CUTTING, CHIPPING OR ROUGH FORGING. MACHINE OPERATIONS ARE NOT REQUIRED AS APPEARANCE IS NOT OBJECTIONABLE. THIS SURFACE, RARELY SPECIFIED, IS SUITABLE FOR UNMACHINED CLEARANCE AREAS ON ROUGH CONSTRUCTION ITEMS.
500	12.5	ROUGH, LOW GRADE SURFACE RESULTING FROM HEAVY CUTS AND COARSE FEEDS IN MILLING TURNING, SHAPING, BORING, AND ROUGH FILING, DISC GRINDING AND SNAGGING. IT IS SUITABLE FOR CLEARANCE AREAS ON MACHINERY, JIGS, AND FIXTURES. SAND CASTING OR ROUGH FORGING PRODUCES THIS SURFACE.
250	6.3	COARSE PRODUCTION SURFACES, FOR UNIMPORTANT CLEARANCE AND CLEANUP OPERATIONS, RESULTING FROM COARSE SURFACE GRIND, ROUGH FILE, DISC GRIND, RAPID FEEDS IN TURNING, MILLING, SHAPING, DRILLING, BORING, GRINDING, ETC., WHERE TOOL MARKS ARE NOT OBJECTIONABLE. THE NATURAL SURFACES OF FORGINGS, PERMANENT MOLD CASTINGS, EXTRUSIONS, AND ROLLED SURFACES ALSO PRODUCE THIS ROUGHNESS. IT CAN BE PRODUCED ECONOMICALLY AND IS USED ON PARTS WHERE STRESS REQUIREMENTS, APPEARANCE, AND CONDITIONS OF OPERATIONS AND DESIGN PERMIT.
125	3.2	THE ROUGHEST SURFACE RECOMMENDED FOR PARTS SUBJECT TO LOADS, VIBRATION, AND HIGH STRESS. IT IS ALSO PERMITTED FOR BEARING SURFACES WHEN MOTION IS SLOW AND LOADS LIGHT OR INFREQUENT. IT IS A MEDIUM COMMERCIAL MACHINE FINISH PRODUCED BY RELATIVELY HIGH SPEEDS AND FINE FEEDS TAKING LIGHT CUTS WITH SHARP TOOLS. IT MAY BE ECONOMICALLY PRODUCED ON LATHES, MILLING MACHINES, SHAPERS, GRINDERS, ETC., OR ON PERMANENT MOLD CASTINGS, DIE CASTINGS, EXTRUSION, AND ROLLED SURFACES.
63	1.6	A GOOD MACHINE FINISH PRODUCED UNDER CONTROLLED CONDITIONS USING RELATIVELY HIGH SPEEDS AND FINE FEEDS TO TAKE LIGHT CUTS WITH SHARP CUTTERS. IT MAY BE SPECIFIED FOR CLOSE FITS AND USED FOR ALL STRESSED PARTS, EXCEPT FAST ROTATING SHAFTS, AXLES, AND PARTS SUBJECT TO SEVERE VIBRATION OR EXTREME TENSION. IT IS SATISFACTORY FOR BEARING SURFACES WHEN MOTION IS SLOW AND LOADS LIGHT OR INFREQUENT. IT MAY ALSO BE OBTAINED ON EXTRUSIONS, ROLLED SURFACES, DIE CASTINGS AND PERMANENT MOLD CASTINGS WHEN RIGIDLY CONTROLLED.
32	0.8	A HIGH-GRADE MACHINE FINISH REQUIRING CLOSE CONTROL WHEN PRODUCED BY LATHES, SHAPERS, MILLING MACHINES, ETC., BUT RELATIVELY EASY TO PRODUCE BY CENTERLESS, CYLINDRICAL OR SURFACE GRINDERS. ALSO, EXTRUDING, ROLLING, OR DIE CASTING MAY PRODUCE A COMPARABLE SURFACE WHEN RIGIDLY CONTROLLED. THIS SURFACE MAY BE SPECIFIED IN PARTS WHERE STRESS CONCENTRATION IS PRESENT. IT IS USED FOR BEARINGS WHEN MOTION IS NOT CONTINUOUS AND LOADS ARE LIGHT. WHEN FINER FINISHES ARE SPECIFIED, PRODUCTION COSTS RISE RAPIDLY; THEREFORE, SUCH FINISHES MUST BE ANALYZED CAREFULLY.
16	0.4	A HIGH QUALITY SURFACE PRODUCED BY FINE CYLINDRICAL GRINDING, EMERY BUFFING, COARSE HONING OR LAPPING. IT IS SPECIFIED WHERE SMOOTHNESS IS OF PRIMARY IMPORTANCE, SUCH AS RAPIDLY ROTATING SHAFT BEARINGS, HEAVILY LOADED BEARINGS AND EXTREME TENSION MEMBERS.
8	0.2	A FINE SURFACE PRODUCED BY HONING, LAPPING, OR BUFFING. IT IS SPECIFIED WHERE PACKINGS AND RINGS MUST SLIDE ACROSS THE DIRECTION OF THE SURFACE GRAIN, MAINTAINING OR WITHSTANDING PRESSURES, OR FOR INTERIOR HONED SURFACES OF HYDRAULIC CYLINDERS. IT MAY ALSO BE REQUIRED IN PRECISION GAGES AND INSTRUMENT WORK, OR SENSITIVE VALUE SURFACES, OR ON RAPIDLY ROTATING SHAFTS AND ON BEARINGS WHERE LUBRICATION IS NOT DEPENDABLE.
4	0.1	A COSTLY REFINED SURFACE PRODUCED BY HONING, LAPPING, AND BUFFING. IT IS SPECIFIED ONLY WHEN THE REQUIREMENTS OF DESIGN MAKE IT MANDATORY. IT IS REQUIRED IN INSTRUMENT WORK, GAGE WORK, AND WHERE PACKINGS AND RINGS MUST SLIDE ACROSS THE DIRECTION OF SURFACE GRAIN SUCH AS ON CHROME–PLATED PISTON RODS, ETC., WHERE LUBRICATION IS NOT DEPENDABLE.
2 / 1	0.05 / 0.025	COSTLY REFINED SURFACES PRODUCED ONLY BY THE FINEST OF MODERN HONING, LAPPING, BUFFING, AND SUPERFINISHING EQUIPMENT. THESE SURFACES MAY HAVE A SATIN OR HIGHLY POLISHED APPEARANCE DEPENDING ON THE FINISHING OPERATION AND MATERIAL. THESE SURFACES ARE SPECIFIED ONLY WHEN DESIGN REQUIREMENTS MAKE IT MANDATORY. THEY ARE SPECIFIED ON FINE OR SENSITIVE INSTRUMENT PARTS OR OTHER LABORATORY ITEMS, AND CERTAIN GAGE SURFACES, SUCH AS ON PRECISION GAGE BLOCKS.

Fig. 7-7-7 Typical surface roughness height applications.

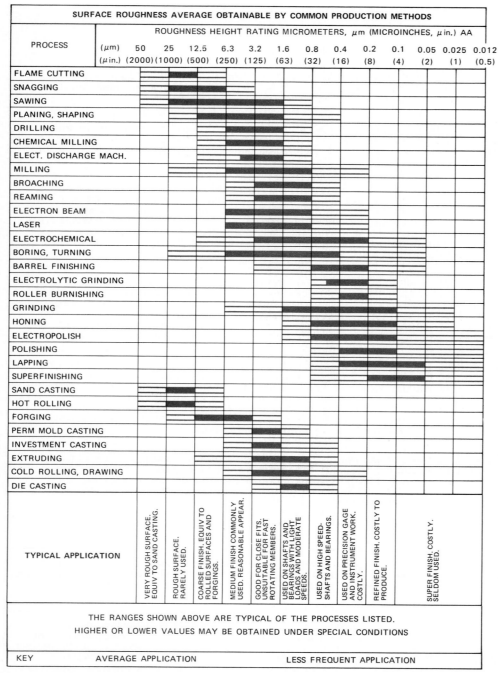

Fig. 7-7-8 Surface roughness range for common production.

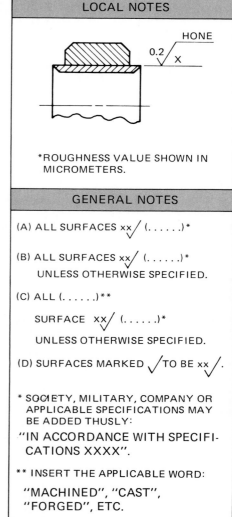

Fig. 7-7-9 Surface texture notes.

MACHINED SURFACES

In preparing working drawings or parts to be cast, molded, or forged, the drafter must indicate the surfaces on the drawing which will require machining or finishing. The symbol √ identifies those surfaces which are produced by machining operations. See Fig. 7-7-12. It indicates that material is to be provided for removal by machining. Where all the surfaces are to be machined, a general note such as FINISH ALL OVER may be used, and the symbols on the drawing may be omitted. Where space is restricted, the machining symbol may be placed on an extension line.

Machining symbols, like dimensions, are not normally duplicated. They should be used on the same view as the dimensions that give the size or location of the surfaces concerned. The symbol is placed on the line representing the surface or, where desirable, on the extension line locating the surface. Figures 7-7-13 and 7-7-14 show examples of the use of machining symbols.

Material Removal Allowance

When it is desirable to indicate the amount of material to be removed, the

SYMBOL	DESIGNATION	EXAMPLE
=	LAY PARALLEL TO THE LINE REPRE-SENTING THE SURFACE TO WHICH THE SYMBOL IS APPLIED.	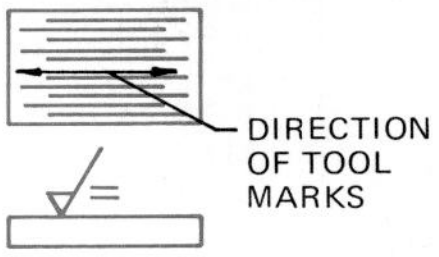 DIRECTION OF TOOL MARKS
⊥	LAY PERPENDICULAR TO THE LINE REPRESENTING THE SURFACE TO WHICH THE SYMBOL IS APPLIED.	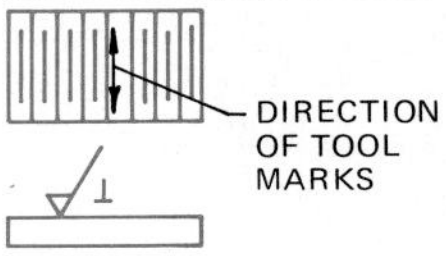 DIRECTION OF TOOL MARKS
X	LAY ANGULAR IN BOTH DIRECTIONS TO LINE REPRESENTING THE SURFACE TO WHICH SYMBOL IS APPLIED.	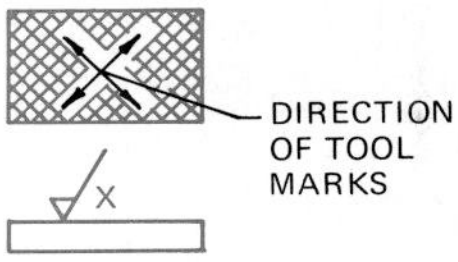 DIRECTION OF TOOL MARKS
M	LAY MULTIDIRECTIONAL.	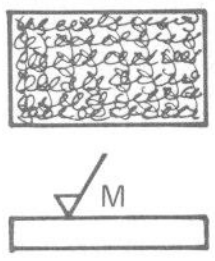
C	LAY APPROXIMATELY CIRCULAR RELATIVE TO THE CENTER OF THE SURFACE TO WHICH THE SYMBOL IS APPLIED.	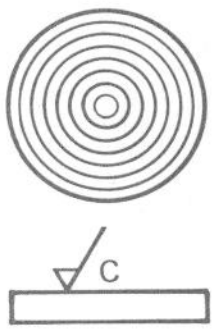
R	LAY APPROXIMATELY RADIAL RELA-TIVE TO THE CENTER OF THE SUR-FACE TO WHICH THE SYMBOL IS APPLIED.	
P	LAY NONDIRECTIONAL, PITTED OR PROTUBERANT.	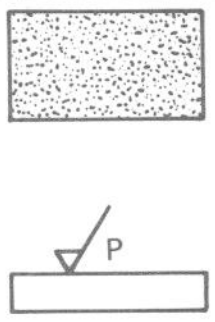

Fig. 7-7-10 Lay symbols.

REMOVAL OF MATERIAL BY MACHINING IS

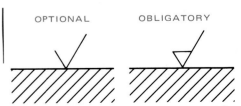

OPTIONAL   OBLIGATORY

Fig. 7-7-12 Indicating the removal of material on the surface texture symbol.

PRESENT SYMBOL	FORMER SYMBOL
LAY SYMBOLS	

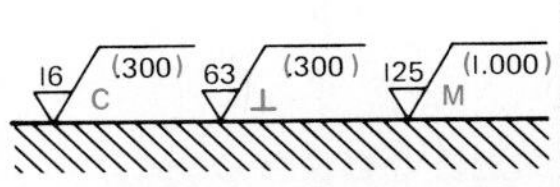

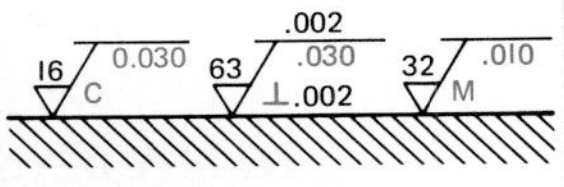

STANDARD ROUGHNESS SAMPLING LENGTH VALUES

INCHES	MILLIMETERS
.003	0.08
.010	0.25
.030	0.8
.100	2.54
.300	8
1.000	25.4

Fig. 7-7-11 Lay and roughness sampling length specifications.

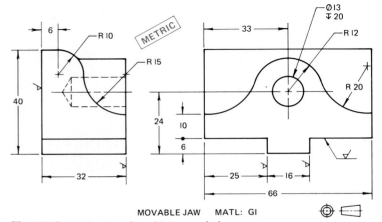

MOVABLE JAW   MATL: GI

Fig. 7-7-13 Application of machining symbols.

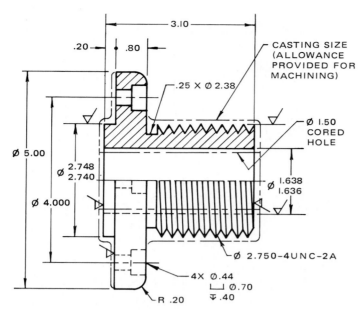

Fig. 7-7-14 **Extra metal allowance for machined surfaces.**

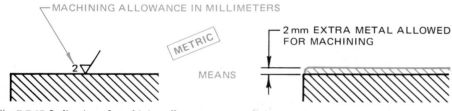

Fig. 7-7-15 **Indication of machining allowance.**

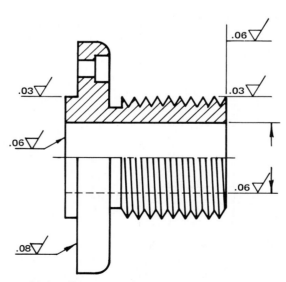

Fig. 7-7-16 **Indicating machining allowance on drawings.**

amount of material in inches or millimeters is shown to the left of the symbol. Illustrations showing material removal allowance are shown in Figs. 7-7-15 and 7-7-16.

Material Removal Prohibited

When it is necessary to indicate that a surface must be produced without material removal, the machining prohibited symbol shown in Fig. 7-7-17 must be used.

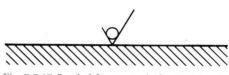

Fig. 7-7-17 Symbol for removal of material not permitted.

Former Machining Symbols

Former machining symbols, as shown in Fig. 7-7-18, may be found on many drawings in use today. When called upon to make changes or revisions to a drawing already in existence, a drafter must adhere to the drawing conventions shown on that drawing.

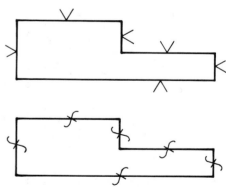

Fig. 7-7-18 Former machining symbols.

References and Source Materials

1. ANSI Y14.36, *Surface Texture Symbols*.
2. GAR.
3. General Motors.

ASSIGNMENTS

See Assignments 21 through 24 for Unit 7-7 starting on page 154.

ASSIGNMENTS FOR CHAPTER 7

Assignments for Unit 7-1, Basic Dimensioning

1. Select one of the template drawings (Fig. 7-1-A or 7-1-B) and on a B (A3) size sheet make a one-view drawing, complete with dimensions, of the part. Scale 1:1.

2. Select one of the parts shown in Figs. 7-1-C to 7-1-F and on a B (A3) size sheet make a three-view drawing, complete with dimensions, of the part. Scale 1:1.

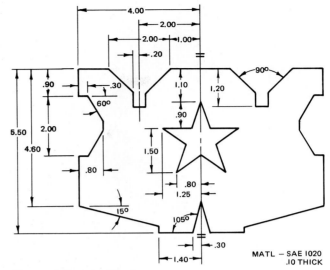

Fig. 7-1-A Template No. 1.

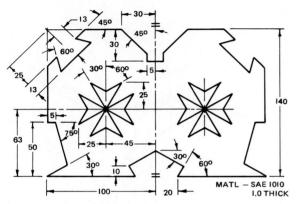

Fig. 7-1-B Template No. 2.

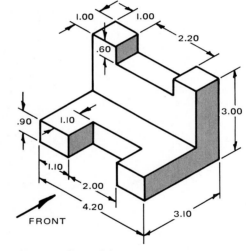

Fig. 7-1-C Cross slide.

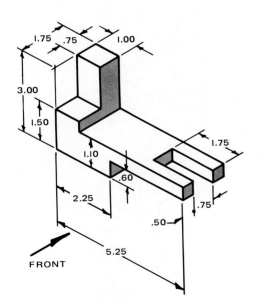

Fig. 7-1-D Notched block.

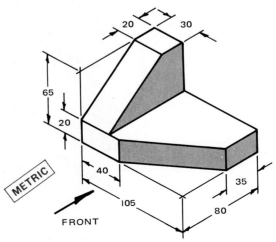

Fig. 7-1-E Angle plate.

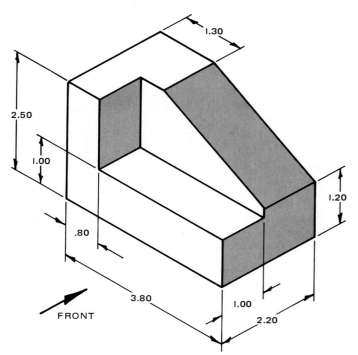

Fig. 7-1-F Guide stop.

3. Select one of the parts shown in Figs. 7-1-G to 7-1-J and on a C (A2) size sheet make a three-view drawing, complete with dimensions, of the part. Show the dimensions with the view which best shows the shape of the part or feature. Scale 1:1.

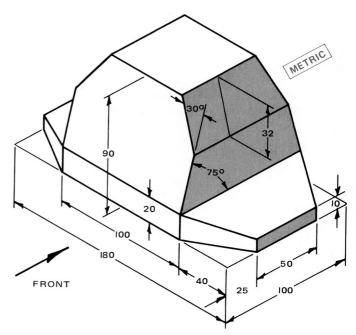

Fig. 7-1-G Base.

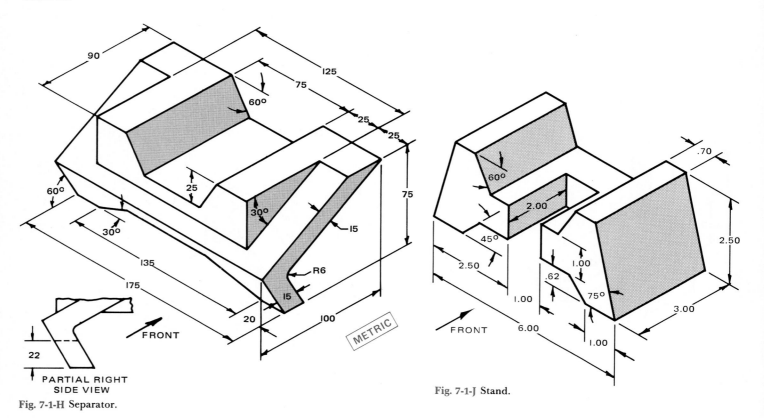

Fig. 7-1-H Separator.

Fig. 7-1-J Stand.

Assignments for Unit 7-2, Dimensioning Circular Features

4. Select one of the problems shown in Figs. 7-2-A and 7-2-B. On a B (A3) size sheet make a one-view drawing, complete with dimensions, of the part. Scale 1:1.

5. Select one of the parts shown in Figs. 7-2-C to 7-2-G and on a B (A3) size sheet make a three-view drawing, complete with dimensions, of the part. Scale 1:1.

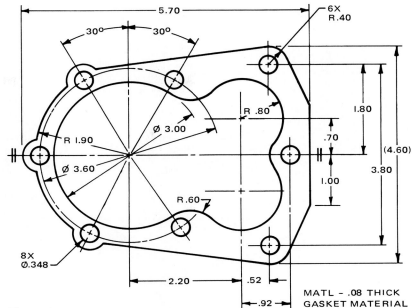

Fig. 7-2-A Gasket.

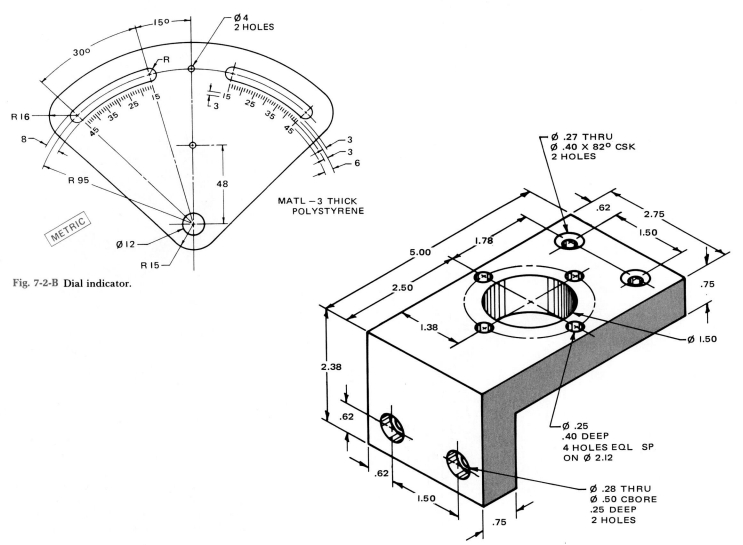

Fig. 7-2-B Dial indicator.

Fig. 7-2-C Guide block.

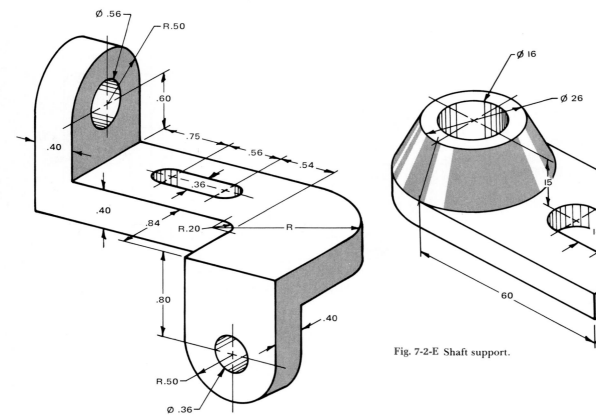

Fig. 7-2-D Bracket.

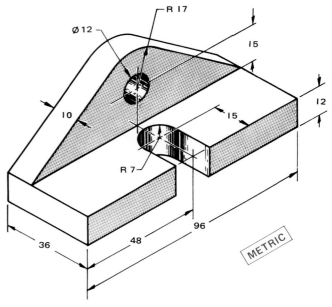

Fig. 7-2-E Shaft support.

Fig. 7-2-G Yoke.

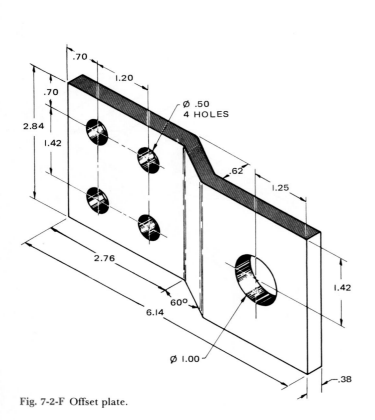

Fig. 7-2-F Offset plate.

6. Select one of the parts shown in Fig. 7-2-H and on a B (A3) size sheet redraw the part and add dimensions. Select one of the scales shown to scale the drawing. Scale 1:1.

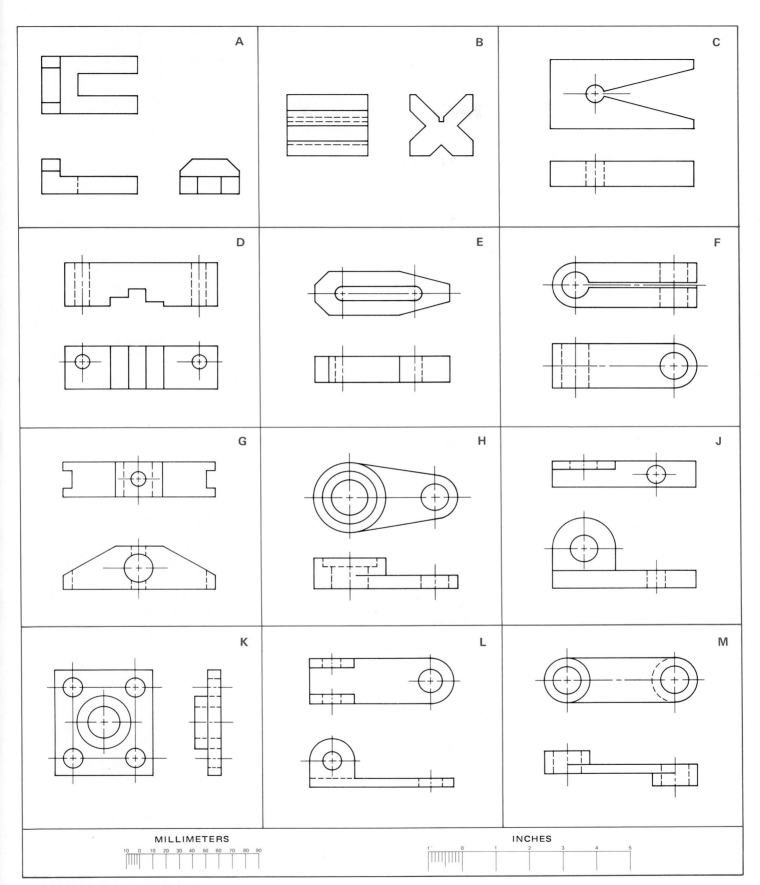

Fig. 7-2-H Problems in dimensioning practice.

Assignments for Unit 7-3, Dimensioning Common Features

7. Redraw the handle shown in Fig. 7-3-A on a B (A3) size sheet. Scale 1:1. The following features are to be added and dimensioned:

(a) 45° × .10 chamfer

(b) 33DP diamond knurl for 1.20 in. starting .80 in. from left end

(c) 1:8 circular taper for 1.20 in. length on right end of Ø1.25

(d) .16 × Ø.54 in. undercut on Ø.75

(e) Ø.189 × .25 in. deep, 4 holes equally spaced

(f) 30° × .10 chamfer. The .10 in. dimension taken horizontally along the shaft.

8. On a B (A3) size sheet redraw the selector shaft shown in Fig. 7-3-B and dimension. Scale the drawing for sizes.

9. On a B (A3) size sheet make a one-view drawing (plus a partial view of the blade), with dimensions, of the screwdriver shown in Fig. 7-3-C. Scale 1:1.

10. On a B (A3) size sheet make a one-view drawing with dimensions of the indicator rod shown in Fig. 7-3-D. Scale 1:1.

11. Make a half-view drawing of one of the parts shown in Figs. 7-3-E to 7-3-G. Add the symmetry symbol to the drawing and dimension using symbolic dimensioning wherever possible. Use the MIRROR command to create the view if CAD is used. Scale 1:1 for Fig. 7-3-E, 10:1 for Fig. 7-3-F, and 1:5 for Fig. 7-3-G.

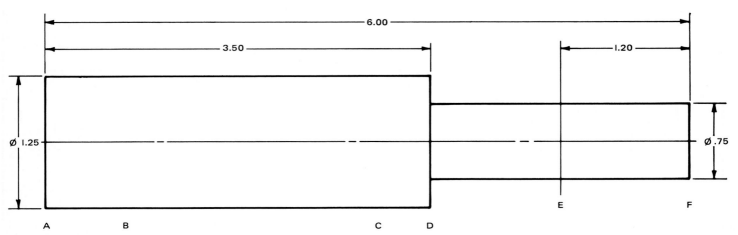

Fig. 7-3-A Handle.

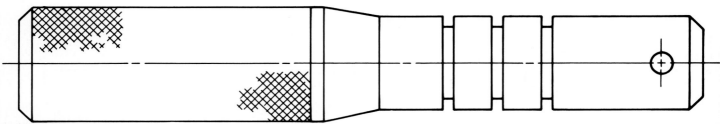

Fig. 7-3-B Selector shaft.

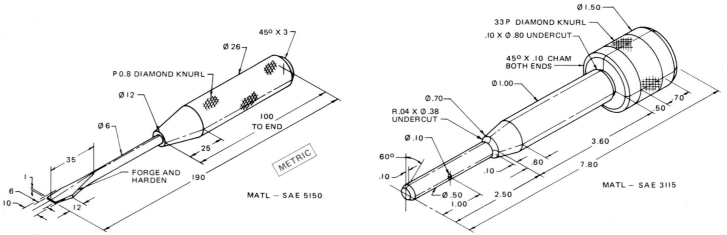

Fig. 7-3-C Screwdriver.

Fig. 7-3-D Indicator rod.

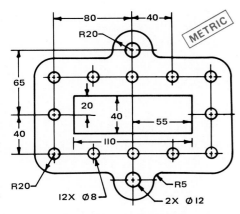

MATL — 2mm FLUORCARBON PLASTIC

Fig. 7-3-E Gasket.

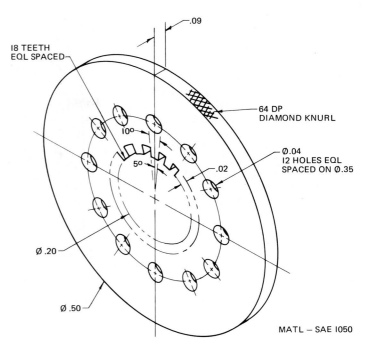

Fig. 7-3-F Adjusting locking plate.

MATL — SAE 1050

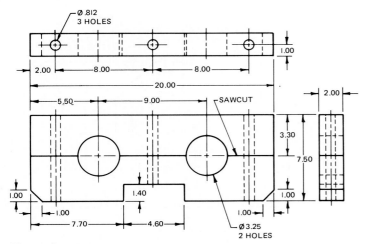

Fig. 7-3-G Tube support.

Assignments for Unit 7-4, Dimensioning Methods

12. Select one of the problems shown in Figs. 7-4-A and 7-4-B, and on a C (A2) size sheet make a working drawing of the part. The arrowless dimensioning shown is to be replaced with rectangular coordinate dimensioning and has the following dimensioning changes.

For Fig. 7-4-A
- Holes A, E, and D are located from the zero coordinates.
- Holes B are located from center of hole E.
- Holes C are located from center of hole D.

For Fig. 7-4-B
- Holes E and D are located from left and bottom edges.
- Holes A and C are located from center of hole D.
- Holes B are located from center of hole E.

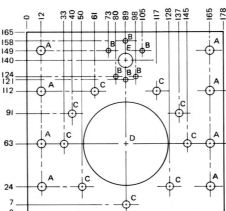

HOLE	DIA
A	8
B	4
C	5
D	76
E	12

MATL — SAE 1006
3mm THICK

METRIC

Fig. 7-4-A Cover plate.

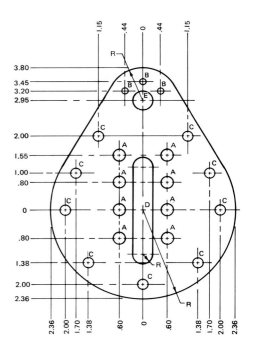

HOLE	SIZE
A	.30
B	.16
C	.24
D	.40 X 2.80
E	.50

MATL — SAE 1008
.12 THICK

Fig. 7-4-B Transmission cover.

13. Divide a B (A3) size sheet into four quadrants by bisecting the vertical and horizontal sides. In each quadrant draw the adapter plate shown in Fig. 7-4-C. Different methods of dimensioning are to be used for each drawing. The methods are rectangular coordinate, chordal, arrowless, and tabular. Scale 1:1.

14. ON a B (A3) size sheet, redraw one of the parts shown in Figs. 7-4-D or 7-4-E. Use arrowless or tabular dimensioning. For Fig. 7-4-D use the bottom and left-hand edge for the data surfaces. For Fig. 7-4-E use the bottom and the center of the part for the data. Use the MIRROR command to create the views if CAD is used. Scale 1:2.

15. On a B (A3) size sheet, redraw the terminal board shown in Fig. 7-4-F using tabular dimensioning. Use the bottom and left-hand edge for the datum surfaces. Scale 1:1.

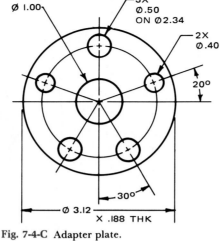

Fig. 7-4-C Adapter plate.

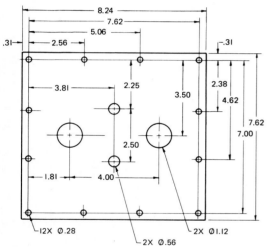

Fig. 7-4-D Cover plate.

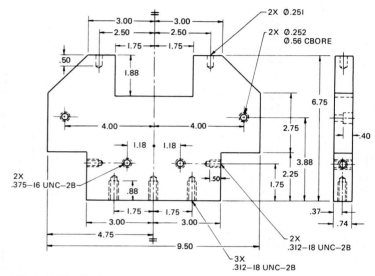

Fig. 7-4-E Back plate.

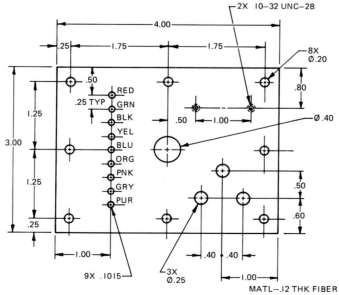

Fig. 7-4-F Terminal board.

148 BASIC DRAWING DESIGN

Assignment for Unit 7-5, Limits and Tolerances

16. Calculate the sizes and tolerances for one of the drawings shown in Fig. 7-5-A or 7-5-B.

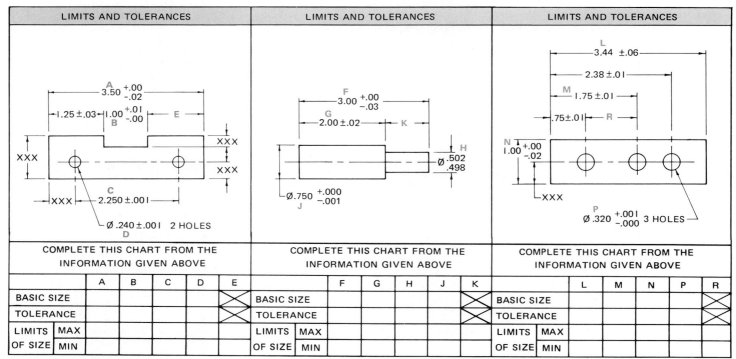

		A	B	C	D	E			F	G	H	J	K			L	M	N	P	R
BASIC SIZE						╳	BASIC SIZE							╳	BASIC SIZE					╳
TOLERANCE						╳	TOLERANCE							╳	TOLERANCE					╳
LIMITS	MAX						LIMITS	MAX							LIMITS	MAX					
OF SIZE	MIN						OF SIZE	MIN							OF SIZE	MIN					

Fig. 7-5-A Inch limits and tolerances.

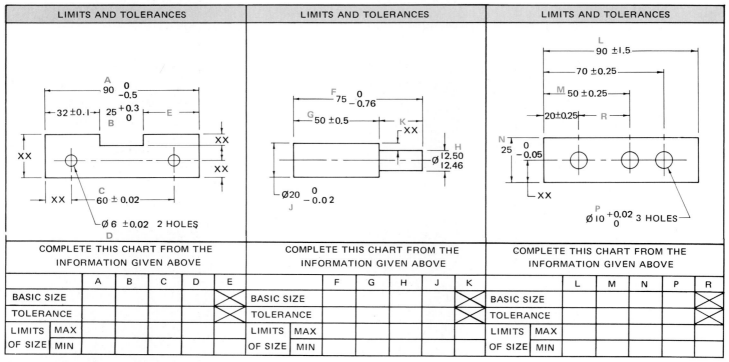

		A	B	C	D	E			F	G	H	J	K			L	M	N	P	R
BASIC SIZE						╳	BASIC SIZE							╳	BASIC SIZE					╳
TOLERANCE						╳	TOLERANCE							╳	TOLERANCE					╳
LIMITS	MAX						LIMITS	MAX							LIMITS	MAX					
OF SIZE	MIN						OF SIZE	MIN							OF SIZE	MIN					

Fig. 7-5-B Metric limits and tolerances.

Assignments for Unit 7-6, Fits and Allowances

17. Using the tables of fits located in the Appendix, calculate the missing dimensions in any of the four charts shown in Figs. 7-6-A to 7-6-D.

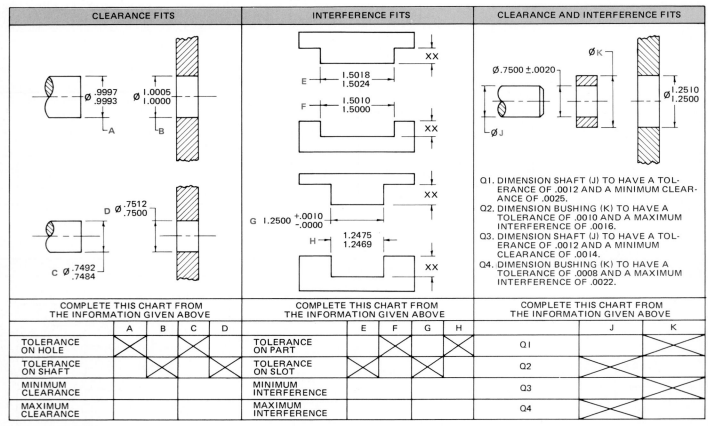

CLEARANCE FITS					INTERFERENCE FITS					CLEARANCE AND INTERFERENCE FITS		
COMPLETE THIS CHART FROM THE INFORMATION GIVEN ABOVE					COMPLETE THIS CHART FROM THE INFORMATION GIVEN ABOVE					COMPLETE THIS CHART FROM THE INFORMATION GIVEN ABOVE		
	A	B	C	D		E	F	G	H		J	K
TOLERANCE ON HOLE					TOLERANCE ON PART					Q1		
TOLERANCE ON SHAFT					TOLERANCE ON SLOT					Q2		
MINIMUM CLEARANCE					MINIMUM INTERFERENCE					Q3		
MAXIMUM CLEARANCE					MAXIMUM INTERFERENCE					Q4		

Q1. DIMENSION SHAFT (J) TO HAVE A TOLERANCE OF .0012 AND A MINIMUM CLEARANCE OF .0025.
Q2. DIMENSION BUSHING (K) TO HAVE A TOLERANCE OF .0010 AND A MAXIMUM INTERFERENCE OF .0016.
Q3. DIMENSION SHAFT (J) TO HAVE A TOLERANCE OF .0012 AND A MINIMUM CLEARANCE OF .0014.
Q4. DIMENSION BUSHING (K) TO HAVE A TOLERANCE OF .0008 AND A MAXIMUM INTERFERENCE OF .0022.

Fig. 7-6-A Inch fits.

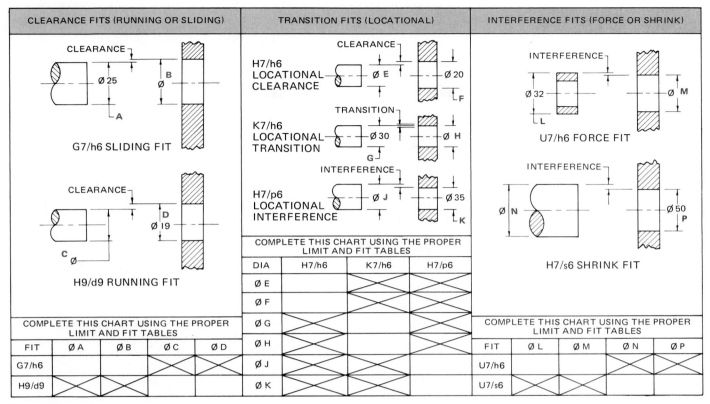

Fig. 7-6-B Metric fits.

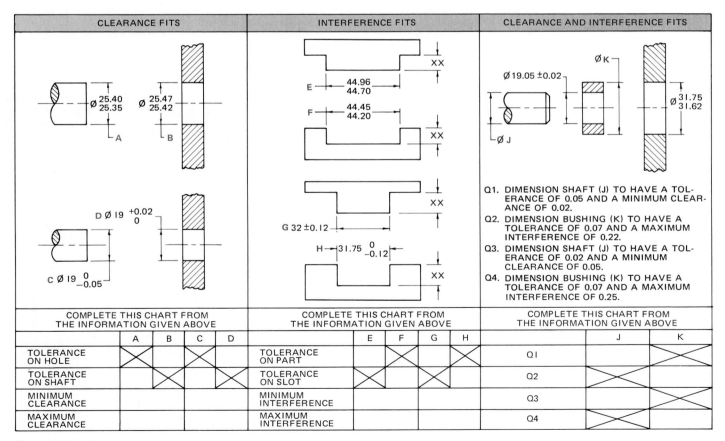

Fig. 7-6-C Inch fits.

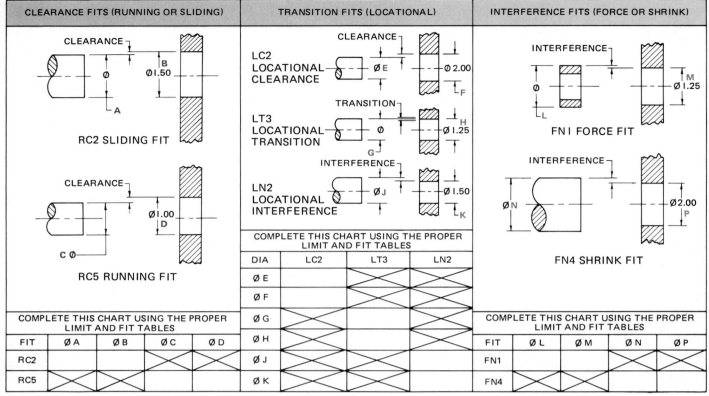

CLEARANCE FITS (RUNNING OR SLIDING) | TRANSITION FITS (LOCATIONAL) | INTERFERENCE FITS (FORCE OR SHRINK)

RC2 SLIDING FIT

RC5 RUNNING FIT

LC2 LOCATIONAL CLEARANCE

LT3 LOCATIONAL TRANSITION

LN2 LOCATIONAL INTERFERENCE

FN1 FORCE FIT

FN4 SHRINK FIT

COMPLETE THIS CHART USING THE PROPER LIMIT AND FIT TABLES

DIA	LC2	LT3	LN2
Ø E		✕	✕
Ø F		✕	✕
Ø G	✕		✕
Ø H	✕		✕
Ø J	✕	✕	
Ø K	✕	✕	

COMPLETE THIS CHART USING THE PROPER LIMIT AND FIT TABLES

FIT	Ø A	Ø B	Ø C	Ø D
RC2				
RC5	✕	✕		

COMPLETE THIS CHART USING THE PROPER LIMIT AND FIT TABLES

FIT	Ø L	Ø M	Ø N	Ø P
FN1			✕	✕
FN4	✕	✕		

Fig. 7-6-D Inch fits.

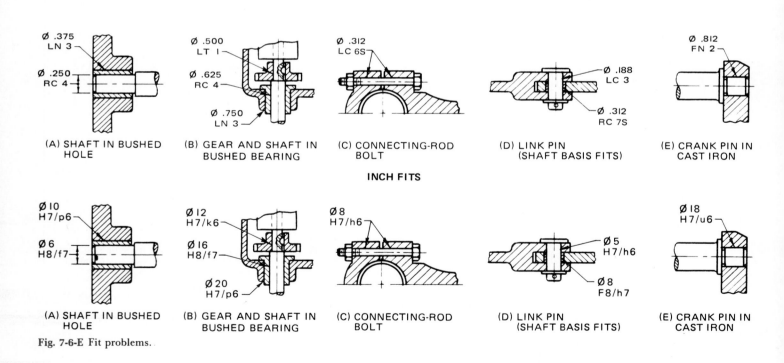

(A) SHAFT IN BUSHED HOLE

(B) GEAR AND SHAFT IN BUSHED BEARING

(C) CONNECTING-ROD BOLT

(D) LINK PIN (SHAFT BASIS FITS)

(E) CRANK PIN IN CAST IRON

INCH FITS

(A) SHAFT IN BUSHED HOLE

(B) GEAR AND SHAFT IN BUSHED BEARING

(C) CONNECTING-ROD BOLT

(D) LINK PIN (SHAFT BASIS FITS)

(E) CRANK PIN IN CAST IRON

Fig. 7-6-E Fit problems.

DESIGN SKETCH	BASIC DIAMETER SIZE IN. [mm]	SYMBOL	BASIS	FEATURE	LIMITS OF SIZE		CLEARANCE OR INTERFERENCE	
					MAX	MIN	MAX	MIN
A	.375 [10]		HOLE	HOLE				
				SHAFT				
A	.250 [6]		HOLE	HOLE				
				SHAFT				
B	.500 [12]		HOLE	HOLE				
				SHAFT				
B	.625 [16]		HOLE	HOLE				
				SHAFT				
B	.750 [20]		HOLE	HOLE				
				SHAFT				
C	.312 [8]		SHAFT	HOLE				
				SHAFT				
D	.188 [5]		HOLE	HOLE				
				SHAFT				
D	.312 [8]		SHAFT	HOLE				
				SHAFT				
E	.812 [18]		HOLE	HOLE				
				SHAFT				

Fig. 7-6-E (continued)

18. Using the fit tables in the Appendix, complete the table shown in Fig. 7-6-E using either U.S. Customary or metric sizes.

19. On B (A3) size sheet make a detail drawing of the roller guide base shown in Fig. 7-6-F. Use one of the scales shown on the drawing to scale the part for sizes. Other considerations are:

(a) Keyseat to be for standard square key and limits on the hole controlled by either an H9/d9 (metric) or RC6 (inch) fit.

(b) Control critical machine surfaces to 0.8 μm or 32 μin.

(c) Dimension in metric or decimal inch.

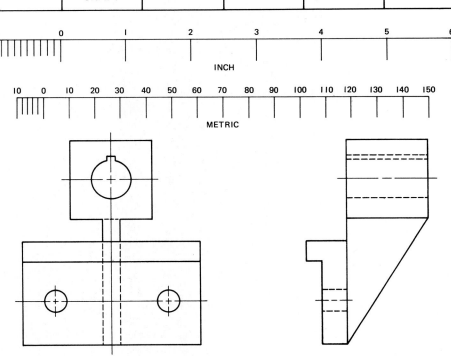

Fig. 7-6-F Roller guide base.

20. On an A (A4) size sheet make a detail drawing of the spindle shown in Fig. 7-6-G. Scale the part for sizes using one of the scales shown with the drawing. Other considerations are:

(a) "A" diameter to have an LC3 (inch) or H7/h6 (metric) fit.
(b) "B" diameter requires a 96 diamond knurl or its equivalent.
(c) "C" diameter to have an LT3 (inch) or H7/K6 (metric) fit.
(d) "D" diameter to be a minimum relief (undercut).
(e) "E" to be a standard No. 807 Woodruff key in center of segment and the diameter to be controlled by an RC3 (inch) or H7/g6 (metric) fit.
(f) "F" to be undercut for a standard retaining ring and controlled form to manufacturer's specifications.
(g) Dimension in decimal inch or metric.

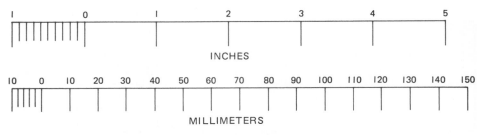

SCALES FOR FIGURES 7–6–F AND 7–6–G

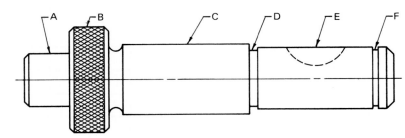

Fig. 7-6-G Spindle.

Assignments for Unit 7-7, Surface Texture

21. On a B (A3) size sheet make a working drawing of the cross slide shown in Fig. 7-7-A. Scale 1:1. The following surface texture information is to be added to the drawing:

- The dovetail slot is to have a maximum roughness value of 3.2 μm and a machining allowance of 2 mm.
- The ends of the shaft support are to have maximum and minimum roughness values of 1.6 and 0.8 μm and a machining allowance of 2 mm.
- The hole is to have an H8 tolerance.

22. On a B (A3) size sheet make a working drawing of the column bracket shown in Fig. 7-7-B. Scale 1:1. The following surface texture information is to be added to the drawing:

- The bottom of the base is to have a maximum roughness value of 125 μin. and a machining allowance of .06 in.
- The tops of the bosses are to have a maximum roughness value of 250 μin. and a machining allowance of .04 in.
- The end surfaces of the hubs supporting the shafts are to have maximum and minimum roughness values of 63 and 32 μin. and a machining allowance of .04 in.
- The large hole is to be dimensioned for an RC4 fit. The small hole is to be dimensioned for an LN3 fit for plain bearings.

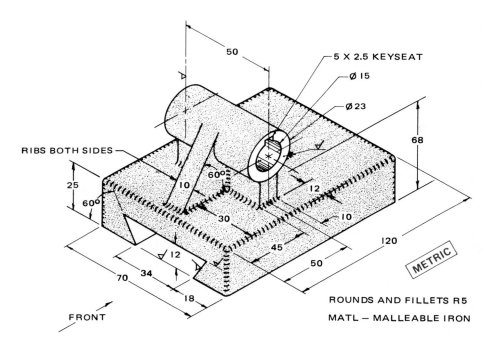

Fig. 7-7-A Cross slide.

23. On a B (A3) size sheet make a working drawing of the adjustable base plate shown in Fig. 7-7-C. The amount of material to be removed on the surfaces requiring machining is 2 mm. The center hole is to be dimensioned having an H8 tolerance. Scale 1:1.

24. On a B (A3) size sheet make a working drawing of the link shown in Fig. 7-7-D. The amount of material to be removed from the end surfaces of the hub is .09 in. and .06 in. on the bosses and bottom of the vertical hub. The two large holes are to have an LN3 fit for journal bearings. Scale 1:1.

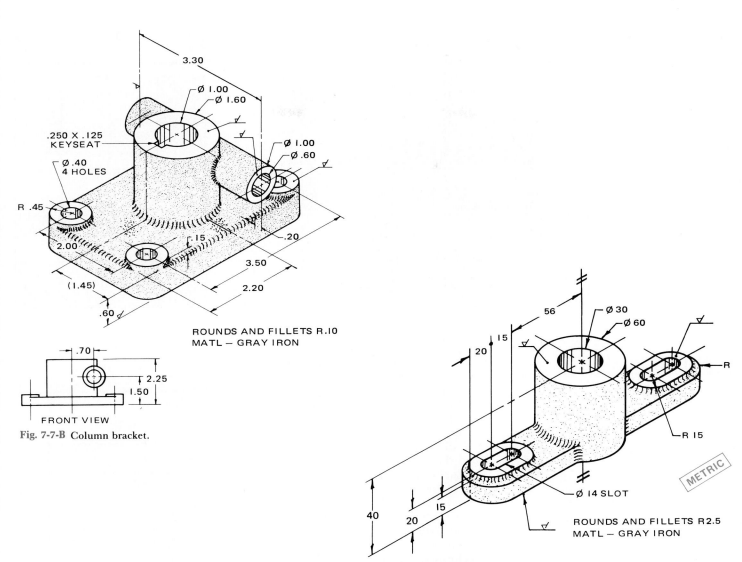

.250 X .125
KEYSEAT

Ø 1.00
Ø 1.60

3.30

Ø 1.00
Ø .60

Ø .40
4 HOLES

R .45

.15

.20

2.00

3.50

2.20

(1.45)

.60

ROUNDS AND FILLETS R.10
MATL — GRAY IRON

.70

2.25

1.50

FRONT VIEW

Fig. 7-7-B Column bracket.

56

Ø 30
Ø 60

15

20

R

R 15

40

20

15

Ø 14 SLOT

METRIC

ROUNDS AND FILLETS R2.5
MATL — GRAY IRON

Fig. 7-7-C Adjustable base plate.

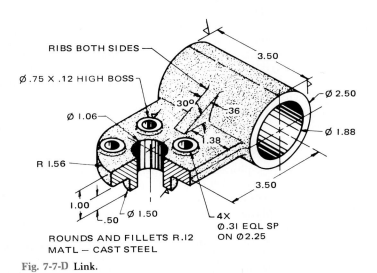

RIBS BOTH SIDES

3.50

Ø .75 X .12 HIGH BOSS

Ø 2.50

Ø 1.06

30°

36

Ø 1.88

1.38

R 1.56

3.50

1.00

.50

Ø 1.50

4X
Ø .31 EQL SP
ON Ø 2.25

ROUNDS AND FILLETS R.12
MATL — CAST STEEL

Fig. 7-7-D Link.

Sections and Conventions

UNIT 8-1
Sectional Views

Sectional views, commonly called *sections*, are used to show interior detail that is too complicated to be shown clearly by regular views containing many hidden lines. For some assembly drawings, they show a difference in materials. A sectional view is obtained by supposing the nearest part of the object to be cut or

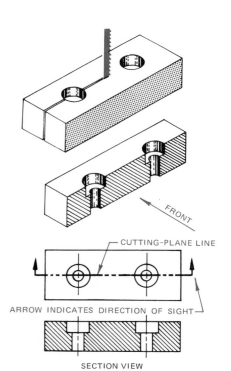

Fig. 8-1-1 A full-section drawing.

broken away on an imaginary cutting plane. The exposed or cut surfaces are identified by section lining or cross-hatching. Hidden lines and details behind the cutting-plane line are usually omitted unless they are required for clarity or dimensioning. It should be understood that only in the sectional view is any part of the object shown as having been removed.

A sectional view frequently replaces one of the regular views. For example, a regular front view is replaced by a front view in section, as shown in Fig. 8-1-1.

Whenever practical, except for revolved sections, sectional views should be projected perpendicular to the cutting plane and be placed in the normal position for third-angle projection.

When the preferred placement is not practical, the sectional view may be moved to some other convenient position on the drawing, but it must be clearly identified, usually by two capital letters, and labeled.

Cutting-Plane Lines

Cutting-plane lines (Fig. 8-1-2) are used to indicate the location of cutting planes for sectional views. Two forms of cutting-plane lines are approved for general use.

The first form consists of evenly spaced, thick dashes with arrowheads. The second form consists of alternating long dashes and pairs of short dashes. The long dashes may vary in length, depending on the size of the drawing.

Both forms of lines should be drawn to

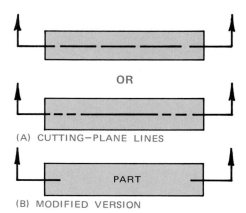

Fig. 8-1-2 Cutting-plane lines.

stand out clearly on the drawing. The ends of the lines are bent at 90° and terminated by bold arrowheads to indicate the direction of sight for viewing the section.

The cutting-plane line can be omitted when it corresponds to the center line of the part and it is obvious where the cutting plane lies. On drawings with a high density of line work, and for offset sections (see Unit 8-6), cutting-plane lines may be modified by omitting the dashes between the line ends for the purpose of obtaining clarity, as shown in Fig. 8-1-2B.

Full Sections

When the cutting plane extends entirely through the object in a straight line and the front half of the object is theoretically removed, a full section is obtained. See Figs. 8-1-3 and 8-1-4. This

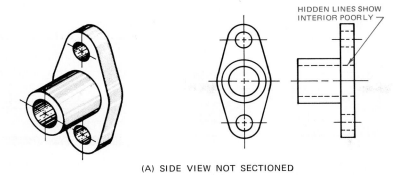

(A) SIDE VIEW NOT SECTIONED

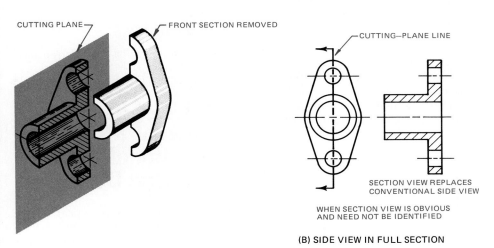

HIDDEN LINES SHOW INTERIOR POORLY

CUTTING PLANE — FRONT SECTION REMOVED

CUTTING–PLANE LINE

SECTION VIEW REPLACES CONVENTIONAL SIDE VIEW

WHEN SECTION VIEW IS OBVIOUS AND NEED NOT BE IDENTIFIED

SECTION A–A

WHEN SECTION VIEW MUST BE IDENTIFIED

(B) SIDE VIEW IN FULL SECTION

Fig. 8-1-3 Full-section view.

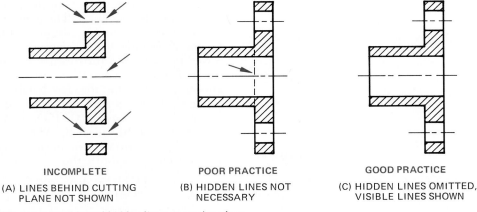

INCOMPLETE

(A) LINES BEHIND CUTTING PLANE NOT SHOWN

POOR PRACTICE

(B) HIDDEN LINES NOT NECESSARY

GOOD PRACTICE

(C) HIDDEN LINES OMITTED, VISIBLE LINES SHOWN

Fig. 8-1-4 Visible and hidden lines on section views.

type of section is used for both detail and assembly drawings. When the section is on an axis of symmetry, it is not necessary to indicate its location. See Fig. 8-1-5. However, it may be identified and indicated in the normal manner to increase clarity, if so desired.

Section Lining

Section lining, sometimes referred to as cross-hatching, can serve a double purpose. It indicates the surface that has

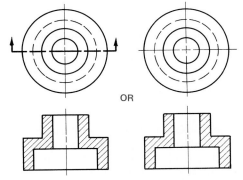

OR

Fig. 8-1-5 Cutting-plane line may be omitted when it corresponds with the center line.

been theoretically cut and makes it stand out clearly, thus helping the observer to understand the shape of the object. Section lining may also indicate the material from which the object is to be made, when the lining symbols shown in Fig. 8-1-6 are used.

CAD systems provide the option to use different section lining. The selection available on a particular system, however, will not be exactly the same as shown in Fig. 8-1-6.

Section Lining for Detail Drawings

Since the exact material specifications for a part are usually given elsewhere on the drawing, the general-purpose section lining symbol is recommended for most detail drawings. An exception may be made for wood when it is desirable to show the direction of the grain.

The lines for section lining are thin and are usually drawn at an angle of 45° to the major outline of the object. The same angle is used for the whole "cut" surface of the object. If the part shape would cause section lines to be parallel, or nearly so, to one of the sides of the part, then some angle other than 45° should be chosen. See Fig. 8-1-7. The spacing of the hatching lines should be

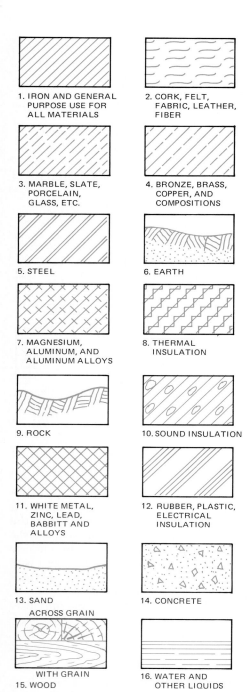

1. IRON AND GENERAL PURPOSE FOR ALL MATERIALS

2. CORK, FELT, FABRIC, LEATHER, FIBER

3. MARBLE, SLATE, PORCELAIN, GLASS, ETC.

4. BRONZE, BRASS, COPPER, AND COMPOSITIONS

5. STEEL

6. EARTH

7. MAGNESIUM, ALUMINUM, AND ALUMINUM ALLOYS

8. THERMAL INSULATION

9. ROCK

10. SOUND INSULATION

11. WHITE METAL, ZINC, LEAD, BABBITT AND ALLOYS

12. RUBBER, PLASTIC, ELECTRICAL INSULATION

13. SAND

14. CONCRETE

15. WOOD
ACROSS GRAIN
WITH GRAIN

16. WATER AND OTHER LIQUIDS

Fig. 8-1-6 Symbolic section lining.

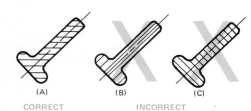

(A) CORRECT (B) INCORRECT (C)

Fig. 8-1-7 Direction of section lining.

reasonably uniform to give a good appearance to the drawing. The pitch, or distance between lines, normally varies between .03 and .12 in. (1 and 3 mm)

depending on the size of the area to be sectioned.

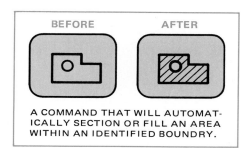

BEFORE   AFTER

A COMMAND THAT WILL AUTOMATICALLY SECTION OR FILL AN AREA WITHIN AN IDENTIFIED BOUNDRY.

Fig. 8-1-8 CAD cross-hatching command.

As a cost reduction, in manual drafting, large areas need not be entirely section-lined. See Fig. 8-1-9. Section lining around the outline will usually be sufficient, providing clarity is not sacrificed.

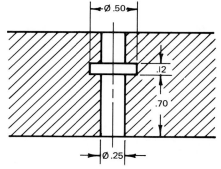

Fig. 8-1-9 Outline section lining.

Dimensions or other lettering should not be placed in sectioned areas. When this is unavoidable, the section lining should be omitted for the numerals or lettering. See Fig. 8-1-10.

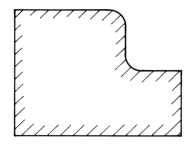

Ø.50 .12 .70 Ø.25

Fig. 8-1-10 Section lining omitted to accommodate dimensions.

Sections which are too thin for effective section lining, such as sheet-metal items, packing, and gaskets, may be shown without section lining, or the area may be filled in completely. See Fig. 8-1-11.

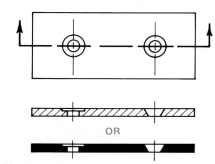

OR

Fig. 8-1-11 Thin parts in section.

ASSIGNMENT

See Assignment 1 for Unit 8-1 on page 169.

UNIT 8-2
Two or More Sectional Views on One Drawing

If two or more sections appear on the same drawing, the cutting-plane lines are identified by two identical large Gothic letters, one at each end of the line, placed behind the arrowhead so that the arrow points away from the letter. Normally, begin alphabetically with A-A, then B-B, and so on. The identification letters should not include I, O, Q, and Z. See Fig. 8-2-1.

Sectional view subtitles are given when identification letters are used and appear directly below the view, incorporating the letters at each end of the cutting-plane line thus: SECTION A-A, or abbreviated, SECT. B-B. When the scale is different from the main view, it is stated below the subtitle thus

SECTION A-A
SCALE 1:10

ASSIGNMENT

See Assignment 2 for Unit 8-2 on page 170.

UNIT 8-3
Half Sections

A *half section* is a view of an assembly or object, usually symmetrical, showing one-half of the view in section. See Figs. 8-3-1 and 8-3-2. Two cutting-plane lines, perpendicular to each other, extend half-

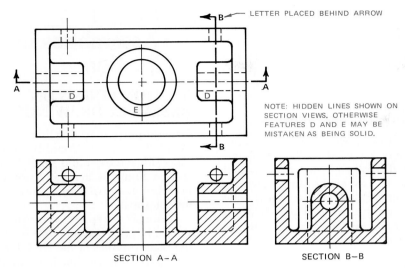

NOTE: HIDDEN LINES SHOWN ON SECTION VIEWS. OTHERWISE FEATURES D AND E MAY BE MISTAKEN AS BEING SOLID.

SECTION A-A   SECTION B-B

Fig. 8-2-1 Detail drawing having two section views.

way through the view, and one-quarter of the view is considered removed with the interior exposed to view.

Similar to the practice followed for full-section drawings, the cutting-plane line need not be drawn for half sections when it is obvious where the cutting took place. Instead, center lines may be used. When a cutting-plane is used, the common practice is to show only one end of the cutting-plane line, terminating with an arrow to show the direction of sight for viewing the section.

On the sectional view a center line or a visible object line may be used to divide the sectioned half from the unsectioned half of the drawing. This type of sectional drawing is best suited for assembly drawings where both internal and external construction is shown on one view and where only overall and center-

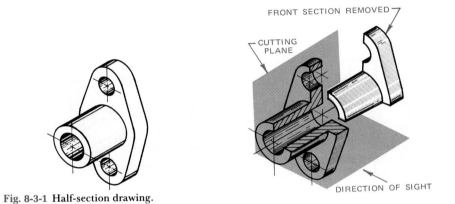

Fig. 8-3-1 Half-section drawing.

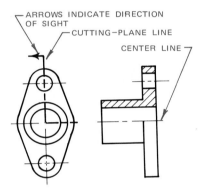

to-center dimensions are required. The main disadvantage of using this type of sectional drawing for detail drawings is the difficulty in dimensioning internal features without adding hidden lines. However, hidden lines may be added for dimensioning, as shown in Fig. 8-3-3.

ASSIGNMENT

See Assignment 3 for Unit 8-3 on page 170.

CENTER LINES OR CUTTING-PLANE LINES ARE USED ON VIEWS WHICH ARE NOT SECTIONED.

OR

OR

A CENTER LINE OR A VISIBLE OBJECT LINE MAY BE USED TO DIVIDE THE SECTIONED HALF FROM THE UNSECTIONED HALF.

Fig. 8-3-2 Half-section views.

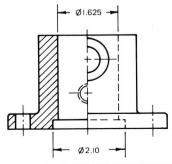

HIDDEN LINES ADDED FOR DIMENSIONING
Fig. 8-3-3 Dimensioning half-section view.

SECTIONS AND CONVENTIONS **159**

U N I T 8 - 4
Threads in Section

True representation of a screw thread is seldom provided on working drawings because it would require very laborious and accurate drawing involving repetitious development of the helix curve of the thread. A symbolic representation of threads is now standard practice.

Three types of conventions are in general use for screw thread representation. See Fig. 8-4-1. These are known as detailed, schematic, and simplified representations. Simplified representation should be used whenever it will clearly portray the requirements. Schematic and detailed representations require more drafting time, but are sometimes necessary to avoid confusion with other parallel lines or to more clearly portray particular aspects of the threads.

Threaded Assemblies

Any of the thread conventions shown here may be used for assemblies of threaded parts, and two or more methods may be used on the same drawing, as shown in Fig. 8-4-2. In sectional views, the externally threaded part is always shown covering the internally threaded part, as illustrated in Fig. 8-4-3.

ASSIGNMENT

See Assignment 4 for Unit 8-4 on page 170.

U N I T 8 - 5
Assemblies in Section

SECTION LINING ON ASSEMBLY DRAWINGS

General-purpose section lining is recommended for most assembly drawings, especially if the detail is small. Symbolic section lining is generally not recommended for drawings that will be microformed.

General-purpose section lining should be drawn at an angle of 45° with the main outlines of the view. On adjacent parts, the section lines should be drawn in the opposite direction, as shown in Figs. 8-5-1 and 8-5-2.

For additional adjacent parts, any suitable angle may be used to make each part stand out separately and clearly.

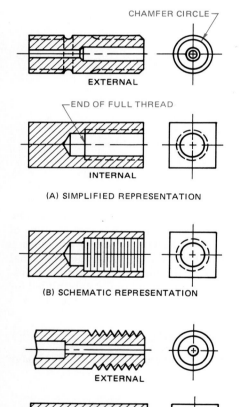

CHAMFER CIRCLE

EXTERNAL

END OF FULL THREAD

INTERNAL

(A) SIMPLIFIED REPRESENTATION

(B) SCHEMATIC REPRESENTATION

EXTERNAL

PREFERRED

ALTERNATE

(C) DETAILED REPRESENTATION

Fig. 8-4-1 Threads in section.

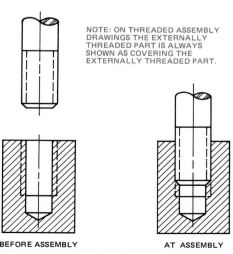

Fig. 8-4-2 Threaded assembly.

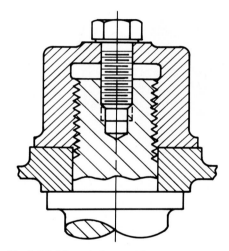

NOTE: ON THREADED ASSEMBLY DRAWINGS THE EXTERNALLY THREADED PART IS ALWAYS SHOWN AS COVERING THE EXTERNALLY THREADED PART.

BEFORE ASSEMBLY   AT ASSEMBLY

Fig. 8-4-3 Drawing threads in assembly drawings.

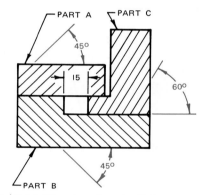

PART A   PART C

45°

15

60°

45°

PART B

Fig. 8-5-1 Direction of section lining.

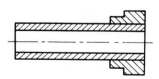

(A) ADJACENT PARTS

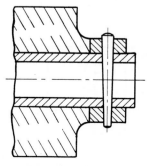

ANGLE AND SPACING OF SECTION LINING

Fig. 8-5-2 Arrangement of section lining.

STEEL PLATES

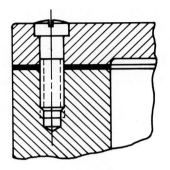

GASKETS

Fig. 8-5-3 Assembly of thin parts in section.

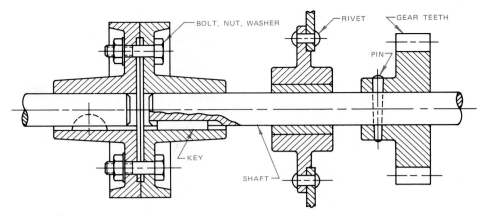

Fig. 8-5-4 Parts that are not section-lined even though the cutting plane passes through them.

Section lines should not be purposely drawn to meet at common boundaries.

When two or more thin adjacent parts are filled in, a space is left between them, as shown in Fig. 8-5-3.

Symbolic section lining is used on special-purpose assembly drawings such as illustrations for parts catalogs, display assemblies, and promotional materials, when it is desirable to distinguish between different materials (Figs. 8-1-6).

All assemblies and subassemblies pertaining to one particular set of drawings should use the same symbolic conventions.

Shafts, Bolts, Pins, Keyseats, and Similar Solid Parts, in Section
Shafts, bolts, nuts, rods, rivets, keys, pins, and similar solid parts, the axes of which lie in the cutting-plane, should not be sectioned except that a broken-out section of the shaft may be used to describe more clearly the key, keyseat, or pin. See Fig. 8-5-4.

ASSIGNMENT

See Assignment 5 for Unit 8-5 starting on page 171.

UNIT 8-6
Offset Sections

In order to include features that are not in a straight line, the cutting-plane may be offset or bent, so as to include several

planes or curved surfaces. See Figs. 8-6-1 and 8-6-2.

An offset section is similar to a full section in that the cutting-plane line extends through the object from one side

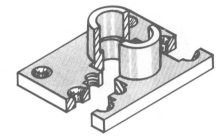

Fig. 8-6-1 An offset section.

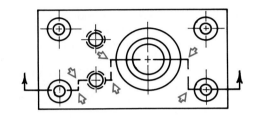

NOTE: CHANGE IN DIRECTION OF CUTTING-PLANE LINE NOT SHOWN IN SECTION VIEW

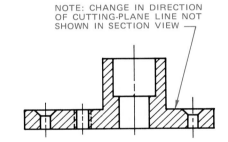

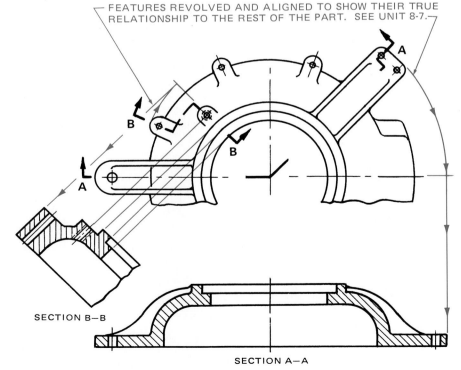

FEATURES REVOLVED AND ALIGNED TO SHOW THEIR TRUE RELATIONSHIP TO THE REST OF THE PART. SEE UNIT 8-7.

SECTION B—B

SECTION A—A

Fig. 8-6-2 Positioning offset sections.

to the other. The change in direction of the cutting-plane line is not shown in the sectional view.

ASSIGNMENT

See Assignment 6 for Unit 8-6 starting on page 172.

UNIT 8-7
Ribs, Holes, and Lugs in Section

RIBS IN SECTIONS

A true-projection sectional view of a part, such as shown in Fig. 8-7-1, would be misleading when the cutting plane passes longitudinally through the center of the rib. To avoid this impression of solidity, a section not showing the ribs section-lined is preferred. When there is an odd number of ribs, such as those shown in Fig. 8-7-1B, the top rib is aligned with the bottom rib to show its true relationship with the hub and flange. If the rib is not aligned or revolved, it would appear distorted on the sectional view and is therefore misleading.

At times it may be necessary to use an alternative method of identifying ribs in a sectional view. Figure 8-7-2 shows a base and a pulley in section. If rib A of the base was not sectioned as previously mentioned, it would appear exactly like rib B in the sectional view and would be misleading. Similarly, ribs C shown on the pulley may be overlooked. To clearly show the relationship of the ribs with the other solid features on the base and pulley, alternate section lining on the ribs is used. The line between the rib and solid portions is shown as a broken line.

HOLES IN SECTIONS

Holes, like ribs, are aligned as shown in Fig. 8-7-1 to show their true relationship to the rest of the part.

LUGS IN SECTION

Lugs, like ribs and spokes, are also aligned to show their true relationship to the rest of the part, because true projection may be misleading. Figure 8-7-3 shows several examples of lugs in section. Note how the cutting-plane line is bent or offset so that the features may be clearly shown in the sectional view.

Some lugs are shown in section, and some are not. When the cutting plane passes through the lug crosswise, the lug is sectioned; otherwise, the lugs are treated in the same manner as ribs.

ASSIGNMENT

See Assignment 7 for Unit 8-7 on page 173.

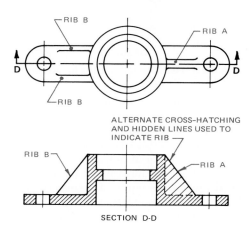

ALTERNATE CROSS-HATCHING AND HIDDEN LINES USED TO INDICATE RIB

SECTION D-D

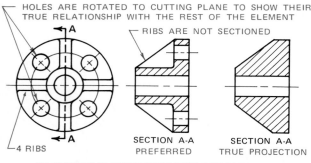

HOLES ARE ROTATED TO CUTTING PLANE TO SHOW THEIR TRUE RELATIONSHIP WITH THE REST OF THE ELEMENT

RIBS ARE NOT SECTIONED

SECTION A-A PREFERRED

SECTION A-A TRUE PROJECTION

4 RIBS

(A) CUTTING PLANE PASSING THROUGH BOTH RIBS

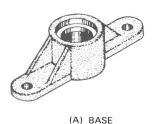

(A) BASE

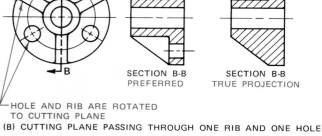

REVOLVE RIB

TRUE PROJECTION GIVES A DISTORTED IMPRESSION

SECTION B-B PREFERRED

SECTION B-B TRUE PROJECTION

HOLE AND RIB ARE ROTATED TO CUTTING PLANE

(B) CUTTING PLANE PASSING THROUGH ONE RIB AND ONE HOLE

Fig. 8-7-1 Preferred and true projection through ribs and holes.

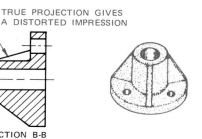

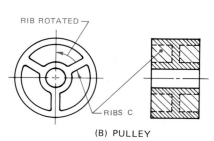

RIB ROTATED

RIBS C

(B) PULLEY

Fig. 8-7-2 Alternate method of showing ribs in section.

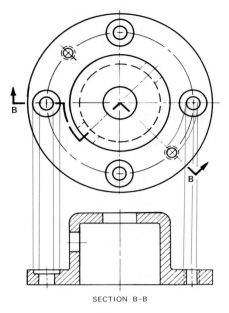

SECTION B-B

(1) HOLES ALIGNED

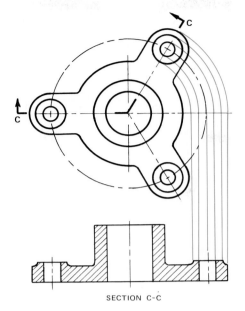

SECTION C-C

(2) LUGS ALIGNED AND SECTIONED

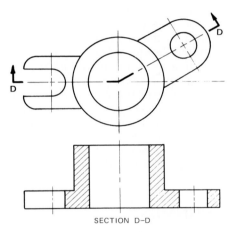

SECTION D-D

(3) LUGS ALIGNED AND SECTIONED

Fig. 8-7-3 Lugs in section.

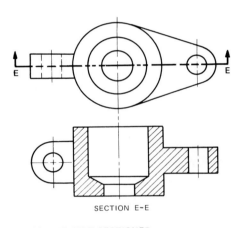

SECTION E-E

(4) LUG NOT SECTIONED

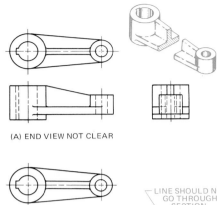

(A) END VIEW NOT CLEAR

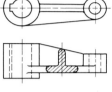

(B) REVOLVED SECTION

LINE SHOULD NOT GO THROUGH SECTION

AVOID

(C) PARTIAL VIEW SHOWING REVOLVED SECTION

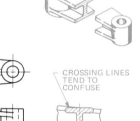

(D) REVOLVED SECTION WITH MAIN VIEW BROKEN FOR CLARITY

CROSSING LINES TEND TO CONFUSE

AVOID

(E) PARTIAL VIEW SHOWING REVOLVED SECTION

Fig. 8-8-1 Revolved sections.

U N I T 8 - 8
Revolved and Removed Sections

Revolved and removed sections are used to show the cross-sectional shape of ribs, spokes, or arms when the shape is not obvious in the regular views. See Figs. 8-8-1 to 8-8-3. Often end views are not needed when a revolved section is used. For a revolved section, draw a center line through the shape on the plane to be described, imagine the part to be rotated 90°, and superimposed on the view the shape that would be seen when rotated. See Figs. 8-8-1 and 8-8-2. If the revolved

section does not interfere with the view on which it is revolved, then the view is not broken unless it would provide for clearer dimensioning. When the revolved section interferes or passes through lines on the view on which it is revolved, then the general practice is to break the view (Fig. 8-8-2). Often the break is used to shorten the length of the object. In no circumstances should the lines on the view pass through the section. When superimposed on the view, the outline of the revolved section is a thin, continuous line.

The removed section differs in that the section, instead of being drawn right on the view, is removed to an open area on the drawing (Fig. 8-8-3). Frequently the

removed section is drawn to an enlarged scale for clarification and easier dimensioning. Removed sections of symmetrical parts should be placed, whenever possible, on the extension of the center line (Fig. 8-8-3B)

If CAD is being used, the BLOCK, ROTATE, SCALE, and MOVE com-

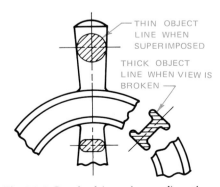

THIN OBJECT LINE WHEN SUPERIMPOSED

THICK OBJECT LINE WHEN VIEW IS BROKEN

Fig. 8-8-2 Revolved (superimposed) sections.

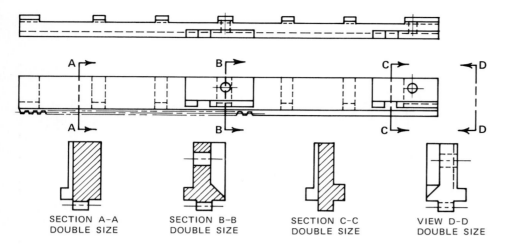

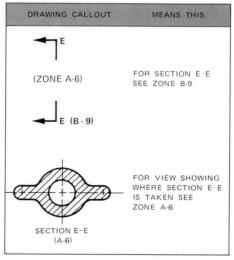

Fig. 8-8-3 Removed sections.

(A) REMOVED SECTIONS AND REMOVED VIEW

(B) CRANE HOOK

KNURL

(C) NUT

ENLARGED DETAIL OF TEETH
SCALE 8:1

mands will assist in the preparation of removed sections.

On complicated drawings where the placement of the removed view may be some distance from the cutting-plane, auxiliary information, such as the reference zone location (Fig. 8-8-4), may be helpful.

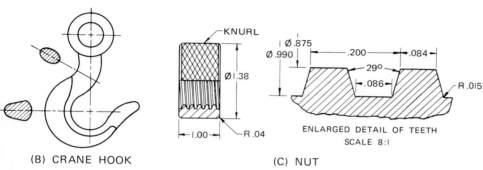

Fig. 8-8-4 Reference zone location.

PLACEMENT OF SECTIONAL VIEWS

Whenever practical, except for revolved sections, sectional views should be projected perendicular to the cutting-plane and be placed in the normal position for third-angle projection. See Fig. 8-8-5.

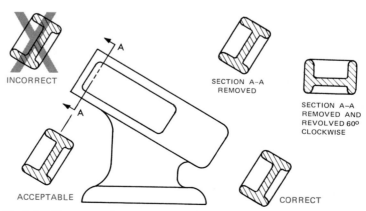

Fig. 8-8-5 Placement of sectional views.

When the preferred placement is not practical, the sectional view may be removed to some other convenient position on the drawing, but it must be clearly identified, usually by two capital letters and be labeled.

ASSIGNMENT

See Assignment 8 for Unit 8-8 on page 174.

UNIT 8-9
Spokes and Arms in Section

A comparison of the true projection of a wheel with spokes and a wheel with a web is made in Figs. 8-9-1A and B. This comparison shows that a preferred section for the wheel and spokes is desirable so that it will not appear to be a wheel with a solid web. In preferred sectioning, any part that is not solid or continuous around the hub is drawn without the section lining, even though the cutting-plane passes through it. When there is an odd number of spokes, as shown in Fig. 8-9-1C, the bottom spoke is aligned with the top spoke to show its true relationship to the wheel and to the hub. If the spoke was not revolved or aligned, it would appear distorted in the sectional view.

ASSIGNMENT

See Assignment 9 for Unit 8-9 on page 174.

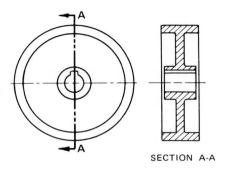

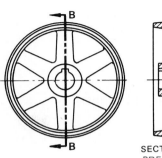
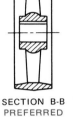
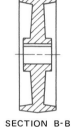

SECTION A-A

SECTION B-B
PREFERRED

SECTION B-B
TRUE PROJECTION

(A) FLAT PULLEY WITH WEB

(B) HANDWHEEL WITH EVEN NUMBER OF SPOKES

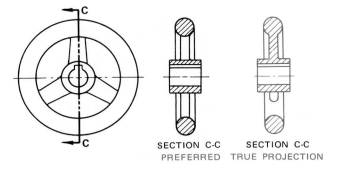

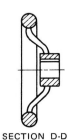
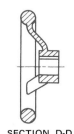

SECTION C-C
PREFERRED

SECTION C-C
TRUE PROJECTION

SECTION D-D
PREFERRED

SECTION D-D
TRUE PROJECTION

(C) HANDWHEEL WITH ODD NUMBER OF SPOKES

(D) HANDWHEEL WITH ODD NUMBER OF OFFSET SPOKES

Fig. 8-9-1 Preferred and true projection through spokes.

UNIT 8-10
Partial or Broken-Out Sections

Where a sectional view of only a portion of the object is needed, partial sections may be used. See Fig. 8-10-1. An irregular break line is used to show the extent of the section. With this type of section, a cutting-plane line is not required.

ASSIGNMENT

See Assignment 10 for Unit 8-10 on page 175.

UNIT 8-11
Phantom or Hiddent Sections

A phantom section is used to show the typical interior shapes of an object in one view when the part is not truly symmetrical in shape, as well as to show mating parts in an assembly drawing. See Fig. 8-11-1. It is a sectional view

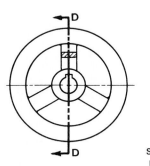

EXAMPLE 1

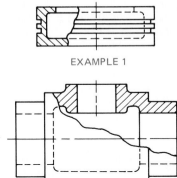

EXAMPLE 2

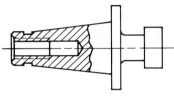

EXAMPLE 3

Fig. 8-10-1 Broken-out or partial sections.

superimposed on the regular view without the removal of the front portion of the object. The section lining used for phantom sections consists of thin, evenly spaced, broken lines.

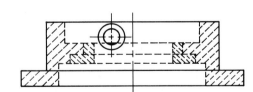

Fig. 8-11-1 Phantom or hidden sections.

ASSIGNMENT

See Assignment 11 for Unit 8-11 on page 176.

UNIT 8-12
Sectional Drawing Review

In Units 8-1 through 8-11 the different types of sectional views have been explained and drawing problems have been assigned with each type of section drawing.

In the drafting office, it is the drafter who must decide which views are required to fully explain the part to be made. In addition, the drafter must select the proper scale(s) which will show the features clearly.

This unit has been designed to review the sectional-view options open to the drafter.

ASSIGNMENT

See Assignment 12 for Unit 8-12 starting on page 177.

UNIT 8-13

Conventional Representation of Common Features

To simplify the representation of common features, a number of conventional drafting practices are used. Many conventions are deviations from true projection for the purpose of clarity; others are used to save drafting time. These conventions must be executed carefully, for clarity is even more important than speed.

Many drafting conventions such as those used on thread, gear, and spring drawings appear in various chapters throughout the text. Only the conventions not described in those chapters appear here.

Repetitive Details

Repetitive features, such as gear and spline teeth, are shown by drawing a partial view, showing two or three of these features, with a phantom line or lines to indicate the extent of the remaining features. See Figs. 8-13-1A and B. Alternatively, gears and splines may be shown with a solid thick line representing the basic outline of the part and a thin line representing the root of the teeth. This is essentially the same convention that is used for screw threads. The pitch line may be added by using the standard center line.

Knurls

Knurling is an operation which puts patterned indentations in the surface of a metal part to provide a good finger grip. See Figs. 8-13-1C and D. Commonly used types of knurls are straight, diagonal, spiral, convex, raised diamond, depressed diamond, and radial. The pitch refers to the distance between corresponding indentations, and it may be a straight pitch, a circular pitch, or a diametral pitch. For cylindrical surfaces, the latter is preferred. The pitch of the

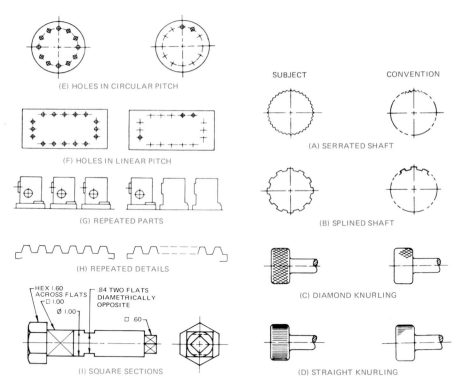

(E) HOLES IN CIRCULAR PITCH

(F) HOLES IN LINEAR PITCH

(G) REPEATED PARTS

(H) REPEATED DETAILS

HEX 1.60 ACROSS FLATS
□ 1.00
.84 TWO FLATS DIAMETRICALLY OPPOSITE
Ø 1.00
□ .60

(I) SQUARE SECTIONS

SUBJECT   CONVENTION

(A) SERRATED SHAFT

(B) SPLINED SHAFT

(C) DIAMOND KNURLING

(D) STRAIGHT KNURLING

Fig. 8-13-1 Conventional representation of common features.

teeth for coarse knurls (measured parallel to the axis of the work) is 14 teeth per inch (TPI) or about 2 mm; for medium knurls, 21 TPI or about 1.2 mm; and for fine knurls, 33 TPI or 0.8 mm. The medium-pitch knurl is the most commonly used.

As a time-saver, the knurl symbol is shown on only a part of the surface being knurled.

Holes

A series of similar holes is indicated by drawing one or two holes and showing only the center for the others. See Fig. 8-13-1E and F.

Repetitive Parts

Repetitive parts, or intricate features, are shown by drawing one in detail and the others in simple outline only. A covering note is added to the drawing. See Figs. 8-13-1G and H.

Repetitive parts or features can readily be added to a drawing when CAD is used. There are several commands that can perform this function, one being the ARRAY command. See Fig. 8-13-2.

Square Sections

Square sections on shafts and similar parts may be illustrated by thin, crossed, diagonal lines, as shown in Fig. 8-13-1J.

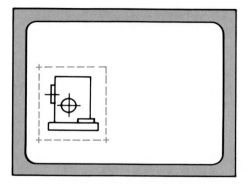
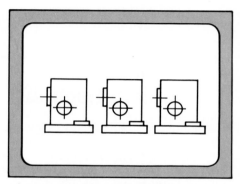

A COPY FUNCTION THAT ALLOWS OBJECT TO BE REPEAT COPIED BY KEYING IN A SPECIFIED X AND Y DISTANCE (OR ANGLE) AND NUMBER OF COPIES. THE PATTERN MAY BE EITHER RECTANGULAR OR CIRCULAR (POLAR).

Fig. 8-13-2 CAD array (minsert, multiple copies) command.

ASSIGNMENT

See Assignment 13 for Unit 8-13 starting on page 178.

UNIT 8-14
Conventional Breaks

Long, simple parts such as shafts, bars, tubes, and arms need not be drawn to their entire length. Conventional breaks located at a convenient position may be used and the true length indicated by a dimension. Often a part can be drawn to a larger scale to produce a clearer drawing if a conventional break is used. Two types of break lines are used for general use as shown in Fig. 8-14-1A. Thick freehand lines are used for short breaks. The thin line with freehand zig-zag is recommended for long breaks and may be used for solid details or for assemblies containing open spaces.

Special break lines, shown in Fig. 8-14-1B, are used when it is desirable to indicate the shape of the features.

ASSIGNMENT

See Assignment 14 for Unit 8-14 on page 179.

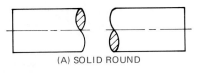

(A) SOLID ROUND

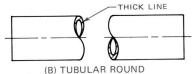

THICK LINE

(B) TUBULAR ROUND

(C) SOLID RECTANGULAR

(D) TUBULAR RECTANGULAR

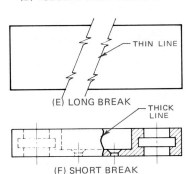

THIN LINE

(E) LONG BREAK

THICK LINE

(F) SHORT BREAK

Fig. 8-14-1 Conventional breaks.

UNIT 8-15
Materials of Construction

Symbols used to indicate materials in sectional views are shown in Fig. 8-1-6. Those shown for concrete, wood, and transparent materials are also suitable for outside views. Other symbols which may be used to indicate areas of different materials are shown in Fig. 8-15-1. It is not necessary to cover the entire area affected with such symbolic lining, as long as the extent of the area is shown on the drawing.

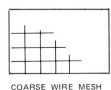

COARSE WIRE MESH

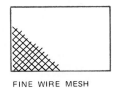

FINE WIRE MESH

MARBLE

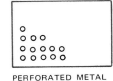

PERFORATED METAL

Fig. 8-15-1 Symbols to indicate materials of construction.

Transparent Materials

These should generally be treated in the same manner as opaque materials; i.e., details behind them are shown with hidden lines if such detail is necessary.

ASSIGNMENT

See Assignment 15 for Unit 8-15 on page 180.

UNIT 8-16
Cylindrical Intersections

The intersections of rectangular and circular contours, unless they are very large, are shown conventionally as in Figs. 8-16-1 and 8-16-2. The same convention may be used to show the intersection of two cylindrical contours, or the curve of intersection may be shown as a circular arc.

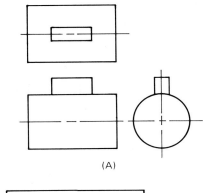

(A)

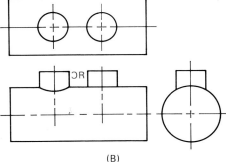

(B)

Fig. 8-16-1 Conventional representation of external intersection.

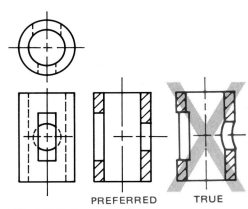

PREFERRED   TRUE

Fig. 8-16-2 Conventional representation of cylindrical intersection.

ASSIGNMENT

See Assignment 16 for Unit 8-16 starting on page 180.

UNIT 8-17

Foreshortened Projection

When the true projection of a feature would result in confusing foreshortening, it should be rotated until it is parallel to the line of the section or projection. See Fig. 8-17-1.

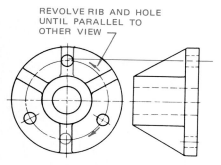

(A) ALIGNMENT OF RIB AND HOLES

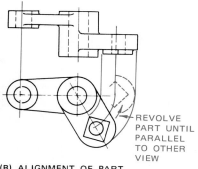

(B) ALIGNMENT OF PART

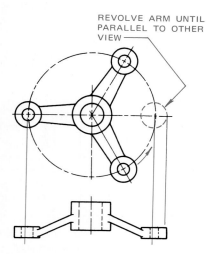

(C) ALIGNMENT OF ARM

Fig. 8-17-1 Alignment of parts and holes to show true relationship.

Holes Revolved to Show True Distance from Center

Drilled flanges in elevation or section should show the holes at their true distance from the center, rather than the true projection.

ASSIGNMENT

See Assignment 17 for Unit 8-17 on page 181.

UNIT 8-18

Intersections of Unfinished Surfaces

The intersections of unfinished surfaces that are rounded or filleted may be indicated conventionally by a line coinciding with the theoretical line of intersection. The need for this convention is demonstrated by the examples shown in Fig. 8-18-1, where the upper top views are shown in true projection. Note that in each example the true projection would be misleading. In the case of the large radius, such as shown in Fig. 8-18-1D, no line is drawn. Members such as ribs and arms that blend into other features terminate in curves called *runouts*. Small runouts are usually drawn freehand. Large runouts are drawn with an irregular curve, template, or compass. See Fig. 8-18-2.

ASSIGNMENT

See Assignment 18 for Unit 8-18 on page 182.

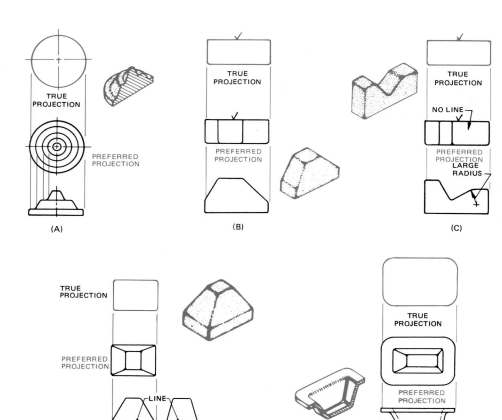

Fig. 8-18-1 Conventional representation of rounds and fillets.

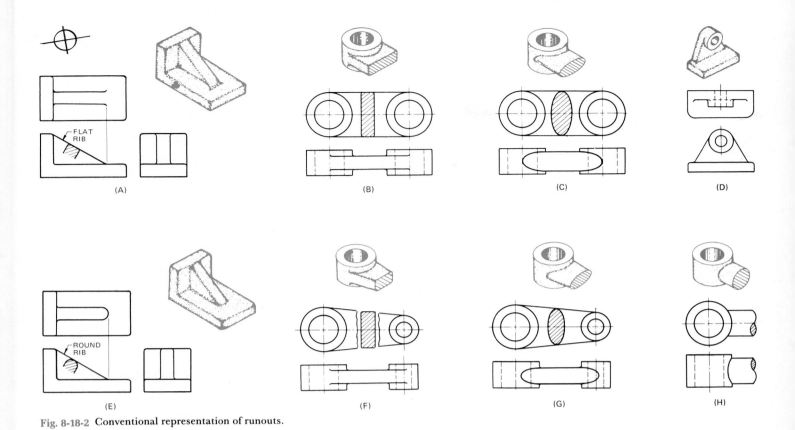

Fig. 8-18-2 Conventional representation of runouts.

ASSIGNMENTS FOR CHAPTER 8

Assignment for Unit 8-1, Sectional Views

1. Select one of the problems shown in Fig. 8-1-A or 8-1-B, and on a B (A3) size sheet make a two-view working drawing of the part, showing one of the views in full section. Use symbolic dimensioning wherever possible. Scale 1:1

Fig. 8-1-A Shaft base.

Ø .500
2 HOLES

R .76   R .50

2.25   .50

Ø .34
Ø .70 SF
6 HOLES

R 1.30

R 1.80

1.60   .44

ROUNDS AND FILLETS R .10   MATL — MALLEABLE IRON

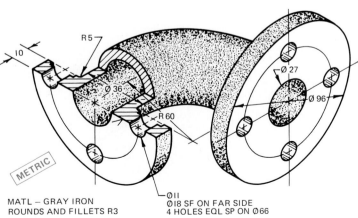

R5
10
Ø 36
R 60
Ø 27
Ø 96
METRIC

MATL – GRAY IRON
ROUNDS AND FILLETS R3

Ø11
Ø18 SF ON FAR SIDE
4 HOLES EQL SP ON Ø66

Fig. 8-1-B Flanged elbow.

Assignment for Unit 8-2, Two or More Sectional Views on One Drawing

2. Select one of the problems shown in Fig. 8-2-A or 8-2-B and on a B (A3) size sheet make a working drawing of the part showing the appropriate views in sections. Refer to the Appendix for taper sizes. Use symbolic dimensioning wherever possible. Scale 1:1.

Assignment for Unit 8-3, Half Sections

3. Select one of the problems shown in Fig. 8-3-A or 8-3-B. On a B (A3) size sheet, make a two-view working drawing of the part, showing the side view in half section. Redimension the keyseat as per Chap. 10. Scale 1:1.

Assignment for Unit 8-4, Threads in Section

4. Select one of the problems shown in Fig. 8-4-A or 8-4-B. On a B (A3) size sheet make a working drawing of the part. Determine the number of views and the best type of section which will clearly describe the part. Use symbolic dimensioning wherever possible and add undercut sizes. Scale 1:1

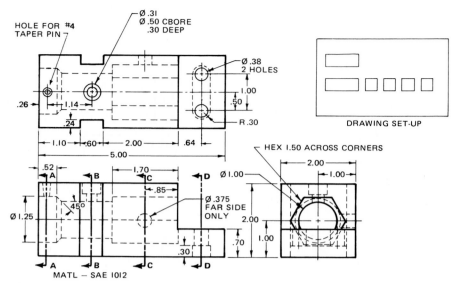

Fig. 8-2-A Casing.

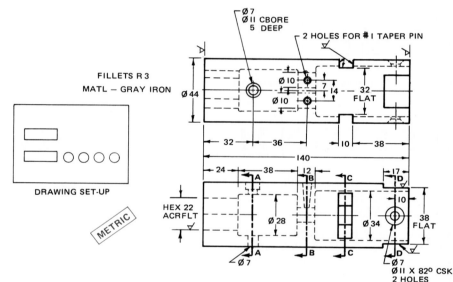

Fig. 8-2-B Housing.

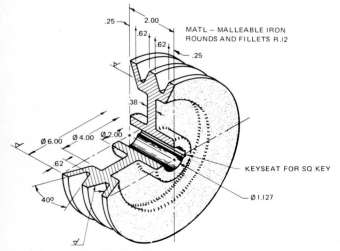

Fig. 8-3-A Double-V pulley.

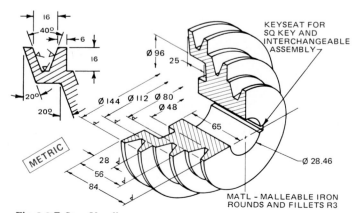

Fig. 8-3-B Step-V pulley.

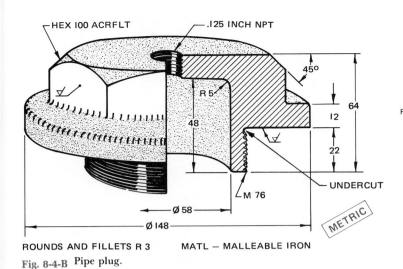

ROUNDS AND FILLETS R 3   MATL — MALLEABLE IRON

Fig. 8-4-B Pipe plug.

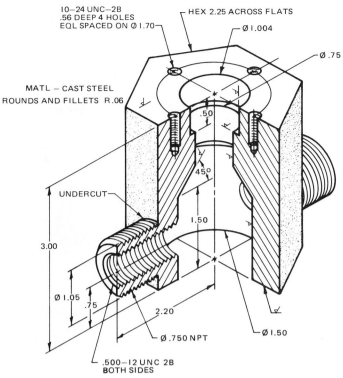

MATL — CAST STEEL
ROUNDS AND FILLETS R.06

Fig. 8-4-A Valve body.

Assignment for Unit 8-5, Assemblies in Section

5. On a B (A3) size sheet, make a one-view section assembly drawing of one of the problems shown in Fig. 8-5-A or 8-5-B. Include on your drawing an item list and identify the parts on the assembly. Assuming that this drawing will be used in a catalog, place on the drawing the dimensions and information required by the potential buyer. Scale 1:1.

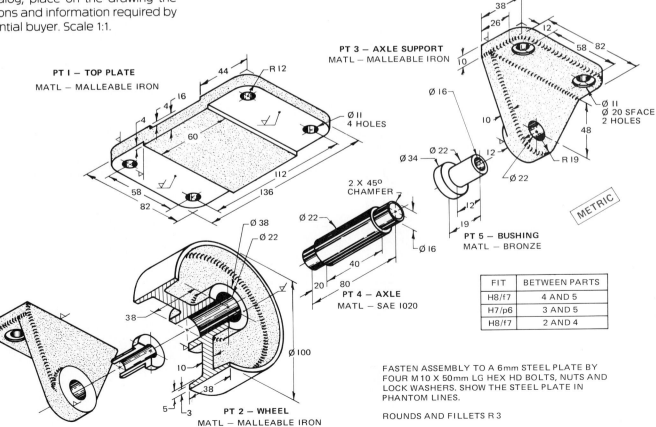

FIT	BETWEEN PARTS
H8/f7	4 AND 5
H7/p6	3 AND 5
H8/f7	2 AND 4

FASTEN ASSEMBLY TO A 6mm STEEL PLATE BY FOUR M10 X 50mm LG HEX HD BOLTS, NUTS AND LOCK WASHERS. SHOW THE STEEL PLATE IN PHANTOM LINES.

ROUNDS AND FILLETS R 3

Fig. 8-5-A Caster.

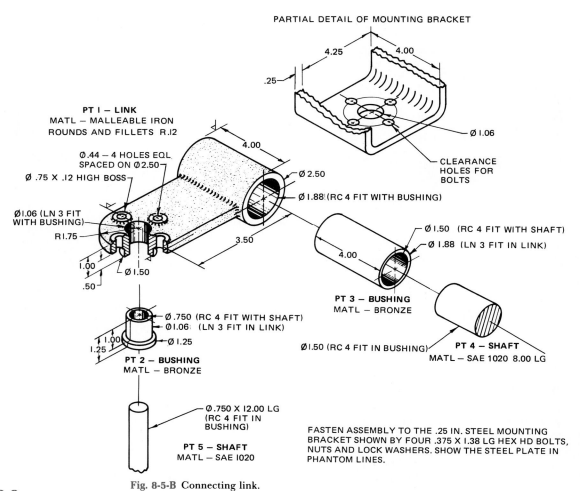

PARTIAL DETAIL OF MOUNTING BRACKET

4.25   4.00
.25

Ø1.06

CLEARANCE HOLES FOR BOLTS

PT I — LINK
MATL — MALLEABLE IRON
ROUNDS AND FILLETS R.12

Ø.44 — 4 HOLES EQL SPACED ON Ø2.50
Ø .75 X .12 HIGH BOSS

4.00

Ø 2.50
Ø 1.88 (RC 4 FIT WITH BUSHING)

Ø1.06 (LN 3 FIT WITH BUSHING)
R1.75

3.50

Ø 1.50   (RC 4 FIT WITH SHAFT)
Ø 1.88 (LN 3 FIT IN LINK)
4.00

PT 3 — BUSHING
MATL — BRONZE

1.00
.50
Ø1.50

Ø.750 (RC 4 FIT WITH SHAFT)
Ø1.06 (LN 3 FIT IN LINK)
Ø1.25

1.00
1.25

PT 2 — BUSHING
MATL — BRONZE

Ø1.50 (RC 4 FIT IN BUSHING)

PT 4 — SHAFT
MATL — SAE 1020 8.00 LG

Ø.750 X 12.00 LG (RC 4 FIT IN BUSHING)

PT 5 — SHAFT
MATL — SAE 1020

FASTEN ASSEMBLY TO THE .25 IN. STEEL MOUNTING BRACKET SHOWN BY FOUR .375 X 1.38 LG HEX HD BOLTS, NUTS AND LOCK WASHERS. SHOW THE STEEL PLATE IN PHANTOM LINES.

Fig. 8-5-B Connecting link.

Assignment for Unit 8-6, Offset Sections

6. Select one of the problems shown in Fig. 8-6-A or 8-6-B. On a B (A3) size sheet, make a working drawing of the part. Scale 1:1.

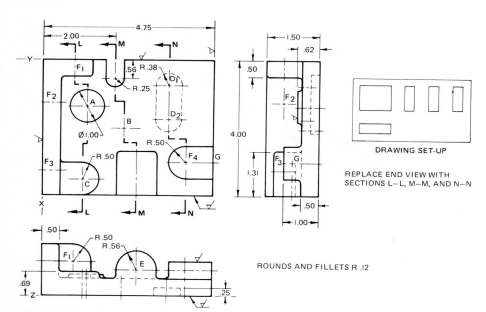

2.00   4.75
.56   R.38
R.25
F_1
F_2 D_1
A
D_2
Ø1.00
B
F_3 R.50
R.50 F_4 G
C

1.50
.62
.50
F_2
4.00
F_3 G
1.31
.50
1.00

DRAWING SET-UP

REPLACE END VIEW WITH SECTIONS L–L, M–M, AND N–N

.50
R.50
R.56
F_1 E
.69
.25

ROUNDS AND FILLETS R .12

DRAW TOP, FRONT AND 3 SECTION VIEWS
MATL — MALLEABLE IRON

HOLE	HOLE SIZE	LOCATION		
		X	Y	Z
A	.500-13UNC-2B	1.25	1.38	
B	Ø.281 CSK Ø.50 X 82°	2.25	1.94	
C	Ø.281 CBORE Ø.50 X .25 DEEP	1.12	3.50	
D_1	Ø.31	3.50	.75	
D_2	Ø.31	3.50	1.75	
E	.500-13UNC-2B X .75 DEEP	2.62		.75
F_1	Ø.50	.88		1.00
F_2	Ø.50		1.25	1.00
F_3	Ø.50		3.25	1.00
F_4	Ø.50	4.00	3.00	
G	Ø.12 THROUGH		3.00	.75

Fig. 8-6-A Base plate.

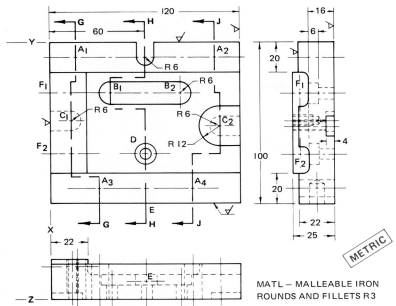

HOLE	HOLE SIZE	LOCATION		
		X	Y	Z
A1	Ø 12	16	9	
A2	Ø 12	100	9	
A3	Ø 12	30	92	
A4	Ø 12	87	92	
B1	Ø 8	38	32	
B2	Ø 8	80	32	
C1	M6 X 12 DEEP	12	50	
C2	M6	104	52	
D	Ø 6 CBORE Ø 12 X 6 DEEP	58	70	
E	Ø 10 X 12 DEEP	58		11
F1	Ø 6		32	20
F2	Ø 6		70	20

METRIC

MATL — MALLEABLE IRON
ROUNDS AND FILLETS R 3

Fig. 8-6-B Mounting plate.

REPLACE END VIEW WITH
SECTIONS G—G, H—H, AND J—J.

Assignment for Unit 8-7, Ribs, Holes, and Lugs in Section

7. Select one of the problems shown in Fig. 8-7-A or 8-7-B. On a B (A3) size sheet make a three-view working drawing of the part showing the front and side views in section. Both parts are to be used here and abroad, so a dual dimensioning system must be used. Scale 1:1.

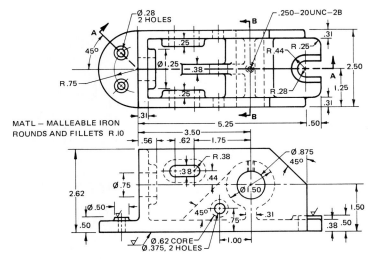

MATL — MALLEABLE IRON
ROUNDS AND FILLETS R.10

Fig. 8-7-A Bracket bearing.

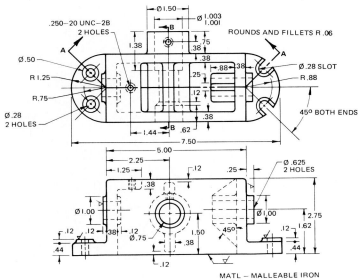

MATL — MALLEABLE IRON

Fig. 8-7-B Shaft support.

Assignment for Unit 8-8, Revolved and Removed Sections

8. Select one of the problems shown in Fig. 8-8-A or 8-8-B and on a B (A3) size sheet make a working drawing of the part. For clarity it is recommended that an enlarged removed view be used to show the detail of the inclined hole. Use symbolic dimensioning wherever possible. Scale 1:1.

Assignment for Unit 8-9, Spokes and Arms in Section

9. Select one of the problems shown in Fig. 8-9-A or 8-9-B and on a B (A3) size sheet make a two-view working drawing. Draw the side view in full section, and show a revolved section of the spoke in the front view. Scale 1:1.

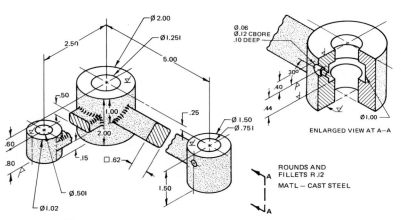

Fig. 8-8-A Idler support.

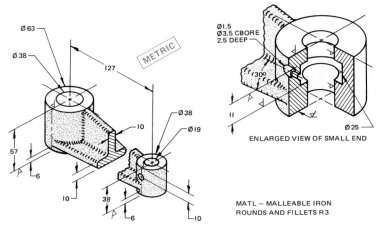

Fig. 8-8-B Shaft support.

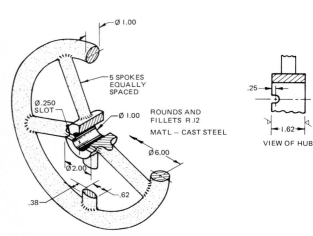

Fig. 8-9-A Handwheel.

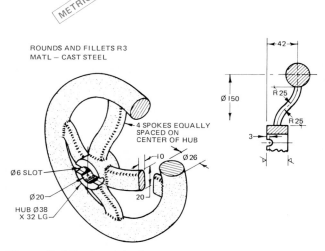

Fig. 8-9-B Offset handwheel.

174 BASIC DRAWING DESIGN

Assignment for Unit 8-10, Partial or Broken-Out Sections

10. Select one of the problems shown in Fig. 8-10-A or 8-10-B and make a two-view working drawing on a B (A3) size sheet. Use partial sections where clarity of drawing can be achieved. Scale 1:1.

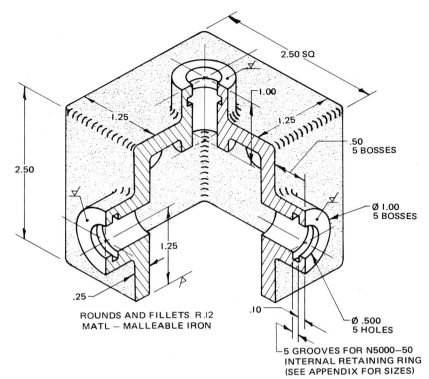

2.50 SQ

1.00

1.25

1.25

.50
5 BOSSES

Ø 1.00
5 BOSSES

2.50

1.25

.25

Ø .500
5 HOLES

.10

5 GROOVES FOR N5000—50
INTERNAL RETAINING RING
(SEE APPENDIX FOR SIZES)

ROUNDS AND FILLETS R .12
MATL — MALLEABLE IRON

Fig. 8-10-A Tumble box.

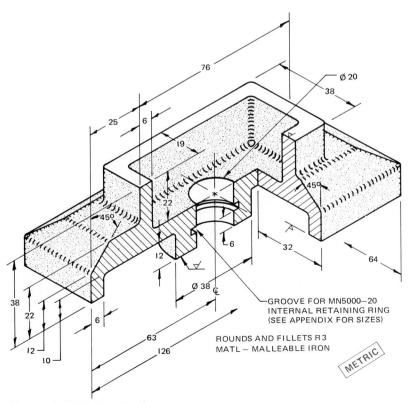

76

Ø 20

38

25

6

19

Ø 20

450

22

45⁰

6

12

32

64

38

22

12

10

6

Ø 38

GROOVE FOR MN5000—20
INTERNAL RETAINING RING
(SEE APPENDIX FOR SIZES)

63

126

ROUNDS AND FILLETS R3
MATL — MALLEABLE IRON

METRIC

Fig. 8-10-B Hold-down bracket.

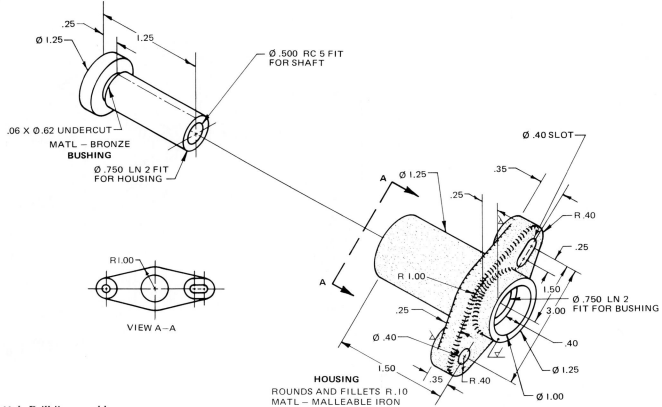

.25
Ø 1.25
1.25
Ø .500 RC 5 FIT
FOR SHAFT

.06 X Ø.62 UNDERCUT
MATL – BRONZE
BUSHING

Ø .750 LN 2 FIT
FOR HOUSING

Ø .40 SLOT
.35
R .40
.25
A
Ø 1.25
.25
R 1.00
1.50
Ø .750 LN 2
FIT FOR BUSHING
3.00
.40
A
Ø .40
.25
Ø 1.25
1.50
.35
R .40
Ø 1.00

HOUSING
ROUNDS AND FILLETS R .10
MATL – MALLEABLE IRON

R1.00

VIEW A–A

Fig. 8-11-A Drill-jig assembly.

Assignment for Unit 8-11, Phantom or Hidden Sections

11. On a B (A3) size sheet, make a two-view assembly drawing of one of the assemblies shown in Fig. 8-11-A or 8-11-B. The front view is to be drawn as a phantom section drawing. Show only the hole and shaft sizes for the fits shown. Scale 1:1.

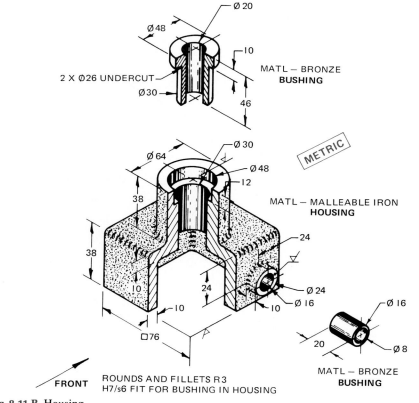

Ø 20
Ø 48
10
MATL – BRONZE
BUSHING
2 X Ø26 UNDERCUT
Ø30
46

Ø 30
Ø 64
Ø 48
12
METRIC
MATL – MALLEABLE IRON
HOUSING
38
24
38
10
24
Ø 24
10
Ø 16
Ø 16
10
20
Ø 8
☐76
MATL – BRONZE
BUSHING

FRONT
ROUNDS AND FILLETS R3
H7/s6 FIT FOR BUSHING IN HOUSING

Fig. 8-11-B Housing.

Assignment for Unit 8-12, Sectional Drawing Review

12. On a B (A3) size sheet, make a three-view working drawing of one of the parts shown in Fig. 8-12-A to 8-12-E. From the information on section drawings found in Units 8-1 to 8-11, select appropriate sectional views which will improve the clarity of the drawing. Scale 1:1.

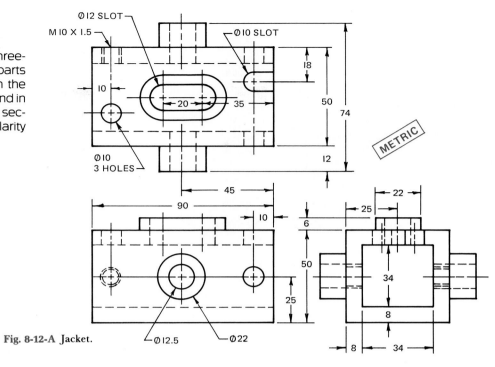

Fig. 8-12-A Jacket.

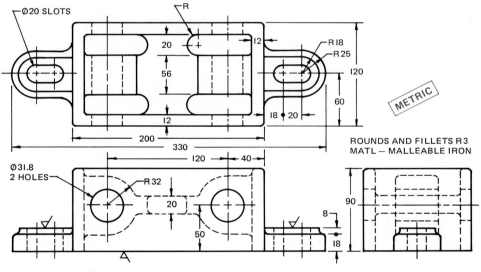

Fig. 8-12-B Guide block.

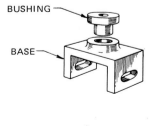

Fig. 8-12-C Connecting link.

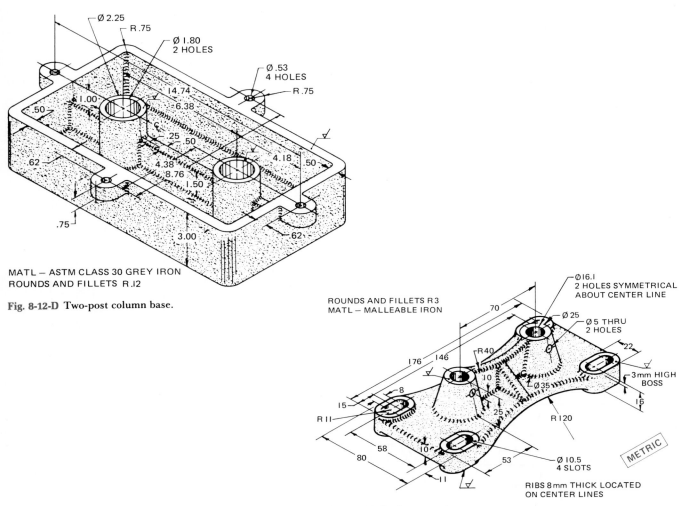

MATL — ASTM CLASS 30 GREY IRON
ROUNDS AND FILLETS R.12

Fig. 8-12-D Two-post column base.

ROUNDS AND FILLETS R 3
MATL — MALLEABLE IRON

Ø16.1
2 HOLES SYMMETRICAL
ABOUT CENTER LINE

Ø 25
Ø 5 THRU
2 HOLES

3mm HIGH
BOSS

METRIC

Ø 10.5
4 SLOTS

RIBS 8mm THICK LOCATED
ON CENTER LINES

Fig. 8-12-E Shaft support base.

Assignment for Unit 8-13, Conventional Representation of Common Features

13. On a B (A3) size sheet, make a working drawing of one of the parts shown in Fig. 8-13-A or 8-13-B. Wherever possible, simplify the drawing by using conventional representation of features and symbolic dimensioning wherever possible (including symmetry). Scale 10:1.

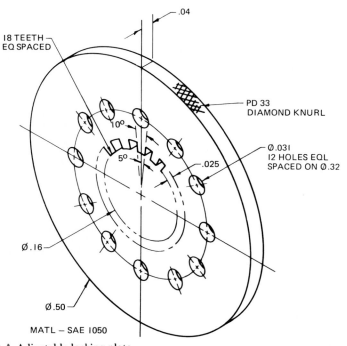

18 TEETH
EQ SPACED

.04

PD 33
DIAMOND KNURL

Ø.031
12 HOLES EQL
SPACED ON Ø.32

10°

5°

.025

Ø.16

Ø.50

MATL — SAE 1050

Fig. 8-13-A Adjustable locking plate.

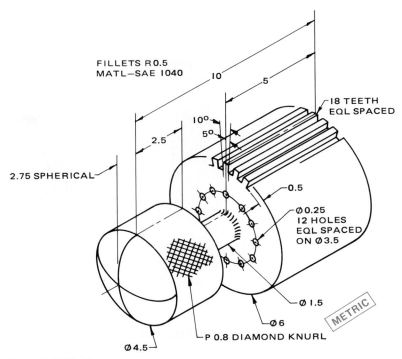

FILLETS R0.5
MATL—SAE 1040

10

5

2.5

10°
5°

18 TEETH
EQL SPACED

2.75 SPHERICAL

0.5

Ø0.25
12 HOLES
EQL SPACED
ON Ø3.5

Ø1.5

Ø6

METRIC

Ø4.5

P 0.8 DIAMOND KNURL

Fig. 8-13-B Clock stem.

Assignment for Unit 8-14, Conventional Breaks

14. On a B (A3) size sheet, make a working drawing of one of the parts shown in Fig. 8-14-A or 8-14-B. Use conventional breaks to shorten the length of the part. An enlarged view is also recommended where the detail cannot be clearly shown at full scale. Apply the symmetry symbol and use symbolic dimensioning wherever possible. Scale 1:1.

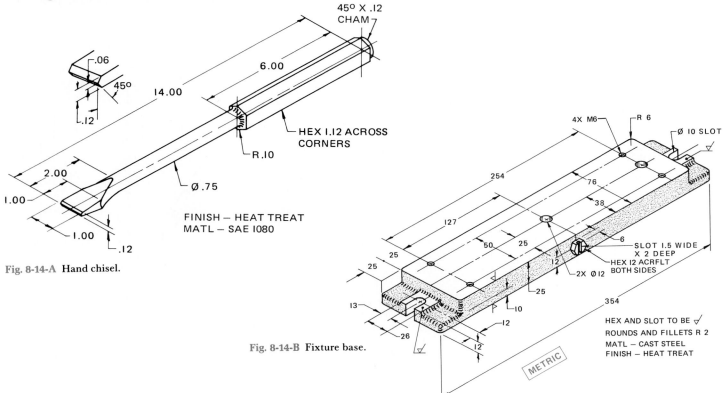

45° X .12
CHAM

.06

45°

.12

14.00

6.00

HEX 1.12 ACROSS
CORNERS

R.10

2.00

Ø.75

1.00

1.00

.12

FINISH — HEAT TREAT
MATL — SAE 1080

Fig. 8-14-A Hand chisel.

4X M6

R 6

Ø 10 SLOT

254

76

38

127

50

25

6

SLOT 1.5 WIDE
X 2 DEEP

25

25

12

HEX 12 ACRFLT
BOTH SIDES

2X Ø12

25

13

10

354

12

HEX AND SLOT TO BE ∀
ROUNDS AND FILLETS R 2
MATL — CAST STEEL
FINISH — HEAT TREAT

26

12

METRIC

Fig. 8-14-B Fixture base.

Assignment for Unit 8-15, Materials of Construction

15. On a B (A3) size sheet, make a detailed assembly drawing of one of the assemblies shown in Fig. 8-15-A or 8-15-B. Enlarged details are recommended for the steel mesh and joints. Use conventional breaks to shorten the length. Scale 1:5.

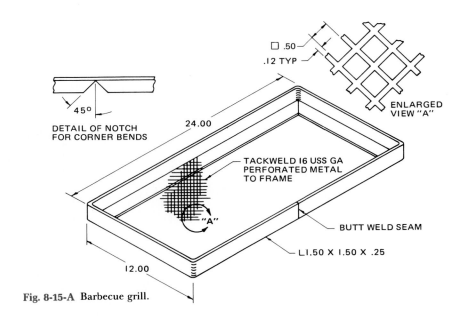

Fig. 8-15-A Barbecue grill.

Assignment for Unit 8-16, Cylindrical Intersections

16. On a B (A3) size sheet, make a working drawing of one of the parts shown in Fig. 8-16-A or 8-16-B. For 8-16-A a bushing is to be pressed (H7/s6) into the large hole and the stepped smaller hole is to have a running fit (H8/f7) with its respective shaft. These sizes are to be given as limit dimensions. All other finished surfaces are to have a 3.2 μm or equivalent finish. For Fig. 8-16-B an LN3 fit is required for the two large holes. Finished surfaces are to have a 63 μin. finish with an .06 in. material-removal allowance. Use your judgment in selecting the number of views required and deciding whether some form of sectional view would be desirable to improve the readability of the drawing. Scale 1:1.

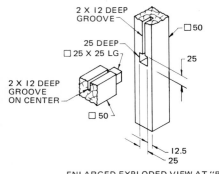

ENLARGED EXPLODED VIEW AT "B"

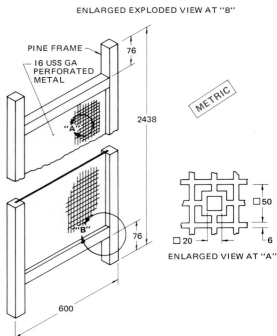

Fig. 8-15-B Room divider.

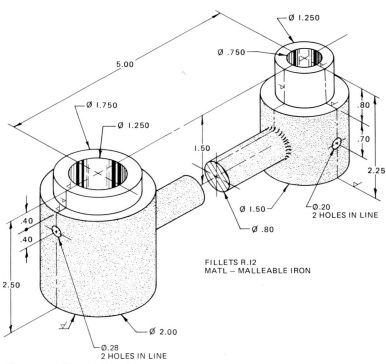

Fig. 8-16-A Steering knuckle.

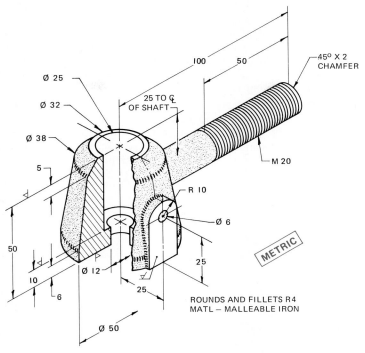

Fig. 8-16-B Shift lever.

Assignment for Unit 8-17, Foreshortened Projection

17. On a B (A3) size sheet, make a detail drawing of one of the parts shown in Fig. 8-17-A or 8-17-B. All surface finishes are to be 1.6 μm or 63 μin. Keyed holes will have H9/d9 or RC6 fits with shafts. Where required, rotate the features to show their true distances from the centers and edges. To show the true shape of the ribs or arms, a revolved section is recommended. Dimension the keyseat as per Chap. 10 and the Appendix. Scale 1:1.

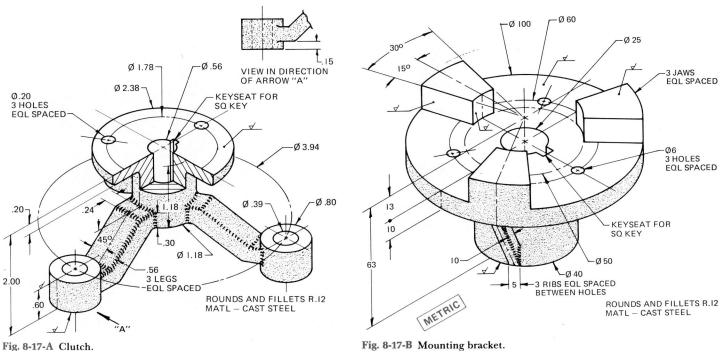

Fig. 8-17-A Clutch.

Fig. 8-17-B Mounting bracket.

Assignment for Unit 8-18, Intersections of Unfinished Surfaces

18. On a B (A3) size sheet, make a three-view detail drawing of one of the problems shown in Fig. 8-18-A or 8-18-B. Scale 1:1. Surface finish requirements are essential for all parts. Use symbolic dimensioning wherever possible. For Fig. 8-18-A the T slot surfaces should have a maximum roughness of 0.8 μm and a maximum waviness of 0.05 mm for a 25-mm length. The back surface should have a maximum roughness of 3.2 μm with no restrictions on waviness. For Fig. 8-18-B the back surface and notch should have an equivalent control as the T slot in Fig. 8-18-A. The faces on the boss should have a maximum roughness of 125 μin. with no restrictions on waviness.

19. On a C (A2) size sheet, make a detail drawing of the pipe vise base shown in Fig. 8-18-C. To clearly show the features, a section and bottom view should also be drawn in addition to the top, front, and side views. The recommended drawing layout is shown. Add hidden lines only where necessary to improve the clarity of the drawing.

The slots are to have a surface finish of 3.2 μm and a machining allowance of 2 mm. The base is to have the same surface finish but with a 3-mm machining allowance.

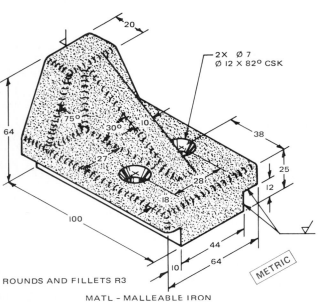

ROUNDS AND FILLETS R3

MATL – MALLEABLE IRON

Fig. 8-18-A Cut-off stop.

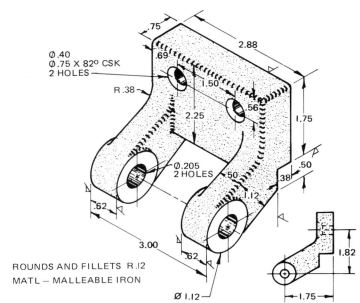

ROUNDS AND FILLETS R.12
MATL – MALLEABLE IRON

Fig. 8-18-B Sparker bracket.

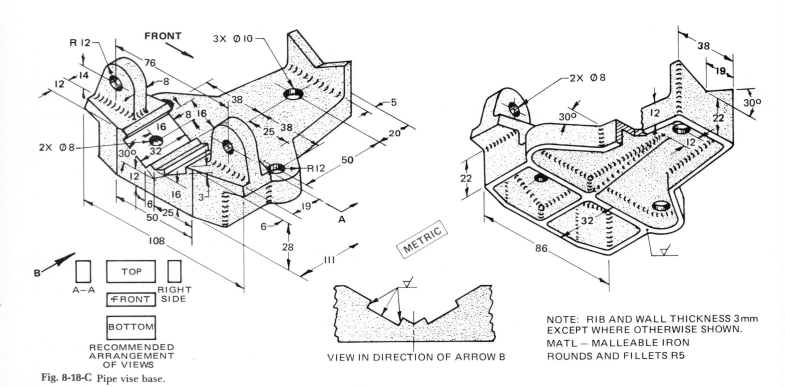

RECOMMENDED ARRANGEMENT OF VIEWS

Fig. 8-18-C Pipe vise base.

VIEW IN DIRECTION OF ARROW B

NOTE: RIB AND WALL THICKNESS 3mm EXCEPT WHERE OTHERWISE SHOWN.
MATL – MALLEABLE IRON
ROUNDS AND FILLETS R5

Fasteners, Materials, and Forming Processes

Threaded Fasteners

UNIT 9-1
Simplified Thread Representation

Fastening devices are important in the construction of manufactured products, in the machines and devices used in manufacturing processes, and in the construction of all types of buildings. Fastening devices are used in the smallest watch to the largest ocean liner. See Fig. 9-1-1.

There are two basic kinds of fasteners: *permanent* and *removable*. Rivets and welds are permanent fasteners. Bolts, screws, studs, nuts, pins, rings, and keys are removable fasteners. As industry progressed, fastening devices became standardized, and they developed definite characteristics and names. A thorough knowledge of the design and graphic representation of the more common fasteners is an essential part of drafting.

The cost of fastening, once considered only incidental, is fast becoming recognized as a critical factor in total product cost. "It's the in-place cost that counts, not the fastener cost" is an old saying of fastener design. The art of holding down fastener cost is not learned simply by scanning a parts catalog. More subtly, it entails weighing such factors as standardization, automatic assembly, tailored fasteners, and joint preparation.

Standardization A favorite cost-reducing method, standardization, not

Fig. 9-1-1 Fasteners. (Industrial Fasteners Institute)

only cuts the cost of parts but also reduces paperwork and simplifies inventory and quality control. By standardizing on type and size, it may be possible to reach the level of usage required to make power tools or automatic assembly feasible.

SCREW THREADS

A *screw thread* is a ridge of uniform section in the form of a helix on the external or internal surface of a cylinder. See Fig. 9-1-2. The helix of a square thread is shown in Fig. 9-1-3.

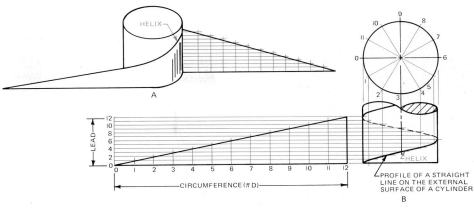

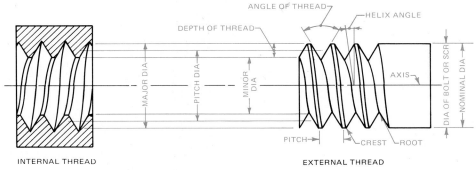

Fig. 9-1-2 The helix.

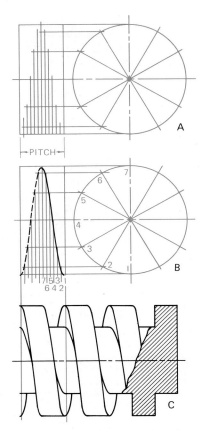

Fig. 9-1-3 The helix of a square thread.

The *pitch* of a thread P is the distance from a point on the thread form to the corresponding point on the next form, measured parallel to the axis. See Fig. 9-1-4. The *lead* L is the distance the threaded part would move parallel to the axis during one complete rotation in relation to a fixed mating part (the distance a screw would enter a threaded hole in one turn).

THREAD FORMS

Figure 9-1-5 shows some of the more common thread forms in use today. The

Fig. 9-1-4 Screw thread terms.

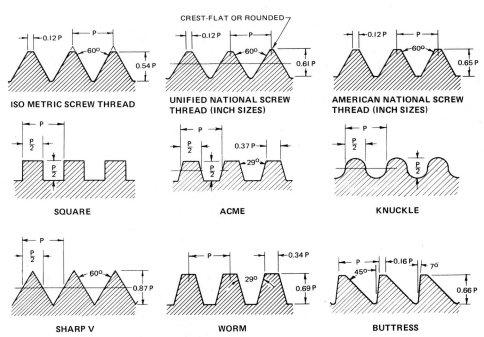

Fig. 9-1-5 Common thread forms and proportions.

ISO metric thread will eventually replace all the V-shaped metric and inch threads. As for the other thread forms shown, the proportions will be the same for both metric- and inch-size threads.

The knuckle thread is usually rolled or cast. A familiar example of this form is seen on electric light bulbs and sockets. See Fig. 9-1-6. The square and acme forms are designed to transmit motion or

Fig. 9-1-6 Application of a knuckle thread.

power, as on the lead screw of a lathe. The buttress thread takes pressure in only one direction—against the surface perpendicular to the axis.

THREAD REPRESENTATION

True representation of a screw thread is seldom used on working drawings. Symbolic representation of threads is now standard practice. There are three types of conventions in general use for screw thread representation. These are known as *simplified*, *detailed*, and *schematic*. See Fig. 9-1-7. Simplified representation should be used whenever it will clearly portray the requirements. Detailed representation is used to show the detail of a screw thread, especially for dimensioning in enlarged views, layouts, and assemblies. The schematic representation is nearly as effective as the detailed representation and is much easier to draw when manual drafting is used.

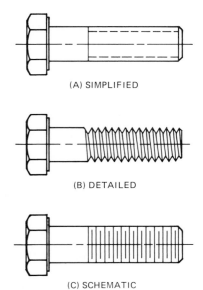

(A) SIMPLIFIED

(B) DETAILED

(C) SCHEMATIC

Fig. 9-1-7 Symbolic thread representation.

This representation has given way to the simplified representation, and as such, has been discarded as a thread symbol by most countries.

RIGHT- AND LEFT-HAND THREADS

Unless designated otherwise, threads are assumed to be right-hand. A bolt being threaded into a tapped hole would be turned in a right-hand (clockwise) direction. See Fig. 9-1-8. For some special applications, such as turnbuckles, left-hand threads are required. When such a thread is necessary, the letters LH are added after the thread designation.

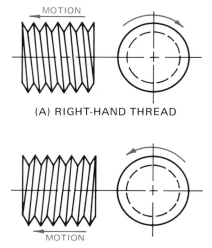

(A) RIGHT-HAND THREAD

(B) LEFT-HAND THREAD

Fig. 9-1-8 Right- and left-hand threads.

SINGLE AND MULTIPLE THREADS

Most screws have single threads. It is understood that unless the thread is designated otherwise, it is a single thread. The single thread has a single ridge in the form of a helix. See Fig. 9-1-9. The lead of a thread is the distance traveled parallel to the axis in one rotation of a part in relation to a fixed mating part (the distance a nut would travel along the axis of a bolt with one rotation of the nut). In single threads, the lead is equal to the pitch. A double thread has two ridges, started 180° apart, in the form of helices; and the lead is twice the pitch. A triple thread has three ridges, started 120° apart, in the form of helices; and the lead is 3 times the pitch. Multiple threads are used where fast movement is desired with a minimum number of rotations, such as on threaded mechanisms for opening and closing windows.

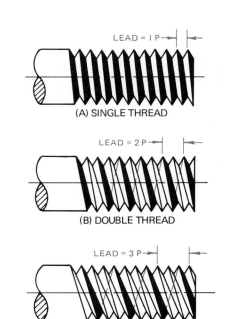

(A) SINGLE THREAD

(B) DOUBLE THREAD

(C) TRIPLE THREAD

Fig. 9-1-9 Single and multiple threads.

SIMPLIFIED THREAD REPRESENTATION

In this system the thread crests, except in hidden views, are represented by a thick outline and the thread roots by a thin broken line. See Fig. 9-1-10. The end of the full-form thread is indicated

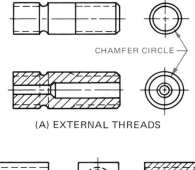

CHAMFER CIRCLE

(A) EXTERNAL THREADS

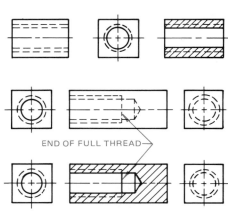

END OF FULL THREAD

(B) INTERNAL THREADS

Fig. 9-1-10 Simplified thread convention.

by a thick line across the part, and imperfect or runout threads are shown beyond this line by running the root line at an angle to meet the crest line. If the length of runout threads is unimportant, this portion of the convention may be omitted.

THREADED ASSEMBLIES

For general use, the simplified representation of threaded parts is recommended for assemblies. See Fig. 9-1-11. In sectional views, the externally threaded part is always shown covering the internally threaded part.

INCH THREADS

In the United States a great number of threaded assemblies are still designed using inch-sized threads. In this system the *pitch* is equal to

$$\frac{1}{\text{Number of threads per inch}}$$

The number of threads per inch is set for different diameters in what is called a thread *series*. For the Unified National system, there is the coarse-threaded series (UNC) and the fine-thread series (UNF). See Table 11 of Appendix 2.

In addition, there is an extra-fine-thread series (UNEF) for use where a small pitch is desirable, such as on thin-walled tubing. For special work and for diameters larger than those specified in the coarse and fine series, the Unified Thread system has three series that provide for the same number of threads per inch regardless of the diameter. These are the 8-thread series, the 12-thread series, and the 16-thread series. These are called constant-pitch threads.

Thread Class

Three classes of external thread (classes 1A, 2A, and 3A) and three classes of internal thread (classes 1B, 2B, and 3B) are provided. These classes differ in the amount of allowances and tolerances provided in each class.

The general characteristics and uses of the various classes are as follows.

Classes 1A and 1B These classes produce the loosest fit, that is, the greatest amount of play in assembly. They are useful for work where ease of assembly and disassembly is essential, such as for stove bolts and other rough bolts and nuts.

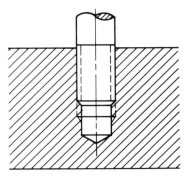

Fig. 9-1-11 Simplified representation of threads in assembly drawings.

Classes 2A and 2B These classes are designed for the ordinary good grade of commercial products, such as machine screws and fasteners, and for most interchangeable parts.

Classes 3A and 3B These classes are intended for exceptionally high-grade commercial products, where a particularly close or snug fit is essential and the high cost of precision tools and machines is warranted.

Thread Designation

Thread designation for inch threads, whether external or internal, is expressed in this order: diameter (nominal or major diameter in decimal form with a minimum of three, or maximum of four decimal places), number of threads per inch, thread form and series, and class of fit (number and letter). See Fig. 9-1-12.

METRIC THREADS

Metric threads are grouped into diameter-pitch combinations distinguished from one another by the pitch applied to specific diameters. See Fig. 9-1-13.

The *pitch* for metric threads is the distance between corresponding points on adjacent teeth. In addition to a coarse- and fine-pitch series, a series of constant pitches is available. See Table 12 of Appendix 2.

Coarse-Thread Series This series is intended for use in general engineering work and commercial applications.

Fine-Thread Series The fine-thread series is for general use where a finer thread than the coarse-thread series is desirable. In comparison with a coarse-thread screw, the fine-thread screw is stronger in both tensile and torsional strength and is less likely to loosen under vibration.

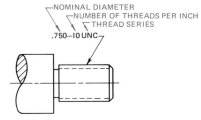

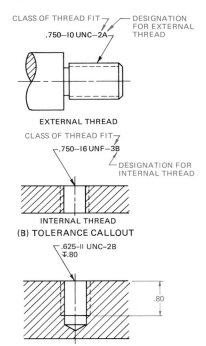

(A) BASIC THREAD CALLOUT

(B) TOLERANCE CALLOUT

(C) BLIND HOLE

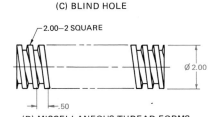

(D) MISCELLANEOUS THREAD FORMS

Fig. 9-1-12 Thread specifications for inch-size threads.

Thread Grades and Classes The *fit* of a screw thread is the amount of clear-

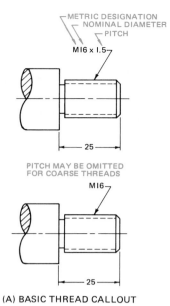

METRIC DESIGNATION
NOMINAL DIAMETER
PITCH
M16 x 1.5

25

PITCH MAY BE OMITTED
FOR COARSE THREADS

M16

25

(A) BASIC THREAD CALLOUT

PITCH DIAMETER TOLERANCE SYMBOL { TOLERANCE POSITION
TOLERANCE GRADE

CREST DIAMETER TOLERANCE SYMBOL { TOLERANCE POSITION
TOLERANCE GRADE

M16 x 1.5 — 5g6g

25

(B) TOLERANCE CALLOUT

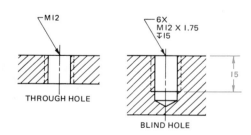

M12

THROUGH HOLE

6X
M12 X 1.75
↧15

15

BLIND HOLE

(C) INTERNAL THREAD CALLOUT

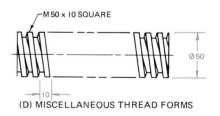

M 50 x 10 SQUARE

Ø50

10

(D) MISCELLANEOUS THREAD FORMS

Fig. 9-1-13 Thread specifications for metric threads.

ance between the internal and external threads when they are assembled.

For each of the two main thread elements—pitch diameter and crest diameter—a number of tolerance grades have been established. The number of the tolerance grades reflects the size of the tolerance. For example, grade 4 tolerances are smaller then grade 6 tolerances, and grade 8 tolerances are larger than grade 6 tolerances.

In each case, grade 6 tolerances should be used for medium-quality length-of-engagement applications. The tolerance grades below grade 6 are intended for applications involving fine quality and/or short lengths of engagement. Tolernce grades above grade 6 are intended for coarse quality and/or long lengths of engagement.

In addition to the tolerance grade, a positional tolerance is required. This tolerance defines the maximum-material limits of the pitch and crest diameters of the external and internal threads and indicates their relationship to the basic profile.

In conformance with current coating (or plating) thickness requirements and the demand for ease of assembly, a series of tolerance positions reflecting the application of varying amounts of allowance has been established as follows:

For external threads:

- Tolerance position e (large allowance)
- Tolerance position g (small allowance)
- Tolerance position h (no allowance)

For internal threads:

- Tolerance position G (small allowance)
- Tolerance position H (no allowance)

Thread Designation

ISO metric screw threads are defined by the nominal size (basic major diameter) and pitch, both expressed in millimeters. An "M" specifying as ISO metric screw thread precedes the nominal size, and an "×" separates the nominal size from the pitch. See Fig. 9-1-13. For the coarse-thread series only, the pitch is not shown unless the dimension for the length of the thread is required. In specifying the length of thread, a × is used to separate the length of thread from the rest of the designations. For external threads, the length or depth of

thread may be given as a dimension on the drawing.

For example, a 10 mm diameter, 1.25 pitch, fine-thread series is expressed as M10 × 1.25. A 10 mm diameter, 1.5 pitch, coarse-thread series is expressed as M10; the pitch is not shown unless the length of thread is required. If the latter thread were 25 mm long and this information was required on the drawing, then the thread callout would be M10 × 1.5 × 25.

A complete designation for an ISO metric screw thread comprises, in addition to the basic designation, an identification for the tolerance class. The tolerance class designation is separated from the basic designation by a dash and includes the symbol for the pitch diameter tolerance followed immediately by the symbol for crest diameter tolerance. Each of these symbols consists of a numeral indicating the grade tolerance followed by a letter representing the tolerance position (a capital letter for internal threads and a lowercase letter for external threads). Where the pitch and crest diameter symbols are identical, the symbol need be given only once.

For external threads, the length of thread may be given as a dimension on the drawing. The length given is to be the minimum length of full thread. For threaded holes that go all the way through the part, the term THRU is sometimes added to the note. If no depth is given, the hole is assumed to go all the way through. For threaded holes that do not go all the way through, the depth (in conjunction with the depth symbol or word) is given in the note, for example, M12 × 1.75 × 20 DEEP. The depth given is the minimum depth of full thread.

Neither the chamfer shown at the beginning of a thread, nor the undercut at the end of a thread where a small diameter meets a larger diameter is required to be dimensioned, as shown in Fig. 9-1-14. A comparison of customary

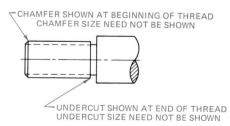

CHAMFER SHOWN AT BEGINNING OF THREAD
CHAMFER SIZE NEED NOT BE SHOWN

UNDERCUT SHOWN AT END OF THREAD
UNDERCUT SIZE NEED NOT BE SHOWN

Fig. 9-1-14 Omission of thread information on detail drawings.

and metric thread sizes is shown in Fig. 9-1-15.

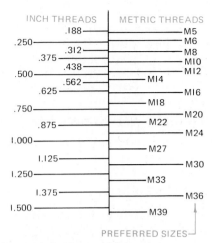

Fig. 9-1-15 Comparison of thread sizes.

PIPE THREADS

The pipe universally used is the inch-sized pipe. When pipe is ordered, the outside diameter and wall thickness (in inches or millimeters) are given. In calling for the size of thread, the note used is similar to that for screw threads. When calling for a pipe thread on a metric drawing, the abbreviation "IN" follows the pipe size. See Fig. 9-1-16.

EXAMPLE 1

$$4 \times 8NPT$$

EXAMPLE 2

$$4 \times 8NPS$$

where 4 = nominal diameter of pipe, in inches
8 = number of threads per inch
N = American Standard
P = pipe
S = straight pipe thread
T = taper pipe thread

When calling for a pipe thread on a metric drawing, the abbreviation "IN" follows the pipe size.

Reference and Source Material

1. ANSI Y14.6 *Screw Thread Representation.*

ASSIGNMENTS

See Assignments 1 through 4 for Unit 9-1 starting on page 201.

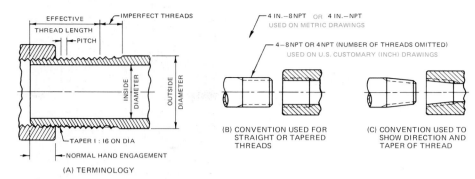

Fig. 9-1-16 Pipe thread terminology and conventions.

UNIT 9-2
Detailed and Schematic Thread Representation

DETAILED REPRESENTATION

Detailed representation of threads is a close approximation of the actual appearance of a screw thread. The form of the thread is simplified by showing the helices as straight lines and the truncated crests and roots as sharp Vs. It is used when a more realistic thread representation is required. See Fig. 9-2-1.

(A) EXTERIOR THREADS

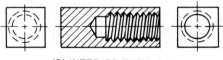

(B) INTERIOR THREADS

Fig. 9-2-1 Detailed representation of threads.

Detailed Representation of V Threads The detailed representation for V-shaped threads uses the sharp-V profile.

The order of drawing the screw threads is shown in Fig. 9-2-2. The pitch is seldom drawn to scale; generally it is approximated. Lay off the pitch P and the half-pitch $P/2$, as shown in Fig. 9-2-2A. Add the crest lines. At B add the V profile for one thread, top and bottom, locating the root diameter. Add construction lines for the root diameter. At C add one side of the remaining Vs (thread profile), then add the other side of the Vs, completing the thread profile. At D, add the root lines, which complete the detailed representation of the threads.

Detailed Representation of Square Threads The depth of the square thread is one-half the pitch. In Fig. 9-2-3A, lay off spaces equal to $P/2$ along the diameter and add construction lines to locate the depth (root dia.) of thread. At B add the crest lines. At C add the

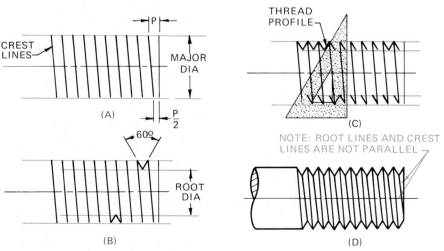

Fig. 9-2-2 Steps in drawing detailed representation of screw threads.

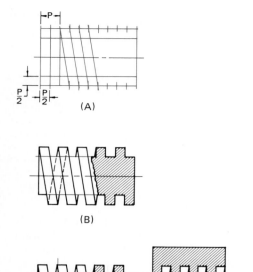

(A)

(B)

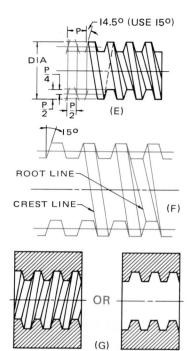

14.5° (USE 15°)

(E)

15°

ROOT LINE

CREST LINE

(F)

(C)   (D)

SQUARE THREADS

(G)

OR

ACME THREADS

Fig. 9-2-3 Steps in drawing detailed representation of square and acme threads.

SCHEMATIC THREAD REPRESENTATION

The staggered lines, symbolic of the thread roots and crests, normally are perpendicular to the axis of the thread. The spacing between the root and crest lines and the length of the root lines are drawn to any convenient size. See Fig. 9-2-5. At one time the root line was shown as a thick line.

Reference and Source Material

1. ANSI Y14.6 *Screw Threads Representation.*

ASSIGNMENTS

See Assignments 5 and 6 for Unit 9-2 starting on page 202.

root lines, as shown. At D the internal square thread is shown in section. Note the reverse direction of the crest and root lines.

Detailed Representation of Acme Threads The depth of the acme thread is one-half the pitch. See Figs. 9-2-3E through G. The stages in drawing acme threads are shown at E. For drawing purposes locate the pitch diameter midway between the outside diameter and the root diameter. The pitch diameter locates the pitch line. On the pitch line, lay off half-pitch spaces and the root lines to complete the view. The construction shown at F is enlarged.

Sectional views of an internal acme thread are shown at G. Showing the root and crest lines beyond the cutting plane on sectional views is optional.

THREADED ASSEMBLIES

It is often desirable to show threaded assembly drawings in detailed form, e.g., in presentation or catalog drawings. Hidden lines are normally omitted on these drawings, as they do nothing to add to the clarity of the drawing. See Fig. 9-2-4. One type of thread representation is generally used within any one drawing. When required, however, all three methods may be used.

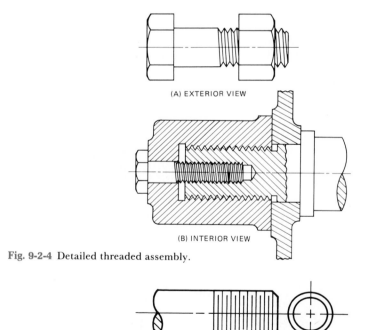

(A) EXTERIOR VIEW

(B) INTERIOR VIEW

Fig. 9-2-4 Detailed threaded assembly.

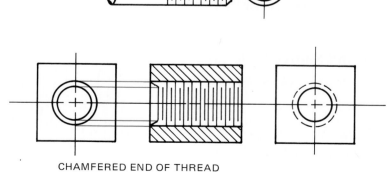

CHAMFERED END OF THREAD

Fig. 9-2-5 Schematic representation of threads.

UNIT 9-3
Common Threaded Fasteners

FASTENER SELECTION

Fastener manufacturers agree that product selection must begin at the design stage. For it is here, when a product is still a figment of someone's imagination, that the best interests of the designer, production manager, and purchasing agent can be served. Designers, naturally, want optimum performance; production people are interested in the ease and economics of assembly; purchasing agents are keen to minimize intial costs and stocking costs.

The answer, pure and simple, is to determine the objectives of the particular fastening job and then consult fastener suppliers. These technical experts can often shed light on the situation and then recommend the right item at the best in-place cost.

Machine screws are among the most common fasteners in industry. See Figs. 9-3-1 and 9-3-2. They are the easiest to install and remove. They are also among the least understood. To obtain maximum machine-screw efficiency, thorough knowledge of the properties of both the screw and the materials to be fastened together is required.

For a given application, a designer should know the load which the screw must withstand, whether the load is one of tension or shear, and whether the assembly will be subject to impact shock or vibration. Once these factors have been determined, then the size, strength, head shape, and thread type can be selected.

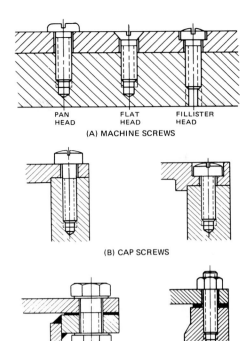

Fig. 9-3-2 Fastener applications.

FASTENER DEFINITIONS

Machine Screws *Machine screws* have either fine or coarse threads and are available in a variety of heads. They may be used in tapped holes as shown in Fig. 9-3-2A, or with nuts.

Cap Screws A *cap screw* is a threaded fastener which joins two or more parts by passing through a clearance hole in one part and screwing into a tapped hole in the other, as in Fig. 9-3-2B. A cap screw is tightened or released by torquing the head. Cap screws range in size

from .25 in. (6 mm) in diameter and are available in five basic types of head.

Captive Screws *Captive screws* are those that remain attached to the panel or parent material even when disengaged from the mating part. They are used to meet military requirements, to prevent screws from being lost, to speed assembly and disassembly operations, and to prevent damage from loose screws falling into moving parts or electrical circuits.

Tapping Screws *Tapping screws* cut or form a mating thread when driven into preformed holes.

Bolts A *bolt* is a threaded fastener which passes through clearance holes in assembled parts and threads into a nut, Fig. 9-3-2C. Bolts and nuts are available in a variety of shapes and sizes. The square and hexagon head are the two most popular designs.

Studs *Studs* are shafts threaded at both ends, and they are used in assemblies. One end of the stud is threaded into one of the parts being assembled; and the other assembly parts, such as washers and covers, are guided over the studs through clearance holes and are held together by means of a nut which is threaded over the exposed end of the stud. See Fig. 9-3-2D.

Explanatory Data

A bolt is designed for assembly with a nut. A screw has features in its design which make it capable of being used in a tapped or other preformed hole in the work. Because of basic design, it is possible to use certain types of screws in combination with a nut.

THE CHANGE TO METRIC FASTENERS

In the United States, the Industrial Fasteners Institute (IFI) has undertaken a major compilation of standards in its *Metric Fastener Standards* book.

FASTENER CONFIGURATION
Head Styles

Which of the various head configurations to specify depends on the type of driving equipment used (screwdriver, socket wrench, etc.), the type of joint load, and the external appearance desired. The head styles shown in Fig.

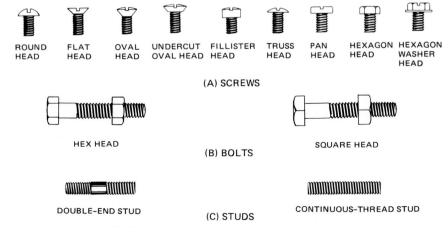

Fig. 9-3-1 Common threaded fasteners.

9-3-3 can be used for both bolts and screws but are most commonly identified with the fastener category called machine screw or cap screws.

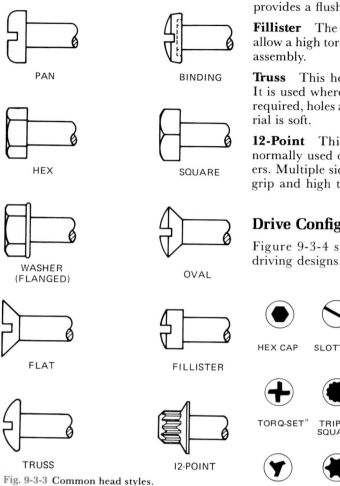

Fig. 9-3-3 Common head styles.

Hex and Square   The hex head is the most commonly used head style. The hex head design offers greater strength, ease of torque input, and area than the square head.

Pan   This head combines the qualities of the truss, binding, and round head types.

Binding   This type of head is commonly used in electrical connections because its undercut prevents fraying of stranded wire.

Washer (flanged)   This configuration eliminates the need for a separate assembly step when a washer is required, increases the bearing areas of the head, and protects the material finish during assembly.

Oval   Characteristics of this head type are similar to those of the flat head but it

is sometimes preferred because of its neat appearance.

Flat   Available with various head angles, this fastener centers well and provides a flush surface.

Fillister   The deep slot and small head allow a high torque to be applied during assembly.

Truss   This head covers a large area. It is used where extra holding power is required, holes are oversize, or the material is soft.

12-Point   This twelve-sided head is normally used on aircraft-grade fasteners. Multiple sides allow for a very sure grip and high torque during assembly.

Drive Configurations

Figure 9-3-4 shows sixteen different driving designs.

Fig. 9-3-4 Drive configurations.

Shoulders and Necks

The shoulder of a fastener is the enlarged portion of the body of a threaded fastener or the shank of an unthreaded fastener. See Fig. 9-3-5.

Point Styles

The point of a fastener is the configuration of the end of the shank of a headed or headless fastener. Standard point styles are shown in Fig. 9-3-6.

OVAL SHOULDER   ROUND SHOULDER

FIN NECK   SQUARE (CARRIAGE) NECK

Fig. 9-3-5 Shoulder and necks.

CUP   FLAT   CONE

OVAL   HALF DOG

Fig. 9-3-6 Point styles.

Cup   Most widely used when the cutting-in action of point is not objectionable.

Flat   Used when frequent resetting of a part is required. Particularly suited for use against hardened steel shafts. This point is preferred where walls are thin or the threaded member is a soft material.

Cone   Used for permanent location of parts. Usually spotted in a hole to half its length.

Oval   Used when frequent adjustment is necessary or for seating against angular surfaces.

Half Dog   Normally applied where permanent location of one part in relation to another is desired.

PROPERTY CLASSES OF FASTENERS

Inch Fasteners

The strength of customary fasteners for most common uses is determined by the size of the fastener and the material from which it is made. Property classes are defined by the Society of Automotive Engineers (SAE), or the American Society for Testing and Materials (ASTM).

HEAD DESIGNATION / GRADE	GRADES 0, I, 2	GRADE 3	GRADE 5	GRADE 7	GRADE 8
MINIMUM TENSILE STRENGTH KIPS	0—NO REQUIREMENTS 1—55 2—69 64 55	110 100	120 115 105	133	150

Fig. 9-3-7 Mechanical requirements for inch-size threaded fasteners.

PROPERTY CLASS	IDENTIFICATION SYMBOL	
	BOLTS, SCREWS AND STUDS	STUDS SMALLER THAN MI2
4.6	4.6	–
4.8	4.8	–
5.8	5.8	–
8.8 (I)	8.8	O
9.8 (I)	9.8	+
10.9 (I)	10.9	□
12.9	12.9	△

NOTE I: PRODUCTS MADE OF LOW CARBON MARTENSITE STEEL SHALL BE ADDITIONALLY IDENTIFIED BY UNDERLINING THE NUMERALS.

Fig. 9-3-9 Metric property class identification symbols for bolts, screws, and studs.

Figure 9-3-7 lists the mechanical requirements of inch-sized fasteners and their identification patterns.

Metric Fasteners

For mechanical and material requirements, metric fasteners are classified under a number of property classes. Bolts, screws, and studs have seven property classes of steel suitable for general engineering applications. The property classes are designated by numbers where increasing numbers generally represent increasing tensile strengths. The designation symbol consists of two parts: the first numeral of a two-digit symbol or the first two numerals of a three-digit symbol approximates one-hundredth of the minimum tensile strength in megapascals; and the last numeral approximates one-tenth of the ratio expressed as a percentage of minimum yield strength and minimum tensile strength.

EXAMPLE 1 A property class 4.8 fastener (see Fig. 9-3-8) has a minimum tensile strength of 420 MPa and a minimum yield strength of 340 MPa. One percent of 420 is 4.2. The first digit is 4. The minimum yield strength of 340 MPa is equal to approximately 80 percent of the minimum tensile strength of 420 MPa. One-tenth of 80 percent is 8. The last digit of the property class is 8.

PROPERTY CLASS ∠	NOMINAL DIAMETER	MINIMUM TENSILE STRENGTH MPa	MINIMUM YIELD STRENGTH MPa
4.6	M5 THRU M36	400	240
4.8	M1.6 THRU M16	420	340
5.8	M5 THRU M24	520	420
8.8	M16 THRU M36	830	660
9.8	M1.6 THRU M16	900	720
10.9	M5 THRU M36	1040	940
12.9	M1.6 THRU M36	1220	1100

Fig. 9-3-8 Mechanical requirements for metric bolts, screws, and studs.

EXAMPLE 2 A property class 10.9 fastener (see Fig. 9-3-8) has a minimum tensile strength of 1040 MPa and a minimum yield strength of 940 MPa. One percent of 1040 is 10.4. The first two numerals of the three-digit symbol are 10. The minimum yield strength of 940 MPa is equal to approximately 90 percent of the minimum tensile strength of 1040 MPa. One-tenth of 90 percent is 9. The last digit of the property class is 9.

Machine screws are normally available only in classes 4.8 and 9.8; other bolts, screws, and studs are available in all classes within the specified product size limitations given in Fig. 9-3-7.

For guidance purposes only, to assist designers in selecting a property class, the following information may be used.

- Class 4.6 is approximately equivalent to SAE grade 1 and ASTM A 307, grade A.
- Class 5.8 is approximately equivalent to SAE grade 2.
- Class 8.8 is approximately equivalent to SAE grade 5 and ASTM A 449.
- Class 9.8 has properties approximately 9 percent stronger than SAE grade 5 and ASTM A 449.
- Class 10.9 is approximately equivalent to SAE grade 8 and ASTM A 354 grade BD.

Fastener Markings

Slotted and crossed recessed screws of all sizes and other screws and bolts of sizes .25 in. or M4 and smaller need not be marked. All other bolts and screws of sizes .25 in. or M5 and larger are marked to identify their strength. The property class symbols for metric fasteners are shown in Fig. 9-3-9. The symbol is located on the top of the bolt head or screw. Alternatively, for hex-head products, the markings may be indented on the side of the head.

All studs of size .25 in. or M5 and larger are identified by the property class symbol. The marking is located on the extreme end of the stud. For studs with an interference fit thread the markings are located at the nut end. Studs smaller than .50 in. or M12 use different identification symbols.

Nuts

The customary terms *regular* and *thick* for describing nut thicknesses have been replaced by the terms *style 1* and *style 2* for metric nuts. The design of style 1 and 2 steel nuts shown in Fig. 9-3-10 is based on providing sufficient nut strength to

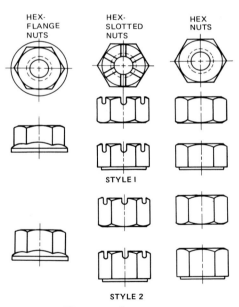

Fig. 9-3-10 Hex-nut styles.

reduce the possibility of thread stripping. There are three property classes of steel nuts available. See Fig. 9-3-11.

Hex Flanged Nuts These nuts are intended for general use in applications requiring a large bearing contact area. The two styles of flanged hex nuts differ

PROPERTY CLASS	NOMINAL NUT SIZE	SUGGESTED PROPERTY CLASS OF MATING BOLT, SCREW OR STUD
5	M5 THRU M36	4.6, 4.8, 5.8
9	M5 THRU MI6	5.8, 9.8
	M20 THRU M36	5.8, 8.8
I0	M6.3 THRU M36	10.9

Fig. 9-3-11 **Metric nut selection for bolts, screws, and studs.**

dimensionally in thickness only. The standard property classes for hex flanged nuts are identical to the hex nuts. All metric nuts are marked to identify their property class.

DRAWING A BOLT AND NUT

Bolts and nuts are not normally drawn on detail drawings unless they are of a special size or have been modified. On some assembly drawings it may be necessary to show a nut and bolt. Approximate nut and bolt sizes are shown in Fig. 9-3-12. Actual sizes are found in the Appendix. Nut and bolt templates are also available and are recommended as a cost-saving device for manual drafting. Conventional drawing practice is to show the nuts and bolt heads in the across-corners position in all views.

STUDS

Studs, as shown in Fig. 9-3-13, are still used in large quantities to best fulfill the needs of certain design functions and for overall economy.

(A) DOUBLE END

(B) CONTINUOUS THREAD

Fig. 9-3-13 **Studs.**

Double-End Studs These studs are designated in the following sequence: type and name; nominal size; thread information; stud length; material, including grade identification; and finish (plating or coating) if required.

EXAMPLE
TYPE 2 DOUBLE-END STUD, .500—13 UNC—2A × 4.00, CADMIUM PLATED

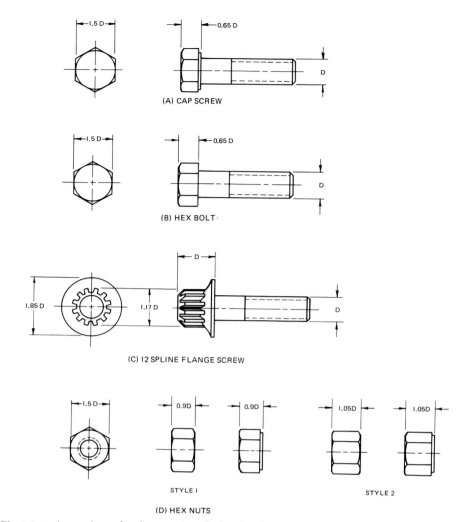

Fig. 9-3-12 **Approximate head proportions for hex-head cap screws, bolts, and nuts.**

Continuous-Thread Studs These studs are designated in the following sequence: product name, nominal size, thread information, stud length, material and finish (plating or coating) if required.

EXAMPLE
TYPE 3 CONTINUOUS THREAD STUD, M24 × 3 × 200, STEEL CLASS 8.8, ZINC PHOSPHATE AND OIL

WASHERS

Washers are one of the most common forms of hardware and perform many varied functions in mechanically fastened assemblies. They may only be required to span an oversize clearance hole, to give better bearing for nuts or screw faces, or to distribute loads over a greater area. Often, they serve as locking devices for threaded fasteners. They are also used to maintain a spring-resistance

pressure, to guard surfaces against marring, and to provide a seal.

Classification of Washers

Washers are commonly the elements which are added to screw systems to keep them tight, but not all washers are locking types. Many washers serve other functions, such as surface protection, insulation, sealing, electrical connection, and spring-tension take-up devices.

Flat Washers Plain, or flat, washers are used primarily to provide a bearing surface for a nut or a screw head, to cover large clearance holes, and to distribute fastener loads over a large area—particularly on soft materials such as aluminum or wood. See Fig. 9-3-14.

Conical Washers These washers are used with screws to effectively add spring take-up to the screw elongation.

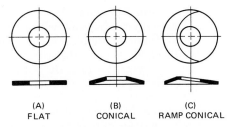

(A) FLAT (B) CONICAL (C) RAMP CONICAL

Fig. 9-3-14 Flat and conical washers.

Helical Spring Washers These washers are made of slightly trapezoidal wire formed into a helix of one coil so that the free height is approximately twice the thickness of the washer section. See Fig. 9-3-15.

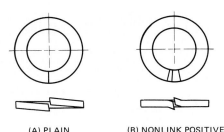

(A) PLAIN (B) NONLINK POSITIVE

Fig. 9-3-15 Helical spring washers.

Tooth Lock Washers Made of hardened carbon steel, a tooth lock washer has teeth that are twisted or bent out of the plane of the washer face so that sharp cutting edges are presented to both the workpiece and the bearing face of the screw head or nut. See Fig. 9-3-16.

Spring Washers There are no standard designs for spring washers. See Fig. 9-3-17. They are made in a great variety of sizes and shapes and are usually selected from a manufacturer's catalog for some specific purpose.

Special-Purpose Washers Molded or stamped nonmetallic washers are available in many materials and may be used as seals, electrical insulators, or for protection of the surface of assembled parts.

Many plain, cone, or tooth washers are available with special mastic sealing compounds firmly attached to the washer. These washers are used for sealing and vibration isolation in high-production industries.

Terms Related to Threaded Fasteners

The *tap drill size* for a threaded (tapped) hole is a diameter equal to the minor

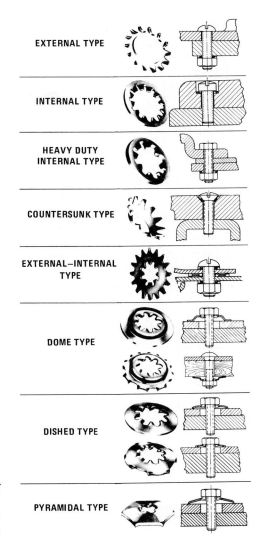

EXTERNAL TYPE

INTERNAL TYPE

HEAVY DUTY INTERNAL TYPE

COUNTERSUNK TYPE

EXTERNAL—INTERNAL TYPE

DOME TYPE

DISHED TYPE

PYRAMIDAL TYPE

Fig. 9-3-16 Tooth lock washers.

diameter of the thread. The *clearance drill size*, which permits the free passage of a bolt, is a diameter slightly greater than the major diameter of the bolt. See Fig. 9-3-18. A *counterbored hole* is a circular, flat-bottomed recess that permits the head of a bolt or cap screw to rest below the surface of the part. A *countersunk hole* is an angular-sided recess that accommodates the shape of a flat-head cap screw or machine screw or an oval-head machine screw. *Spot-facing* is a machine operation that provides a smooth, flat surface where a bolt head or a nut will rest.

SPECIFYING FASTENERS

In order for the purchasing department to properly order the fastening device which has been selected in the design, the following information is required. (*Note:* The information listed will not apply to all types of fasteners):

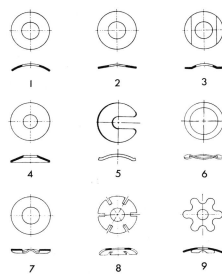

1 2 3

4 5 6

7 8 9

Fig. 9-3-17 Typical spring washer.

1. Type of fastener
2. Thread specifications
3. Fastener length
4. Material
5. Head style
6. Type of driving recess
7. Point type (setscrews only)
8. Property class
9. Finish

EXAMPLES

.375—16 UNC—2A × 4.00
HEX BOLT, ZINC PLATED

M10 × 1.5 × 50, 9.8 12-SPLINE
FLANGE SCREW, CADMIUM
PLATED

TYPE 2 DOUBLE-END STUD, M10
× 1.5 × 100, STEEL CLASS 9.8,
CADMIUM PLATED

NUT, HEX, STYLE 1, .500 UNC
STEEL

MACH SCREW, PHILLIPS ROUND
HD, 8—32 UNC × 1.00, BRASS

WASHER, FLAT 8.4 ID × 17 OD ×
2 THK, STEEL HELICAL SPRING

References and Source Material

1. *Machine Design*, Fastening and joining reference issue, Nov. 1981.
2. *Design Engineering* and Staff of Stelco's "Fastener Facts."

ASSIGNMENTS

See Assignments 7 and 8 for Unit 9-3 starting on page 203.

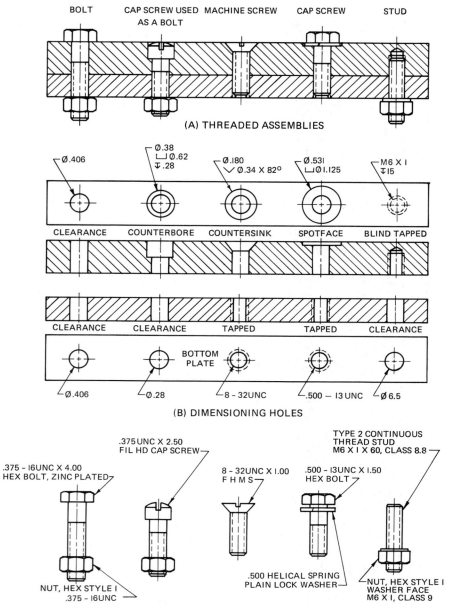

Fig. 9-3-18 Specifying threaded fasteners and holds.

U N I T 9 - 4
Special Fasteners

SETSCREWS

Setscrews are used as semipermanent fasteners to hold a collar, sheave, or gear on a shaft against rotational or translational forces. In contrast to most fastening devices, the setscrew is essentially a compression device. Forces developed by the screw point on tightening produce a strong clamping action that resists relative motion between assembled parts. The basic problem in setscrew

selection is to find the best combination of setscrew form, size, and point style that provides the required holding power.

Setscrews can be categorized in two ways: by their head style and by the point style desired. See Fig. 9-4-1. Each setscrew style is available in any one of five point styles.

The conventional approach to selecting the setscrew diameter is to make it roughly equal to one-half the shaft diameter. This rule of thumb often gives satisfactory results, but its range of usefulness is limited.

Setscrews and Keyseats When a setscrew is used in combination with a key, the screw diameter should be equal to the width of the key. In this combination the setscrew is locating the parts in an axial direction only. The torsional load on the parts is carried by the key.

The key should be tight-fitting, so that no motion is transmitted to the screw. Key design is covered in Chap. 10.

KEEPING FASTENERS TIGHT

Fasteners are inexpensive, but the cost of installing them can be substantial. Prob-

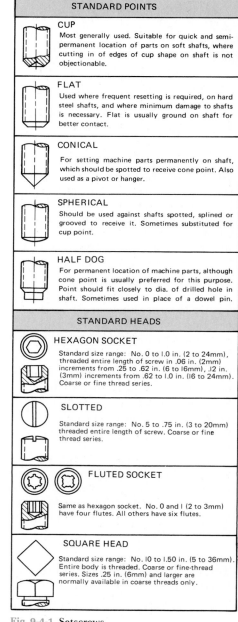

Fig. 9-4-1 Setscrews.

ably the simplest way to cut assembly costs is to make sure that, once installed, fasteners stay tight.

The American National Standards Institute has identified three basic locking methods: free-spinning, prevailing-torque, and chemical locking. Each has its own advantages and disadvantages. See Fig. 9-4-2.

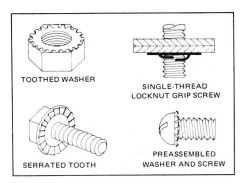

TOOTHED WASHER

SINGLE-THREAD LOCKNUT GRIP SCREW

SERRATED TOOTH

PREASSEMBLED WASHER AND SCREW

(A) FREE-SPINNING

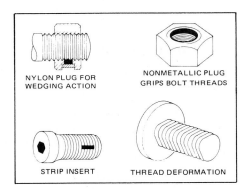

NYLON PLUG FOR WEDGING ACTION

NONMETALLIC PLUG GRIPS BOLT THREADS

STRIP INSERT

THREAD DEFORMATION

(B) PREVAILING TORQUE

Fig. 9-4-2 Typical setscrew installation locking fasteners.

Free-spinning devices include toothed and spring lockwashers and screws and bolts with washerlike heads. With these arrangements, the fasteners spin free in the clamping direction, which makes them easy to assemble, and the break-loose torque is greater than the seating torque. However, once break-loose torque is exceeded, free-spinning washers have no prevailing torque to prevent further loosening.

Prevailing-torque methods make use of increased friction between nut and bolt. Metallic types usually have deformed threads or contoured thread profiles that jam the threads on assembly. Nonmetallic types make use of nylon or polyester insert elements that produce interference fits on assembly.

Chemical locking is achieved by coating the fastener with an adhesive.

(A) PREVAILING TORQUE LOCKNUTS

NONMETALLIC COLLAR CLAMPED IN THE TOP OF THIS NUT PRODUCES LOCKING ACTION.

THREADED ELLIPTICAL SPRING-STEEL INSERT GRIPS THE BOLT AND PREVENTS TURNING.

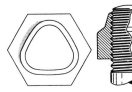

SLOTTED SECTION OF THIS PREVAILING-TORQUE NUT FORMS BEAMS WHICH ARE DEFLECTED INWARD AND GRIP THE BOLT.

THREE SECTORS OF TAPERED CONE, PREFORMED INWARDLY, ARE ELASTICALLY RETURNED TO CIRCU-FORM WHEN THE NUT IS APPLIED.

(B) FREE-SPINNING LOCKNUTS

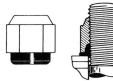

DEFORMED BEARING SURFACE. TEETH ON THE BEARING SURFACE "BITE" INTO WORK TO PROVIDE A RATCHET LOCKING ACTION.

NYLON INSERT FLOWS AROUND THE BOLT RATHER THAN BEING CUT BY THE BOLT THREADS TO PROVIDE LOCKING ACTION AND AN EFFECTIVE SEAL.

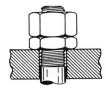

JAM NUT, APPLIED UNDER A LARGE REGULAR NUT, IS ELASTICALLY DEFORMED AGAINST BOLT THREADS WHEN THE LARGE NUT IS TIGHTENED.

NUT WITH A CAPTIVE-TOOTHED WASHER. WHEN TIGHTENED, THE CAPTIVE WASHER PROVIDES THE LOCKING MEANS WITH SPRING ACTION BETWEEN THE NUT AND WORKING SURFACE.

(C) OTHER TYPES

SLOTTED NUT USES A COTTER PIN THROUGH A HOLE IN THE BOLT FOR LOCKING ACTION.

SINGLE-THREAD LOCKNUT, WHICH IS SPEEDILY APPLIED, LOCK BY GRIP OF ARCHED PRONGS WHEN BOLT OR SCREW IS TIGHTENED.

Fig. 9-4-3 Locknuts.

LOCKNUTS

A *locknut* is a nut with special internal means for gripping a threaded fastener to prevent rotation. Generally it has the dimensions, mechanical requirements, and other specifications of a standard nut, but with a locking feature added.

Locknuts are divided into three general classifications: prevailing-torque, free-spinning and other types. These are shown in Figs. 9-4-3 and 9-4-4.

Prevailing-Torque Locknuts

Prevailing-torque locknuts spin freely for a few turns, and then must be wrenched to final position. The maximum holding and locking power is reached as soon as the threads and the locking feature are engaged. Locking action *is* maintained until the nut is removed. Prevailing-torque locknuts are classified by basic design principles:

1. Thread deflection causes friction to develop when the threads are mated; thus the nut resists loosening.
2. The out-of-round top portion of the tapped nut grips the bolt threads and resists rotation.
3. The slotted section of locknut is pressed inward to provide a spring frictional grip on the bolt.

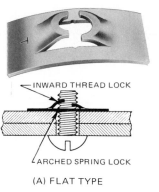

(A) FLAT TYPE

(B) FLAT-TYPE CONICAL THREAD

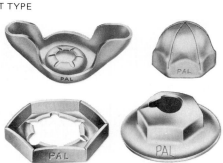

(C) SPIRAL-FORMED THREAD

Fig. 9-4-4 Single-thread engaging nuts.

4. Inserts, either nonmetallic or of soft metal, are plastically deformed by the bolt threads to produce a frictional interference fit.
5. A spring wire or pin engages the bolt threads to produce a wedging or ratchet-locking action.

Free-Spinning Locknuts

Free-spinning locknuts are free to spin on the bolt until seated. Additional tightening locks the nut.

Since most free-spinning locknuts depend on clamping force for their locking action, they are usually not recom-

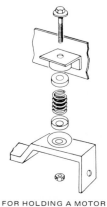

FOR HOLDING A MOTOR MOUNTING SECURELY IN POSITION.

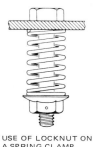

USE OF LOCKNUT ON A SPRING CLAMP.

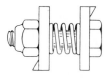

USE OF LOCKNUT FOR TUBULAR FASTENING.

FOR RUBBER-INSULATED AND CUSHION MOUNTINGS WHERE THE NUT MUST REMAIN STATIONARY.

USE OF LOCKNUT ON A BOLTED CONNECTION THAT REQUIRES PREDETERMINED PLAY.

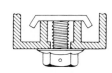

FOR AN EXTRUDED PART ASSEMBLY.

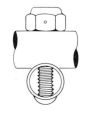

FOR SPRING-MOUNTED CONNECTIONS WHERE THE NUT MUST REMAIN STATIONARY OR IS SUBJECT TO ADJUSTMENT.

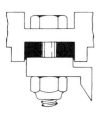

USE OF LOCKNUT WHERE ASSEMBLY IS SUBJECTED TO VIBRATORY OR CYCLIC MOTIONS THAT COULD CAUSE LOOSENING.

Fig. 9-4-5 Typical locknut applications.

mended for joints that might relax through plastic deformation or for fastening materials that might crack or crumble.

Other Locknut Types

Jam nuts are thin nuts used under full-sized nuts to develop locking action. The large nut has sufficient strength to elastically deform the lead threads of the bolt and jam nut. Thus, a considerable resistance against loosening is built up. The use of jam nuts is decreasing; a one-piece, prevailing-torque locknut usually is used instead at a savings in assembled cost.

Slotted and castle nuts have slots which receive a cotter pin that passes through a drilled hole in the bolt and thus serves as the locking member.

Castle nuts differ from slotted nuts in that they have a circular crown of a reduced diameter.

Single-thread locknuts are spring steel fasteners which may be speedily applied. Locking action is provided by the grip of the thread-engaging prongs and the reaction of the arched base. Their use is limited to nonstructural assemblies and usually to screw sizes below 6 mm in diameter. See Fig. 9-4-5.

CAPTIVE OR SELF-RETAINING NUTS

Captive or self-retaining nuts provide a permanent, strong, multiple-thread fastener for use on thin materials. See Fig. 9-4-6. They are especially good where there are blind locations, and they can generally be attached without damaging finishes. Methods of attaching these types of nuts vary and tools required for assembly are generally uncomplicated and inexpensive. The self-retained nuts are grouped according to four means of attachment:

1. Plate or anchor nuts: These nuts have mounting lugs which can be riveted, welded, or screwed to the part.
2. Caged nuts: A spring-steel cage retains a standard nut. The cage snaps into a hole or clips over an edge to hold the nut in position.
3. Clinch nuts: They are specially designed nuts with pilot collars

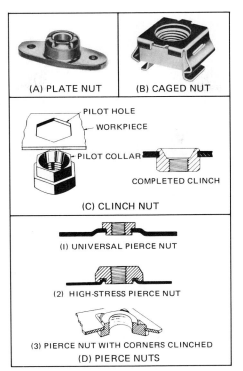

Fig. 9-4-6 Captive or self-retaining nuts.

which are clinched or staked into the parent part through a precut hole.
4. Self-piercing nuts: A form of clinch nut that cuts its own hole.

INSERTS

Inserts are a special form of nut designed to serve the function of a tapped hole in blind or through-hole locations. See Fig. 9-4-7.

SEALING FASTENERS

Fasteners hold two or more parts together, but they can perform other functions as well. One important auxiliary function is that of sealing gases and liquids against leakage.

Two types of sealed-joint constructions are possible with fasteners. See Fig. 9-4-8. In one approach, the fasteners enter the sealed medium and are separately sealed.

The second approach uses a separate sealing element which is held in place by the clamping forces produced by conventional fasteners, such as rivets or bolts.

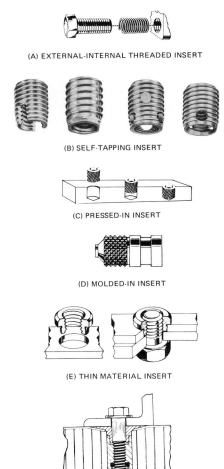

Fig. 9-4-7 Inserts.

Fig. 9-4-8 Types of sealed-joint construction.

There are many methods of obtaining a seal using sealing fasteners, as shown in Fig. 9-4-9.

Reference and Source Material

1. *Machine Design*, Fastening and joining reference issue, Nov. 1981.

ASSIGNMENTS

See Assignments 9 and 10 for Unit 9-4 starting on page 204.

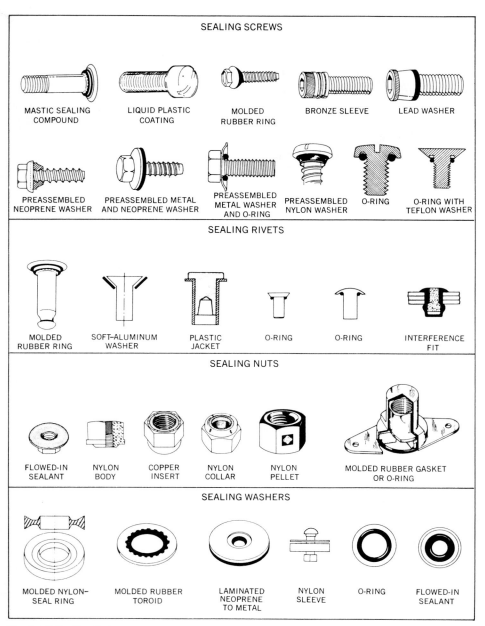

SEALING SCREWS

MASTIC SEALING COMPOUND

LIQUID PLASTIC COATING

MOLDED RUBBER RING

BRONZE SLEEVE

LEAD WASHER

PREASSEMBLED NEOPRENE WASHER

PREASSEMBLED METAL AND NEOPRENE WASHER

PREASSEMBLED METAL WASHER AND O-RING

PREASSEMBLED NYLON WASHER

O-RING

O-RING WITH TEFLON WASHER

SEALING RIVETS

MOLDED RUBBER RING

SOFT–ALUMINUM WASHER

PLASTIC JACKET

O-RING

O-RING

INTERFERENCE FIT

SEALING NUTS

FLOWED-IN SEALANT

NYLON BODY

COPPER INSERT

NYLON COLLAR

NYLON PELLET

MOLDED RUBBER GASKET OR O-RING

SEALING WASHERS

MOLDED NYLON– SEAL RING

MOLDED RUBBER TOROID

LAMINATED NEOPRENE TO METAL

NYLON SLEEVE

O-RING

FLOWED-IN SEALANT

Fig. 9-4-9 Sealing fasteners.

U N I T 9 - 5

Fasteners for Light-Gage Metal, Plastic, and Wood

TAPPING SCREWS

Tapping screws cut or form a mating thread when driven into drilled or cored holes. These one-piece fasteners permit rapid installation, since nuts are not used and access is required from only one side of the joint. The mating thread produced by the tapping screw fits the screw threads closely, and no clearance

is necessary. This close fit usually keeps the screw tight, even under vibrating conditions. See Fig. 9-5-1.

Tapping screws are practically all case-hardened and, therefore, can be driven tight and have a relatively high ultimate torsional strength. The screws are used in steel, aluminum (cast, extruded, rolled, or die-formed) die cast-

TYPE AB   TYPE B   TYPE F   TYPE U

Fig. 9-5-1 Self-tapping screws.

ings, cast iron, forgings, plastics, reinforced plastics, asbestos, and resin-impregnated plywood. See Fig. 9-5-2 on the facing page.

Types C, D, F, G, and T tapping screws are available in both coarse- and fine-thread series. Coarse threads should be used with weak materials.

Self-drilling tapping screws have special points for drilling and then tapping their own holes. See Fig. 9-5-3. These eliminate drilling or punching, but they must be driven by a power screwdriver.

SPECIAL TAPPING SCREWS

Typical special tapping screws are the self-captive screws and double-thread combinations for limited drive. Self-captive screws combine a coarse-pitched starting thread (similar to type B) with a finer pitch (machine-screw thread) farther along the screw shank.

Sealing tapping screws, with preassembled washers or O-rings (Fig. 9-5-4B) are available in a variety of styles.

.09   .02   .07

Fig. 9-5-3 Self-drilling tapping screws.

(A) TAPPING SCREWS WITH PREASSEMBLED WASHERS.

(B) TAPPING SCREWS WITH PREASSEMBLED SEALING WASHERS OR COMPOUNDS.

Fig. 9-5-4 Special tapping screws.

Reference and Source Material

1. *Machine Design*, Fastening and joining reference issue, Nov. 1981.

ASSIGNMENT

See Assignment 11 for Unit 9-5 on page 205.

HEAVY GAGE SHEET METAL AND STRUCTURAL STEEL
USE TYPES B, U, F.

Holes may be drilled or clean-punched.

Two parts may have pierced holes to nest burrs. This results in a stronger joint.

Use a pierced hole in workpiece if clearance hole is needed in part to be fastened.

Extruded hole may also be used in workpiece if clearance hole is needed in fastened part.

LIGHT GAGE SHEET METAL
USE TYPES AB, B.

Holes may be drilled or clean-punched the same size in both sheet metal parts. For thicker sheet metal and structural steel, a clearance hole should be provided in the part to be fastened. Hole size depends on thickness of the workpiece.

Notes: 1. Use hex-head on Type B screws.

2. With Type U screws, material should be thick enough to permit sufficient thread engagement—at least one screw diameter.

PLASTICS
USE TYPES B, U, F.

Screw holes may be molded or drilled. If material is brittle or friable, molded holes should be formed with a rounded chamfer, and drilled holes should be machine chamfered. Provide a clearance in the part to be fastened. Depth of penetration should be held within the "minimum and maximum" limits recommended. The hole should be deeper than the screw penetration to allow for chip clearance.

CASTINGS AND FORGINGS
USE TYPES B, U, F.

Holes may be cored if it is practical to maintain close tolerances. Otherwise blind drill holes to recommended hole size. Provide a clearance hole for screw in the part to be fastened. The hole in the casting, if it is a blind hole, should be deeper than the screw penetration to allow for chip clearance.

Notes: 1. Hole in fastened part may be the same size as workpiece hole for Type U screws.

2. Type B is only suitable for use in nonferrous castings.

Fig. 9-5-2 Tapping-screws application chart.

ASSIGNMENTS FOR CHAPTER 9

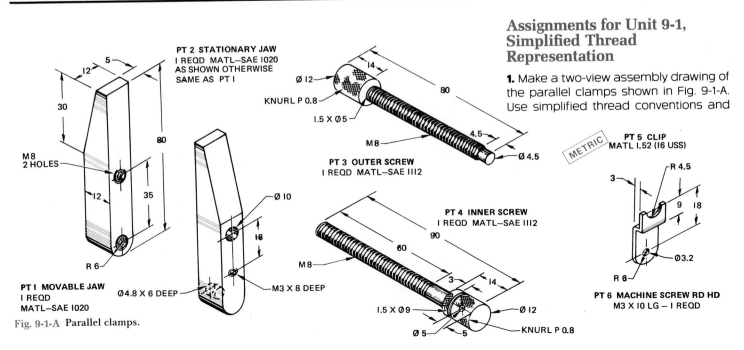

Assignments for Unit 9-1, Simplified Thread Representation

1. Make a two-view assembly drawing of the parallel clamps shown in Fig. 9-1-A. Use simplified thread conventions and

PT 2 STATIONARY JAW
I REQD MATL—SAE 1020
AS SHOWN OTHERWISE
SAME AS PT I

Ø 12
KNURL P 0.8
1.5 X Ø 5
M8

PT 3 OUTER SCREW
I REQD MATL—SAE 1112

PT 4 INNER SCREW
I REQD MATL—SAE 1112

M8
Ø 10
Ø 12
KNURL P 0.8
Ø 5
1.5 X Ø9

M8
2 HOLES
R 6
Ø 4.8 X 6 DEEP
M3 X 8 DEEP

PT I MOVABLE JAW
I REQD
MATL—SAE 1020

Fig. 9-1-A Parallel clamps.

METRIC **PT 5 CLIP**
MATL 1.52 (16 USS)

R 4.5
Ø3.2
R 6

PT 6 MACHINE SCREW RD HD
M3 X 10 LG — I REQD

THREADED FASTENERS **201**

include an item list calling for all the parts. The only dimension required on the drawing is the maximum opening of the jaws. Identify the parts on the assembly. Scale 1:1.

2. Make detail drawings of the parts shown in Fig. 9-1-A. Scale 1:1. Use your judgment for the number of views required for each part.

3. Make a one-view assembly drawing of the turnbuckle shown in Fig. 9-1-B. Show the assembly in its shortest length and also include the maximum position shown in phantom lines. The only dimensions required are the minimum and maximum distances between the eye centers. Scale 1:1.

4. Make detail drawings of the parts shown in Fig. 9-1-B. Scale 1:1.

Assignments for Unit 9-2, Detailed and Schematic Thread Representation

5. Make one-view drawings of the parts shown in Fig. 9-2-A or 9-2-B. Use detailed representation for the threads. Use conventional breaks to shorten the lengths of the guide rod and jack screw.

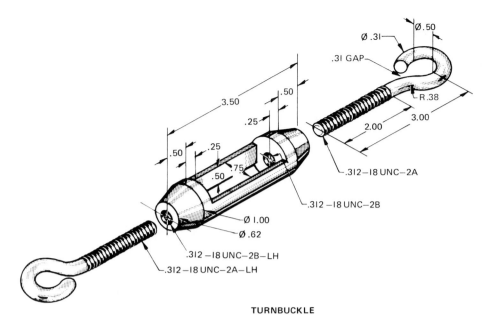

TURNBUCKLE

Fig. 9-1-B Turnbuckle.

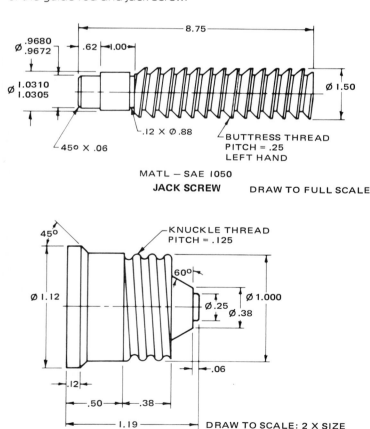

MATL — SAE 1050

JACK SCREW   DRAW TO FULL SCALE

FUSE

Fig. 9-2-A Jack screw and fuse.

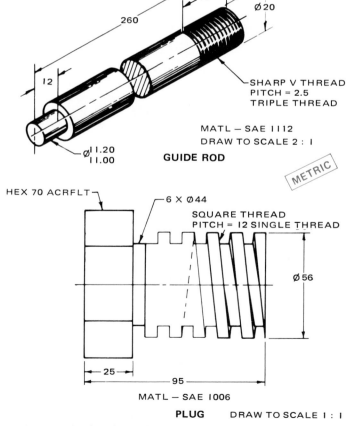

MATL — SAE 1112
DRAW TO SCALE 2 : 1

GUIDE ROD

METRIC

PLUG   DRAW TO SCALE 1 : 1

MATL — SAE 1006

Fig. 9-2-B Guide rod and plug.

6. Make an assembly drawing for either Figure 9-2-C or 9-2-D showing the connector and end rods. The end rods are to be drawn in section. Scale 1:1.

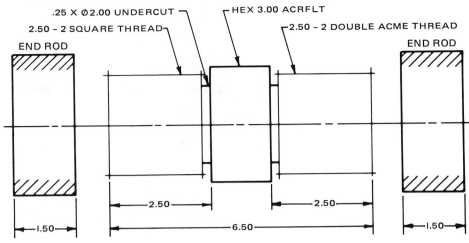

Fig. 9-2-C Connector and supports.

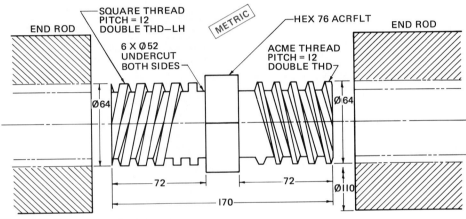

Fig. 9-2-D Connector and supports.

Assignments for Unit 9-3, Common Threaded Fasteners

7. Prepare a full-section assembly drawing of the four fastener assemblies shown in Fig. 9-3-A. Dimension both the clearance and threaded holes. A top view may be shown, if required. Scale 1:1.

8. Prepare full-section assembly drawings of the five fastener assemblies shown in Fig. 9-3-B. Dimension both the clearance and threaded holes. A top view of the fastener may be shown if required. Scale 1:1.

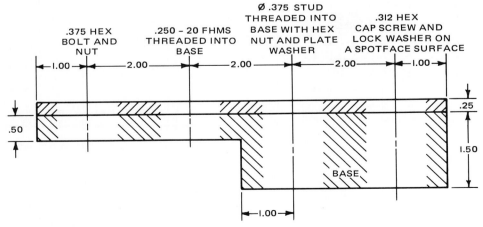

Fig. 9-3-A Threaded fasteners.

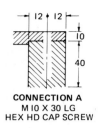

CONNECTION A
M I0 X 30 LG
HEX HD CAP SCREW

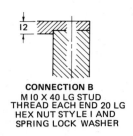

CONNECTION B
M I0 X 40 LG STUD
THREAD EACH END 20 LG
HEX NUT STYLE I AND
SPRING LOCK WASHER

CONNECTION C
M I0 X 30 LG
FL HD CAP SCREW

CONNECTION D
M I0 X 1.25 X 25 LG
SOCKET HEAD CAP SCREW
AND SPRING LOCK WASHER

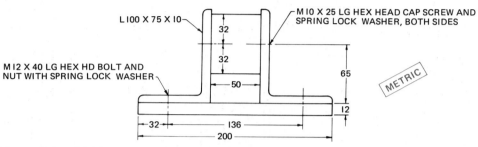

Fig. 9-3-B Standard fasteners.

Assignments for Unit 9-4, Special Fasteners

9. Make a one-view assembly drawing of the flexible coupling shown in Fig. 9-4-A. The shafts, which are coupled, are 1.50 in. in diameter and are to be shown in the assembly. They are to extend beyond the coupling for approximately 2.00 in. and end with a conventional break. Show the setscrews and keys in position. Scale 1:1.

10. Make a one-view assembly drawing of the adjustable shaft support shown in Fig. 9-4-B. Show the base in full section. A broken out section is recommended to clearly show the setscrews in the yoke. Add part numbers to the assembly drawing and include an item list. Do not dimension. Scale 1:1.

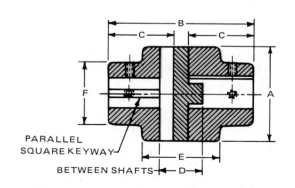

Fig. 9-4-A Flexible coupling.

MAXIMUM BORE	A	B	C	D	E	F
.9375	3.00	3.75	1.75	.88	1.50	2.38
1.1875	3.50	4.69	2.19	1.06	1.81	2.75
1.4375	4.00	5.62	2.62	1.25	2.12	3.12
1.6875	5.00	6.56	3.06	1.44	2.44	3.50
1.9375	5.50	7.50	3.50	1.50	2.50	4.00
2.1875	6.00	8.44	3.94	1.81	3.06	4.38

DIMENSIONS SHOWN ARE IN INCHES

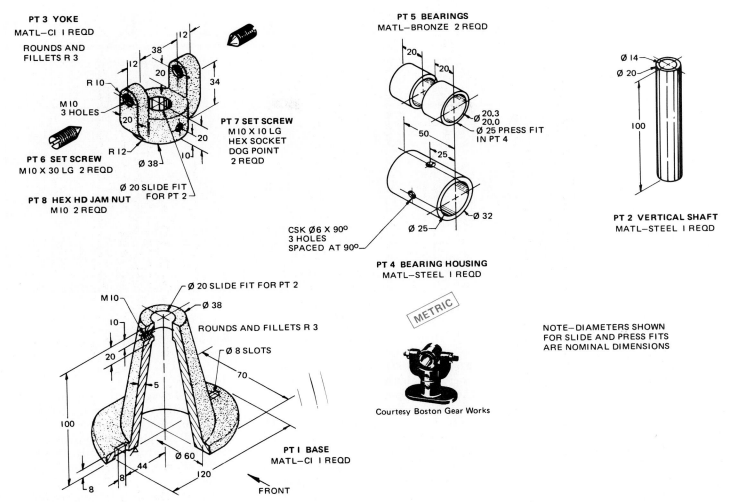

PT 3 YOKE
MATL-CI I REQD

ROUNDS AND
FILLETS R 3

R 10

M 10
3 HOLES

PT 6 SET SCREW
M 10 X 30 LG 2 REQD

R 12

Ø 38

Ø 20 SLIDE FIT
FOR PT 2

PT 8 HEX HD JAM NUT
M 10 2 REQD

PT 7 SET SCREW
M 10 X 10 LG
HEX SOCKET
DOG POINT
2 REQD

PT 5 BEARINGS
MATL-BRONZE 2 REQD

Ø 20.3
20.0
Ø 25 PRESS FIT
IN PT 4

Ø 32

Ø 25

CSK Ø6 X 90°
3 HOLES
SPACED AT 90°

PT 4 BEARING HOUSING
MATL-STEEL I REQD

Ø 14
Ø 20

100

PT 2 VERTICAL SHAFT
MATL-STEEL I REQD

NOTE—DIAMETERS SHOWN
FOR SLIDE AND PRESS FITS
ARE NOMINAL DIMENSIONS

METRIC

Ø 20 SLIDE FIT FOR PT 2
Ø 38
ROUNDS AND FILLETS R 3
Ø 8 SLOTS

M 10

70

5

100

44

120

8

8

Ø 60

FRONT

PT I BASE
MATL-CI I REQD

Courtesy Boston Gear Works

Fig. 9-4-B Adjustable shaft support.

Assignment for Unit 9-5, Fasteners for Light-Gage Metal, Plastic, and Wood

11. Make a drawing of the assembly shown in Fig. 9-5-A. Either inch or metric fasteners may be used. The steel post is fastened to the panel by two rows of tapping screws. The steel strap is held to the post by a single tapping screw which has the equivalent strength (body area) of at least three of the other tapping screws. Dimension the holes and fastener sizes. Scale to suit.

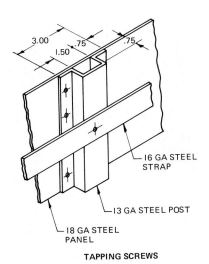

3.00
1.50
.75
.75

16 GA STEEL
STRAP

13 GA STEEL POST

18 GA STEEL
PANEL

TAPPING SCREWS

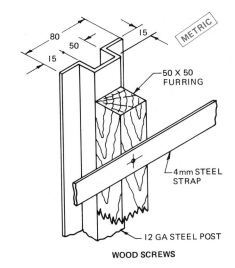

80
50
15
15

METRIC

50 X 50
FURRING

4mm STEEL
STRAP

12 GA STEEL POST

WOOD SCREWS

Fig. 9-5-A Special fasteners.

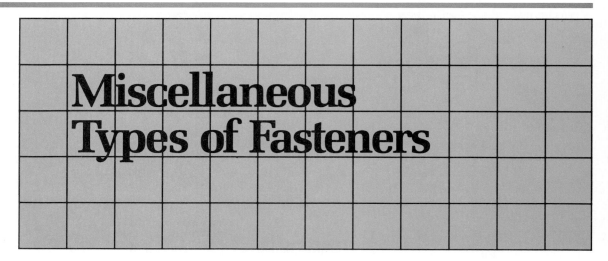

Miscellaneous Types of Fasteners

UNIT 10-1
Keys, Splines, and Serrations

KEYS

A *key* is a piece of steel lying partly in a groove in the shaft and extending into another groove in the hub. These grooves are called keyseats. See boxes 6 and 8 in Fig. 10-1-1. A key is used to secure gears, pulleys, cranks, handles, and similar machine parts to shafts, so that the motion of the part is transmitted to the shaft, or the motion of the shaft to the part, without slippage. The key may also act in a safety capacity; its size is generally calculated so that when overloading takes place, the key will shear or break before the part or shaft breaks.

There are many kinds of keys. The most common types are shown in Fig. 10-1-2. Square and flat keys are widely used in industry. The width of the square and flat key should be approximately one-quarter the shaft diameter, but for proper key selection refer to the Appendix. These keys are also available with a 1:100 taper on their top surfaces and are then known as *square taper* or *flat tapered* keys. The keyseat in the hub is tapered to accommodate the taper on the key.

The gibhead key is the same as the square or flat tapered key but has a head added for easy removal.

The Pratt and Whitney key is rectangular with rounded ends. Two-thirds of this key sits in the shaft, one-third sits in the hub.

The Woodruff key is semicircular and fits into a semicircular keyseat in the shaft and a rectangular keyseat in the hub. The width of the key should be approximately one-quarter the diameter of the shaft, and its diameter should approximate the diameter of the shaft. Half the width of the key extends above the shaft and into the hub. Refer to the Appendix for exact sizes. Woodruff keys are identified by a number which gives the nominal dimensions of the key. The numbering system, which originated many years ago, is identified with the fractional-inch system of measurement. The last two digits of the number give the normal diameter in eighths of an inch, and the digits preceding the last two give the nominal width in thirty-seconds of an inch. For example, a No. 1210 Woodruff key indicates a key $\frac{12}{32} \times \frac{10}{8}$ in., or a $\frac{3}{8} \times 1\frac{1}{4}$ in. key.

In calling up keys on the item list, only the information shown in the column "Specifications" in Fig 10-1-2 need be given.

Dimensioning of Keyseats

Keyseats are dimensioned by width,

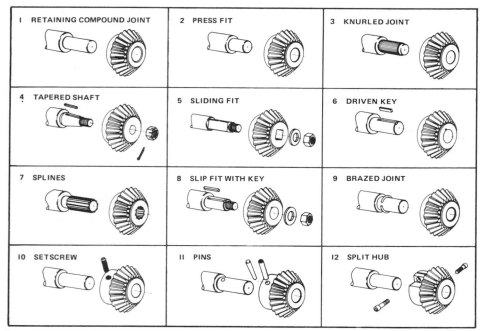

1 RETAINING COMPOUND JOINT	2 PRESS FIT	3 KNURLED JOINT
4 TAPERED SHAFT	5 SLIDING FIT	6 DRIVEN KEY
7 SPLINES	8 SLIP FIT WITH KEY	9 BRAZED JOINT
10 SETSCREW	11 PINS	12 SPLIT HUB

Fig. 10-1-1 Miscellaneous types of fasteners.

TYPE OF KEY	ASSEMBLY SHOWING KEY, SHAFT AND HUB	SPECIFICATION
SQUARE		.25 SQUARE KEY, 1.25 LG OR .25 SQUARE TAPERED KEY, 1.25 LG
FLAT		.188 X .125 FLAT KEY, 1.00 LG OR .188 X .125 FLAT TAPERED KEY, 1.00 LG
GIB-HEAD		.375 SQUARE GIB-HEAD KEY, 2.00 LG
PRATT AND WHITNEY		NO. 15 PRATT AND WHITNEY KEY
WOODRUFF		NO. 1210 WOODRUFF KEY

Fig. 10-1-2 Common keys.

depth, location, and, if required, length. The depth is dimensioned from the opposite side of the shaft or hole. See Fig. 10-1-3.

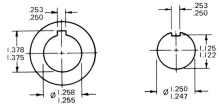

Fig. 10-1-3 Dimensioning keyseats.

Tapered Keyseats. The depth of tapered keyseats in hubs, which is shown on the drawing, is the nominal depth $H/2$ minus an allowance. This is always the depth at the large end of the tapered keyseat and is indicated on the drawing by the abbreviation LE.

The radii of fillets, when required, must be dimensioned on the drawing.

Since standard milling cutters for Woodruff keys have the same appropri-

ate number, it is possible to call for a Woodruff keyseat by the number only.

Where it is desirable to detail Woodruff keyseats on a drawing, all dimensions are given in the form of a note in the following order: width, depth, and radius of cutter.

Woodruff keyseats may alternately be dimensioned in the same manner as for square and flat keys, specifying first the width and then the depth. See Fig. 10-1-4.

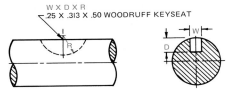

Fig. 10-1-4 Alternative method of detailing a Woodruff keyseat.

Splines and Serrations

A *splined shaft* is a shaft having multiple grooves, or keyseats, cut around its cir-

cumference for a portion of its length, in order that a sliding engagement may be made with corresponding internal grooves of a mating part.

Splines are capable of carrying heavier loads than keys, permit lateral movement of a part while maintaining positive rotation, and allow the attached part to be indexed or changed to another angular position.

Splines have either straight-sided teeth or curved-sided teeth. The latter type is known as an *involute spline*.

Involute Splines The splines are similar in shape to involute gear teeth but have pressure angles of 30, 37.5, or 45°. There are two types of fits, the *side fit* and the *major-diameter fit*. See Fig. 10-1-5.

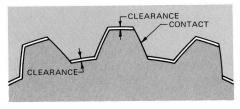

(A) SIDE FIT

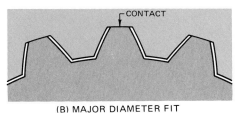

(B) MAJOR DIAMETER FIT

Fig. 10-1-5 Involute splines.

Straight-Side Splines The most popular are the SAE straight-side splines, as shown in Fig. 10-1-6. They have been used in many applications in the automotive and machine industry.

Serrations are shallow, involute splines with 45° pressure angles. They are primarily used for holding parts, such as plastic knobs, on steel shafts.

Drawing Data

It is essential that a uniform system of drawing and specifying splines and serrations be used on drawings. The conventional method of showing and calling out of splines on a drawing is shown in Figs. 10-1-7 and 10-1-8. Distance L does not include the cutter runout. The drawing callout shows the symbol indicating the type of spline followed by the type of fit, the pitch diameter, number of teeth

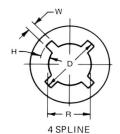

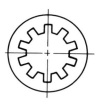

4 SPLINE   6 SPLINE   10 SPLINE

NUMBER OF SPLINES	W FOR ALL FITS	PERMANENT FIT		TO SLIDE WITHOUT LOAD		TO SLIDE UNDER LOAD	
		H	R	H	R	H	R
4	0.241 D	0.075 D	0.85 D	0.125 D	0.75 D		
6	0.250 D	0.050 D	0.90 D	0.075 D	0.85 D	0.100 D	0.80 D
10	0.156 D	0.045 D	0.91 D	0.070 D	0.86 D	0.095 D	0.81 D
16	0.098 D	0.045 D	0.91 D	0.070 D	0.86 D	0.095 D	0.81 D

Fig. 10-1-6 Sizes of SAE parallel-side splines.

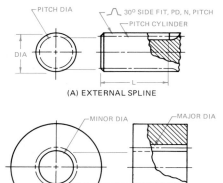

(A) EXTERNAL SPLINE

(B) INTERNAL SPLINE

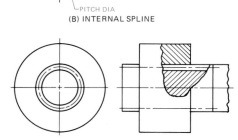

(C) ASSEMBLY DRAWINGS

Fig. 10-1-7 Callout and representation of involute splines.

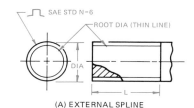

(A) EXTERNAL SPLINE

(B) INTERNAL SPLINE

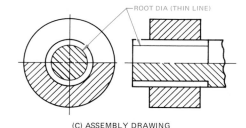

(C) ASSEMBLY DRAWING

Fig. 10-1-8 Callout and representation of straight-sided splines.

and pitch for involute splines, and number of teeth and outside diameter for straight-sided teeth.

ASSIGNMENT

See Assignment 1 for Unit 10-1 starting on page 220.

UNIT 10-2
Pin Fasteners

Pin fasteners are an inexpensive and effective approach to assembly where loading is primarily in shear. They can be separated into two groups: *semipermanent* and *quick-release*.

SEMIPERMANENT PINS

Semipermanent pin fasteners require application of pressure or the aid of tools for installation or removal. The two basic types are machine pins and radial-locking pins.

The following general design rules apply to all types of semipermanent pins:

- Avoid conditions where the direction of vibration parallels the axis of the pin.
- Keep the shear plane of the pin a minimum distance of one diameter from the end of the pin.
- In applications where engaged length is at a minimum and appearance is not critical, allow pins to protrude the length of the chamfer at each end for maximum locking effect.

Machine Pins

Four types are generally considered to be most important: *hardened and ground dowel pins and commercial straight pins, taper pins, clevis pins,* and *standard cotter pins*. Descriptive data and recommended assembly practices for these four traditional types of machine pins are presented in Fig. 10-2-1. For proper size selection of cotter pins, refer to Fig. 10-2-2.

Radial Locking Pins

Two basic pin forms are employed: *solid with grooved surfaces* and *hollow spring pins,* which may be either slotted or spiral-wrapped.

Grooved Straight Pins Locking action of the grooved pin is provided by parallel, longitudinal grooves uniformly spaced around the pin surface. Rolled or pressed into solid pin stock, the grooves expand the effective pin diameter. When the pin is driven into a drilled hole corresponding in size to nominal pin diameter, elastic deformation of the raised groove edges produces a secure force-fit with the hole wall. Figure 10-2-3 shows six of the grooved-pin constructions that have been standardized. For typical grooved pin applications and size selection, refer to Figs. 10-2-4 and 10-2-5.

Hollow Spring Pins Resilience of hollow cylinder walls under radial compression forces is the principle of spiral-wrapped and slotted tubular pins. Both pin forms are made to controlled diameters greater than the holes into which they are pressed. Compressed when

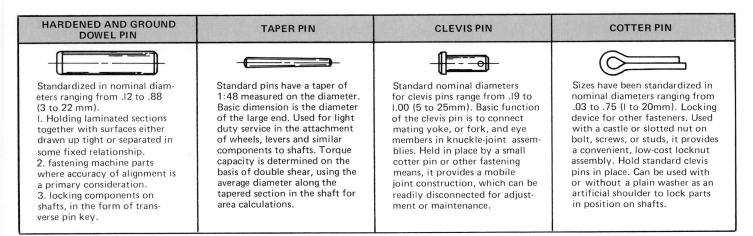

HARDENED AND GROUND DOWEL PIN	TAPER PIN	CLEVIS PIN	COTTER PIN
Standardized in nominal diameters ranging from .12 to .88 (3 to 22 mm). 1. Holding laminated sections together with surfaces either drawn up tight or separated in some fixed relationship. 2. fastening machine parts where accuracy of alignment is a primary consideration. 3. locking components on shafts, in the form of transverse pin key.	Standard pins have a taper of 1:48 measured on the diameter. Basic dimension is the diameter of the large end. Used for light duty service in the attachment of wheels, levers and similar components to shafts. Torque capacity is determined on the basis of double shear, using the average diameter along the tapered section in the shaft for area calculations.	Standard nominal diameters for clevis pins range from .19 to 1.00 (5 to 25mm). Basic function of the clevis pin is to connect mating yoke, or fork, and eye members in knuckle-joint assemblies. Held in place by a small cotter pin or other fastening means, it provides a mobile joint construction, which can be readily disconnected for adjustment or maintenance.	Sizes have been standardized in nominal diameters ranging from .03 to .75 (1 to 20mm). Locking device for other fasteners. Used with a castle or slotted nut on bolt, screws, or studs, it provides a convenient, low-cost locknut assembly. Hold standard clevis pins in place. Can be used with or without a plain washer as an artificial shoulder to lock parts in position on shafts.

Fig. 10-2-1 **Machine pins.**

NOMINAL THREAD SIZE	NOMINAL COTTER PIN SIZE	COTTER PIN HOLE	END CLEARANCE*
.250 (6)	.062 (1.5)	.078 (1.9)	.11 (3)
.312 (8)	.078 (2)	.094 (2.4)	.11 (3)
.375 (10)	.094 (2.5)	.109 (2.8)	.14 (4)
.500 (12)	.125 (3)	.141 (3.4)	.17 (5)
.625 (14)	.156 (3)	.172 (3.4)	.23 (5)
.750 (20)	.156 (4)	.172 (4.5)	.27 (7)
1.000 (24)	.188 (5)	.203 (5.6)	.31 (8)
1.125 (27)	.188 (5)	.203 (5.6)	.39 (8)
1.250 (30)	.219 (6)	.234 (6.3)	.41 (10)
1.375 (36)	.219 (6)	.234 (6.3)	.44 (11)
1.500 (42)	.250 (6)	.266 (6.3)	.48 (12)
1.750 (48)	.312 (8)	.312 (8.5)	.55 (14)
*DISTANCE FROM EXTREME POINT OF BOLT OR SCREW TO CENTER OF COTTER PIN HOLE.			

Fig. 10-2-2 **Recommended cotter pin sizes.**

driven into the hole, the pins exert spring pressure against the hole wall along their entire engaged length to develop locking action.

Standard slotted tubular pins are designed so that several sizes can be used inside one another. In such combinations, shear strengths of the individual pins are additive. For spring pin application refer to Fig. 10-2-6.

QUICK-RELEASE PINS

Commercially available quick-release pins vary widely in head styles, types of locking and release mechanisms, and range of pin lengths. See Fig. 10-2-7.

Quick-release pins may be divided into two basic types: *push-pull* and *positive-locking* pins. The positive-locking pins can be further divided into three categories: heavy-duty cotter pins, single-acting pins, and double-acting pins.

Push-Pull Pins

These pins are made with either a solid or a hollow shank, containing a detent

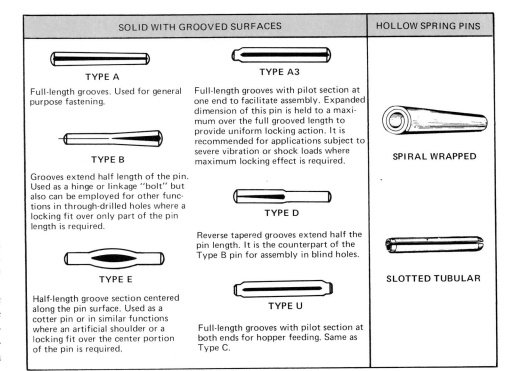

SOLID WITH GROOVED SURFACES		HOLLOW SPRING PINS
TYPE A — Full-length grooves. Used for general purpose fastening.	**TYPE A3** — Full-length grooves with pilot section at one end to facilitate assembly. Expanded dimension of this pin is held to a maximum over the full grooved length to provide uniform locking action. It is recommended for applications subject to severe vibration or shock loads where maximum locking effect is required.	**SPIRAL WRAPPED**
TYPE B — Grooves extend half length of the pin. Used as a hinge or linkage "bolt" but also can be employed for other functions in through-drilled holes where a locking fit over only part of the pin length is required.	**TYPE D** — Reverse tapered grooves extend half the pin length. It is the counterpart of the Type B pin for assembly in blind holes.	**SLOTTED TUBULAR**
TYPE E — Half-length groove section centered along the pin surface. Used as a cotter pin or in similar functions where an artificial shoulder or a locking fit over the center portion of the pin is required.	**TYPE U** — Full-length grooves with pilot section at both ends for hopper feeding. Same as Type C.	

Fig. 10-2-3 **Grooved radical locking pins.**

assembly in the form of a locking lug, button, or ball backed up by some type of resilient core, plug, or spring. The detent member projects from the surface of the pin body until sufficient force is applied in assembly or removal to cause it to retract against the spring action of the resilient core and release the pin for movement.

Positive-Locking Pins

For some quick-release fasteners, the locking action is independent of insertion and removal forces. As in the case of push-pull pins, these pins are primarily suited for shear-load applications. How-

SHAFT DIA		TRANSVERSE KEY		LONGITUDINAL KEY	
		PIN DIA	TAPER PIN NO.	PIN DIA	
in.	(mm)	in. (mm)		in. (mm)	
.375	(10)	.125 (3)	3/0	.094 (2.5)	
.438	(12)	.156 (4)	0	.125 (3)	
.500	(14)	.156 (4)	0	.125 (3)	
.562	(16)	.188 (5)	2	.156 (4)	
.625	(18)	.188 (5)	2	.156 (5)	
.750	(20)	.250 (6)	4	.156 (5)	
.875	(22)	.250 (6)	4	.219 (6)	
1.000	(24)	.312 (8)	6	.250 (6)	
1.062	(26)	.312 (8)	6	—	—
1.125	(28)	.375 (10)	7	—	—
1.188	(30)	.375 (10)	7	—	—
1.250	(32)	.375 (10)	7	.312 (8)	
1.375	(34)	.438 (11)	7	.375 (10)	
1.438	(36)	.438 (11)	7	—	—
1.500	(38)	.500 (12)	8	.438 (11)	

Fig. 10-2-4 **Recommended groove pin sizes.**

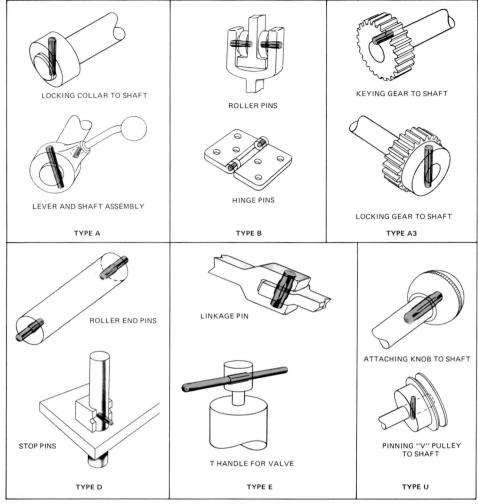

Fig. 10-2-5 Groove pin applications.

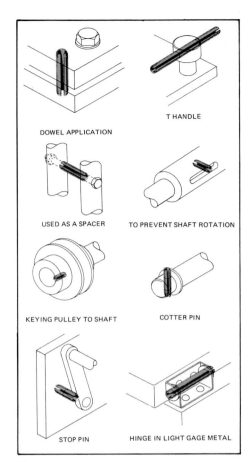

Fig. 10-2-6 Spring pin applications. (Drive Lok)

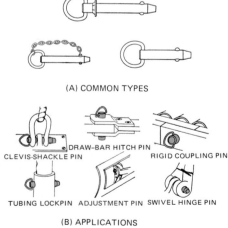

(A) COMMON TYPES

CLEVIS-SHACKLE PIN   DRAW-BAR HITCH PIN   RIGID COUPLING PIN

TUBING LOCKPIN   ADJUSTMENT PIN   SWIVEL HINGE PIN

(B) APPLICATIONS

Fig. 10-2-7 Quick-release pins.

ever, some degree of tension loading usually can be tolerated without affecting the pin function.

Reference and Source Material

1. *Machine Design,* Fastening and joining reference issue, November 1981.

ASSIGNMENTS

See Assignment 2 through 5 for Unit 10-2 starting on page 221.

UNIT 10-3
Retaining Rings

Retaining rings, or *snap rings,* are used to provide a removable shoulder to accurately locate, retain, or lock components on shafts and in bores of housings. See Fig. 10-3-1. They are easily installed and removed, and since they are usually made of spring steel, retaining rings have a high shear strength and impact capacity. In addition to fastening and positioning, a number of rings are designed for taking up end play caused by accumulated tolerances or wear in the parts being retained. In general, these devices can be placed into three categories which describe the type and method

of fabrication: stamped retaining rings, wire-formed rings, and spiral-wound retaining rings.

STAMPED RETAINING RINGS

Stamped retaining rings, in contrast to wire-formed rings with their uniform cross-sectional area, have a tapered radial width which decreases symmetrically from the center section to the free ends. The tapered construction permits the rings to remain circular when they are expanded for assembly over a shaft or contracted for insertion into a bore or housing. This constant circularity ensures maximum contact surface with the bottom of the groove.

Stamped retaining rings can be classified into three groups: axially assembled rings, radially assembled rings, and self-locking rings which do not require grooves. *Axially assembled rings* slip over the ends of shafts or down into bores, while *radially assembled rings* have side openings which permit the rings to be snapped directly into grooves on a shaft.

Commonly used types of stamped

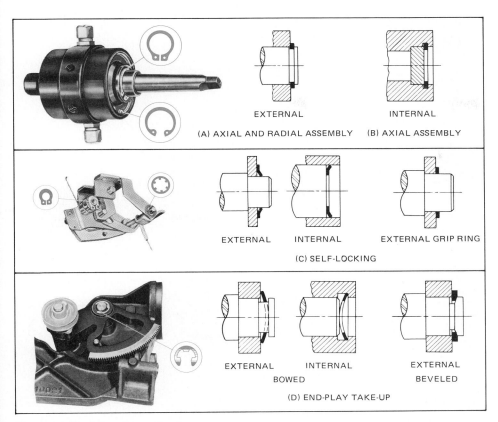

EXTERNAL   INTERNAL

(A) AXIAL AND RADIAL ASSEMBLY   (B) AXIAL ASSEMBLY

EXTERNAL   INTERNAL   EXTERNAL GRIP RING

(C) SELF-LOCKING

EXTERNAL   INTERNAL   EXTERNAL

BOWED   BEVELED

(D) END-PLAY TAKE-UP

Fig. 10-3-1 Retaining ring applications.

retaining rings are illustrated and compared in Fig. 10-3-2.

WIRE-FORMED RETAINING RINGS

The *wire-formed retaining ring* is a split ring formed and cut from spring wire of uniform cross-sectional size and shape. The wire is cold-drawn or rolled into shape from a continuous coil or bar. Then the gap ends are cut into various configurations for ease of application and removal.

Rings are available in many cross-sectional shapes, but the most commonly used are the rectangular and circular cross sections.

SPIRAL-WOUND RETAINING RINGS

Spiral-wound retaining rings consist of two or more turns of rectangular material, wound on edge to provide continuous crimped or uncrimped coil.

Reference and Source Material

1. *Machine Design,* Fastening and joining reference issue, Nov. 1981.

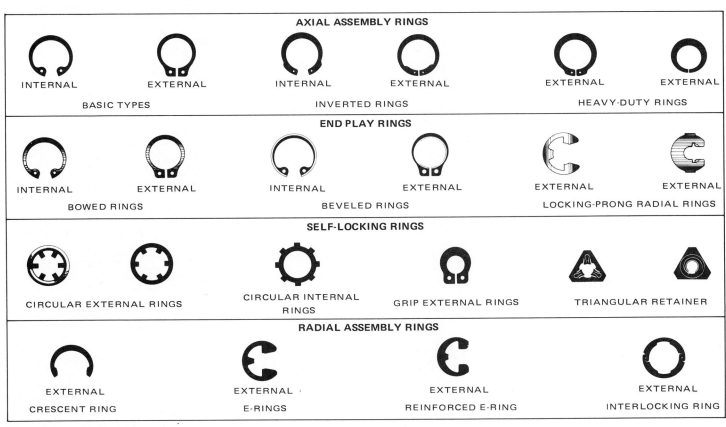

AXIAL ASSEMBLY RINGS

INTERNAL   EXTERNAL   INTERNAL   EXTERNAL   EXTERNAL   EXTERNAL

BASIC TYPES   INVERTED RINGS   HEAVY-DUTY RINGS

END PLAY RINGS

INTERNAL   EXTERNAL   INTERNAL   EXTERNAL   EXTERNAL   EXTERNAL

BOWED RINGS   BEVELED RINGS   LOCKING-PRONG RADIAL RINGS

SELF-LOCKING RINGS

CIRCULAR EXTERNAL RINGS   CIRCULAR INTERNAL RINGS   GRIP EXTERNAL RINGS   TRIANGULAR RETAINER

RADIAL ASSEMBLY RINGS

EXTERNAL   EXTERNAL   EXTERNAL   EXTERNAL

CRESCENT RING   E-RINGS   REINFORCED E-RING   INTERLOCKING RING

Fig. 10-3-2 Stamped retaining rings.

UNIT 10-4
Springs

Springs may be classified into three general groups according to their application.

Controlled Action Springs Controlled action springs have a well-defined function, or a constant range of action for each cycle of operation. Examples are valve, die, and switch springs.

Variable-Action Springs Variable-action springs have a changing range of action because of the variable conditions imposed upon them. Examples are suspension, clutch, and cushion springs.

Static Springs Static springs exert a comparatively constant pressure or tension between parts. Examples are packing or bearing pressure, anti-rattle, and seal springs.

TYPES OF SPRINGS

The type or name of a spring is determined by characteristics such as function, shape of material, application, or design. Figure 10-4-1 illustrates common springs in use. Figure 10-4-2 designates spring nomenclature.

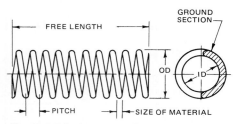

Fig. 10-4-2 Spring nomenclature.

Compression Springs

A *compression spring* is an open-coil, helical spring that offers resistance to a compressive form.

Compression Spring Ends Figure 10-4-3A shows the ends commonly used on compression springs.

Plain open ends are produced by straight cutoff with no reduction of helix angle. See Fig. 10-4-4. The spring should

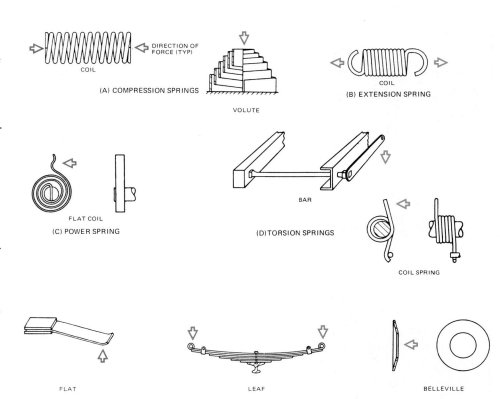

Fig. 10-4-1 Types of springs.

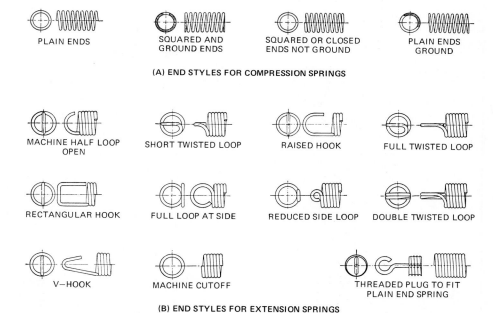

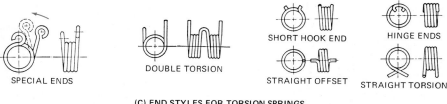

Fig. 10-4-3 End styles for helical springs.

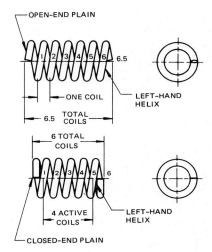

Fig. 10-4-4 Coil definitions.

be guided on a rod or in a hole to operate satisfactorily.

Ground open ends are produced by parallel grinding of open-end coil springs. Advantages of this type of end are improved stability and a larger number of total coils.

Plain closed ends are produced with a straight cutoff and with reduction of helix angle to obtain closed-end coils, resulting in a more stable spring.

Ground closed ends are produced by parallel grinding of closed-end coil springs, resulting in maximum stability.

Extension Springs

An *extension spring* is a close-coiled helical spring that offers resistance to a pulling force. It is made from round or square wire.

Extension Spring Ends The end of an extension spring is usually the most highly stressed part. Thus, proper consideration should be given to its selection. The types of ends shown in Fig. 10-4-3B are most commonly used on extension springs. Different types of ends can be used on the same spring.

Torsion Springs

Springs exerting pressure along a path which is a circular arc, or, in other words, providing a torque, are called torsion springs, motor springs, power springs, etc. The term *torsion spring* is usually applied to a helical spring of round, square, or rectangular wire, loaded by torque.

The variation in ends used is almost limitless, but a few of the more common types are illustrated in Fig. 10-4-3C.

Torsion Bar Springs A torsion bar spring is a relatively straight bar anchored at one end, on which a torque may be exerted at the other end, thus tending to twist it about its axis.

Power Springs

Clock or Motor Type A *flat coil spring*, also known as a clock or motor spring, consists of a strip of tempered steel wound on an arbor and usually confined in a case or drum.

Flat Springs

Flat springs are made of flat material formed in such a manner as to apply force in the desired direction when deflected in the opposite direction.

Leaf Springs A *leaf spring* is composed of a series of flat springs nested together and arranged to provide approximately uniform distribution of stress throughout its length. Springs may be used in multiple arrangements, as shown in Fig. 10-4-5.

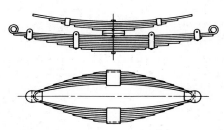

Fig. 10-4-5 Leaf springs—multiple arrangements.

Belleville Springs Belleville springs are washer-shaped, made in the form of a short, truncated cone.

Belleville washers may be assembled in series to accommodate greater deflections, in parallel to resist greater forces, or in combination of series and parallel, as shown in Fig. 10-4-6.

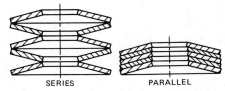

SERIES   PARALLEL

Fig. 10-4-6 Belleville spring applications.

SPRING DRAWINGS

On working drawings, a schematic drawing of a helical spring is recommended to save drafting time. See Fig.

10-4-7. As in screw-thread representation, straight lines are used in place of the helical curves. On assembly drawings, springs are normally shown in section, and either cross-hatching lines or solid black shading is recommended, depending on the size of the wire's diameter. See Fig. 10-4-8.

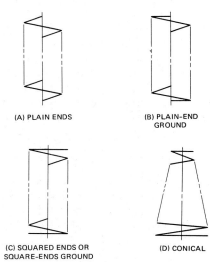

(A) PLAIN ENDS   (B) PLAIN-END GROUND

(C) SQUARED ENDS OR SQUARE-ENDS GROUND   (D) CONICAL

Fig. 10-4-7 Schematic drawing of compression springs.

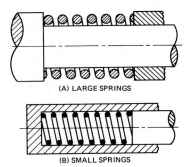

(A) LARGE SPRINGS

(B) SMALL SPRINGS

Fig. 10-4-8 Showing helical springs on assembly drawings.

Dimensioning Springs The following information should be given on a drawing of a spring:

- Size, shape, and kind of material used in the spring
- Diameter (outside or inside)
- Pitch or number of coils
- Shape of ends
- Length
- Load and rate (not covered in this text)

EXAMPLE
ONE HELICAL TENSION SPRING 3.00 LG (OR NUMBER OF COILS),

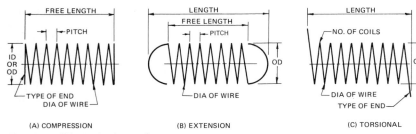

Fig. 10-4-9 Dimensioning springs.

.50 ID, PITCH .25, 18 B & S GA
SPRING BRASS WIRE

In using single-line representation, the dimensions should state the applicable size of material required to ensure correct interpretation on such features as inside diameter, outside diameter, and end loops. See Fig. 10-4-9.

SPRING CLIPS

Spring clips are a relatively new class of industrial fasteners. They perform multiple functions, eliminate the handling of several small parts, and thus reduce assembly costs. See Fig. 10-4-10.

The *spring clip* is generally self-retaining, requiring only a flange, panel edge, or mounting hole to clip to.

Basically, spring clips are light-duty fasteners and serve the same function as small bolts and nuts, self-tapping screws, clamps, spot welding, and formed retaining plates.

Dart-Type Spring Clips Dart-shaped panel retaining elements have hips to engage within panel or component holes. The top or arms of the fastener can be formed in any shape to perform unlimited fastening functions.

Stud Receiver Clips There are three basic types of stud receivers: push-ons, tubular types, and self-threading fasteners. All are designed to make attachments to unthreaded studs, rivets, pins, or rods of metal or plastic.

Cable, Wire, and Tube Clips These fasteners incorporate self-retaining elements for engaging panel holes or mounting on panel edges and flanges.

Spring-clip cable, wire, and tubing fasteners are front-mounting devices, requiring no access to the back of the panel.

Spring Molding Clips Molding retaining clips are formed with legs that hold the clips to a panel and arms that positively engage the flanges of various sizes and shapes of trim molding and pull the molding tightly to the attaching panel.

U-, S-, and C-Shaped Spring Clips These spring clips get their names from their shapes. The fastening function is accomplished by using inward compressive spring force to secure assembly components or provide self-retention after installation.

References and Source Material

1. General Motors Corporation.
2. The Wallace Barnes Company Limited.
3. *Machine Design*, Fastening and joining reference issue, Nov. 1981.

ASSIGNMENT

See Assignment 7 for Unit 10-4 starting on page 223.

UNIT 10-5
Rivets

STANDARD RIVETS

Riveting is a popular method of fastening and joining, primarily because of its simplicity, dependability, and low cost. A myriad of manufactured products and structures, both small and large, are held together by these fasteners. Rivets are classified as permanent fastenings, as distinguished from removable fasteners, such as bolts and screws.

Basically, a *rivet* is a ductile metal pin which is inserted through holes in two or more parts, and the ends are formed over to securely hold the parts.

Another important reason for riveting is versatility, with respect to both the properties of rivets as fasteners and the method of clinching.

- Part materials: Rivets can be used to join dissimilar materials, metallic or nonmetallic, in various thicknesses.
- Multiple functions: Rivets can serve as fasteners, pivot shafts, spacers, electric contacts, stops, or inserts.
- Fastening finished parts: Rivets can be used to fasten parts that have already received a final painting or other finishing.

Riveted joints are neither watertight nor airtight, although such a joint may

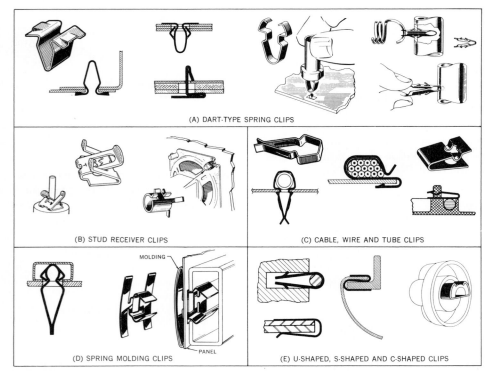

Fig. 10-4-10 Spring clips.

be attained at some added cost by using a sealing compound. The riveted parts cannot be disassembled for maintenance or replacement without knocking the rivet out and clinching a new one in place for reassembly. Common riveted joints are shown in Fig. 10-5-1.

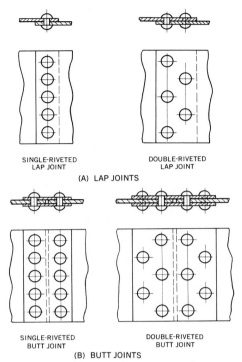

Fig. 10-5-1 Common riveted joints.

Large Rivets

Large rivets are used in structural work of buildings and bridges. Today, however, high-strength bolts have almost completely replaced rivets in field connections because of cost, strength, and the noise factor. Rivet joints are of two types: butt and lapped. The more common types of large rivets are shown in Fig. 10-5-2. In order to show the dif-

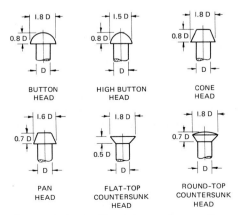

Fig. 10-5-2 Approximate sizes and types of large rivets .50 in. (12 mm) and up.

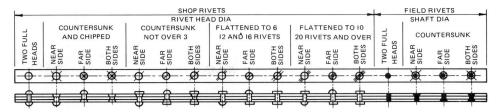

Fig. 10-5-3 Conventional rivet symbols.

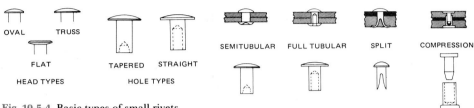

Fig. 10-5-4 Basic types of small rivets.

ference between *shop rivets* (rivets that are put in the structure at the shop) and *field rivets* (rivets that are used on the site), two types of symbols are used. In drawing shop rivets, the diameter of the rivet head is shown on the drawings. For field rivets, the shaft diameter is used. Figure 10-5-3 shows the conventional rivet symbols adopted by the American and Canadian Institutes of Steel Construction.

Small Rivets

Design of small rivet assemblies is influenced by two major considerations:

1. The joint itself, its strength, appearance, and configuration
2. The final riveting operation, in terms of equipment capabilities and production sequence

Types of Small Rivets (Fig. 10-5-4)

Semitubular This is the most widely used type of small rivet. The depth of the hole in the rivet, measured along the wall, does not exceed 112 percent of the mean shank diameter. The hole may be extruded (straight or tapered) or drilled (straight), depending on the manufacturer and/or rivet size.

Full Tubular This rivet has a drilled shank with a hole depth more than 112 percent of the mean shank diameter. It can be used to punch its own hole in fabric, some plastic sheets, and other soft materials, eliminating a preliminary punching or drilling operation.

Bifurcated (Split) The rivet body is sawed or punched to produce a pronged

shank that punches its own hole through fiber, wood, or plastic.

Compression This rivet consists of two elements: the solid or blank rivet and the deep-drilled tubular member. Pressed together, these form an interference fit.

Design Recommendations (Fig. 10-5-5)

Select the Right Rivets Basic types are covered in Fig. 10-5-4. Rivet standards for all types but compression rivets have been published by the Tubular and Split Rivet Council.

Rivet Diameters The optimum rivet diameter is determined not by performance requirements but by economics—the costs of the rivet and the labor to install it. The rivet length-to-diameter ratio should not exceed 6:1.

Rivet Positioning The location of the rivet in the assembled product influences both joint strength and clinching requirements. The important dimensions are edge distance and pitch distance.

Edge distance is the interval between the edge of the part and the center line of the rivet.

The recommended edge distance for plastic materials, either solid or laminated, is between 2 and 3 diameters, depending on the thickness and inherent strength of the material.

Pitch distance—the interval between center lines of adjacent rivets—should not be too small. Unnecessarily high stress concentrations in the riveted material and buckling at adjacent empty

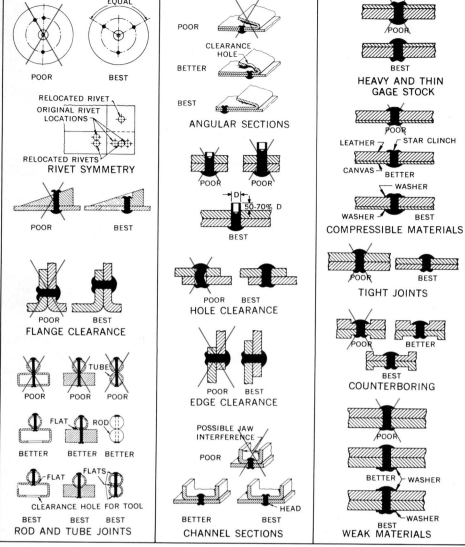

Fig. 10-5-5 Small rivet design data.

holes can result if the pitch distance is less than 3 times the diameter of the largest rivet in the assembly (metal parts) or 5 times the diameter (plastic parts).

BLIND RIVETS

Blind riveting is a technique for setting a rivet without access to the reverse side of the joint. However, blind rivets may also be used in applications in which both sides of the joint are actually accessible.

Blind rivets are classified according to the methods with which they are set: pull-mandrel, drive-pin, and chemically expanded. See Fig. 10-5-6.

Design Considerations (Fig. 10-5-7)

Type of Rivet Selection depends on a number of factors, such as speed of assembly, clamping capacity, available sizes, adaptability to the assembly, ease of removal, cost, and structural integrity of the joint.

Joint Design Factors that must be known include allowable tolerances of rivet length versus assembly thickness, hole clearance, joint configuration, and type of loading.

Speed of Installation The fastest, most efficient installation is done with power tools—air, hydraulic, or electric. Manual tools, such as special pliers, can be used efficiently with practically no training.

In-Place Costs Blind rivets often have lower in-place costs than solid rivets or tapping screws.

Loading A blind-rivet joint is usually in compression or shear.

Material Thickness Some rivets can be set in materials as thin as .02 in. (0.5 mm). Also, if one component is of compressible material, rivets with extra-large head diameter should be used.

Edge Distance The average recommended edge distance (Fig. 10-5-7, top) is twice the diameter of the rivet.

Spacing Rivet pitch (Fig. 10-5-7, top) should be 3 times the diameter of the rivet.

Length The amount of length needed (Fig. 10-5-7, top) for clinching action varies greatly. Most rivet manufacturers provide data on grip ranges of their rivets.

Backup Clearance Full entry of the rivet is essential for tightly clinched joints. Sufficient backup clearance (Fig. 10-5-7, top) must be provided to accommodate the full length of the unclinched rivet, A.

Blind Holes or Slots A useful application of a blind rivet is in fastening members in a blind hole (Fig. 10-5-7,

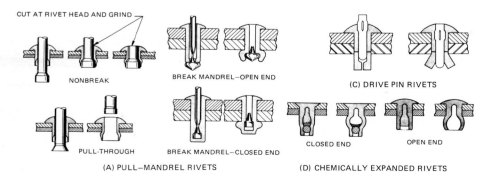

Fig. 10-5-6 Basic types of blind rivets and methods of setting. (Machine Design, Vol. 53, No. 26)

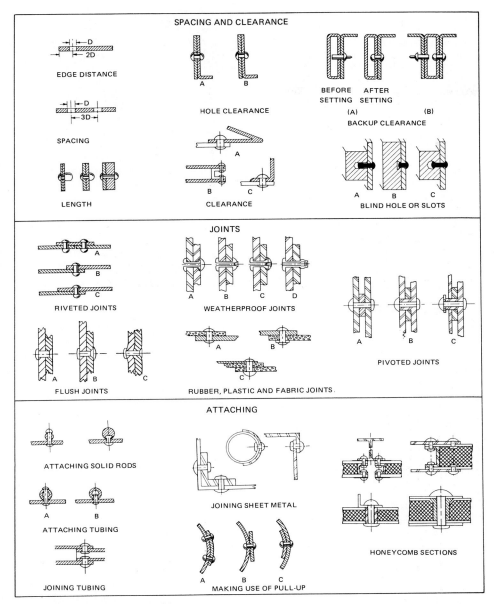

Fig. 10-5-7 Blind-rivet design data.

top). At A, the formed head bears against the side of the hole only. This joint is not as strong as the other two (B and C).

Riveted Joints Riveted cleat or batten holds a butt joint, A. The simple lap joint, B, must have sufficient material beyond the hole for strength. Excessive material beyond rivet hole C may curl up or vibrate or cause interference problems, depending on the installation (Fig. 10-5-7, middle).

Flush Joints Generally, flush joints are made by countersinking one of the sections and using a rivet with a countersunk head, A.

Weatherproof Joints A hollow-core rivet can be sealed by capping it, A; by plugging it, B; or by using both a cap and a plug, C. To obtain a true seal, however, a gasket or mastic should be used between the sections and perhaps under the rivet head. An ideal solution is to use a closed-end rivet, D (Fig. 10-5-7, middle).

Rubber, Plastic, and Fabric Joints Some plastics, such as reinforced molded fiberglass, or polystyrene, which are reasonably rigid, present no problem for most small rivets. However, when the material is very flexible or is a fabric, set the rivet as shown at A or B, with the upset head against the solid member. If this practice is not possible, use a backup strip as shown at C (Fig. 10-5-7, middle).

Pivoted Joints There are a number of ways of producing a pivoted assembly. Three are shown in Fig. 10-5-7, middle.

Attaching Solid Rod When attaching a rod to other members, the usual practice is to pass the rivet completely through the rod (Fig. 10-5-7, bottom).

Attaching Tubing Attaching tubing is an application for which the blind rivet is ideally suited.

Joining Tubing This tubing joint is a common form of blind riveting, used for both structural and low-cost power transmission assemblies (Fig. 10-5-7, bottom).

Making Use of Pull-up By judicious positioning of rivets and parts that are to be assembled with rivets, the setting force can sometimes be used to pull together unlike parts (Fig. 10-5-7, bottom).

Honeycomb Sections Inserts should be employed to strengthen the section and provide a strong joint.

Reference and Source Material

1. *Machine Design*, Fastening and joining reference issue, Nov. 1981.

ASSIGNMENT

See Assignment 8 for Unit 10-5 starting on page 224.

U N I T   1 0 - 6
Welded Fasteners

The most common forms of welded fasteners are screws and nuts. In this unit, welded fasteners are grouped into *resistance-welded* threaded fasteners and *arc-welded* studs.

RESISTANCE-WELDED FASTENERS

Simply defined, a *resistance-welded fastener* is an externally or internally threaded metal part designed to be fused permanently in place by standard production welding equipment. Two methods of resistance welding are used to attach these fasteners (Fig. 10-6-1): projection welding and spot welding.

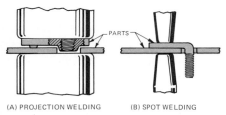

Fig. 10-6-1 Basic methods of attaching resistance-welded fasteners.

Design Considerations

Before fasteners can be used, three basic requirements must be met. See Figs. 10-6-2 to 10-6-4.

1. The materials to be joined, both part and fastener, must be suitable for resistance welding.

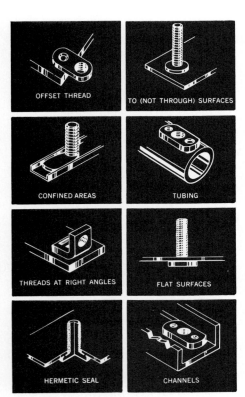

Fig. 10-6-2 Application of resistance-welded fasteners. (Ohio Nut and Bolt Co.)

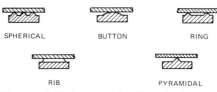

Fig. 10-6-3 Basic types of weld projection.

Use Projection-Weld Fasteners When	Use Spot-Weld Fasteners When
• Suitable projection welding equipment is available. • Appearance is an important consideration. Projection welding does not mark the surface on the opposite side of the weld. • Simultaneous welding of multiple fasteners is required. • Spacing between fasteners must be kept close. • Fasteners must be welded to part sections of varying thicknesses. • Fasteners must be welded to parts of unusual shape or a watertight weld joint is required. • Welding fixtures can be used for easier locating or automatic feeding. • Length of production run without maintenance is critical.	• Suitable rocker-arm welding equipment is available. • Appearance of the part surface opposite the weld is not critical. Spot welding leaves a slight indentation from the electrode tips. • Other spot welds are being performed on parts of the assembly. • Length of production run without maintenance is not too important. Spot-welding electrode tips will mushroom to some extent in production welding. Shorter runs before refacing or redressing must be expected. • Dissimilar materials, such as aluminum, copper, or magnesium, are being welded. • Shape, size, or space requirements do not permit use of projection welded fasteners.

Fig. 10-6-4 Guide to weld fastener selection.

2. The parts to be welded must be portable enough to be carried to the welder.
3. Production volume should be great enough to justify tooling costs.

Figure 10-6-5 shows typical resistance-welded fasteners.

ARC-WELDED STUDS

There are two basic stud welding processes: *electric-arc* and *capacitor-discharge*.

Electric-Arc Stud Welding The more widely used stud welding process is a semiautomatic electric-arc process. To avoid burn-through, the plate thickness should be at least one-fifth the weld base diameter.

Capacitor-Discharge Stud Welding This stud welding process derives its heat from an arc produced by a rapid discharge of stored electrical energy.

Design Considerations

In most instances, the thickness of the plate for stud attachment will determine the stud welding process. Electric-arc stud welding is generally used for fasteners .32 in. (8 mm) and larger.

Reference and Source Material

1. *Machine Design*, Fastening and joining reference issue, Nov. 1981.

ASSIGNMENT

See Assignment 9 for Unit 10-6 starting on page 225.

U N I T   1 0 - 7
Adhesive Fastenings

Industrial designers and manufacturers are relying on adhesives more than ever before. They allow greater versatility in design, styling, and materials. They can also cut costs. However, as with any engineering tool, there are limitations as well as advantages.

Adhesion Versus Stress

Adhesion is the force that holds materials together.

Stress, on the other hand, is the force pulling materials apart. See Fig. 10-7-1. The basic types of stress in adhesives are:

1. *Tensile*. Pull is exerted equally over the entire joint. Pull direction is straight and away from the adhesive bond. All adhesive contributes to bond strength.
2. *Shear*. Pull direction is across the adhesive bond. The bonded materials are being forced to slide over one another.
3. *Cleavage*. Pull is concentrated at one edge of the joint and exerts a prying force on the bond. The other edge of the joint is theoretically under zero stress.
4. *Peel*. One surface must be flexible. Stress is concentrated along a thin line at the edge of the bond.

Resistance to stress is one reason for the rapid increase in the use of adhesives for product assembly. The following points elaborate on stress resistance and the other advantages of adhesives.

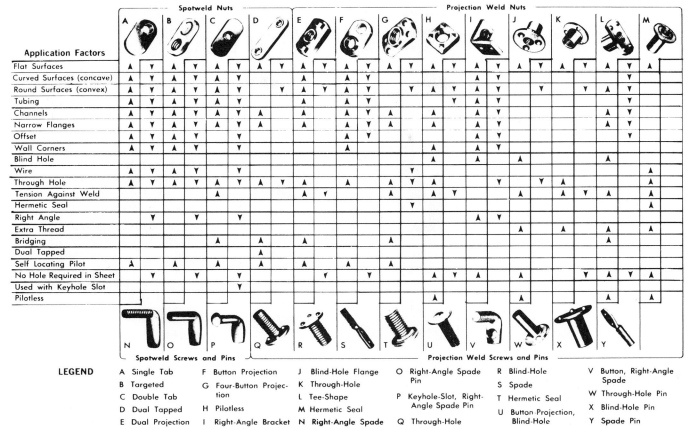

Fig. 10-6-5 Resistance-welded fasteners.

LEGEND

A Single Tab
B Targeted
C Double Tab
D Dual Tapped
E Dual Projection
F Button Projection
G Four-Button Projection
H Pilotless
I Right-Angle Bracket
J Blind-Hole Flange
K Through-Hole
L Tee-Shape
M Hermetic Seal
N Right-Angle Spade
O Right-Angle Spade Pin
P Keyhole-Slot, Right-Angle Spade Pin
Q Through-Hole
R Blind-Hole
S Spade
T Hermetic Seal
U Button-Projection, Blind-Hole
V Button, Right-Angle Spade
W Through-Hole Pin
X Blind-Hole Pin
Y Spade Pin

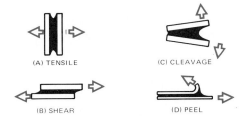

Fig. 10-7-1 Basic types of stress on bonded joints. (3M Co.)

Advantages

1. Adhesives allow uniform distribution of stress over the entire bond area. See Fig. 10-7-2. This eliminates stress concentration caused by rivets, bolts, spot welds, and similar fastening techniques. Lighter, thinner materials can be used without sacrificing strength.
2. Adhesives can effectively bond dissimilar materials.
3. Continuous contact between mating surfaces effectively bonds and seals against many environmental conditions.
4. They eliminate holes needed for mechanical fasteners and surface marks resulting from spot welding, brazing, etc.

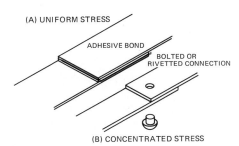

Fig. 10-7-2 Stress caused by fasteners.

Limitations

1. Adhesive bonding can be slow or require critical processing. This is particularly true in mass production. Some adhesives require heat and pressure, or special jigs and fixtures, to establish the bond.
2. Adhesives are sensitive to surface conditions. Special surface preparation may be required.
3. Some adhesive solvents present hazards. Special ventilation may be required to protect employees from toxic vapors.
4. Environmental conditions can reduce bond strength of some adhesives. Some do not hold well when exposed to low temperatures, high humidity, severe heat, chemicals, water, etc.

JOINT DESIGN

Joints should be specifically designed for use with structural adhesives. First, the joint should be designed so that all the bonded area shares the load equally. Second, the joint configuration should be designed so that basic stress is primarily in shear or tensile, with cleavage and peel minimized or eliminated.

The following structural joints and their advantages and disadvantages illustrate some typical design alternatives. See Fig. 10-7-3.

Lap Joints Lap joints are most practical and applicable in bonding thin materials. The *simple lap joint* is offset. This can result in cleavage and peel

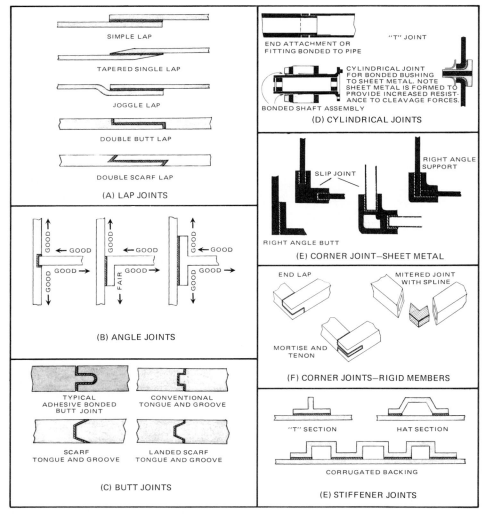

Fig. 10-7-3 Adhesive joint design.

stress under load when thin materials are used. A *tapered single lap joint* is more efficient than a simple lap joint. The tapered edge allows bending of the joint edge under stress. The *joggle lap joint* gives more uniform stress distribution than either the simple or tapered lap joint.

The *double-butt lap joint* gives more uniform stress distribution in the load-bearing area than the above joints. This type of joint, however, requires machining which is not always feasible with thinner-gage metals. *Doublescarf lap joints* have better resistance to bending forces than double-butt joints.

Angle Joints Angle joints give rise to either peel or cleavage stress depending on the gage of the metal. Typical approaches to the reduction of cleavage are illustrated.

Butt Joints The following recessed butt joints are recommended: landed scarf tongue and groove, conventional tongue and groove, and scarf tongue and groove.

Cylindrical Joints The *T joint* and *overlap slip joint* are typical for bonding cylindrical parts such as tubing, bushings, and shafts.

Corner Joints—Sheet Metal Corner joints can be assembled with adhesives by using simple supplementary attachments. This permits joining and sealing in a single operation. Typical designs are right-angle butt joints, slip joints, and right-angle support joints.

Corner Joints—Rigid Members Corner joints, as in storm doors or decorative frames, can be adhesive-bonded. End lap joints are the simplest design type, although they require machining. Mortise and tenon joints are excellent from a design standpoint, but they also require machining. The mitered joint with an insert is best if both members are hollow extrusions.

Stiffener Joints Deflection and flutter of thin metal sheets can be minimized with adhesive-bonded stiffeners.

Reference and Source Material

1. 3M Company.

ASSIGNMENT

See Assignment 10 for Unit 10-7 on page 226.

ASSIGNMENTS FOR CHAPTER 10

Assignment for Unit 10-1, Keys, Splines, and Serrations

1. Lay out the two fastener assemblies shown in Fig. 10-1-A or 10-1-B. The following fasteners are used:

For Fig. 10-1-A
- *Assembly A:* flat key
- *Assembly B:* serrations

For Fig. 10-1-B
- *Assembly A:* square key
- *Assembly B:* Woodruff key

Refer to the Appendix and manufacturers' catalogs for sizes and use your judgment for dimensions not shown. Show the dimensions for the keyseats and serrations. Scale 1:1.

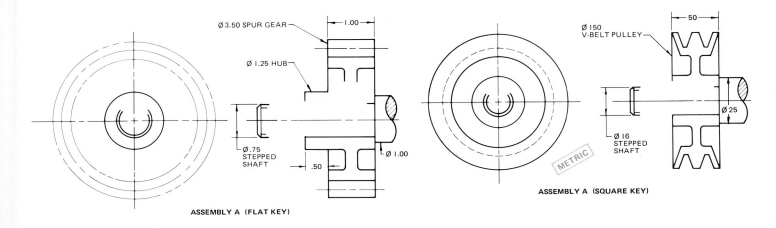

ASSEMBLY A (FLAT KEY)

ASSEMBLY A (SQUARE KEY)

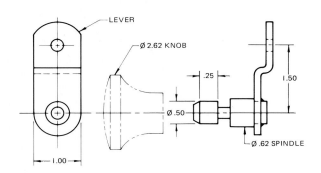

ASSEMBLY B (SERRATIONS)

Fig. 10-1-A Key and serration fasteners.

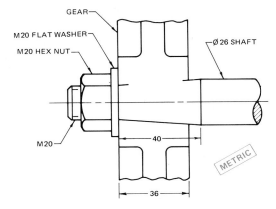

ASSEMBLY B (WOODRUFF KEY)

Fig. 10-1-B Key fasteners.

Assignments for Unit 10-2, Pin Fasteners

2. Complete the pin assemblies shown in Fig. 10-2-A or 10-2-B, given the following information:

For Fig. 10-2-A
- *Assembly A.* Slotted tubular spring pins are used to fasten the cap and handle to the shaft. Scale 1:2.
- *Assembly B.* A clevis pin whose area is equal to the four rivets is used to fasten the trailer hitch to the tractor draw bar. Scale 1:2.

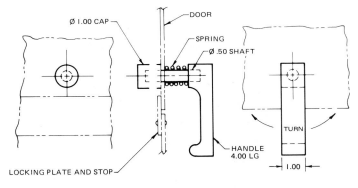

ASSEMBLY A (CABINET HANDLE)

Fig. 10-2-A Pin fasteners.

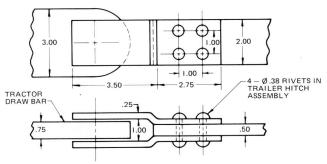

ASSEMBLY B (DRAW BAR HITCH)

MISCELLANEOUS TYPES OF FASTENERS **221**

For Fig. 10-2-B

- *Assembly A.* A type E grooved pin holds the roller to the bracket. A washer and cotter pin are used to fasten the bracket to the pushrod. Scale 1:1.
- *Assembly B.* A type A3 grooved pin holds the V-belt pulley to the shaft. Scale 1:1.

Refer to manufacturers' catalogs for pin sizes and provide the complete information to order each fastener.

3. Prepare detail drawings of the parts shown in Fig. 10-2-C. Use your judgment for the scale and selection of views. Include an item list.

4. Make a two-view assembly drawing of the crane hook shown in Fig. 10-2-D. The hook is to be held to the U-frame with a slotted locknut. A spring pin is inserted through the locknut slots to prevent the nut from turning. A clevis pin with washer and cotter pin holds the pulley to the frame. Include on the drawing an item list. Scale 1:1.

5. Prepare detail drawings of the parts in Assignment 4. Use your judgment for the scale and selection of views.

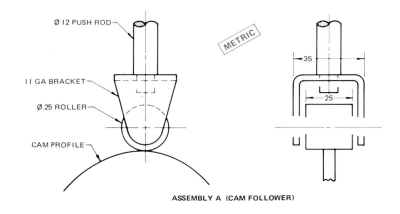

ASSEMBLY A (CAM FOLLOWER)

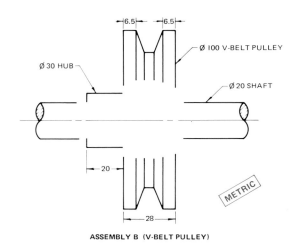

ASSEMBLY B (V-BELT PULLEY)

Fig. 10-2-B Pin fasteners.

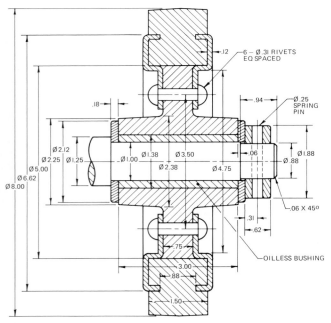

Fig. 10-2-C Wheel assembly.

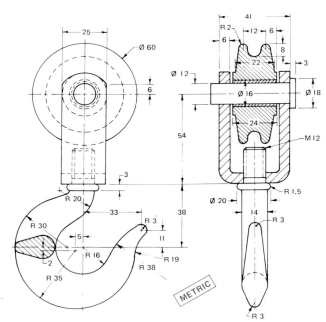

Fig. 10-2-D Crane hook.

Assignments for Unit 10-3, Retaining Rings

6. Complete the assemblies shown in Fig. 10-3-A or 10-3-B by adding suitable retaining rings as per the information supplied below. Refer to manufacturers' catalogs and show on the drawing the catalog number for the retaining ring. Add ring and groove sizes. Scale 1:1. Use your judgment for dimensions not shown.

For Fig. 10-3-A
- *Assembly A.* An external radial retaining ring mounted on the shaft is to act as a shoulder for the shaft support. An external axial retaining ring is required to hold the gear on the shaft.
- *Assembly B.* The plunger and punch are held into the punch holder by internal retaining rings.

For Fig. 10-3-B
- *Assembly A.* External self-locking retaining rings hold the roller shaft in position on the bracket.
- *Assembly B.* An external self-locking ring holds the plastic housing to the viewer case. An internal self-locking ring holds the lens in position.

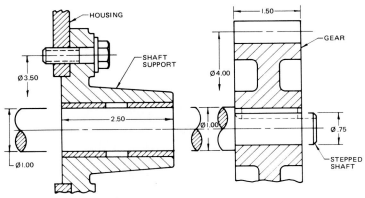

ASSEMBLY A (EXTERNAL RETAINING RINGS)

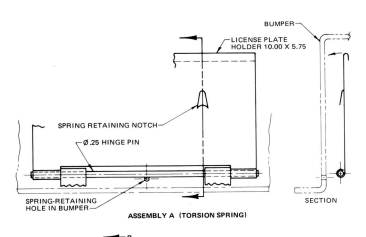

ASSEMBLY A
(EXTERNAL SELF-LOCKING)

ASSEMBLY B
(EXTERNAL AND INTERNAL SELF-LOCKING)

Fig. 10-3-B Retaining ring fasteners.

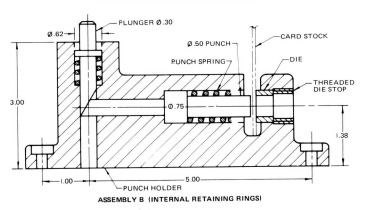

ASSEMBLY B (INTERNAL RETAINING RINGS)

Fig. 10-3-A Retaining ring fasteners.

Assignments for Unit 10-4, Springs

7. Lay out the two assembly drawings as shown in Figs. 10-4-A or 10-4-B. Complete the drawings from the information supplied below, and make detail drawings of the springs. Use your judgment for sizes not given.

For Fig. 10-4-A
- *Assembly A.* The license plate holder is held to the frame of the car by a hinge. A torsion spring is required to keep the plate holder in position. The torsion spring is slipped over the hinge pin during assembly, and one end of the spring passes through the hole in the

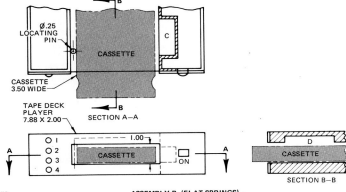

ASSEMBLY A (TORSION SPRING)

ASSEMBLY B (FLAT SPRINGS)

Fig. 10-4-A Spring fasteners.

bumper. The other end of the spring is locked into the spring-retaining notch in the license plate holder. Scale 1:2.

- *Assembly B.* Flat springs are positioned in openings C and D in the tape deck player. These springs hold the cassette against the bottom, and the locating pin positioned in the left side of the tape deck. Scale 1:2.

For Fig. 10-4-B

- *Assembly A.* An extension spring controls the lever. The spring is fastened to the neck in the pin and through the hole in the lever. Scale 1:1.
- *Assembly B.* A compression spring mounted on the shaft of the handle provides sufficient pressure to hold the lever in position, thus maintaining the door against the panel. To open the door, the handle is pushed in and turned. This action compresses the spring and forces the lever away from the notch in the panel edge, thus permitting the lever to turn. Scale 1:1.

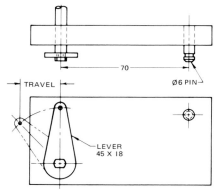

ASSEMBLY A (EXTENSION SPRING)

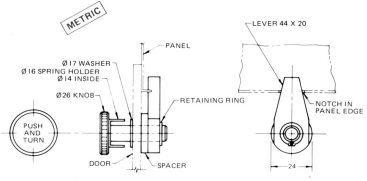

ASSEMBLY B (COMPRESSION SPRING)

Fig. 10-4-B Spring fasteners.

Assignment for Unit 10-5, Rivets

8. Complete the two assembly drawings shown in Fig. 10-5-A or 10-5-B from the information supplied below. Refer to manufacturers' catalogs for rivet type and sizes, and on each assembly show the callout for the rivets. Use your judgment for sizes not given.

For Fig. 10-5-A

- *Assembly A.* Padlock brackets are riveted to the locker door and door frame with two blind rivets in each bracket. Scale 1:1.
- *Assembly B.* The roof truss is assembled in the shop with five evenly spaced Ø .50-in. (12-mm) rivets in each angle. Scale 1:4.

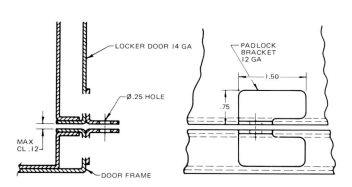

ASSEMBLY A (BLIND RIVETS)

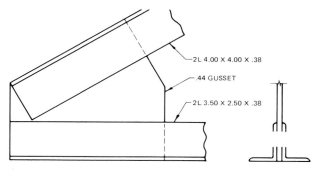

ASSEMBLY B (LARGE STRUCTURAL RIVETS)

Fig. 10-5-A Rivet fasteners.

For Fig. 10-5-B

- *Assembly A.* The grill is held to the panel by four truss-head full tubular rivets. Scale 1:1.
- *Assembly B.* The support is held to the plywood panel by drive rivets uniformly spaced on the gage lines. Two rivets hold the bracket to the support. Scale 1:1.

Assignment for Unit 10-6, Welded Fasteners

9. Complete the two assemblies shown in Fig. 10-6-A or 10-6-B. Refer to manufacturers' catalogs and the Appendix for standard fastener components. Complete the drawings from the information supplied below. Use your judgment for sizes not given. Scale 1:1.

For Fig. 10-6-A

- *Assembly A.* Two resistance-welded threaded fasteners, one on each side of the pipe, are required. The bracket drops over the fasteners, and lockwashers and nuts secure the bracket to the pipe.
- *Assembly B.* A leakproof attaching method (stud welding) is required to hold the adapter to the panel.

For Fig. 10-6-B

- *Assembly A.* A spot-weld nut is to be attached to the panel. A hole in the clamp permits a machine screw to fasten the pipe clamp to the nut.
- *Assembly B.* A right-angle bracket is to be fastened (projection welding) to the bottom plate. The vertical plate is secured to the bracket by a machine screw and lock washer.

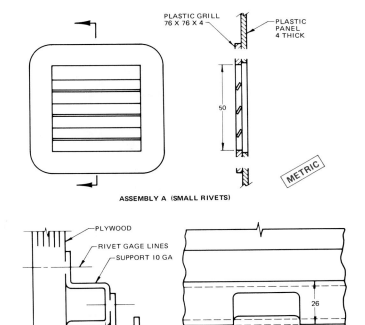

Fig. 10-5-B Rivet fasteners.

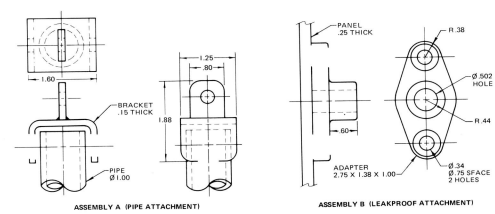

Fig. 10-6-A Welded fasteners.

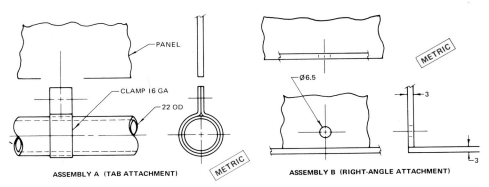

Fig. 10-6-B Welded fasteners.

Assignment for Unit 10-7, Adhesive Fastenings

10. Complete the two adhesive-bonded assemblies shown in Fig. 10-7-A or 10-7-B from the information supplied below and the adhesive chart in the Appendix. List the adhesive product number and state the method of application you would recommend. Use your judgment for sizes not shown, and dimension the joint. Scale is to suit.

For Fig. 10-7-A
- *Assembly A.* The riveted joint shown is to be replaced by a joggle lap joint. It must be fast-drying.
- *Assembly B.* The sheet-metal corner joint shown is to be replaced by a slip joint. It must be water-resistant.

For Fig. 10-7-B
- *Assembly A.* Three pieces of wood are to be assembled into the shape shown. Joint design has not been shown.
- *Assembly B.* The riveted joint shown is to be replaced by a joggle lap joint. Must meet specification requirements by MMM-A-121.

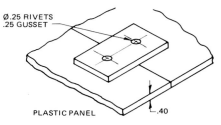

ASSEMBLY A (BUTT JOINT)

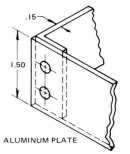

ASSEMBLY B (SLIP JOINT)

Fig. 10-7-A Adhesive fastenings.

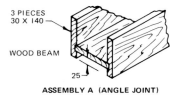

ASSEMBLY A (ANGLE JOINT)

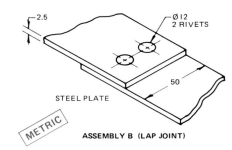

ASSEMBLY B (LAP JOINT)

Fig. 10-7-B Adhesive fastenings.

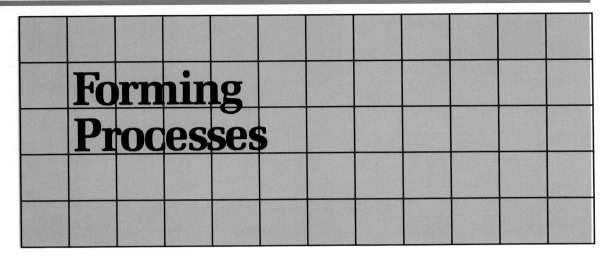

Forming Processes

UNIT 11-1
Metal Castings

FORMING PROCESSES[1]

When a component of a machine takes shape on the drawing board or CAD monitor of the designer, the method of its manufacture may still be entirely open. The number of possible manufacturing processes is increasing day by day, and the optimum process is found only by carefully weighing technological advantages and drawbacks in relation to the economy of production.

The choice of the manufacturing process depends on the size, shape, and quantity of the component. Manufacturing processes are therefore important to the engineer and drafter in order to properly design a part. They must be familiar with the advantages, disadvantages, costs, and machines necessary for manufacturing. Since the cost of the part is influenced by the production method, such as welding or casting, the designer must be able to choose wisely the method which will reduce the cost. In some cases it may be necessary to recommend the purchase of a new or different machine in order to produce the part at a competitive price.

This means the designer should design the part for the process as well as for the function. Most of all, unnecessarily close tolerances on non-functional dimensions should be avoided.

This chapter covers the following manufacturing processes: casting, forging, and powder metallurgy. Forming by means of welding and stamping (punches and dies) is covered in Chaps. 17 and 30, respectively.

CASTING PROCESSES[2,3]

Casting is the process whereby parts are produced by pouring molten metal into a mold. A typical cast part is shown in Fig. 11-1-1. Casting processes for metals can be classified by either the type of mold or pattern or the pressure or force used to fill the mold. Conventional sand, shell, and plaster molds utilize a permanent pattern, but the mold is used only once. Permanent molds and die-casting dies are machined in metal or graphite sections and are employed for a large number of castings. Investment casting and the relatively new full-mold process involve both an expendable mold and an expendable pattern.

Fig. 11-1-1 Typical cast part.

Casting metals are usually alloys or compounds of two or more metals. They are generally classed as ferrous or nonferrous metals. *Ferrous* metals are those which contain iron, the most common being gray iron, steel, and malleable iron. *Nonferrous* alloys, which contain no iron, are those containing metals such as aluminum, magnesium, and copper.

Sand Mold Casting

The most widely used casting process for metals uses a permanent pattern of metal or wood that shapes the mold cavity when loose molding material is compacted around the pattern. This material consists of a relatively fine sand, which serves as the refractory aggregate, plus a binder.

A typical sand mold, with the various provisions for pouring the molten metal and compensating for contraction of the solidifying metal, and a sand core for forming a cavity in the casting are shown in Fig. 11-1-2. Sand molds consist of two or more sections: bottom (*drag*), top (*cope*), and intermediate sections (*cheeks*) when required. The sand is contained in flasks equipped with pins and plates to ensure the alignment of the cope and drag.

Molten metal is poured into the sprue, and connecting runners provide flow channels for the metal to enter the mold cavity through gates. Riser cavities are located over the heavier sections of the casting. A vent is usually added to permit the escape of gases which are formed during the pouring of metal.

When a hollow casting is required, a form called a *core* is usually used. Cores occupy that part of the mold which is intended to be hollow in the casting. Cores, like molds, are formed of sand and placed in the supporting impressions or *core prints* in the molds. The core

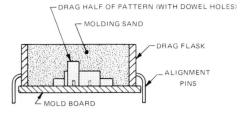

(A) STARTING TO MAKE THE SAND MOLD

DRAG HALF OF PATTERN (WITH DOWEL HOLES)
MOLDING SAND
DRAG FLASK
ALIGNMENT PINS
MOLD BOARD

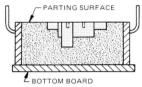

PARTING SURFACE
BOTTOM BOARD

(B) AFTER ROLLING OVER THE DRAG

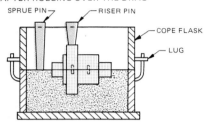

SPRUE PIN
RISER PIN
COPE FLASK
LUG

(C) PREPARING TO RAM MOLDING SAND IN COPE

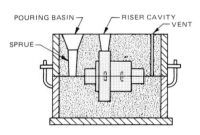

POURING BASIN
RISER CAVITY
VENT
SPRUE

(D) REMOVING RISER AND GATE SPRUE PINS AND ADDING POURING BASIN

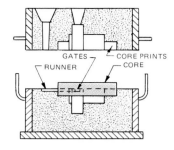

GATES
RUNNER
CORE PRINTS
CORE

(E) PARTING FLASKS TO REMOVE PATTERN AND TO ADD CORE AND RUNNER

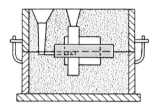

(F) SAND MOLD READY FOR POURING

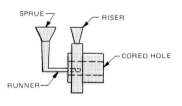

SPRUE
RISER
CORED HOLE
RUNNER

SPRUE, RISER, AND RUNNER TO BE REMOVED FROM CASTING.

(G) CASTING AS REMOVED FROM THE MOLD

Fig. 11-1-2 Sequence in preparing a sand casting.

prints ensure positive location of the core in the mold and, as such, should be placed so that they support the mass of the core uniformly to prevent shifting or sagging. Metal core supports called *chaplets*, which are used in the mold cavity and which fuse into the casting, are sometimes used by the foundry in addition to core prints. See Fig. 11-1-3. Chaplets and their locations are not usually specified on drawings.

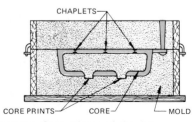

CHAPLETS
CORE PRINTS
CORE
MOLD

Fig. 11-1-3 Core prints and chaplets. (General Motors Corp.)

In producing sand molds, a metal or wooden pattern must first be made. The pattern, normally made in two parts, is slightly larger in every dimension than the part to be cast, to allow for shrinkage when the casting cools. This is known as *shrinkage allowance*, and the pattern maker allows for it by using a shrink rule for each of the cast metals.

Drafts or slight tapers are also placed on the pattern to allow for easy withdrawal from the sand mold. The parting line location and amount of draft are very important considerations in the design process.

In the construction of patterns for castings in which various points on the surface of the casting must be machined, sufficient excess metal should be provided for all machined surfaces. Allowance depends on the metal used, the shape and size of the part, the tendency

to warp, the machining method, and setup.

The molten metal is poured into the pouring basin and runs down the sprue to a runner and into the mold cavity.

When the metal has hardened, the sand mold is broken and the casting removed. Next the excess metal, gates, and risers are removed and remelted.

Shell Mold Casting[2]

The refractory sand used in shell molding is bonded by a thermostable resin that forms a relatively thin shell mold. A heated, reusable metal pattern plate (Fig. 11-1-4) is used to form each half of the mold by either dumping a sand-resin mixture on top of the heated pattern or by blowing resin-coated sand under air pressure against the pattern.

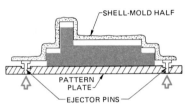

SHELL-MOLD HALF
PATTERN PLATE
EJECTOR PINS

Fig. 11-1-4 Shell mold being stripped from pattern.

Plaster Mold Casting

Plaster of paris and fillers are mixed with water and setting-control agents to form a *slurry*. This slurry is poured around a reusable metal or rubber pattern and sets to form a gypsum mold. See Fig. 11-1-5. The molds are then dried, assembled, and filled with molten (nonferrous) metals. Plaster mold casting is ideal for producing thin, sound walls. As in sand mold casting, a new mold is required for each casting. Castings made by this process have smoother finish, finer detail, and greater dimensional accuracy than sand castings.

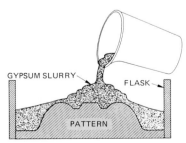

GYPSUM SLURRY
FLASK
PATTERN

Fig. 11-1-5 Pouring slurry over a plaster mold pattern.

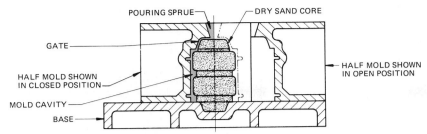

Fig. 11-1-6 Permanent mold. (General Motors Corp.)

Permanent Mold Casting

This process makes use of a metal mold, similar to a die, which is utilized to produce many castings from each mold. See Fig. 11-1-6. It is used to produce some ferrous alloy castings, but due to rapid deterioration of the mold caused by the high pouring temperatures of these alloys, and the high mold cost, the process is confined largely to production of nonferrous alloy castings.

Investment Mold Casting

Investment castings have been better known in the past by the term *lost wax castings.* The term *investment* refers to the refractory material used to encase the wax patterns.

This process uses both an expendable pattern and an expendable mold. Patterns of wax, plaster, or frozen mercury are cast in metal dies. The molds are formed either by pouring a slurry of a refractory material around the pattern positioned in a flask or by building a thick layer of shell refractory on the pattern by repeated dipping into slurries and drying. The arrangement of the wax patterns in the flask method is shown in Fig. 11-1-7.

Full Mold Casting

The characteristic feature of the full mold process is the use of consumable patterns made of foamed plastic. These are not extracted from the mold, but are vaporized by the molten metal.

The full mold process is suitable for individual castings. The advantages it offers are obvious: it is very economical and reduces the delivery time required for prototypes, articles urgently needed for repair jobs, or individual large machine parts.

Centrifugal Casting

In the centrifugal casting process, commonly applied to cylindrical casting of either ferrous or nonferrous alloys, a permanent mold is rotated rapidly about the axis of the casting while a measured amount of molten metal is poured into the mold cavity. See Fig. 11-1-8. The centrifugal force is used to hold the metal against the outer walls of the mold with the volume of metal poured determining the wall thickness of the casting. Rotation speed is rapid enough to form the central hole without a core. Castings made by this method are smooth, sound, and clean on the outside.

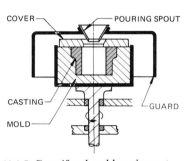

Fig. 11-1-8 Centrifugal mold equipment. (General Motors Corp.)

Continuous Casting

Continuous casting produces semi-finished shapes such as uniform section rounds, ovals, squares, rectangles, and plates. These shapes are cast from nearly all ferrous and nonferrous metals by continuously pouring the molten metal into a water-jacketed mold. The metal solidifies in the mold, and the solid billet exits continuously into a water spray. These sections are processed further by rolling, drawing, or extruding into smaller, more intricate shapes. Iron bars cast by this process are finished by machining.

Die Casting

One of the least expensive, fastest, and most efficient processes used in the production of metal parts is die casting. Die castings are made by forcing molten metal into a die or mold. Large quantities, accurately cast, can be produced with a die-casting die, thus eliminating or reducing machining costs. Many parts are completely finished when taken from a die. Since die castings can

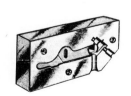

(1) THE DIE

(2) THE WAX PATTERN

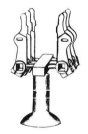

(3) THE CLUSTER ASSEMBLY

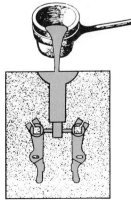

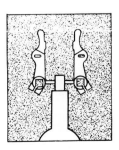

(4) REFRACTORY MOLD

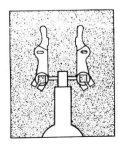

(5) FIRED MOLD

(6) THE CASTING

Fig. 11-1-7 Investment mold casting.

be accurate to within .001 in. (0.02 mm) of size, internal and external threads, gear teeth, and lugs can readily be cast.

Die casting has its limitations. Only nonferrous alloys can be die-cast economically because of the lack of a suitable die material to withstand the higher temperatures required for steel and iron.

Die-casting machines are two types: the *submerged-plunger* type for low-melting alloys containing zinc, tin, lead, etc., and the *cold-chamber* type for high-melting nonferrous alloys containing aluminum and magnesium. See Fig. 11-1-9.

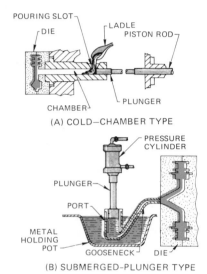

Fig. 11-1-9 Die-casting machines.

SELECTION OF PROCESS

Selection of the most feasible casting process for a given part requires an evaluation of the type of metal, the number of castings required, their shape and size, the dimensional accuracy required, and the casting finish required. When the casting can be produced by a number of methods, selection of the process is based on the most economical production of the total requirement. Since final cost of the part, rather than price of the rough casting, is the significant factor, the number of finishing operations necessary on the casting is also considered. Those processes that provide the closest dimensions, the best surface finish, and the most intricate detail generally require the smallest number of finishing operations.

A direct comparison of the capabilities, production characteristics, and limitations of several processes is indicated in Fig. 11-1-10.

DESIGN CONSIDERATIONS[3]

The advantages of using castings for engineering components are well appreciated by designers. Of major importance is the fact that they can produce shapes of any degree of complexity and of virtually any size.

Solidification of Metal in a Mold

While this is not the first step in the sequence of events, it is of such fundamental importance that it forms the most logical point to begin understanding the making of a casting.

Consider a few simple shapes transformed into mold cavities and filled with molten metal.

In a sphere, heat dissipates from the surface through the mold while solidification commences from the outside and proceeds progressively inward, in a series of layers. As liquid metal solidifies, it contracts in volume, and unless feed metal is supplied, a shrinkage cavity may form in the center. See Fig. 11-1-11.

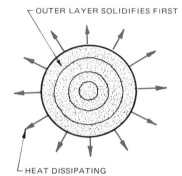

Fig. 11-1-11 Circular mold cavity filled with molten metal.

PROCESS	METALS CAST	USUAL WEIGHT (MASS) RANGE	MINIMUM PRODUCTION QUANTITIES	RELATIVE SETUP COST	CASTING DETAIL FEASIBLE	MINIMUM THICKNESS in. (mm)	DIMENSIONAL TOLERANCES in. (mm)	SURFACE FINISH, RMS (μin.)
SAND (Green, Dry, and Core) CO$_2$ Sand	All ferrous and nonferrous	Less than I lb. (0.5 kg) to several tons	3, without mechanization	Very low to high depending on mechanization	Fair	.12 to .25 (3 to 6) / .10 to .25 (2.5 to 6)	± .03 (0.8) / ± .02 (0.5)	350 / 250
SHELL	All ferrous and nonferrous	0.5 to 30 lb. (0.2 to 15 kg)	50	Moderate to high depending on mechanization	Fair to good	.03 to .10 (0.8 to 2.5)	± .015 (0.4)	200
PLASTER	Al, Mg, Cu, and Zn alloys	Less than I lb. to 3000 lb. (0.5 to 1350 kg)	1	Moderate	Excellent	.03 to .08 (0.8 to 2)	± .01 (0.2)	100
INVESTMENT	All ferrous and nonferrous	Less than I oz. to 50 lb. (30 g to 25 kg)	25	Moderate	Excellent	0.2 to .06 (0.5 to 1.5)	± .005 (0.1)	80
PERMANENT MOLD Metal Mold	Nonferrous and cast iron	I to 40 lb. (0.5 to 20 kg)	100	Moderate to high	Poor	.18 to .25 (4.5 to 6)	± .02 (0.5)	200
Graphite Mold	Steel	5 to 300 lb. (2 to 150 kg)	100			.25 (6)	± .03 (0.8)	200
DIE	Sn, Pb, Zn, Al, Mg, and Cu alloys	Less than I lb. to 20 lb. (0.5 to 10 kg)	1000	High	Excellent	.05 to .08 (1.2 to 2)	± .002 (0.5)	60

* Values listed are primarily for aluminum alloys, but data applies generally to other metals also.

† Depends on surface area.   Double if dimension is across parting line.

Fig. 11-1-10 General characteristics of casting processes.

The designer must realize that a shrinkage problem exists and that the foundry worker must attach risers to the casting or resort to other means to overcome it.

When the simple sphere has solidified further, it continues to contract in volume, so that the final casting is smaller than the mold cavity.

Consider a shape with a square cross section such as the one shown in Fig. 11-1-12A. Here again, cooling proceeds at right angles to the surface and is necessarily faster at the corners of the casting. Thus, solidification proceeds more rapidly at the corners.

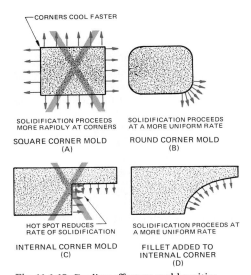

Fig. 11-1-12 Cooling effect on mold cavities filled with molten metal. (Mechanite Metal Corp.)

The resulting hot spot prolongs solidification, promoting solidification shrinkage and lack of density in this area. The only logical solution, from the designer's viewpoint, is the provision of very generous fillets or radii at the corners. Additionally, the relative size or shape of the two sections forming the corner is of importance. If they are materially different, as in Fig. 11-1-12D, contraction in the lighter member will occur at a different rate from that in the heavier member. Differential contraction is the major cause of casting stress, warping and cracking.

General Design Rules

Design for Casting Soundness
See Fig. 11-1-13. Most metals and alloys shrink when they solidify. Therefore, the design must be such that all members of the parts increase in dimension progressively to one or more suitable loca-

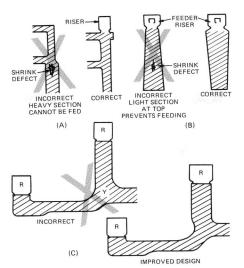

Fig. 11-1-13 Design members so that all parts increase progressively to feeder risers. (Mechanite Metal Corp.)

tions where feeder heads can be placed to offset liquid shrinkage.

Fillet All-Sharp Angles
See Fig. 11-1-14. Fillets have three functional purposes: to reduce stress concentration in the casting in service; to eliminate cracks, tears, and draws at reentry angles; and to make corners more moldable to eliminate hot spots.

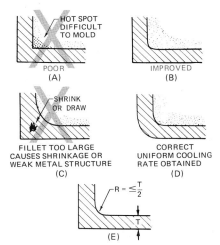

Fig. 11-1-14 Fillet all sharp angles. (Mechanite Metal Corp.)

Bring the Minimum Number of Adjoining Sections Together
See Fig. 11-1-15. A well-designed casting brings the minimum number of sections together and avoids acute angles.

Design All Sections as Nearly Uniform in Thickness as Possible
Shrink defects and casting strains existed in the casting illustrated in Fig. 11-1-16. Redesigning eliminated excessive metal and resulted in a casting that

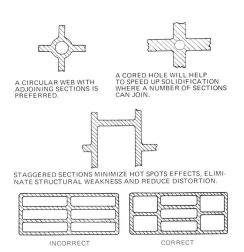

Fig. 11-1-15 Bringing the minimum number of adjoining sections together. (Mechanite Metal Corp.)

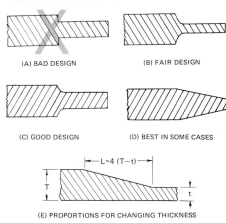

Fig. 11-1-16 Design all sections as nearly uniform in thickness as possible. (Mechanite Metal Corp.)

was free from defects, was lighter in weight (mass), and prevented the development of casting strains in the light radial veins.

Avoid Abrupt Section Changes—Eliminate Sharp Corners at Adjoining Sections
See Fig. 11-1-17. The difference in the relative thickness of adjoining sections should be minimum and not exceed a 2:1 ratio.

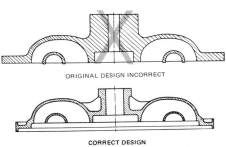

Fig. 11-1-17 Avoid abrupt changes. (Mechanite Metal Corp.)

When a change of thickness must be less than 2:1, it may take the form of a fillet; where the difference must be greater, the form recommended is that of a wedge.

Wedge-shaped changes in wall thickness are to be designated with a taper not exceeding 1 in 4.

Design Ribs for Maximum Effectiveness See Fig. 11-1-18. Ribs have two functions: to increase stiffness and to reduce the mass. If too shallow in depth or too widely spaced, they are ineffectual.

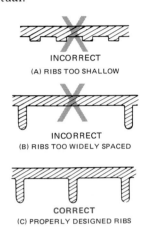

INCORRECT
(A) RIBS TOO SHALLOW

INCORRECT
(B) RIBS TOO WIDELY SPACED

CORRECT
(C) PROPERLY DESIGNED RIBS

INCORRECT
(D) THIN RIBS SHOULD BE AVOIDED WHEN JOINED TO A HEAVY SECTION. OTHERWISE, THEY WILL LEAD TO HIGH STRESSES AND CRACKING.

INCORRECT
(E) AS FAR AS POSSIBLE, JUNCTION BETWEEN RIBS AND MAIN CASTING SHOULD PREVENT ANY LOCAL ACCUMULATION OF METAL.

CORRECT
(F) RIBS SHOULD SOLIDIFY BEFORE THE CASTING SECTION THEY ADJOIN.

(G) T- AND H-SHAPED RIBBED DESIGNS HAVE THE ADVANTAGE OF UNIFORM METAL SECTIONS AND HENCE UNIFORM COOLING.

THICKNESS OF RIBS SHOULD APPROXIMATE 0.8 CASTING THICKNESS.

Fig. 11-1-18 Design ribs for maximum effectiveness. (Mechanite Metal Corp.)

Bosses and Pads Should Not Be Used Unless Absolutely Necessary
Bosses and pads increase metal thickness, create hot spots, and cause open grain or draws. Blend these into the casting by tapering or flattening the fillets. Bosses should not be included in casting design when the surface to support bolts, etc., may be obtained by milling or countersinking.

Spoked Wheels See Fig. 11-1-19. A curved spoke is preferred to a straight one. It will tend to straighten slightly, thereby offsetting the dangers of cracking.

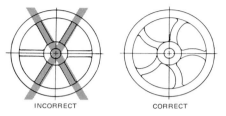

INCORRECT   CORRECT
(A) USE AN ODD NUMBER OF CURVED SPOKES

CAREFULLY BLEND SECTIONS

INCORRECT   CORRECT
(B) AVOID EXCESSIVE SECTION VARIATION

Fig. 11-1-19 Spoked-wheel design. (Mechanite Metal Corp.)

Use an Odd Number of Spokes. A wheel having an odd number of spokes will not have the same direct tensile stress along the arms as one having an even number and will have more resiliency to casting stresses.

Wall Thicknesses Walls should be of minimum thickness, consistent with good foundry practice, and should provide adequate strength and stiffness. Wall thicknesses for different materials are as follows:

1. Walls of gray-iron castings and aluminum sand castings should not be less than .16 in. (4 mm) thick.
2. Walls of malleable iron and steel castings should not be less than .18 in. (5 mm) thick.
3. Walls of bronze, brass, or magnesium castings should not be less than .10 in. (2.4 mm) thick.

Parting Lines A *parting line* is a line along which the pattern is divided for molding, or along which the sections of a mold separate. Selection of a parting line depends on a number of factors:

- Shape of the casting
- Elimination of machining on draft surfaces

- Method of supporting cores
- Location of gates and feeders

Holes in Castings Small holes usually are drilled and not cored.

Drafting Practices[2]

It is important that a detail drawing give complete information on all cast parts, e.g.:

- Machining allowances
- Surface texture
- Draft angles
- Limits on cast surfaces that must be controlled
- Locating points
- Parting lines

On small, simple parts all casting information is included on the finished drawing. See Fig. 11-1-20. On more complicated parts, it may be necessary to show additional casting views and sections to completely illustrate the construction of the casting. These additional views should show the rough casting outline in phantom lines and the finished contour in solid lines.

Material In the selection of material for any particular application, the designer is influenced primarily by the physical characteristics such as strength, hardness, density, resistance to wear, mass, antifrictional properties, conductivity, corrosion resistance, shrinkage, and melting point.

Machining Allowance In the construction of patterns for castings in which various points on the surface of the casting must be machined, sufficient excess metal should be provided for all machined surfaces. Unless otherwise specified, Fig. 11-1-21 may be used as a guide to machine finish allowance.

Fillets and Radii Generous fillets and radii (rounds) should be provided on cast corners and specified on the drawing.

Casting Tolerances A great many factors contribute to the dimensional variations of castings. However, the standard drawing tolerances specified in Fig. 11-1-22 can be satisfactorily attained in the production of castings.

Draft All casting methods require a draft or taper on all surfaces perpendicular to the parting line, to facilitate removal of the pattern and ejection of the casting. The permissible draft must be specified on the drawing, in either

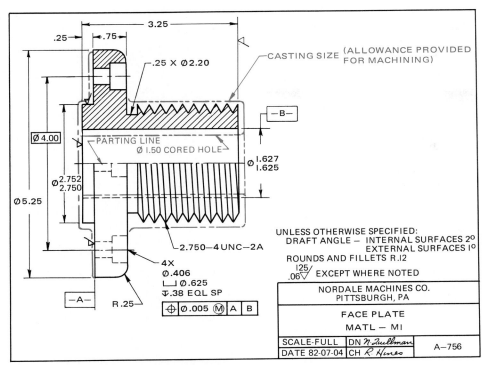

(A) WORKING DRAWING OF A CAST PART

- CASTING SIZE (ALLOWANCE PROVIDED FOR MACHINING)

UNLESS OTHERWISE SPECIFIED:
DRAFT ANGLE — INTERNAL SURFACES 2°
            EXTERNAL SURFACES 1°
ROUNDS AND FILLETS R.12
$\frac{125}{.06}$ EXCEPT WHERE NOTED

NORDALE MACHINES CO.
PITTSBURGH, PA

FACE PLATE
MATL — MI

| SCALE-FULL | DN N. Quillman | A—756 |
| DATE 82-07-04 | CH R. Hines | |

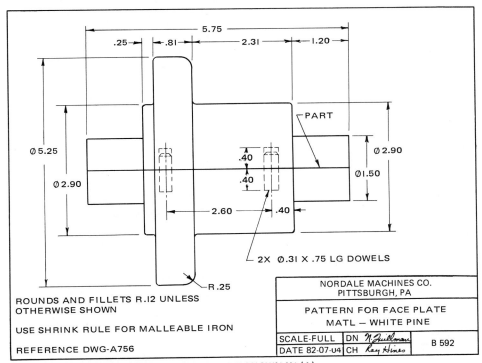

ROUNDS AND FILLETS R.12 UNLESS OTHERWISE SHOWN

USE SHRINK RULE FOR MALLEABLE IRON

REFERENCE DWG-A756

NORDALE MACHINES CO.
PITTSBURGH, PA

PATTERN FOR FACE PLATE
MATL — WHITE PINE

| SCALE-FULL | DN N. Quillman | B 592 |
| DATE 82-07-04 | CH Ray Hines | |

(B) PATTERN DRAWING FOR THE CAST PART SHOWN IN (A)

Fig. 11-1-20 Cast part drawings.

degrees of taper for each surface, inches of taper per inch of length, or millimeters of taper per millimeter of length.

Suitable draft angles for general use, for both sand and die castings, are 1° for external surfaces and 2° for internal surfaces, as shown in Fig. 11-1-23.

The drawing must always clearly indicate whether the draft should be added to, or subtracted from, the casting dimensions.

CASTING DATUMS[2]

It is recognized that in many cases a drawing is made of the fully machined

CASTING ALLOY	DIMENSIONS WITHIN THIS RANGE	CASTING ALLOWANCE
CAST IRON, ALUMINUM, BRONZE, ETC. SAND CASTINGS	UP TO 8.00 8.00 TO 16.00 16.00 TO 24.00 24.00 TO 32.00 OVER 32.00	.06 .09 .12 .18 .25
PEARLITIC, MALLEABLE, AND STEEL SAND CASTINGS	UP TO 8.00 8.00 TO 16.00 16.00 TO 24.00 OVER 24.00	.06 .09 .18 .25
PERMANENT AND SEMIPERMANENT MOLD CASTINGS	UP TO 12.00 12.00 TO 24.00 OVER 24.00	.06 .09 .18
PLASTER MOLD CASTINGS	UP TO 8.00 8.00 TO 12.00 OVER 12.00	.03 .06 .10

Fig. 11-1-21 Guide to machining allowance for castings.

TYPE OF CASTING	DIMENSIONS WITHIN THIS RANGE	STANDARD DRAWING TOLERANCE (±)
IRON AND ALUMINUM SAND CASTINGS	UP TO 8.00 8.00 TO 16.00 16.00 TO 24.00 24.00 TO 32.00 OVER 32.00	.03 .06 .07 .09 .12
PEARLITIC, MALLEABLE IRON AND STEEL SAND CASTINGS	UP TO 8.00 8.00 TO 16.00 16.00 TO 24.00 OVER 24.00	.03 .06 .09 .12
PERMANENT MOLD CASTING (SEMIPERMANENT MOLD CASTING)	UP TO 5.00 5.00 TO 12.00 12.00 TO 24.00 OVER 24.00	.03 .03 .06 .09
PLASTER MOLD CASTING	UP TO 4.00 4.00 TO 8.00 8.00 TO 12.00 OVER 12.00	.02 .02 .03 .06
CENTRIFUGAL PRECISION CASTING	UP TO .50 .50 TO 5.00 OVER 5.00	.02 .02 .02

Fig. 11-1-22 Guide to casting tolerances.

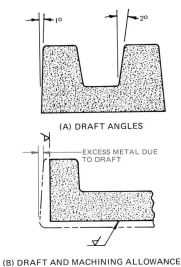

(A) DRAFT ANGLES

EXCESS METAL DUE TO DRAFT

(B) DRAFT AND MACHINING ALLOWANCE

Fig. 11-1-23 Draft for removal of pattern from mold.

end product, and casting dimensions, draft, and machining allowances are left entirely to the patternmaker or foundry worker. However, for mass-production purposes it is generally advisable to make a separate casting drawing, with carefully selected datums, to ensure that parts will fit into machining jigs and fixtures and will meet final requirements after machining. Under these circumstances, dimensioning requires the selection of two sets of datum surfaces, lines, or points—one for the casting and one for the machining—to provide common reference points for measuring, machining, and assembly. To select suitable datums, it will be necessary to know how the casting is to be made, where the parting line or lines are to be, and how the part is going to fit into machining jigs and fixtures.

The first step in dimensioning is to select a primary datum surface for the casting, which is sometimes referred to as the *base* surface, and to identify it as datum A. See Fig. 11-1-24. This primary datum should be a surface which meets the following criteria as closely as possible:

1. It must be a surface, or datum targets on a surface (see Fig. 11-1-25), which can be used as the basis for measuring the casting and which can later be used for mounting and locat-

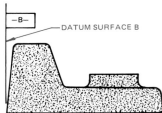

Fig. 11-1-24 Casting datums.

ing the part in a jig or fixture, for the purpose of machining the finished part.

2. It should be a surface which will not be removed by machining, so that control of material to be removed is not lost, and can be checked at final inspection.

3. It should be parallel with the top of the mold, or parting line; that is, a surface which has no draft or taper.

4. It should be integral with the main body of the casting, so that measurements from it to the main surfaces of the casting will be least affected by cored surfaces, parting lines, or gated surfaces.

5. It should be a surface, or target areas on a surface, on which the part can be clamped without causing any

distortion, so that the casting will not be under a distortional stress for the first machining operation.

6. It should be a surface which will provide locating points as far apart as possible, so that the effect of any flatness error will be minimized.

The second step is to select two other planes to serve as secondary and tertiary surfaces. These planes should be at right angles to one another and to the primary datum surface. They probably will not coincide with actual surfaces, because of taper or draft, except at one point, usually a point adjacent to the primary datum surface. These are identified as datum B and datum C, respectively, as shown in Fig. 11-1-26.

In the case of a circular part, the end-view center lines may be selected as sec-

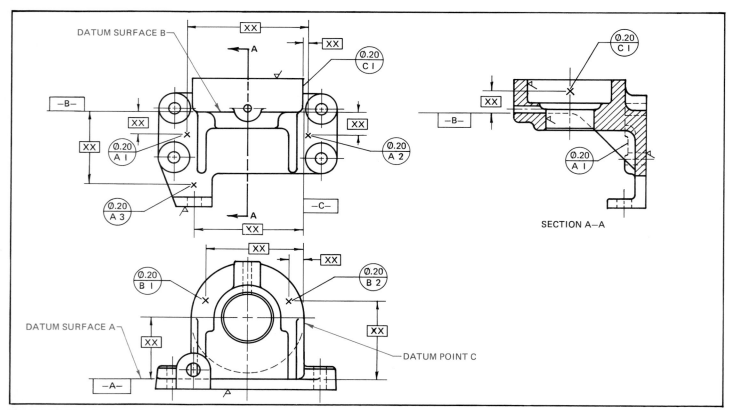

Fig. 11-1-25 Machined cast drawing illustrating datum lines, setup points, and surface finish.

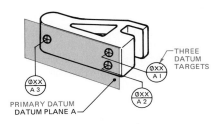

PRIMARY DATUM – PLANE A

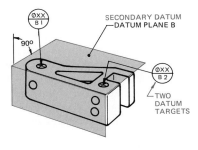

SECONDARY DATUM – PLANE B

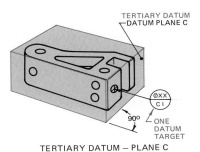

TERTIARY DATUM – PLANE C

Fig. 11-1-26 Datum planes and datum targets.

ondary and tertiary datums, as shown in Fig. 11-1-27. In this case, unless otherwise specified, the center lines represent the center of the outside or overall diameter of the part.

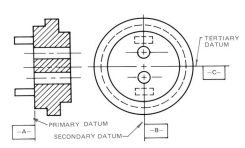

Fig. 11-1-27 Datums for circular casting.

MACHINING DATUMS

The first step in dimensioning the machined or finished part is to select a primary datum surface for machining and to identify it as datum D. This surface is the first surface on the casting to

be machined and is thereafter used as the datum surface for all other machining operations. It should be selected to meet the following criteria:

1. It is generally preferable, though not essential, that it be a surface which is parallel to the primary casting datum surface.
2. It may be a large, flat, machined surface or several small areas of surfaces in the same or parallel planes.
3. If the primary casting datum surface is smooth and does not require machining, as in die castings, or if suitable target areas have been selected, the same surface may be used as the machining datum surface.
4. If the primary casting datum surface of sand castings appears to be the only suitable surface, it is recommended that three or four pads be provided, which can be machined to form the machining datum surface, as shown in Fig. 11-1-28.

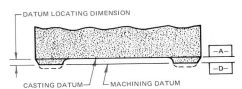

Fig. 11-1-28 Primary machining datums.

5. When pads or small target areas are selected, they should be placed as far apart as possible and located where the part can be readily clamped in jigs or fixtures without distorting it or interfering with other machining operations.

The second step is to select two other surfaces to serve as secondary and tertiary datums. If these datum surfaces are required only for locating and dimensioning purposes, and not for clamping in a jig or fixture, some suitable datums other than flat, machined surfaces may be chosen. These could be the same datums as used for casting, if the locating point in each case is clearly defined and is not removed in machining. For circular parts, a hole drilled in the center, or a turned diameter other than the outside diameter, may provide suitable center lines for use as secondary datum surfaces, as shown in Fig. 11-1-29.

The third step is to specify the *datum-locating dimension*, that is, the dimension between each casting datum surface and the corresponding machining datum surface. See Fig. 11-1-28. There is never

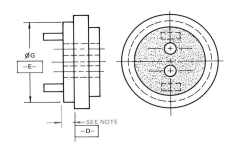

START MACHINING WITH DATUM D TO DIMENSION SHOWN

Fig. 11-1-29 Machining datums for circular parts.

more than one such dimension from each casting datum surface.

Dimensions

When suitable datum surfaces have been selected, with datum-locating dimensions for the machined casting drawing, dimensioning may proceed, with dimensions being specified directly from the datums to all main surfaces. However, where it is necessary to maintain a particular relationship between two or more surfaces or features, regular point-to-point dimensioning is usually the preferred method. This will generally include all such items as thickness of ribs, height of bosses, projections, depth of grooves, most diameters and radii, and center distances between holes and similar features. Whenever possible, specify dimensions to surfaces or surface intersections, rather than to radii centers or nonexistent center lines.

Dimensions given on the casting drawing should not be repeated, except as reference dimensions, on the machined part drawing.

References and Source Material

1. American Iron and Steel Institute, "Principles of Forging Design."
2. General Motors Corporation.
3. Meehanite Metal Corporation.

ASSIGNMENTS

See Assignments 1 through 3 for Unit 11-1 on page 244.

UNIT 11-2
Forgings

Forging consists of plastically deforming, either by a squeezing pressure or sharp blows, a cast or sintered ingot, a wrought bar or billet, or a powder-metal shape, to produce a desired shape with good mechanical properties. Practically all ductile metals can be forged. See Fig. 11-2-1.

(A) BILLET

(B) TONGHOLD IS FIRST FORGED

(C) PREFORMED IMPRESSION

(D) PREFORMED IMPRESSION

(E) PREFORMED IMPRESSION

(F) BLOCKING AND FINISHING

(G) AFTER TRIMMING, CRANKS ARE TWISTED INTO POSITION

Fig. 11-2-1 The forging of a crankshaft. (Wyman-Gordon Co.)

Types of Forgings

Closed-Die Forgings Closed-die forgings are made by hammering or pressing metal until it conforms closely to the shape of the enclosing dies. Grain flow in the closed-die-forged parts can be oriented in the direction requiring greatest strength.

Three-dimensional control of the material to be forged requires a closed die, a simple and common form of which is the impression die.

In the simplest example of impression-die forging (Fig. 11-2-2) the workpiece is cylindrical and is placed in the bottom-half die. On closing the top-half die, the cylinder undergoes elastic compression until its enlarged sides touch the side walls of the die impression. At this point, a small amount of excess material begins to form the flash between the two die faces.

The forging impression die gives control over all three directions, except when the die is similar to that shown in Fig. 11-2-2, and the deforming forging machine tool has an unlimited stroke (e.g., a hammer or hydraulic press). In the latter case, the die must be shaped to allow complete closing of the striking faces at the end of the stroke. See Fig. 11-2-3.

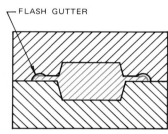

┌─ FLASH GUTTER

Fig. 11-2-3 Forged die with flash gutter.

In practice, *closed-die forging* has become the term applied to all forging operations involving three-dimensional control.

Forging dies can be divided into three main classes: single-impression, double-impression, and interlocking. See Fig. 11-2-4.

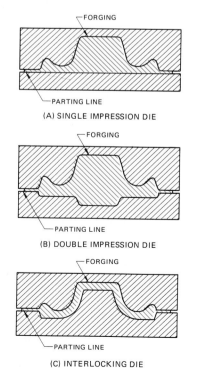

┌─FORGING

└─PARTING LINE

(A) SINGLE IMPRESSION DIE

┌─FORGING

└─PARTING LINE

(B) DOUBLE IMPRESSION DIE

┌─FORGING

└─PARTING LINE

(C) INTERLOCKING DIE

Fig. 11-2-4 Forging dies.

Single-impression dies have the impression of the desired forging entirely in one half of the die.

Double-impression dies have part of the impression of the desired forging sunk in each die in such a manner that no part of the die projects past the parting line into the other die. This type is the most common class of forging.

Trimming

Because the quantity of forging metal is generally in excess of the space in the die cavity, space is provided between the die surfaces for the escape of the excess metal. This space is called the *flash space*, and the excess metal which flows into it is called *flash*. The flash thickness is proportionate to the mass of the forging.

The flash is removed from forgings by trimming dies which are formed to the outline of the part. See Fig. 11-2-5.

BEFORE TRIMMING   FLASH   AFTER TRIMMING

Fig. 11-2-5 Flash trimming.

General Design Rules

Corner and Fillet Radii It is important in forging design to use correct radii where two surfaces meet. Corner and fillet radii on forgings should be sufficient to facilitate the flow of metal.

Stress concentrations resulting from abrupt changes in section thickness or direction are minimized by corner and fillet radii of correct size. Any radius larger than recommended will increase die life. Any radius smaller than recommended will decrease die life. See Fig. 11-2-6 for recommendations.

Sharp fillets cause the formation of cold shuts. In a forging, a *cold shut* is a lap where two surfaces of metal have folded against each other, forming an undesirable flow of metal. A cold shut causes a weak spot that may be opened into a crack by heat treatment. Cold shuts are most likely to form at fillets in deep depressions or in deep sections, especially where the metal is confined. See Fig. 11-2-7. In these cases larger fillets are required as shown in Fig. 11-2-6.

Draft Angle Draft is one of the first factors to be considered in designing a forged part. See Fig. 11-2-8. *Draft* is

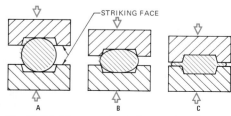

┌─STRIKING FACE

A   B   C

Fig. 11-2-2 Compression in impression dies.

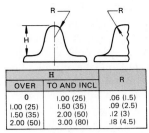

H		R
OVER	TO AND INCL	
0	1.00 (25)	.06 (1.5)
1.00 (25)	1.50 (35)	.09 (2.5)
1.50 (35)	2.00 (50)	.12 (3)
2.00 (50)	3.00 (80)	.18 (4.5)

MIN CORNER RADII

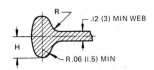

.12 (3) MIN WEB

R .06 (1.5) MIN

H		R
OVER	TO AND INCL	
0	.30 (8)	R = H
.30 (8)	.50 (13)	R = $\frac{3H}{4}$

FILLET RADII FOR SMALL RIBS

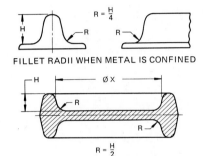

$R = \frac{H}{4}$

FILLET RADII WHEN METAL IS CONFINED

$R = \frac{H}{2}$

DEPTH OF A FORGED RECESS SHOULD NOT EXCEED 0.67 X DIA

FILLET RADII WHEN METAL IS NOT CONFINED

Fig. 11-2-6 Corner and filled radii.

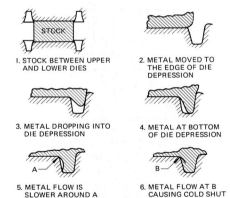

1. STOCK BETWEEN UPPER AND LOWER DIES

2. METAL MOVED TO THE EDGE OF DIE DEPRESSION

3. METAL DROPPING INTO DIE DEPRESSION

4. METAL AT BOTTOM OF DIE DEPRESSION

5. METAL FLOW IS SLOWER AROUND A

6. METAL FLOW AT B CAUSING COLD SHUT

Fig. 11-2-7 Cold shut.

defined as the slope given to the side walls of the die in order to facilitate removal of the forging. Where little or no draft is allowed, stripper or ejection mechanisms must be used. The usual amount of draft for exterior contours is 7° and for interior contours 10°.

Die draft equivalent is the amount of offset that results from draft. Figure

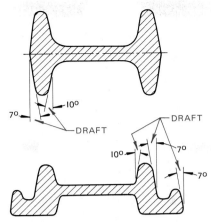

Fig. 11-2-8 Draft application.

11-2-9 shows the draft equivalents for varying angles and depth of draft.

Parting Line The surfaces of dies that meet in forgings are the striking surfaces. The line of meeting is the parting line. The parting line of the forging must be established in order to determine the amount of draft and its location.

The location and the type of parting as applied to simple forgings are shown in Fig. 11-2-10.

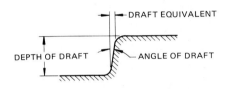

Fig. 11-2-9 Die draft equivalent.

DEPTH OF DRAFT	DRAFT EQUIV FOR ANGLE OF		
	5°	7°	10°
.20 (5)	.018 (0.437)	.024 (0.614)	.035 (0.882)
.40 (10)	.035 (0.875)	.050 (1.228)	.070 (1.763)
.60 (15)	.052 (1.312)	.074 (1.842)	.106 (2.645)
.80 (20)	.070 (1.750)	.100 (2.456)	.140 (3.527)
1.00 (25)	.088 (2.187)	.123 (3.070)	.176 (4.408)

Drafting Practices

In preparing forging drawings, it is important to consider drafting practices which may be peculiar to forgings, such as:

- Draft angles and parting lines
- Corner and fillet radii
- Forging tolerances

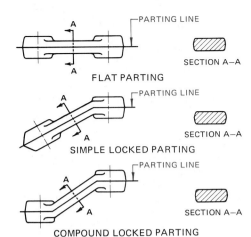

FLAT PARTING

SIMPLE LOCKED PARTING

COMPOUND LOCKED PARTING

Fig. 11-2-10 Parting line application.

- Machining allowances
- Heat treatment
- Location of trademark, part number, and vendor specification

Dimensioning It is generally desirable to apply dimensions to the forged part of the depths of the die impressions. Draft is additive to these dimensions and should be expressed in degrees or linear dimensions.

When the depth of the die impression is located, only one dimension should originate from the parting line. This surface should then be used to establish other dimensions, as shown in Fig. 11-2-11.

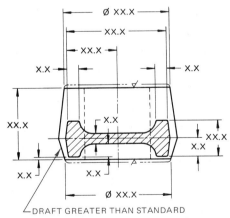

Fig. 11-2-11 Dimensioning a forged drawing.

Allowance for Machining When a forging is to be machined, allowance must be made for metal to be removed.

Composite Drawings Generally, a forged part should be shown on one drawing with the forging outline shown in phantom lines, as in Fig. 11-2-12. Forging outlines for machining allowance should not be dimensioned unless

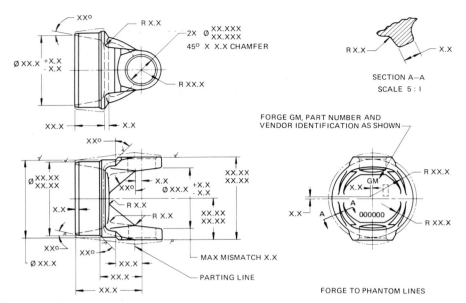

R X.X

SECTION A–A
SCALE 5:1

FORGE GM, PART NUMBER AND
VENDOR IDENTIFICATION AS SHOWN

R XX.X

GM

000000

R XX.X

FORGE TO PHANTOM LINES

MAX MISMATCH X.X

PARTING LINE

Fig. 11-2-12 Composite forging drawing.

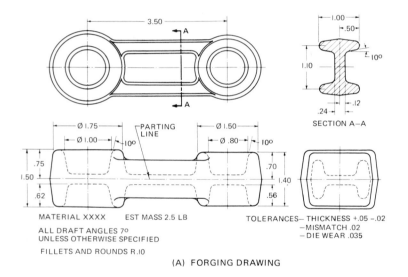

1.00
.50
10°

1.10

.24   .12

SECTION A–A

Ø 1.75
Ø 1.00
PARTING LINE
10°
Ø 1.50
Ø .80   10°

.75
1.50
.62
.70
1.40
.56

MATERIAL XXXX   EST MASS 2.5 LB

ALL DRAFT ANGLES 7°
UNLESS OTHERWISE SPECIFIED

FILLETS AND ROUNDS R.10

TOLERANCES— THICKNESS +.05 –.02
—MISMATCH .02
—DIE WEAR .035

(A) FORGING DRAWING

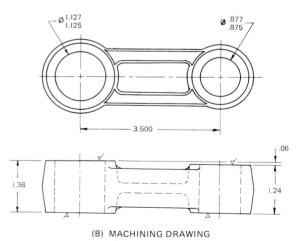

Ø 1.127
1.125

Ø .877
.875

3.500

.06

1.36

1.24

(B) MACHINING DRAWING

Fig. 11-2-13 Separate forging and machining drawings.

the amount of finish cannot be controlled by the machining symbol.

Separate drawings for rough forgings should be made only when the part is complicated and the outline of the rough forging cannot be clearly visualized, or where the outline of the rough forging must be maintained for tooling purposes.

Where both the forging and machining drawings are shown on the same sheet, as in Fig. 11-2-13, place the headings FORGING DRAWING and MACHINING DRAWING directly under the corresponding views.

Reference and Source Material

1. Frank Burbank, "Forging," *Machine Design*, vol. 37, no. 21.

ASSIGNMENTS

See Assignments 4 and 5 for Unit 11-2 on page 245.

UNIT 11-3
Powder Metallurgy

Powder metallurgy is the process of making parts by compressing and sintering various metallic and nonmetallic powders into shape. See Fig. 11-3-1.

Dies and presses known as *briquetting machines* are used to compress the powders into shape. These briquets or compacts are then sintered or heated in an atmosphere-controlled furnace, bonding the powdered materials.

Design Considerations

The following should be considered when powder metal parts are designed in order to realize the maximum benefits from the powder metallurgy process. This process is most applicable to the production of cylindrical, rectangular, or irregular shapes that do not have large variations in cross-sectional dimensions. Splines, gear teeth, axial holes, counterbores, straight knurls, serrations, slots, and keyseats present few problems.

Ejection from the Die The shape of the part must permit ejection from the die. The design requirements for some

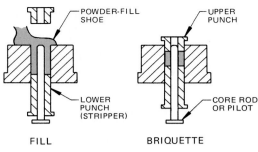

Fig. 11-3-1 Compacting sequence.

parts can be achieved only by subsequent machining, as in some corner relief designs, reverse tapers, holes at right angles to the direction of pressing, diamond knurls, and undercuts.

Axial Variations Slots having a depth greater than one-fourth the axial length of the part require multiple punch action and result in high production costs. See Fig. 11-3-2.

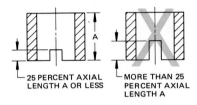

Fig. 11-3-2 Axial variations.

Corner Reliefs Corner reliefs can be molded or machined. A molded corner relief will save machining. See Fig. 11-3-3.

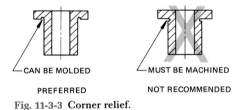

Fig. 11-3-3 Corner relief.

Reverse Tapers Reverse tapers cannot be molded. They must be machined. See Fig. 11-3-4.

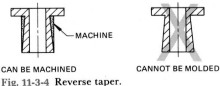

Fig. 11-3-4 Reverse taper.

Holes at Right Angles to the Direction of Pressing Right-angle holes must be machined. See Fig. 11-3-5.

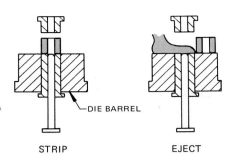

Fig. 11-3-5 Right angle.

Knurls Straight knurls can be molded; diamond knurls cannot. See Fig. 11-3-6.

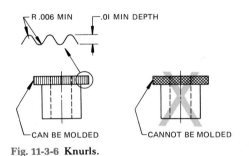

Fig. 11-3-6 Knurls.

Undercuts Undercuts must be machined. See Fig. 11-3-7.

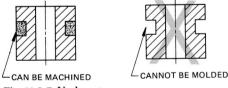

Fig. 11-3-7 Undercuts.

Wall Thickness In general, sidewalls bordering a depression or hole should be a minimum of .03 in. (0.8 mm) thick. See Fig. 11-3-8.

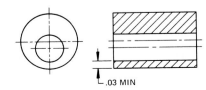

Fig. 11-3-8 Wall thickness.

Corners A fillet radius must be provided under the flange on a flanged part. It allows uniform powder flow in the die and produces a high-strength part. See Fig. 11-3-9.

Fig. 11-3-9 Corners.

Flanges A .06-in. (1.5 mm) minimum flange overhang is desired to provide longer tool life. (See Fig. 11-3-10.)

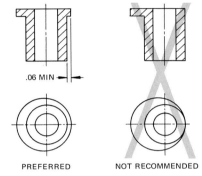

Fig. 11-3-10 Flanges.

Blind Holes If a flange is opposite the blind end of the hole, the part must be modified to allow powder to fill in the die. See Fig. 11-3-11.

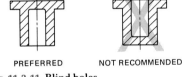

Fig. 11-3-11 Blind holes.

Changes in Cross Section Large changes in cross section should be avoided because they cause density variation. Warping and cracking are likely to occur during sintering. See Fig. 11-3-12.

Fig. 11-3-12 Change in cross section.

Chamfers Care in the design of chamfers minimizes sharp edges on tools and improves tool life. See Fig. 11-3-13.

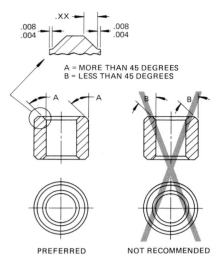

Fig. 11-3-13 Chamfers.

Holes The use of round holes, instead of odd-shaped holes, will simplify tooling, strengthen the part, and reduce costs. See Fig. 11-3-14.

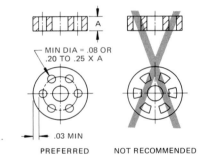

Fig. 11-3-14 Holes.

Reference and Source Material

1. General Motors Corporation.

ASSIGNMENT

See Assignment 6 for Unit 11-3 on page 246.

U N I T 11-4
Plastic Molded Parts

SINGLE PARTS

The design of molded parts involves several factors not normally encountered with machine-fabricated and assembled parts. It is important that designers take these factors into consideration.

Shrinkage *Shrinkage* is defined as the difference between dimensions of the

mold and the corresponding dimensions of the molded part. Normally the mold designer is more concerned with shrinkage than the molded part designer. Shrinkage does, however, affect dimensions, warpage, residual stress, and moldability.

Section Thickness Solidification is a function of heat transfer from or to the mold for both thermoplastics and thermosets. Each material has a fixed rate of heat transfer. Therefore, where section thickness varies, areas within a molded part will solidify at different rates. The varying rates will cause irregular shrinkage, sink marks, additional strain, and warpage. For these reasons, uniform section thickness is important and may be maintained by adding holes or depressions, as shown in Fig. 11-4-1 and 11-4-2.

UNIFORM THICKNESS
PREFERRED

VARYING THICKNESS
NOT RECOMMENDED

Fig. 11-4-1 Section thickness.

MATERIAL	ABSOLUTE MINIMUM		RECOM-MENDED MINIMUM	
	(in.)	(mm)	(in.)	(mm)
THERMOPLASTICS				
ABS	.024	0.6	.064	1.6
ACRYLIC	.024	0.6	.100	2.4
CELLULOSIC	.024	0.6	.076	1.9
FLUOROCARBON	.010	0.2	.010	0.3
POLYAMIDE	.016	0.4	.060	1.5
POLYCARBONATE	.024	0.6	.100	2.4
POLYETHYLENE	.036	0.9	.064	1.6
PLOYSTYRENE	.032	0.8	.064	1.6
POLYVINYL CHLORIDE	.100	2.4	.100	2.4
THERMOSETS				
EPOXY	.064	1.6	.130	3.2
POLYESTER				
GLASS FILLED	.040	1.0	.190	4.7
MINERAL FILLED	.040	1.0	.130	3.2
PHENOLICS				
GENERAL PURPOSE	.050	1.3	.130	3.2
FABRIC FILLED	.064	1.6	.190	4.7
MINERAL FILLED	.130	3.2	.190	4.7
UREA AND MELAMINES				
GENERAL PURPOSE	.036	0.9	.100	2.5
FABRIC FILLED	.050	1.3	.130	3.2
MINERAL FILLED	.040	1.0	.190	4.7

Fig. 11-4-2 Section thickness for various plastics.

Molded Holes A through hole is more advantageous than a blind hole since it is more accurate and economical. Blind holes should not be more than

twice as deep as their diameter, as shown in Fig. 11-4-3. Avoid placing holes at angles other than perpendicular to the flash line. If such holes are necessary, consider using a drilled hole to maintain simple molding. See Fig. 11-4-4.

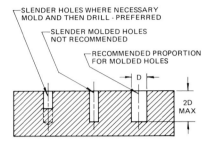

Fig. 11-4-3 Blind holes.

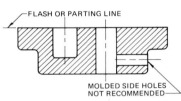

Fig. 11-4-4 Avoiding molded holes which are perpendicular to parting lines.

Threads External and internal threads can be easily molded by means of loose-piece inserts and rotating core pins. External threads may be formed by placing the cavity so that the threads are formed in the mold pattern.

Gates Gate location should be anticipated during the design stage. Avoid gating into areas subjected to high stress levels, fatigue, or impact. To optimize molding, locate gates in the heaviest section of the part. See Fig. 11-4-5.

Fig. 11-4-5 Gating.

Internal and External Draft Draft is necessary on all rigid molded articles to facilitate removal of the part from the mold. Draft may vary from 0.25 to 4° per side, depending upon the length of the vertical wall, surface area, finish, kind of material, and the mold or method of ejection used.

Parting or Flash Line *Flash* is that portion of the molding material which flows or exudes from the mold parting line during molding. Any mold which is made of two or more parts may produce flash at the line of junction of the mold parts. The thickness of flash usually varies between .002 and .016 in. (0.05 and 0.40 mm), depending upon the accuracy of the mold, type of material, and the process used. See Fig. 11-4-6.

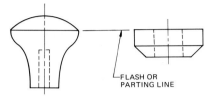

Fig. 11-4-6 Parting or flash line.

Fillets and Radii The principal functions of fillets and radii (rounds) are to ease the flow of plastic within the mold, to facilitate ejection of the part, and to distribute stress in the part in service. During molding, the material is liquefied, but it is a heavy, viscous liquid which does not easily flow around sharp corners. The liquid tends to bend around corners; therefore, rounded corners permit the liquid plastic to flow smoothly and easily through the mold. For recommended radii, see Fig. 11-4-7.

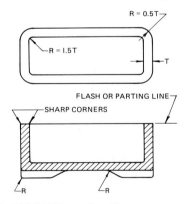

Fig. 11-4-7 Fillets and radii.

Ribs and Bosses Ribs increase rigidity of a molded part without increasing wall thickness and sometimes facilitate flow during molding. Bosses reinforce small, stressed areas, providing sufficient strength for assembly with inserts or screws. Recommended proportions for ribs and bosses are shown in Fig. 11-4-8A.

Undercuts Parts with undercuts should be avoided. Normally, parts with external undercuts cannot be withdrawn from a one-piece mold. Internal undercuts are considered impractical

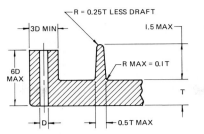

(A) RIB AND BOSS PROPORTION

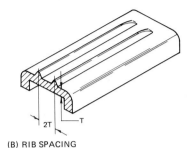

(B) RIB SPACING

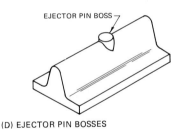

(C) BOSSES

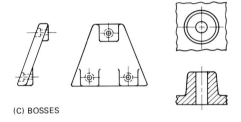

(D) EJECTOR PIN BOSSES

Fig. 11-4-8 Undercuts.

and should be avoided. If an internal undercut is essential, it may be achieved by machining or by use of a flexible mold core material. See Fig. 11-4-9.

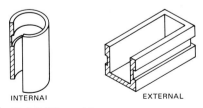

INTERNAL   EXTERNAL

Fig. 11-4-9 Ribs and bosses.

ASSEMBLIES

The design of molded parts which are to be assembled with typical fastening methods involves factors different from those normally encountered with metal.

Holes and Threads

Mechanical fasteners, in general, depend upon a hole of some type. Holes should be designed and located to provide maximum strength and minimum molding problems. Any straight hole, molded or machined, should have between it and an adjacent hole, or side wall, an amount of material equal to or greater than the diameter or width of the hole. Any threaded hole, molded or tapped, should have between it and an adjacent hole, or side wall, an amount of material at least 3 times the outside diameter of the thread. Spacing may be reduced, however, by proper use of bosses.

Drilled holes are often more accurate and easier to produce than molded holes, even though they require a second operation.

Tapped holes provide an economical means of joining a molded part to its assembly. The designer should avoid threads with a pitch of less than .03 in. (0.8 mm). Holes which are to be tapped should be countersunk to prevent chipping when the tap is inserted.

External and internal threads can be molded integrally with the part. Molded threads are generaly more expensive to form than other threads because either a method of unscrewing the part from the mold must be provided or a split mold must be used.

Inserts

After the molding material has been determined, the insert should be designed. The molded part should be designed around the insert.

Inserts of round rod stock, coarse diamond knurled and grooved, provide the strongest anchorage under torque and tension. A large single groove with knurling on each end, as in Fig. 11-4-10,

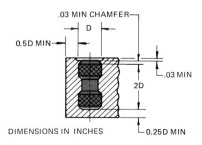

DIMENSIONS IN INCHES

KNURLING DEPTH SHOULD BE ABOUT .01 in. ANGULAR GROOVES GIVE INCREASED AXIAL ANCHORAGE. PLASTIC SHRINKAGE ALONE SHOULD NOT BE RELIED ON TO PROVIDE FIRM SUPPORT FOR INSERTS.

Fig. 11-4-10 Supporting inserts.

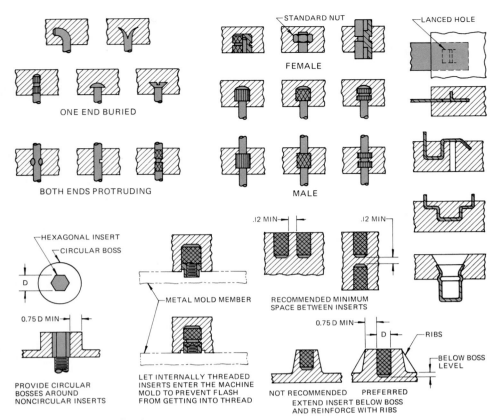

ONE END BURIED

BOTH ENDS PROTRUDING

STANDARD NUT

FEMALE

MALE

LANCED HOLE

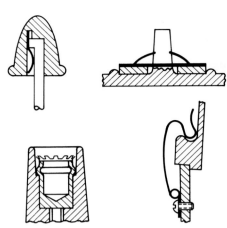

Fig. 11-4-14 Mechanical fasteners.

HEXAGONAL INSERT

CIRCULAR BOSS

D

0.75 D MIN

PROVIDE CIRCULAR
BOSSES AROUND
NONCIRCULAR INSERTS

METAL MOLD MEMBER

LET INTERNALLY THREADED
INSERTS ENTER THE MACHINE
MOLD TO PREVENT FLASH
FROM GETTING INTO THREAD

.12 MIN

.12 MIN

RECOMMENDED MINIMUM
SPACE BETWEEN INSERTS

0.75 D MIN

D

RIBS

BELOW BOSS
LEVEL

NOT RECOMMENDED   PREFERRED
EXTEND INSERT BELOW BOSS
AND REINFORCE WITH RIBS

Fig. 11-4-11 Insert application.

is superior to two or more grooves with smaller knurled surface areas. See also examples of inserts in Fig. 11-4-11.

Press and Shrink Fits

Inserts may be secured by a press fit, or the plastic molding material may be assembled to a larger part by a shrink fit, as shown in Fig. 11-4-12. Both methods rely on shrinkage of the material, which is greatest immediately after removal from the mold.

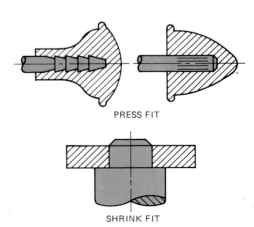

PRESS FIT

SHRINK FIT

Fig. 11-4-12 Press and shrink fits.

Heat Forming and Heat Sealing

Most thermoplastics can be reformed by the application of heat and pressure, as shown in Fig. 11-4-13. This reforming often eliminates the need for other assembly methods, such as adhesive bonding and mechanical fasteners. This method cannot be used with thermosetting materials.

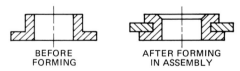

BEFORE
FORMING

AFTER FORMING
IN ASSEMBLY

Fig. 11-4-13 Heat forming.

Mechanical Fastening

Various designs of mechanical fasteners are commercially available. Spring-type metal hinges and clips, speed clips or nuts, and expanding rivets are a few of these designs. Design of the parts for assembly requires that molded parts have sufficient sectional strength to withstand the stresses that will be encountered with fasteners. A strengthening of the area which will receive the

brunt of these applied stresses is usually required. See Fig. 11-4-14.

Rivets

Conventional riveting equipment and procedures can be used with plastics. Care must be exercised to minimize stresses induced during the fastening operation. To do this, the rivet head should be 2.5 to 3 times the shank diameter. Also, rivets should be backed with either plates or washers to avoid high localized stresses. See Fig. 11-4-15.

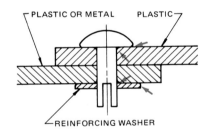

PLASTIC OR METAL   PLASTIC

REINFORCING WASHER

NOTE: BREAK ALL SHARP EDGES ON
RIVET, WASHER AND HOLES.

Fig. 11-4-15 Recommended riveting procedures.

Drilled holes rather than punched holes are preferred for fasteners. If possible, fastener clearance in the hole should be at least .01 in. (0.3 mm) to maintain a plane stress condition at the fastener.

Boss Caps

A boss cap is a cup-shaped metal ring which is pressed onto the boss by hand, with an air cylinder, or with a light-duty press. It is designed to reinforce the boss against the expansion force exerted by tapping screws. See Fig. 11-4-16.

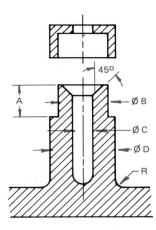

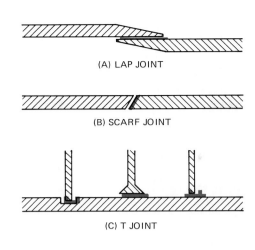

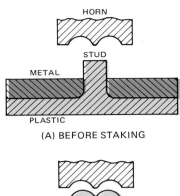

THREAD SIZE	DIMENSIONS				
	A	B	C	D	R
# 6(.138)	.14	.21	.10	.28	.02
# 8(.164)	.16	.25	.13	.34	.02
# 10(.190)	.18	.29	.17	.40	.02

Fig. 11-4-16 Boss cap design.

Adhesive Bonding

When two or more parts are to be joined into an assembly, adhesives permit a strong, durable fastening between similar materials and often are the only fastening method available for joining dissimilar materials. Structural adhesives are made from the same basic resins as many plastics and thus react to their operating environment in a similar manner. In order to provide maximum strength, adhesives must be applied as a liquid to thoroughly wet the surface of the part. The bonding surface must be chemically clean to permit complete wetting. Basic plastics vary in physical properties, so adhesives made from these materials also vary. See Fig. 11-4-17.

Ultrasonic Bonding

Ultrasonic bonding often is used instead of solvent cementing to bond plastic parts. By using this technique, irregularly shaped parts can be bonded in 2 seconds or less. The bonded parts may be handled and used at reasonable temperatures within minutes after joining.

Only one of the mating parts comes in contact with the horn. The part transmits the ultrasonic vibration to small, hidden bonding areas, resulting in fast, perfect welds. Both mating halves remain cool except at the seam, where the energy is quickly dissipated.

(A) LAP JOINT

(B) SCARF JOINT

(C) T JOINT

(D) CORNER JOINT

(E) BUTT JOINTS

Fig. 11-4-17 Adhesive bonding.

This technique is not recommended where high impact strength is required in the bond area. See Fig. 11-4-18.

Fig. 11-4-18 Design joints for ultrasonic bonding.

Ultrasonic Staking

Ultrasonic staking frequently involves the assembly of metal parts. In this technique, a stud molded into the plastic part protrudes through a hole in the metal part. The surface of the stud is vibrated with a horn having high amplitude and a relatively small contact area. The vibration causes the stud to melt and re-form in the configuration of the horn tip. See Fig. 11-4-19.

(A) BEFORE STAKING

(B) AFTER STAKING

Fig. 11-4-19 Typical ultrasonic staking operation.

Friction or Spin Welding

This welding technique is limited to parts with circular joints. It is especially useful for large parts where ultrasonic welding or chemical bonding is impractical.

In friction or spin welding, the faces to be joined are pressed together while one part is spun and the other is held fixed. Frictional heat produces a molten zone that becomes a weld when spinning stops. See Fig. 11-4-20.

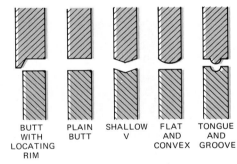

| BUTT WITH LOCATING RIM | PLAIN BUTT | SHALLOW V | FLAT AND CONVEX | TONGUE AND GROOVE |

Fig. 11-4-20 Joint shapes for spin welding.

Drawings

In addition to the usual considerations, the following points should be taken into account when a detail drawing of a plastic part is made

1. Can the part be removed from the mold?
2. Is location of flash line consistent with design requirements?
3. Is section thickness consistent? Are there thick sections? Thin sections? Could greater uniformity of section thickness be maintained?

4. Has the material been correctly specified?

5. Is each feature in accordance with the thinking of competent materials engineers and molders?

6. Have close tolerance requirements been reviewed with responsible engineers?

7. Have marking requirements been specified to inform field service people of the material from which the part is fabricated?

References and Source Material

1. General Motors Corporation.
2. General Electric Company.

ASSIGNMENTS

See Assignments 6 to 8 for Unit 11-4 starting on page 246.

ASSIGNMENTS FOR CHAPTER 11

Assignments for Unit 11-1, Castings

1. Complete the detail drawing of the base for the adjustable shaft support assembly shown in Fig. 11-1-A. Cored holes are to be used for the shaft holes. Scale 1:1.

2. Complete the assembly drawing of the fork for the hinged pipe vise shown in Fig. 11-1-B. Use your judgment for dimensions not given. Scale 1:1.

3. Prepare both the casting and the machining drawings for the connector shown in Fig. 11-1-C. Draw a one-view full section, complete with the necessary dimensions for each drawing. Scale 1:1.

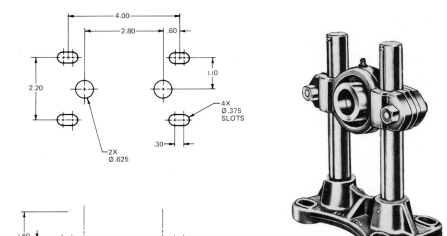

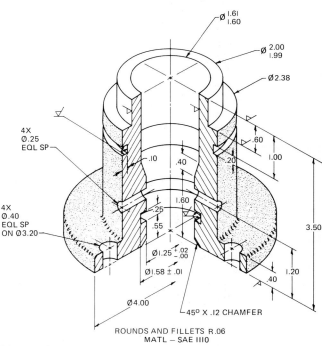

Fig. 11-1-A Adjustable shaft support.

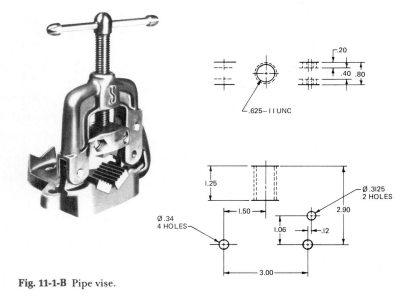

Fig. 11-1-B Pipe vise.

Fig. 11-1-C Connector.

ROUNDS AND FILLETS R.06
MATL — SAE 1110

Assignments for Unit 11-2, Forgings

4. Prepare a forging drawing of one of the parts shown in Fig. 11-2-A or 11-2-B. Scale 1:1.

5. Prepare a forging drawing for the wrench handle shown in Fig. 11-2-C. Scale 1:1.

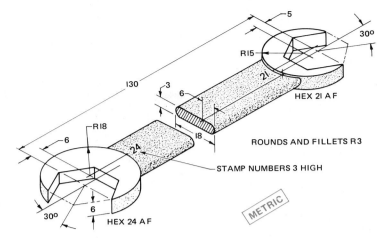

Fig. 11-2-A Open-end wrench.

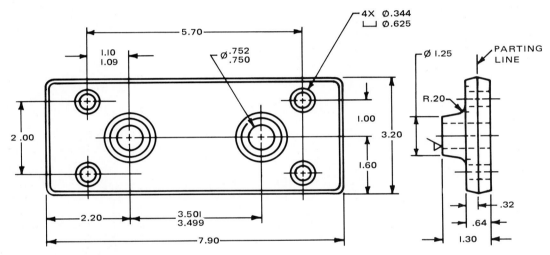

Fig. 11-2-B Bracket.

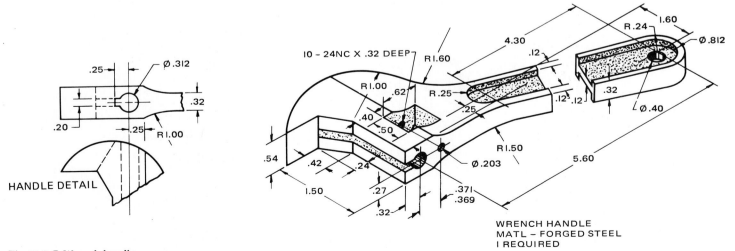

Fig. 11-2-C Wrench handle.

Assignment for Unit 11-3, Powder Metallurgy

6. Prepare two drawings, one for machining the part, the second for the making of the briquet (powder metallurgy) for one of the parts shown in Fig. 11-1-C or 11-3-A. Scale 1:1.

Assignments for Unit 11-4, Plastic Molded Parts

7. Redesign for plastic molding one of the parts shown in Figs. 11-4-A to 11-4-C. Refer to the molding recommendations shown in this unit and indicate the parting line on the drawing. Use your judgment for dimensions not given. Scale 1:1.

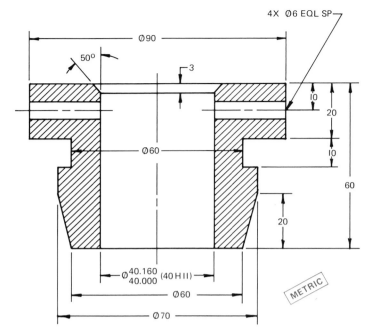

Fig. 11-3-A Bracket.

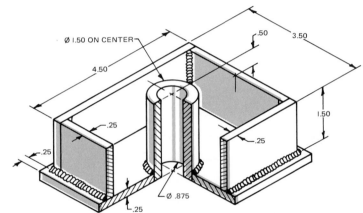

NOTE: BASE EXTENDED BEYOND WALLS FOR WELDING PURPOSES ONLY

Fig. 11-4-A Shaft base.

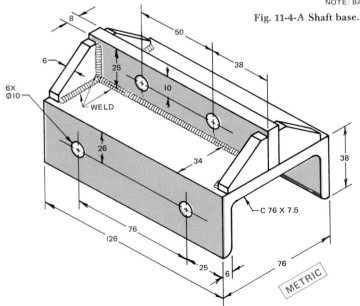

Fig. 11-4-B Caster frame.

Fig. 11-4-C Slide bracket.

8. Using a plastic molding design add threaded inserts to one of the parts shown in Fig. 11-4-D or 11-4-E. Use your judgment for dimensions not shown and the type and number of views required. Scale 2:1.

9. Make a plastic molding assembly drawing of the parts shown in Fig. 11-4-F. The retaining ring is to be positioned in the center of the part and molded into position. Modification to the retaining ring may be required to prevent the ring from turning in the wheel. Scale 5:1. Show a top view and a full-section view. Dimension the finished assembly.

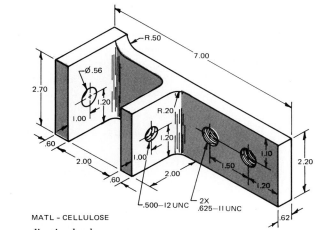

MATL – CELLULOSE

Fig. 11-4-D Lamp adjusting knob.

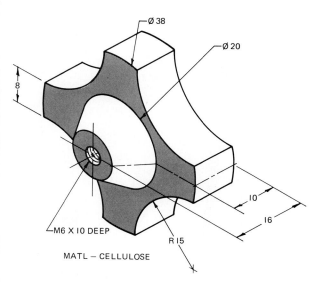

MATL – CELLULOSE

Fig. 11-4-E Connector.

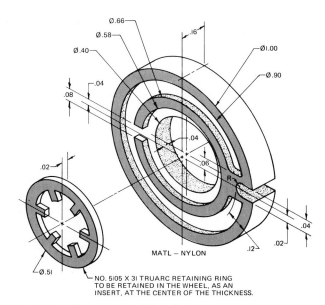

MATL – NYLON

NO. 5I05 X 3I TRUARC RETAINING RING TO BE RETAINED IN THE WHEEL, AS AN INSERT, AT THE CENTER OF THE THICKNESS.

Fig. 11-4-F Cassette-tape drive wheel.

Manufacturing Materials

This chapter is an up-to-date reference on manufacturing materials. It provides the drafter and designer with basic information on materials and their properties to ensure the proper selection of the product material.

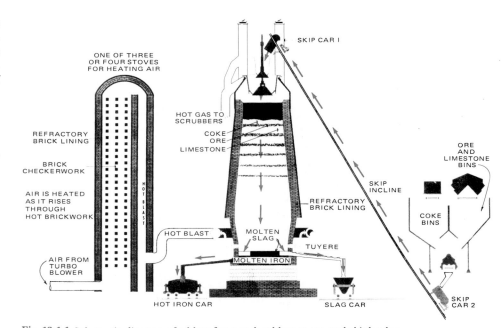

Fig. 12-1-1 Schematic diagram of a blast furnace, hot blast stone, and skiploader. (American Iron and Steel Institute)

U N I T   1 2 - 1
Cast Irons

FERROUS METALS

Iron and the large family of iron alloys called steel are the most frequently specified metals. Iron is abundant (iron ore constitutes about 5 percent of the earth's crust), easy to convert from ore to a useful form, and iron and steel are sufficiently strong and stable for most engineering applications.

All commercial forms of iron and steel contain carbon, which is an integral part of the metallurgy of iron and steel.

CAST IRON

Because of its low cost, cast iron is often considered a simple metal to produce and to specify. Actually, the metallurgy of cast iron is more complex than that of steel and other familiar design materials. Whereas most other metals are usually specified by a standard chemical analysis, the same analysis of cast iron can produce several entirely different types of iron, depending upon rate of cooling, thickness of the casting, and how long the casting remains in the mold. By con-

trolling these variables, the foundry can produce a variety of irons for heat- or wear-resistant uses, or for high-strength components. See Fig. 12-1-1.

Types of Cast Iron

Ductile, or Nodular, Iron Ductile, or nodular, iron is not as available as gray iron, and it is more difficult to control in production. However, ductile iron can be used where higher ductility or strength is required than is available in gray iron. See Fig. 12-1-2.

Ductile iron, sometimes called nodular iron, is used in applications such as crankshafts because of its good machinability, fatigue strength, and high modulus of elasticity; heavy-duty gears because of its high yield strength and wear resistance; and automobile door hinges because of its ductility.

Gray Iron Gray iron is a supersaturated solution of carbon in an iron matrix. The excess carbon precipitates out in the form of graphite flakes. Typical applications of gray iron include

MECHANICAL PROPERTY		DUCTILE				WHITE	GRAY						MALLEABLE						
		50-55-06	60-40-18	100-70-03	120-90-02		20	25	30	40	50	60	32510	35018	40010	45006	50005	70003	90001
Yield strength	10³ lb/in.²	60-75	45-60	75-90	90-125								32	35	40	45	50	70	90
	MPa	410-520	310-410	520-620	620-860		*	*	*	*	*	*	220	240	275	310	345	485	620
Tensile strength	10³ lb/in.²	90-110	60-80	100-120	125-150	20-50	20-25	25-30	30-35	40-48	50-57	60-66	50	53	60	65	70	85	105
	MPa	620-760	410-550	690-825	860-1035	140-345	140-170	170-205	205-240	275-330	345-390	415-455	345	365	415	450	480	585	725
Elongation in 2.00 in. (50 mm)	%	3-10	10-25	6-10	2-7	—	1	1	1	0.8	0.5	0.5	10	18	10	6	5	3	1
Modulus of elasticity	10³ lb/in.²	22-25	22-25	22-25	22-25	—	12	13	15	17	19	20	25	25	26	26	26	26-28	26-28
	10³ MPa	150-170	150-170	150-170	150-170	8 / —	83	90	103	117	131	138	172	172	180	180	180	180-193	180-193

*Yield strength usually about 65–80% of tensile strength.

Fig. 12-1-2 Mechanical properties of cast iron.

automotive blocks, flywheels, brake disks and drums, machine bases, and gears. Generally, gray iron serves well in any machinery applications because of its fatigue resistance.

White Iron White iron is produced by a process called *chilling* which prevents graphite carbon from precipitating out. Either gray or ductile iron can be chilled to produce a surface of white iron. In castings that are white iron throughout, however, the composition of iron is selected according to part size to ensure that the volume of metal involved can chill rapidly enough to produce white iron.

Because of their extreme hardness, white irons are used primarily for applications requiring wear and abrasion resistance such as mill liners and shot-blasting nozzles. Other uses include railroad brake shoes, rolling-mill rolls, clay-mixing and brick-making equipment, and crushers and pulverizers. Generally, plain (unalloyed) white iron costs less than other cast irons.

The principal disadvantage of white iron is that it is very brittle.

High-Alloy Irons High-alloy irons are ductile, gray, or white irons that contain over 3 percent alloy content. These irons have properties that are significantly different from the unalloyed irons and are usually produced by specialized foundries.

Malleable Iron Malleable iron is white iron that has been converted to a malleable condition by a two-stage heat-treating process.

It is a commercial cast material which is similar to steel in many respects. It is strong and ductile, has good impact and fatigue properties, and has excellent machining characteristics.

The two basic types of malleable iron are ferritic and pearlitic. Ferritic grades are more machinable and ductile, whereas the pearlite grades are stronger and harder.

Forming Process

For design information on the preparation of metal castings see Chap. 11, Units 11-1 and 11-3.

Reference and Source Material

1. *Machine Design*, Materials reference issue.

ASSIGNMENT

See Assignment 1 for Unit 12-1 on page 259.

U N I T   1 2 - 2
Carbon Steel

Carbon steel is essentially an iron-carbon alloy with small amounts of other elements (either intentionally added or unavoidably present) such as silicon, magnesium, copper, and sulfur. Steels can be either cast to shape or wrought into various mill forms from which finished parts can be machined, forged, formed, stamped, or otherwise generated.

Wrought steel is either poured into ingots or is sand-cast. After solidification the metal is reheated and hot rolled—often in several steps—into the finished wrought form. Hot-rolled steel is characterized by a scaled surface and a decarburized skin.

CARBON AND LOW-ALLOY CAST STEELS

Carbon and low-alloy cast steels lend themselves to the formation of streamlined, intricate parts with high strength and rigidity.

A number of advantages favor steel casting as a method of construction:

1. The metallographic structure of steel castings is uniform in all directions. It is free from the directional variations in properties of wrought-steel products.
2. Cast steels are available in a wide range of mechanical properties depending on the compositions and heat treatments.
3. Steel castings can be annealed, normalized, tempered, hardened, or carburized.
4. Steel castings are as easy to machine as wrought steels.
5. Most compositions of carbon and low-alloy cast steels are easily welded because their carbon content is under 0.45 percent.

The making of steel is illustrated in Fig. 12-2-1.

HIGH-ALLOY CAST STEELS

The term *high alloy* is applied arbitrarily to steel castings containing a minimum of 8 percent nickel and/or chromium. Such castings are used mostly to resist corrosion or provide strength at temperatures above 1200°F (560°C).

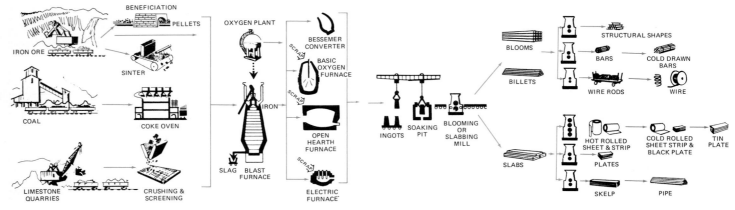

Fig. 12-2-1 Flowchart for steel making.

CARBON STEELS

Carbon steels are the workhorse of product design. They account for over 90 percent of total steel production. More carbon steels are used in product manufacturing than all other metals combined.

A thorough understanding of the selection and specification criteria for all types of steel requires knowledge of what is implied by carbon-steel mill forms, qualities, grades, tempers, finishes, edges, and heat treatments; also how and where these terms relate to dimensions, tolerances, physical and mechanical properties, and manufacturing requirements.

The designer's specification job really begins the instant that molten steel hits the mold. The conditions under which steel solidifies have a significant effect on production and on performance of subsequent mill products.

STEEL SPECIFICATION

Several ways are used to identify a specific steel: by chemical or mechanical properties, by its ability to meet a standard specification or industry-accepted practice, or by its ability to be fabricated into an identified part.

Chemical Composition

The steel producer can be instructed to produce a desired composition in one of three ways:

1. By a maximum limit
2. By a minimum limit
3. By an acceptable range

The following are some commonly specified elements.

Carbon   The principal hardening element in steel. As carbon content is increased to about 0.85 percent, hardness and tensile strength increase, but ductility and weldability decrease.

Manganese   A lesser contributor to hardness and strength. Properties depend on carbon content. Increasing manganese increases the rate of carbon penetration during carburizing.

Phosphorus   Large amounts increase strength and hardness but reduce ductility and impact toughness, particularly in the higher-carbon grades. Phosphorus in low-carbon, free-machining steels improves machinability.

Silicon   A principal deoxidizer in the steel industry. Silicon increases strength and hardness but to a lesser extent than manganese.

Sulfur   Increased sulfur content reduces transverse ductility, notch-impact toughness, and weldability. Sulfur is added to improve machinability of steels.

Copper   Improves atmospheric corrosion resistance when present in excess of 0.15 percent.

Lead   Improves the machinability of steel.

Classification Bodies

The specifications covering the composition of iron and steel have been issued by various classification bodies. They serve as a selection guide and provide a means for the buyer to conveniently specify certain known and recognized requirements. The main classification bodies are:

SAE—Society of Automotive Engineers

AISI—American Iron and Steel Institute   This is an association of steel producers which issues steel specifications for the steel-making industry and cooperates with the SAE in using the same numbers for the same steel.

ASTM—American Society for Testing and Materials   This group is interested in materials of all kinds and writes specifications. The ASTM steel specifications for steel plate and structural shapes are used by all steel makers in North America.

The ASTM has several specifications covering structural steel. Both the AISA and AISC (American Institute of Steel Construction) refer to ASTM specifications.

ASME—American Society of Mechanical Engineers   This group is interested in the steel used in pressure vessels and other mechanical equipment.

SAE AND AISI—SYSTEMS OF STEEL IDENTIFICATION

The specifications for steel bar are based on a code that specifies the composition of each type of steel covered. They include both plain carbon and alloy steels. The code is a four-number system. See Figs. 12-2-2 and 12-2-3. Each figure in the number has the following specific function: the first or left-side figure represents the major class of steel, the second figure represents a subdivision of the major class; for example, the series having *one* (1) as the left-hand figure covers the carbon steels. The second figure breaks this class up into normal low-sulfur steels, the high-sulfur free-machining grades, and another grade having higher than normal manganese.

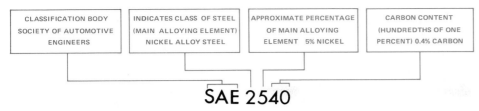

| CLASSIFICATION BODY SOCIETY OF AUTOMOTIVE ENGINEERS | INDICATES CLASS OF STEEL (MAIN ALLOYING ELEMENT) NICKEL ALLOY STEEL | APPROXIMATE PERCENTAGE OF MAIN ALLOYING ELEMENT 5% NICKEL | CARBON CONTENT (HUNDREDTHS OF ONE PERCENT) 0.4% CARBON |

SAE 2540

Fig. 12-2-2 Steel designation system.

TYPE OF CARBON STEEL	NUMBER SYMBOL	PRINCIPAL PROPERTIES	COMMON USES
Plain carbon	10XX		
Low carbon steel (0.06 to 0.20% carbon)	1006 to 1020	Toughness and less strength	Chains, rivets, shafts, and pressed steel products
Medium carbon steel (0.20 to 0.50% carbon)	1020 to 1050	Toughness and strength	Gears, axles, machine parts, forgings, bolts, and nuts
High carbon steel (over 0.50% carbon)	1050 and over	Less toughness and greater hardness	Saws, drills, knives, razors, finishing tools, and music wire
Sulfurized (free-cutting)	11XX	Improves machinability	Threads, splines, and machined parts
Phosphorized	12XX	Increases strength and hardness but reduces ductility	
Manganese steels	13XX	Improves surface finish	

Fig. 12-2-3 Carbon steel designations, properties and uses.

Originally the second figure represented the percentage of the major alloying element present, and this is true of many of the alloy steels. However, this had to be varied in order to account for all the steels that are available.

The third and fourth figures represent carbon content in hundredths of 1 percent, thus the figure xx15 means 0.15 of 1 percent carbon.

EXAMPLE SAE 2335 is a nickel steel containing 3.5 percent nickel and 0.35 of 1 percent carbon.

Typical properties of rolled carbon steels are shown in Fig. 12-2-4.

Carbon-Steel Sheet

Flat-rolled carbon-steel sheets are made from heated slabs that are progressively reduced in size as they move through a series of rolls.

Hot-Rolled Sheets Hot-rolled sheets are produced into three principal qualities, commercial, drawing, and physical.

Cold-Rolled Sheets Cold-rolled sheets are made from hot-rolled coils which are pickled, then cold reduced to the desired thickness. The commercial quality of cold-rolled sheets is normally produced with a matte finish suitable for painting or enameling but not suitable for electroplating.

Carbon-Steel Plates

Carbon-steel plates are produced (in rectangular plates or in coils) by hot rolling directly from the ingot or slab. Plate thickness ranges from .19 in. (4 mm) and thicker for plates up to 48 in. (1200 mm) wide, and from .25 in. (6 mm) and thicker for plates wider than 48 in. (1200 mm). Thickness is specified in millimeters or inches. It can also be specified by weight (lb/ft^2) or mass (kg/m^2).

Carbon-Steel Bars

Hot-Rolled Bars Hot-rolled carbon-steel bars are produced from blooms or billets in a variety of cross sections and sizes. See Figs. 12-2-5 and 12-2-6.

Cold-Finished Bars Cold-finished carbon-steel bars are produced from hot-rolled steel by a cold-finishing process which improves surface finish, dimensional accuracy, and alignment. Cold drawing and cold rolling also increase the yield and tensile strength. For machinability ratings of cold drawn carbon steel see Fig. 12-2-7.

Steel Wire

Steel wire is made from hot-rolled rods produced in continuous-length coils. Most wire is drawn, but some special shapes are rolled.

Pipe and Tubing

Pipe and tubing range from the familiar plumber's black pipe to high-precision mechanical tubing for bearing races. Pipe and tubing may contain fluids, sup-

MECHANICAL PROPERTY		AISI STEEL											
		1015/1020/1022			1035/1040			1045/1050			1095		
		Hot-Rolled	Cold-Drawn	Annealed	Hot-Rolled	Cold-Drawn	Quenched and Tempered	Hot-Rolled	Cold-Drawn	Quenched and Tempered	Hot-Rolled	Cold-Drawn and Annealed	Quenched and Tempered
Yield strength	10^3 lb/in.^{2}	40	51	42	42	71	63–96	49	84	68–117	66	76	80–152
	MPa	270	350	295	290	440	435–660	335	580	470–800	455	525	580–1050
Tensile strength	10^3 lb/in.^{2}	65	61	60	76	85	96–130	90	100	105–137	130	99	130–216
	MPa	450	420	415	525	585	660–895	620	690	725–945	895	680	895–1490
% Elongation in 2.00 in. (50 mm)		25	15	38	18	12	17–24	15	10	25--15	9	13	10–84

Fig. 12-2-4 Typical mechanical properties of rolled carbon steel.

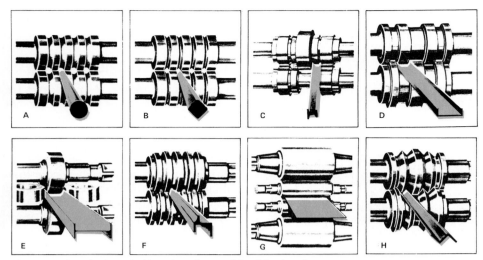

Fig. 12-2-5 Standard stock. (American Iron and Steel Institute)

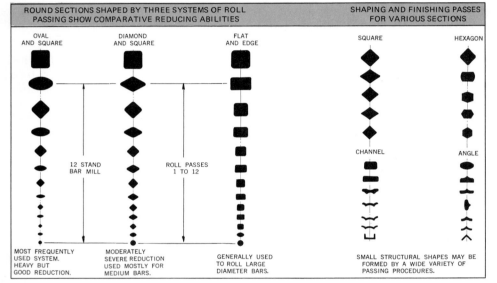

| ROUND SECTIONS SHAPED BY THREE SYSTEMS OF ROLL PASSING SHOW COMPARATIVE REDUCING ABILITIES | | | SHAPING AND FINISHING PASSES FOR VARIOUS SECTIONS | |

Fig. 12-2-6 Bar-mill roll passes. (American Iron and Steel Institute)

AISI No.	RATING*
12L14	195
1213	137
1215	137
1212	100
1211	91
1117	89
1114	85
1137	72
1141	69
1018	66
1045	55

*Based on 1212 = 100%

Fig. 12-2-7 Machinability rating of cold-drawn carbon steel.

port structures, or be a primary shape from which products are fabricated.

Welded Tubular Products   Welded tubular products are made from hot-rolled or cold-rolled flat steel coils.

Pipe   Pipe is produced from carbon or alloy steel to nominal dimensions. Nominal pipe sizes are expressed in inch sizes, but in the metric system the outside diameter and the wall thickness are expressed in millimeters. The outside diameter is often much larger than the nominal size. For example, a .75-in. standard-weight pipe has an outside diameter of 1.050 in. (26.7 mm). The outside diameter of nominal size pipe always remains the same and the mass or wall thickness changes. ANSI B36 has developed 10 different wall thicknesses (schedules) of pipe (See Table 21, Appendix B).

Nominal pipe size designation stops at 12 in. Pipe 14 in. and over is listed on the basis of outside diameter and wall thickness.

Tubing   Tubing is usually specified by a combination of either outside diameter, inside diameter, or wall thickness. Sizes range from approximately .25 to 5.00 in. (6 to 125 mm), in increments of .12 in. (3 mm). Wall thickness is usually specified in inches, millimeters, or by gage numbers.

Structural-Steel Shapes

A large tonnage of structural-steel shapes goes into manufactured products rather than buildings. The frame of a truck, railroad car, or earth-moving equipment is a structural design problem, just as is a high-rise building.

Size Designations   Several ways are used to describe a structural section in a specification, depending primarily on its shape.

1. Beams and channels are measured by the depth of the section in inches (millimeters) and by weight (lb/ft) or mass (kg/m).
2. Angles are described by length of legs and thickness in inches (millimeters), or more commonly, by lengths of legs and weight (lb/ft) or mass (kg/m). The longest leg is always stated first.
3. Tees are specified by width of flange, overall depth of stem, and pounds per foot (kilograms per meter) in that order.
4. Zees are specified by width of flange and thickness in inches (millimeters), or by depth, width across flange, and pounds per foot (kilograms per meter).
5. Wide-flange sections are described by depth, width across flange, and pounds per foot (kilograms per meter).

HIGH-STRENGTH LOW-ALLOY STEELS

The properties of high-strength low-alloy (HSLA) steels generally exceed those of conventional carbon structural steels. These low-alloy steels are usually chosen for their high ratios of yield to tensile strength, resistance to puncturing, abrasion resistance, corrosion resistance, and toughness.

ASTM Specifications

ASTM has six specifications covering high-strength low-alloy steels. These are:

ASTM A94 Used primarily for riveted and bolted structures and for special structural purposes.

ASTM A242 Used primarily for structural members where light weight or low mass and durability are important.

ASTM A374 Used where high strength is required and where resistance to atmospheric corrosion must be at least equal to that of plain copper-bearing steel.

ASTM A375 This specification differs slightly from ASTM A374 in that material can be specified in the annealed or normalized condition.

ASTM A440 This covers high-strength intermediate-manganese steels for nonwelded applications.

ASTM A441 This covers the intermediate manganese HSLA steels which are readily weldable when proper welding procedures are used.

LOW- AND MEDIUM-ALLOY STEELS

There are two basic types of alloy steel: *through hardenable* and *surface hardenable*. Each type contains a broad family of steels whose chemical, physical, and mechanical properties make them suitable for specific product applications. See Fig. 12-2-8.

STAINLESS STEELS

Stainless steels have many industrial uses because of their desirable corrosion resistance and strength properties.

Free-Machining Steels

A whole family of free-machining steels has been developed for fast and economical machining. See Fig. 12-2-9. These steels are available in bar stock in various compositions, some standard and some proprietary. When utilized properly, they lower the cost of machining by reducing metal removal time.

Reference and Source Material

1. *Machine Design*, Materials reference issue.

ASSIGNMENT

See Assignment 2 for Unit 12-2 on page 260.

TYPE OF STEEL	ALLOY SERIES	APPROXIMATE ALLOY CONTENT (%)	PRINCIPAL PROPERTIES	COMMON USES
Manganese steel	13xx	Mn 1.6–1.9	Improve surface finish	
Molybdenum steels	40xx	Mo 0.15–0.3		
	41xx	Cr 0.4–1.1; Mo 0.08–0.35		Axles, forgings, gears, cams, mechanical parts
	43xx	Ni 1.65–2; Cr 0.4–0.9; Mo 0.2–0.3	High strength	
	44xx	Mo 0.45–0.6		
	46xx	Ni 0.7–2; Mo 0.15–0.3		
	47xx	Ni 0.9–1.2; Cr 0.35–0.55; Mo 0.15–0.4		
	48xx	Ni 3.25–3.75; Mo 0.2–0.3		
Chromium steels	50xx	Cr 0.3–0.5	Hardness, great strength and toughness	Gears, shafts, bearings, springs, connecting rods
	51xx	Cr 0.7–1.15		
	E51100	C 1.0; Cr 0.9–1.15		
	E52100	C 1.0; Cr 0.9–1.15		
Chromium vanadium steel	61xx	Cr 0.5–1.1; V 0.1–0.15	Hardness and strength	Punches and dies, piston rods, gears, axles
Nickel-chromium-molybdenum steels	86xx	Ni 0.4–0.7; Cr 0.4–0.6; Mo 0.15–0.25	Rust resistance hardness, and stength	Food containers, surgical equipment
	87xx	Ni 0.4–0.7; Cr 0.4–0.6; Mo 0.2–0.3		
	88xx	Ni 0.4–0.7; Cr 0.4–0.6; Mo 0.3–0.4		
Silicon-manganese steel	92xx	Si 1.8–2.2	Springiness and elasticity	Springs

Fig. 12-2-8 AISI designation system for alloy steel.

UNIT 12-3
Nonferrous Metals

Although ferrous alloys are specified for more engineering applications than all nonferrous metals combined, the large family of nonferrous metals offers a wider variety of characteristics and mechanical properties. For example, the lightest metal is lithium, .02 lb/in.^{3} (0.53 g/cm^3); the heaviest is osmium with a weight of .81 lb/in.^{3} (mass

DESIGNATION		PHOSPHORIZED				SULPHURIZED								
MECHANICAL PROPERTY		12L13/12L14 12L15		1211/1212/ 1213		1117/1118/1119			1137			1141/1144		
		Hot-Rolled	Cold-Drawn	Hot-Rolled	Cold-Drawn	Hot-Rolled	Cold-Drawn	Quenched and Tempered	Hot-Rolled	Cold-Drawn	Quenched and Tempered	Hot-Rolled	Cold-Drawn	Quenched and Tempered
Yield strength	10^3 lb/in.^{2}	34	60–80	33	58	34–46	51–68	50–76	48	82	136	51	90	163
	MPa	235	416–550	225	400	235–315	350–470	345–525	330	565	335	350	620	1120
Tensile strength	10^3 lb/in.^{2}	57	70–90	55	75	62–76	69–78	89–113	88	98	157	94	100	190
	MPa	390	480–620	380	517	425–525	475–535	615–780	605	675	1080	650	690	1310
Elongation in 2.00 in. (50 mm)	%	22	10–18	25	10	23–33	15–20	17–22	15	10	5	15	10	9
Machinability (B1212 = 100%)		195–296		91–137		89–100			71			69		

Fig. 12-2-9 Typical mechanical properties of free-machining carbon steels.

of 22.5 g/cm^3)—nearly twice the weight of lead. Mercury melts at around −38°F (−30°C), while tungsten, the highest-melting metal, liquefies at 6170°F (3410°C).

Availability, abundance, and the cost to convert the metal into useful forms all play an important part in selecting a nonferrous metal. Although nearly 80 percent of all elements are called "metals," only about two dozen of these are used as structural engineering materials. Of the balance, however, many are used as coatings, in electronic devices, as nuclear materials, and as minor constituents in other systems.

One of the most important aspects in selecting a material for a mechanical or structural application is how easily the material can be shaped into the finished part—and how its properties can be either intentionally or inadvertently altered in the process. See Fig. 12-3-1. Frequently, metals are simply cast into the finished part. In other cases, metals are cast into an intermediate form (such as an ingot) then worked or "wrought" by rolling, forging, extruding, or other deformation processes.

MANUFACTURING WITH METALS (Refer to Fig. 12-3-1)

Machining Most metals can be machined. Machinability is best for metals that allow easy chip removal with minimum tool wear.

Powder Metallurgy (PM) Compacting Parts can be made from most metals and alloys by PM compacting, although only a few are economically justified. Iron and iron-copper alloys are most commonly used.

Casting Theoretically, any metal that can be melted and poured can be cast. However, economic limitations usually narrow down the number of ways metals are cast commercially.

Extruding and Forging Metals to be forged or extruded must be ductile and not work-harden at working temperature. Some metals show these characteristics at room temperature and can be cold worked; others must be heated.

Stamping and Forming Most metals, except brittle alloys, can be press worked.

Cold Heading Metals must be ductile and should not work-harden rapidly.

Forming method	Aluminum	Copper	Iron	Lead	Magnesium	Nickel	Silver, Gold Platinum	Molybdenum, Copper Tantalum, Tungsten	Steel	Tin	Titanium	Zinc
Casting												
Centrifugal	✓	✓	✓			✓			✓			
Continuous	✓	✓	✓	✓		✓			✓			
Ceramic mold	✓	✓	✓		✓	✓			✓			✓
Investment	✓	✓	✓		✓	✓	✓		✓			
Permanent mold	✓	✓	✓	✓	✓	✓			✓	✓		✓
Sand	✓	✓	✓	✓	✓	✓			✓	✓		✓
Shell mold	✓	✓	✓		✓	✓			✓			
Die casting	✓	✓			✓	✓				✓		✓
Cold heading	✓	✓		✓		✓	✓		✓			
Deep drawing	✓	✓		✓	✓	✓			✓		✓	✓
Extruding	✓	✓		✓	✓	✓		✓	✓	✓	✓	
Forging	✓	✓			✓	✓		✓	✓		✓	
Machining	✓	✓	✓		✓	✓		✓	✓		✓	✓
PM compacting	✓	✓	✓			✓		✓	✓			
Stamping and forming	✓	✓			✓	✓	✓		✓		✓	✓

Fig. 12-3-1 Common methods of forming metals.

Annealing should restore ductility and softness in cold heading alloys.

Deep Drawing Deep drawing involves severe deformation and the metal is usually stretched over the die.

ALUMINUM

The density of aluminum is about one-third that of steel, brass, nickel, or copper. Yet, some alloys of aluminum are stronger than structural steel. Under most service conditions, aluminum has high resistance to corrosion and forms no colored salts which might stain or discolor adjacent components. See Fig. 12-3-2.

MAJOR ALLOYING ELEMENT	DESIGNATION
Aluminum (99% or more)	1xxx
Copper	2xxx
Manganese	3xxx
Silicon	4xxx
Magnesium	5xxx
Magnesium and silicon	6xxx
Zinc	7xxx
Other elements	8xxx
Unused series	9xxx

Fig. 12-3-2 Wrought aluminum alloy designations.

COPPER

Copper alloys, approximately 250 of them, are fabricated in rod, sheet, tube, and wire form. Each of these alloys has

some property or combination of properties which makes it unique. They can be grouped into several general headings, such as coppers, brasses, leaded brasses, phosphor bronzes, aluminum bronzes, silicon bronzes, beryllium coppers, cupronickels, and nickel silvers. See Fig. 12-3-3.

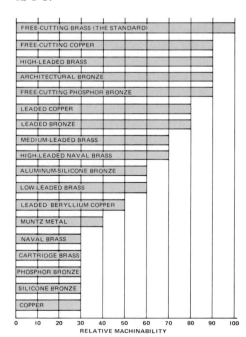

Fig. 12-3-3 Free-machining copper alloys.

Copper alloys are used where one or more of the following properties is needed: thermal or electrical conductivity, corrosion resistance, strength, ease of forming, ease of joining, and color.

The major alloy usages are:

1. Copper in pure form as a conductor in the electrical industry
2. Copper or alloy tubing for water, drainage, air conditioning, and refrigeration lines
3. Brasses, phosphor bronzes, and nickel silvers as springs or in construction of equipment if corrosive conditions are too severe for iron or steel

An advantage of copper and its alloys, offered by no other metals, is the wide range of colors available.

NICKEL

Commercially pure wrought nickel is a grayish-white metal capable of taking a high polish. Because of its combination of attractive mechanical properties, corrosion resistance, and formability, nickel or its alloys is used in a variety of structural applications usually requiring specific corrosion resistance.

MAGNESIUM

Magnesium, with density of only .06 lb/in.^{3} (1.74 g/cm^3), is the world's lightest structural metal. The combination of low density and good mechanical strength makes possible alloys with a high strength-to-weight ratio.

ZINC

Zinc is a relatively inexpensive metal which has moderate strength and toughness and outstanding corrosion resistance in many types of service.

The principal characteristics which influence the selection of zinc alloys for die castings include the dimensional accuracy obtainable, castability of thin sections, smooth surface, dimensional stability, and adaptability to a wide variety of finishes.

TITANIUM

Titanium is a light metal at .16 lb/in.^{3} (4.43 g/cm^3); 60 percent heavier than aluminum but 45 percent lighter than alloy steel. It is the fourth most abundant metallic element in the earth's crust and the ninth most common element.

Titanium-based alloys are much stronger than aluminum alloys and superior in many respects to most alloy steels.

BERYLLIUM

Beryllium has a strength-to-weight ratio comparable to high-strength steel, yet it is lighter than aluminum. Its melting point is 2345°F (1285°C) and it has excellent thermal conductivity. It is nonmagnetic and a good conductor of electricity.

REFRACTORY METALS

Refractory metals are those metals with melting points above 3600°F (2000°C). Among these, the best known and most extensively used are tungsten, tantalum, molybdenum, and niobium.

Refractory metals are characterized by high-temperature strength, corrosion resistance, and high melting points.

Tantalum and Niobium

Tantalum and niobium are usually discussed together, since most of their working operations are identical. Unlike molybdenum and tungsten, tantalum and niobium can be worked at room temperatures. The major differences between tantalum and niobium are in density, nuclear cross section, and corrosion resistance. The density of tantalum is almost twice that of niobium.

Molybdenum

Molybdenum is widely used in missiles, aircraft, industrial furnaces, and nuclear projects. Its melting point is lower than that of tantalum and tungsten. Molybdenum has a high strength-to-weight ratio and a low vapor pressure, is a good conductor of heat and electricity, has a high modulus of elasticity, and a low coefficient of expansion.

Tungsten

Tungsten is the only refractory metal that has the combination of excellent corrosion resistance, good electrical and thermal conductivity, a low coefficient of expansion, and high strength at elevated temperatures.

PRECIOUS METALS

Gold cost over 8000 times more than an equal amount of iron; rhodium costs nearly 32,000 times more than copper. With prices such as these, why are precious metals ever specified?

In some cases, precious metals are used for their unique surface characteristics. They reflect light better than other metals. Gold, for example, is specified as a surface for heat reflectors, insulators, and collectors because of its outstanding ability to reflect ultraviolet radiation.

The family of metals called precious metals can be divided into three subgroups: silver and silver alloys; gold and gold alloys; and the so-called platinum metals, which are platinum, palladium, rhodium, ruthenium, iridium, and osmium.

Reference and Source Material

1. *Machine Design*, Materials reference issue.

ASSIGNMENTS

See Assignments 3 and 4 for Unit 12-3 starting on page 260.

U N I T   12-4
Plastics

This unit will acquaint drafters with the general characteristics of commercially available plastics so that they can make proper use of plastics in products.

Plastics may be defined as nonmetallic materials capable of being formed or molded with the aid of heat, pressure, chemical reactions, or a combination of these. See Fig. 12-4-1.

Plastics are strong, tough, durable materials which solve many problems in machine and equipment design. Metals, it is true, are hard and rigid. This means that they can be machined, to very close tolerances, into cams, bearings, bushings, and gears, which will work smoothly under heavy loads for long periods. Although some come close, no plastic has the hardness and creep resistance of, say, steel. However, metals have many weaknesses which engineering plastics do not. Metals corrode or rust, they must be lubricated, their working surfaces wear readily, they cannot be used as electrical or thermal insulators, they are opaque and noisy, and where they must flex, metals fatigue rapidly.

Plastics can resolve these weaknesses,

Fig. 12-4-1 A variety of plastic parts.

though not necessarily all with one material. The engineering plastics are resistant to most chemicals; fluorocarbon is one of the most chemically inert substances known. None of the engineering plastics corrode or rust; acetal resin and fluorocarbon are unaffected even when continuously immersed in water. Engineering plastics can be run at low speeds and loads, and, without lubrication, are among the world's slipperiest solids, being comparable to ice. Engineering plastics are resilient; therefore, they run more quietly and smoothly than equivalent metal ones, and they are able to stand periodic overloads without harmful effects.

Plastics are a family of materials—not a single material—each member of which has its special advantages.

Being manufactured, plastics raw materials are capable of being variously combined to give almost any property desired in an end product. But these are controlled variations unlike those of nature's products. Some thermoplastics can be sterilized.

The widespread and growing use of plastics in almost every phase of modern living can be credited in large part to their unique combinations of advantages. These advantages are light weight, range of color, good physical properties, adaptability to mass-production methods, and, often, lower cost.

Aside from the range of uses attribut-

able to the special qualities of different plastics, these materials achieve still greater variety through the many forms in which they can be produced.

They may be made into definite shapes like dinnerware and electric switchboxes. They may be made into flexible film and sheeting such as shower curtains and upholstery. Plastics may be made into sheets, rods, and tubes that are later shaped or machined into internally lighted signs or airplane blisters. They may be made into filaments for use in household screening, industrial strainers, and sieves. Plastics may be used as a coating on textiles and paper. They may be used to bind such materials as fibers of glass and sheets of paper or wood to form boat hulls, airplane wing tips, and tabletops.

Plastics are usually classified as either *thermoplastic* or *thermosetting*.

THERMOSETTING PLASTICS

These materials undergo an irreversible chemical change when heat is applied or when a catalyst or reactant is added. They become hard, insoluble, and infusible, and they do not soften upon reapplication of heat. Thermosetting plastics include phenolics, amino plastics (melamine and urea), cold-molded polyesters, epoxies, silicones, alkyds, allylics, and casein. See Fig. 12-4-2.

THERMOPLASTICS

These materials soften, or liquefy, and flow when heat is applied. Removal of the heat causes these materials to set or solidify. They may be reheated and reformed or reused. In this group fall the acrylics, the cellulosics, nylons (polyamides), polyethylene, polystyrene, polyfluorocarbons, the vinyls, polyvinylidene, ABS, acetal resin, polypropylene, and polycarbonates. See Fig. 12-4-3.

MACHINING

Practically all thermoplastics and thermosets can be satisfactorily machined on standard equipment with adequate tooling. The nature of the plastic will determine whether heat should be applied, as in some laminates, or avoided, as in buffing some thermoplastics. Standard machining operations can be used, such as turning, drilling, tapping, milling, blanking, and punching.

MATERIAL SELECTION

One of the first decisions a designer makes is the choice of materials. The choice is influenced by many factors, such as the end use of the product and the properties of the selected material. No attempt is made at this point to discuss the engineering approach to selection of materials.

However, a basic examination and selection of a plastic material at this time will help acquaint the drafter with the wide range of plastics available.

For instance, the preliminary production report for the material selection of the telephone case shown in Fig. 12-4-4 is an example of the type of research required in selecting a material.

Forming Processes

For design information on the preparation of molded plastics, see Chap. 11, Unit 11-4.

References and Source Material

1. The Society of the Plastics Industry, Inc.
2. Crystaplex Plastics.
3. General Motors Corporation.

ASSIGNMENT

See Assignment 5 for Unit 12-4 on page 261.

THERMOSETTING PLASTICS			
NAME OF PLASTIC	PROPERTIES	FORMS AND METHODS OF FORMING	USES
ALKYDS	EXCELLENT DIELECTRIC STRENGTH, HEAT RESISTANCE, AND RESISTANCE TO MOISTURE.	AVAILABLE IN MOLDING POWDER AND LIQUID RESIN. FINISHED MOLDED PRODUCTS ARE PRODUCED BY COMPRESSION MOLDING.	LIGHT SWITCHES, ELECTRIC MOTOR INSULATOR AND MOUNTING CASES, TELEVISION TUNING DEVICES AND TUBE SUPPORTS. ENAMELS AND LACQUERS FOR AUTOMOBILES, REFRIGERATORS, AND STOVES ARE TYPICAL USES FOR THE LIQUID FORM.
ALLYLICS	EXCELLENT DIELECTRIC STRENGTH AND INSULATION RESISTANCE. NO MOISTURE ABSORPTION; STAIN RESISTANCE. FULL RANGE OF OPAQUE AND TRANSPARENT COLORS.	AVAILABLE IN THE FORM OF MONOMERS, PRE-POLYMERS, AND POWDERS. FINISHED ARTICLES MAY BE MADE BY TRANSFER OR COMPRESSION MOLDING, LAMINATION, COATING, OR IMPREGNATION.	ELECTRICAL CONNECTORS, APPLIANCE HANDLES, KNOBS, ETC. LAMINATED OVERLAYS OR COAT-INGS FOR PLYWOOD, HARDBOARD, AND OTHER LAMI-NATED MATERIALS NEEDING PROTECTION FROM MOISTURE.
AMINO (MELAMINE AND UREA)	FULL RANGE OF TRANSLUCENT AND OPAQUE COLORS. VERY HARD, STRONG, BUT NOT UNBREAKABLE. GOOD ELECTRICAL QUALITIES.	AVAILABLE AS MOLDING POWDER OR GRANULES, AS A FOAMED MATERIAL IN SOLUTION, AND AS RESINS. FINISHED PRODUCTS CAN BE MADE BY COMPRES-SION, TRANSFER, PLUNGER MOLDING, AND LAMINATING WITH WOOD, PAPER, ETC.	MELAMINE—TABLEWEAR, BUT-TONS, DISTRIBUTOR CASES, TABLETOPS, PLYWOOD ADHESIVE, AND AS A PAPER AND TEXTILE TREATMENT. UREA—SCALE HOUSING, RADIO CABINETS, ELECTRICAL DEVICES, APPLIANCE HOUSINGS, STOVE KNOBS IN RESIN FORM AS BAK-ING ENAMEL COATINGS, PLYWOOD ADHESIVE AND AS A PAPER AND TEXTILE TREATMENT.
CASEIN	EXCELLENT SURFACE POLISH. WIDE RANGE OF NEAR TRANSPARENT AND OPAQUE COLORS. STRONG, RIGID, AFFECTED BY HUMIDITY AND TEMPERATURE CHANGES.	AVAILABLE IN RIGID SHEETS, RODS, AND TUBES, AS A POWDER AND LIQUID. FINISHED PRODUCTS ARE MADE BY MACHINING OF THE SHEETS, RODS, AND TUBES.	BUTTONS, BUCKLES, BEADS, GAME COUNTERS, KNITTING NEEDLES, TOYS, AND ADHESIVES.
COLD-MOLDED 3 TYPES: BITUMIN PHENOLIC CEMENT-ASBESTOS	RESISTANCE TO HIGH HEAT, SOL-VENTS, WATER, AND OIL.	AVAILABLE IN COMPOUNDS. FINISHED ARTICLES PRODUCED BY MOLDING AND CURING.	SWITCH BASES AND PLUGS, INSULATORS, SMALL GEARS, HAN-DLES AND KNOBS, TILES, JIGS AND DIES, TOY BUILDING BLOCKS.
EPOXY	GOOD ELECTRICAL PROPERTIES; WATER AND WEATHER RESISTANCE.	AVAILABLE AS MOLDING COMPOUNDS, RESINS, FOAMED BLOCKS, LIQUID SOLUTIONS, ADHESIVES, COATINGS, SEALANTS.	PROTECTIVE COATING FOR APPLIANCES, CANS, DRUMS, GYM-NASIUM FLOORS, AND OTHER HARD-TO-PROTECT SURFACES. THEY FIRMLY BOND METALS, GLASS, CERAMICS, HARD RUBBER AND PLASTICS, PRINTED CIRCUITS, LAMINATED TOOLS AND JIGS, AND LIQUID STORAGE TANKS.
PHENOLICS	STRONG AND HARD. HEAT AND COLD RESISTANT; EXCEL-LENT INSULATORS.	CAST AND MOLDED.	RADIO AND TV CABINETS, WASH-ING MACHINE AGITATORS, JUKE BOX HOUSINGS, JEWELRY, PULLEYS, ELECTRICAL INSULATION.
POLYESTERS (FIBERGLASS)	STRONG AND TOUGH, BRIGHT AND PASTEL COLORS, HIGH DIELECTRIC QUALITIES.	PRODUCED AS LIQUIDS, DRY POWDERS, PREMIX MOLDING COMPOUNDS, AND AS CAST SHEETS, RODS, AND TUBES. THEY ARE FORMED BY MOLDING, CASTING, IMPREGNATING, AND PREMIXING.	USED TO IMPREGNATE CLOTH OR MATS OF GLASS FIBERS, PAPER, COTTON, AND OTHER FIBERS IN THE MAKING OF REINFORCED PLASTIC FOR USE IN BOATS, AUTOMBILE BODIES, LUGGAGE.
SILICONES	HEAT RESISTANT, GOOD DIELECTRIC PROPERTIES.	AVAILABLE AS MOLDING COMPOUNDS, RESINS, COATINGS, GREASES, FLUIDS, AND SILICON RUB-BER. FINISHED BY COMPRESSION AND TRANSFER MOLDING, EXTRUSION, COATING, CALENDERING, CASTING, FOAMING, AND IMPREGNATING.	COIL FORMS, SWITCH PARTS, INSULATION FOR MOTORS, AND GENERATOR COILS.

Fig. 12-4-2 Thermosetting plastics. (The Society of Plastics Industry)

THERMOPLASTICS			
NAME OF PLASTIC	PROPERTIES	FORMS AND METHODS OF FORMING	USES
ABS (ACRYLONITRILE BUTADIENE-STYRENE)	STRONG, TOUGH, GOOD ELECTRICAL PROPERTIES	AVAILABLE IN POWDER OR GRANULES FOR INJECTION MOLDING, EXTRUSION, AND CALENDERING AND AS SHEET FOR VACUUM FORMING.	PIPE, WHEELS, FOOTBALL, HELMETS, BATTERY CASES, RADIO CASES, CHILDREN'S SKATES, TOTE BOXES.
ACETAL RESIN	RIGID WITHOUT BEING BRITTLE, TOUGH, RESISTANT TO EXTREME TEMPERATURES, GOOD ELECTRICAL PROPERTIES.	PRODUCED IN POWDER FORM FOR MOLDING AND EXTRUSION, AVAILABLE IN ROD, BAR, TUBE, STRIP, SLAB.	AUTOMOBILE INSTRUMENT CLUSTERS, GEARS, BEARINGS, BUSHINGS, DOOR HANDLES, PLUMBING FIXTURES, THREADED FASTENERS, CAMS.
ACRYLICS	EXCEPTIONAL CLARITY AND GOOD LIGHT TRANSMISSION. STRONG, RIGID, AND RESISTANT TO SHARP BLOWS. EXCELLENT INSULATOR, COLORLESS OR FULL RANGE OF TRANSPARENT, TRANSLUCENT, OR OPAQUE COLORS.	AVAILABLE IN SHEET, ROD, TUBE, AND MOLDING POWDERS, PLASTIC PRODUCTS CAN BE PRODUCED BY FABRICATING OF SHEETS, RODS, AND TUBES, HOT FORMING OF SHEETS, INJECTION AND COMPRESSION MOLDING OF POWDER, EXTRUSION, CASTING.	AIRPLANE CANOPIES AND WINDOWS, TELEVISION AND CAMERA VIEWING LENSES, COMBS, COSTUME JEWELRY, SALAD BOWLS, TRAYS, LAMP BASES, SCALE MODELS, AUTOMOBILE TAIL LIGHTS, OUTDOOR SIGNS.
CELLULOSICS (A) CELLULOSE ACETATE		AVAILABLE IN PELLETS, SHEETS, FILM, RODS, TUBES, STRIPS, COATED CORD. CAN BE MADE INTO PRODUCTS BY INJECTION, COMPRESSION MOLDING, EXTRUSION, BLOW MOLDING, AND VACUUM FORMING, OR SHEETS AND COATING.	SPECTACLE FRAMES, TOYS, LAMP SHADES, COMBS, SHOE HEELS.
(B) CELLULOSE ACETATE BUTYRATE	AMONG THE TOUGHEST OF PLASTICS. RETAINS A LUSTROUS FINISH UNDER NORMAL WEAR. TRANSPARENT, TRANSLUCENT, OR OPAQUE IN WIDE VARIETY OF COLORS AND IN CLEAR TRANSPARENT. GOOD INSULATORS.	AVAILABLE IN PELLETS, SHEETS, RODS, TUBES, STRIPES AND AS A COATING. CAN BE MADE INTO PRODUCTS BY INJECTION, COMPRESSION MOLDING, EXTRUSION, BLOWING AND DRAWING OF SHEET, LAMINATING, COATING.	STEERING WHEELS, RADIO CASES, PIPE AND TUBING, TOOL HANDLES, PLAYING CARDS.
(C) CELLULOSE PROPIONATE		AVAILABLE IN PELLETS FOR INJECTION EXTRUSION OR COMPRESSION MOLDING.	APPLIANCE HOUSING, TELEPHONE HAND SETS, PENS AND PENCILS.
(D) ETHYL CELLULOSE		AVAILABLE IN GRANULES, FLAKE, SHEET, ROD, TUBE, FILM, OR FOIL. CAN BE MADE INTO FINISHED PRODUCTS BY INJECTION, COMPRESSION MOLDING, EXTRUSION, DRAWING.	EDGE MOLDINGS, FLASHLIGHTS, ELECTRICAL PARTS
(E) CELLULOSE NITRATE		AVAILABLE IN RODS, TUBES, SHEETS FOR MACHINING AND AS A COATING.	SHOE HEEL COVERS, FABRIC COATING.
FLUOROCARBONS	LOW COEFFICIENT OF FRICTION, RESISTANT TO EXTREME HEAT AND COLD. STRONG, HARD, AND GOOD INSULATORS.	AVAILABLE AS POWDER AND GRANULES IN RESIN FORM. SHEET, ROD, TUBE, FILM, TAPE, AND DISPERSIONS, MOLDED, EXTRUDED, AND MACHINED.	VALVE SEATS, GASKETS, COATINGS, LININGS, TUBINGS.
NYLON (POLYAMIDES)	RESISTANT TO EXTREME TEMPERATURES. STRONG AND LONG-WEARING RANGE OF SOFT COLORS.	AVAILABLE AS A MOLDING POWDER, IN SHEETS, RODS, TUBES, AND FILAMENTS, INJECTION, COMPRESSION, BLOW MOLDING, AND EXTRUSION.	TUMBLERS, FAUCET WASHERS, GEARS. AS A FILAMENT, IT IS USED AS BRUSH BRISTLES, FISHING LINE.
POLYCARBONATE	HIGH IMPACT STRENGTH, RESISTANT TO WEATHER, TRANSPARENT.	PRIMARILY A MOLDING MATERIAL, MAY TAKE FORM OF FILM, EXTRUSION, COATINGS, FIBERS, OR ELASTOMERS.	PARTS FOR AIRCRAFT, AUTOMOBILES, BUSINESS MACHINES, GAGES, SAFETY-GLASS LENSES.
POLYETHYLENE	EXCELLENT INSULATING PROPERTIES, MOISTURE PROOF, CLEAR, TRANSPARENT, TRANSLUCENT.	AVAILABLE IN PELLET, POWDER, SHEET, FILM, FILAMENT, ROD, TUBE, AND FOAMED. INJECTION, COMPRESSION, BLOW MOLDING, EXTRUSION, COATING, AND CASTING.	ICE CUBE TRAYS, TUMBLERS, DISHES, BOTTLES, BAGS, BALLOONS, TOYS, MOISTURE BARRIERS.
POLYSTYRENE	CLEAR, TRANSPARENT, TRANSLUCENT, OR OPAQUE. ALL COLORS. WATER AND WEATHER RESISTANT, RESISTANCE TO HEAT OR COLD.	AVAILABLE IN MOLDING POWDERS OR GRANULES, SHEETS, RODS, FOAMED BLOCKS, LIQUID SOLUTION, COATINGS, AND ADHESIVES, INJECTION, COMPRESSION MOLDING, EXTRUSION, LAMINATING, MACHINING.	KITCHEN ITEMS, FOOD CONTAINERS, WALL TILE, TOYS, INSTRUMENT PANELS.
POLYPROPYLENES	GOOD HEAT RESISTANCE. HIGH RESISTANCE TO CRACKING. LIGHT RANGE OF COLOR.	PROCESSED BY INJECTION MOLDING, BLOW MOLDING, AND EXTRUSION.	THERMAL DISHWARE, WASHING MACHINE AGITATORS, PIPE AND PIPE FITTINGS, WIRE AND CABLE INSULATION, BATTERY BOXES, PACKAGING FILM AND SHEETS.

Fig. 12-4-3 Thermoplastics. (The Society of Plastics Industry)

THERMOPLASTICS			
NAME OF PLASTIC	**PROPERTIES**	**FORMS AND METHODS OF FORMING**	**USES**
URETHANES	TOUGH AND SHOCK RESISTANT FOR SOLID MATERIALS. FLEXIBLE FOR FOAMED MATERIAL, CAN BE FOAMED IN PLACE.	SOLID TYPE—STARTING TWO REACTANTS, FINAL ARTICLE CAN BE EXTRUDED, MOLDED, CALENDERED, OR CAST. FOAMED TYPE—CAN BE MADE BY EITHER A PREPOLYMER OR ONE-SHOT PROCESS, IN EITHER SLAB STOCK OR MOLDED FORM.	MATTRESSES, CUSHIONING, PADDING, TOYS, RUG UNDERLAYS, CRASH-PADS, SPONGES, MATS, ADHESION, THERMAL INSULATION, INDUSTRIAL TIRES.
VINYLS	STRONG AND ABRASION-RESISTING. RESISTANT TO HEAT AND COLD. WIDE COLOR RANGE.	AVAILABLE IN MOLDING POWDER, SHEET, ROD, TUBE, GRANULES, POWDER. IT CAN BE FORMED BY EXTRUSION, CASTING, CALENDERING, COMPRESSION, AND INJECTION MOLDING.	RAINCOATS, GARMET BAGS, INFLATABLE TOYS, HOSE, RECORDS, FLOOR AND WALL TILE, SHOWER CURTAINS, DRAPERIES, PIPE, PANELING.

Fig. 12-4-3 Continued.

PRODUCTION REPORT					
Part Name	**Material**		**Reason for Selection**	**Machining Required**	**Color**
	1st Choice	**2nd Choice**			
Telephone Case	ABS	Cellulosics	Good impact strength Good range of colors Light mass. Excellent surface finish Good electrical properties Variety of forming methods	None	Green Blue White Tan Red

Fig. 12-4-4 Selection of material.

ASSIGNMENTS FOR CHAPTER 12

Assignment for Unit 12-1, Cast Irons

1. Make a two-view working drawing of one of the parts shown in Fig. 12-1-A or 12-1-B. Use a revolved section to show the center section of the arm. Select a suitable cast iron for the part. Scale 1:1.

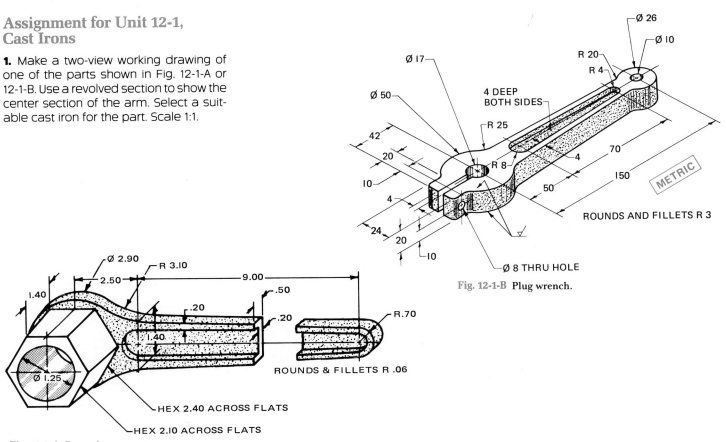

Fig. 12-1-B Plug wrench.

Fig. 12-1-A Door closer arm.

Assignment for Unit 12-2, Carbon Steel

2. Make a working drawing of one of the parts shown in Fig. 12-2-A or 12-2-B. Show the worm threads in pictorial form. Scale 2:1. Select a suitable steel for the part. Conventional breaks may be used to shorten the length of the view.

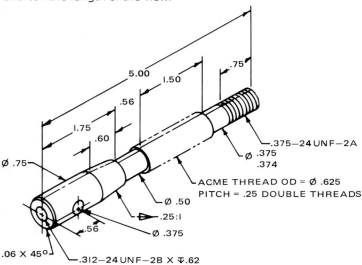

Fig. 12-2-A Raising bar.

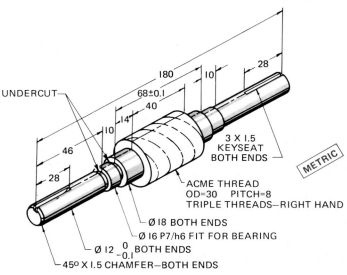

Fig. 12-2-B Worm for gear jack.

Assignments for Unit 12-3, Nonferrous Metals

3. Make a two-view working drawing of the outboard motor clamp shown in Fig. 12-3-A or 12-3-B. Use lines or surfaces marked *A, B,* and *C* as the zero lines and use arrowless dimensioning. Scale 1:2. Select a suitable material noting that the part must be water resistant, have a painted finish, have moderate strength, and have a light weight or mass.

4. Recommend the material for each of the parts shown in Figs. 12-3-C and 12-3-D. State your reasons for each of the materials selected.

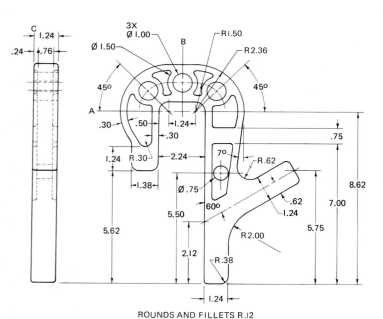

Fig. 12-3-A Outboard motor clamp.

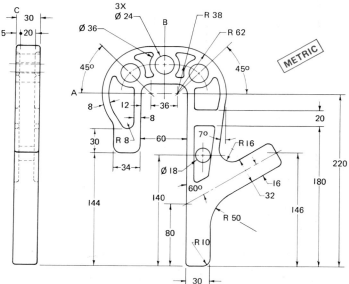

Fig. 12-3-B Outboard motor clamp.

260   FASTENERS, MATERIALS, AND FORMING PROCESSES

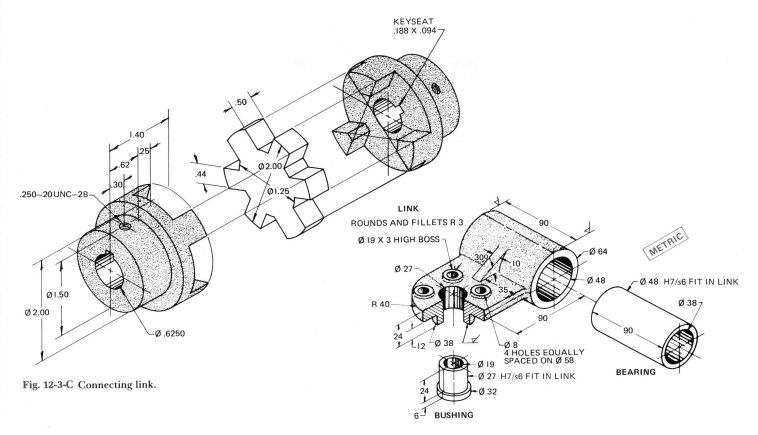

Fig. 12-3-C Connecting link.

KEYSEAT
.188 X .094

LINK
ROUNDS AND FILLETS R 3
Ø 19 X 3 HIGH BOSS

Ø 27
300
10
35
R 40
24
12 Ø 38
Ø 8
4 HOLES EQUALLY
SPACED ON Ø 58

90
Ø 64
Ø 48
90

METRIC

Ø 48 H7/s6 FIT IN LINK
Ø 38
90
BEARING

Ø 19
Ø 27 H7/s6 FIT IN LINK
24
Ø 32
6 BUSHING

Fig. 12-3-D Coupling.

Assignment for Unit 12-4, Plastics

5. Prepare a production report for the selection of materials for the parts shown in Fig. 12-4-A or 12-4-B. Convert as many parts as possible to plastic. The crane hook assembly is to be used in a water dip-tank operation. Assume that the production run is such that all forming processes can be considered. Include with your report an item list.

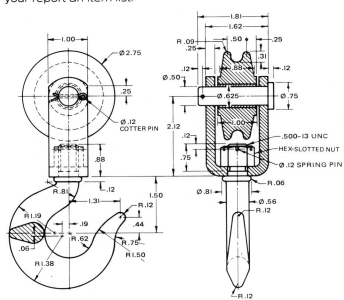

Fig. 12-4-A Crane hook assembly.

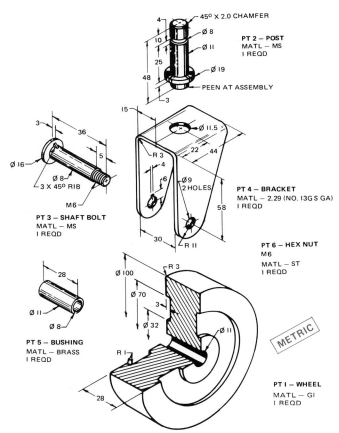

45° X 2.0 CHAMFER
4
10
25
48
3
Ø 8
Ø 11
Ø 19
PEEN AT ASSEMBLY

PT 2 – POST
MATL – MS
I REQD

15
3
36
5
Ø 16
Ø 8
3 X 45° RIB
M6
PT 3 – SHAFT BOLT
MATL – MS
I REQD

Ø 11.5
R 3
4
22 44
6
Ø 9
2 HOLES
58
30
R 11
PT 4 – BRACKET
MATL – 2.29 (NO. 13G S GA)
I REQD

PT 6 – HEX NUT
M6
MATL – ST
I REQD

28
Ø 11
Ø 8
PT 5 – BUSHING
MATL – BRASS
I REQD

R 3
Ø 100
Ø 70
Ø 32
3
R 1
28
Ø 11
METRIC
PT I – WHEEL
MATL – GI
I REQD

Fig. 12-4-B Caster assembly.

P A R T 3

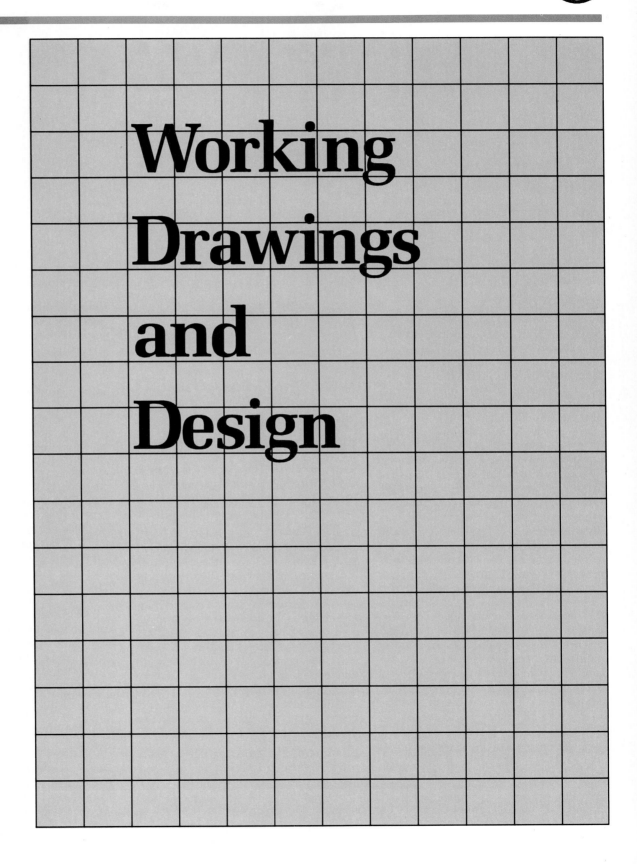

Working Drawings and Design

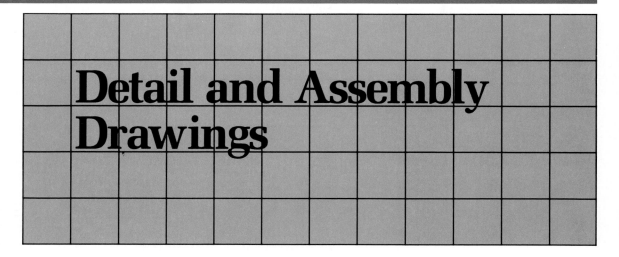

C H A P T E R   1 3

Detail and Assembly Drawings

U N I T   1 3 - 1
Functional Drafting

Since the basic function of the drafting department is to provide sufficient information to produce or assemble parts, functional drafting must embrace every possible means to communicate this information in the least expensive manner. Functional drafting also applies to any method which would lower the cost of producing the part. New technological developments have provided many new ways of producing drawings at lower costs and/or in less time. This means that the drafting office must be prepared to discard some of the old, traditional methods in favor of these newer means of communication.

There are many ways in which to reduce the drafting time in preparing a drawing. These drawing shortcuts, when collectively used, are of prime importance in an effective drafting system.

These newer techniques cannot be blindly applied, however, but must be carefully evaluated to make certain that the benefits outweigh the potential disadvantages. This evaluation should answer the following questions:

- What is its purpose?
- Is it a personal preference disguised as a project requirement?
- Does it meet contractual requirements?
- Will the shortcut increase costs in

other areas such as manufacturing, purchasing, or inspection?
- Is it an effective communication link?
- How much training or education is required to make effective use of it?
- Are facilities available to implement it?
- Does the shortcut bypass a real bottleneck?

As each of these categories is examined, the advantages of the shortcuts will become apparent.

PROCEDURAL SHORTCUTS

There are a number of procedural shortcuts which, if properly applied and carefully managed, can shorten the drawing preparation cycle and result in savings.

Streamlined Approval Requirements
It is obvious that the more signatures required on a drawing, the greater the delays in releasing data. The decision as to who will approve drawings and drawing changes must be carefully considered to make certain that all necessary functions have been taken into account (checkers, responsible engineers, important technical specialists, etc.) without imposing undue restrictions. Project ground rules and contractual requirements also play an important part in this decision.

Eliminating the Drawing Check from the Preparation Cycle One of the most common suggested shortcuts, proposed usually when a project is behind schedule or exceeding its budget or

when experienced personnel are involved, is to eliminate checking from the drawing preparation cycle.

Using Standard and Existing Drawings Every year numerous drawings of parts are prepared which are repetitions of existing drawings. If the drafter were to incorporate in the new design parts that were already drawn, many drawing hours would be saved. Good drawing application records and an efficient multiple-use drawing system can eliminate a great deal of duplication. Standard tabulated drawings may be used to eliminate hundreds of drawings. See Figs. 13-1-1 and 13-1-2.

Standard Drafting Practices Standard drafting practices are obviously the backbone of efficient drafting room operations. The best way to establish and implement these practices is through a good drafting room manual, whose requirements must be strictly observed by all personnel. See Fig. 13-1-3.

The drafting room manual should contain data on the use and preparation of specific types of drawings, drawing and part number requirements, standard and special drafting practices, rules for dimensioning and tolerancing, specifications for associated lists, and company procedures for the preparation, handling, release, and control of drawings.

Team Drafting Many engineering departments have turned out drawings by the method of one drafter to one drawing. Team drafting involves a

QTY	PART	MATL	DESCRIPTION	PT NO.
2	CABLE SUPPORT	MAPLE	A–5374 PT 1	1
2	CABLE SUPPORT	MAPLE	A–5374 PT 2	2
3	CABLE SUPPORT	MAPLE	A–5374 PT 4	3

(A) DRAWING CALLOUT

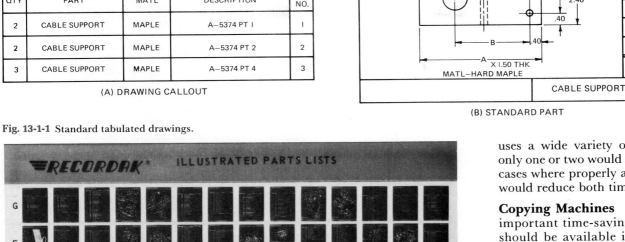

4	5.75	2.40	.80
3	4.00	1.60	.80
2	4.00	1.60	1.00
1	5.00	1.60	1.00
PT	A	B	Ø C

CABLE SUPPORT — A–5374

(B) STANDARD PART

Fig. 13-1-1 Standard tabulated drawings.

Fig. 13-1-2 Standard parts drawings stored on microfilm. (Eastman Kodak Co.)

Fig. 13-1-3 General Motors drawing standards manuals.

number of people producing one drawing. While this may seem uneconomical, it is an expeditious means with visible cost savings over the traditional method.

Some firms are using team drafting because it is a better utilization of skill levels. It is a training program through which drafting skills are taught and semiskilled people are given an opportunity to gain experience.

Data Retrieval The use of microform reader-printers in the drafting room provides quick and ready access to standard drawings and parts. The use of microfiche cards is becoming popular, because they can hold up to 70 pages of information. However, for this method to be effective, a full-time librarian is needed.

Standard Parts and Design Standard Information Encouraging the use of standard parts and standard approaches to design will not only result in drafting time saved, but will also cut costs in areas such as purchasing, material control, manufacturing, etc. The odd-size cutout that requires special tooling, the design that calls for nonstandard hardware, and the equipment that uses a wide variety of fasteners when only one or two would suffice are typical cases where properly applied standards would reduce both time and cost.

Copying Machines One of the most important time-saving devices, which should be available in every drafting area, is a copying machine for reference copies, checking prints of work in preparation, and similar uses. See Fig. 13-1-4. When a drafter needs a copy, work is delayed until the copy is made available. Therefore, a good copying machine will soon pay for itself in drawing hours saved.

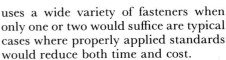

Fig. 13-1-4 Copying machine.

Training Programs To provide drafters with standard procedures and technical information is not enough; they must be trained in their use. New drafters are frequently overwhelmed by a strange environment, while old employees fail to keep up with new requirements or properly use the services available. Training programs for the indoctrination of new personnel and the updating of long-service employees are rewarded by more efficient and versatile operation.

DRAFTING EQUIPMENT AND MATERIALS

The quality of the material and supplies used in the preparation of drawings is as important as the quality of the instruments used in fabrication.

Numerous timesaving devices are available for manual drafting: templates for every application, "pens" for easier linework, and more application of tape to artwork, transfer-type lettering, etc. Since drafting applications vary so widely, only the drafting supervisor can determine which devices will increase the drafting production.

Drafting aids are designed to facilitate the making of drawings by removing or reducing some of the more tedious aspects of drafting.

Templates, such as shown in Fig. 13-1-5, play an important part in functional drafting, for they save a great deal of time in drawing common shapes of details such as rounds, squares, hexagons, and ellipses. In addition to common shapes, templates have been made for standard parts such as nuts, and bolt heads, for electrical symbols, outlines of tools and equipment, and many other outlines which are often repeated.

With CAD, common shapes such as rounds and ellipses can readily be created by commands. Other specialized symbols need only be drawn once and stored in a symbol library. It then can be recalled at any time and placed on the drawing.

REDUCING THE NUMBER OF DRAWINGS REQUIRED

The cost of a project is, to some extent, directly related to the number of drawings which must be prepared. Therefore, careful planning to reduce the number of drawings required can result in significant savings. Some ways to reduce the number of drawings are explained below.

Detail Assembly Drawings Detail assembly drawings, in which parts are detailed in place on the assembly (See Fig. 13-1-6), and multidetail assembly drawings, in which there are separate detail views for the assembly and each of its parts, will reduce the number of drawings required. However, they must be used with extreme care. They can easily become too complicated and confusing to be an effective means of communication. See Unit 13-7.

Selecting the Most Suitable Type of Projection to Describe the Part The selection of the type of projection (orthographic, isometric, or oblique) can greatly increase the ease with which

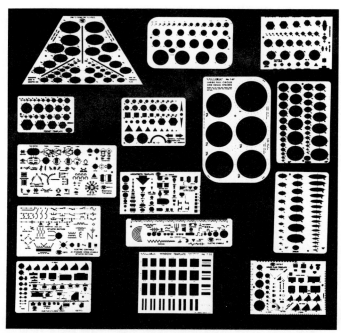

Fig. 13-1-5 Templates are made for many different uses and save a lot of time.

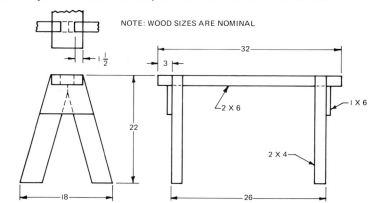

Fig. 13-1-6 Detail assembly drawing of a sawhorse.

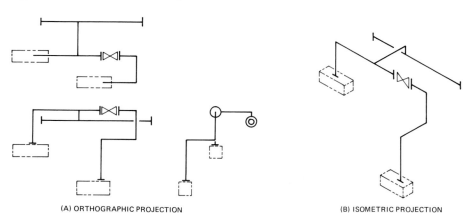

(A) ORTHOGRAPHIC PROJECTION   (B) ISOMETRIC PROJECTION

Fig. 13-1-7 Selecting the most suitable type of projection.

some drawings can be read and, in many cases, reduce drafting time. For example, a single-line piping drawing drawn in isometric projection simplifies an otherwise difficult drawing problem in orthographic projection. See Fig. 13-1-7.

SIMPLIFICATION OF DETAIL DRAWING

1. Complicated parts are best described by means of a drawing.

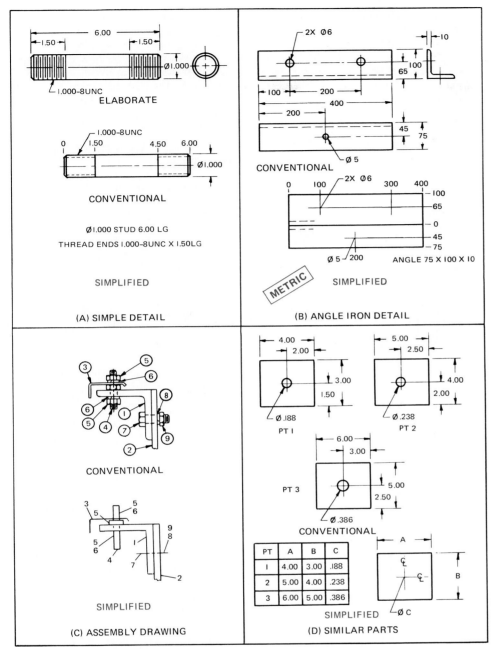

Fig. 13-1-8 Comparison between conventional and simplified drawings.

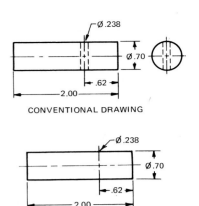

Fig. 13-1-9 Simplified drafting practices for detailed parts.

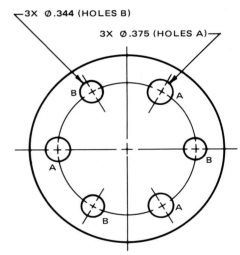

Fig. 13-1-10 Identification of similar sized holes.

However, explanatory notes can complement the drawing, thereby eliminating views that are time-consuming to draw. See Figs. 13-1-8 and 13-1-9.

2. Use simplified drawing practices, as described throughout this text, especially on threads and common features.

3. Avoid unnecessary views. In many cases one or two views are sufficient to explain the part fully.

4. When a number of holes of similar size are to be made in a part, there is a chance that the person producing the part may misinterpret the diameter of some of the holes. In such cases, the identification of similar sized holes should be made clear. See Fig. 13-1-10.

5. The use of the symmetry symbol means that all dimensions are symmetrical about that line.

6. A simplified assembly drawing should be used for assembly purposes only. Some means of simplification are:
   • Standard parts such as nuts, bolts, and washers need not be drawn.
   • Small fillets and rounds on cast parts need not be shown.
   • Phantom outlines of complicated details can often be used.

7. Use templates or symbol libraries.

8. Within limits, a small drawing is made more quickly than a large drawing when manual drafting is used.

9. Eliminate hidden lines which do not add clarification.

10. Show only partial views of symmetrical objects. See Fig. 13-1-11.

11. Avoid the use of elaborate pictorial and repetitive detail when manual drafting is used.

12. Eliminate repetitive data by use of general notes or phantom lines when manual drafting is used.

13. Eliminate views where the shape or

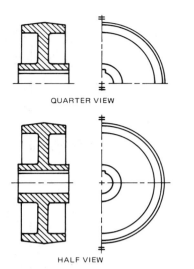

QUARTER VIEW

HALF VIEW

Fig. 13-1-11 Partial views.

dimension can be given by description, for example, ∅, □, HEX, THK, etc.

FREEHAND SKETCHING

Many shops care little whether the drawing is freehand, whether one view is shown, or whether the drawing is to scale, as long as the proportions are approximate. They want the necessary information clearly shown. Freehand sketches and drawings made with instruments can be shown on one sheet. However, it must be clearly understood that the use of freehand sketching does not give the drafter a license to turn out sloppy work.

Savings as high as 30 percent in the preparation of working drawings have been attributed to the use of freehand sketches as opposed to instrument-produced drawings. However, freehand sketching has its limitations. It is highly effective on simple detailed parts, for small radii, such as rounds and fillets, and for small holes. In many cases the term *freehand* is not entirely correct. For instance, templates may be used to draw circles, resistors, or other common features, or a straightedge may be used to produce long lines since it is faster and more accurate than freehand sketching. But short lines are drawn more quickly freehand.

Drawing paper with nonreproducible grid lines is ideal for freehand sketching. See Fig. 13-1-12. For this reason many companies have their drawing paper made with nonreproducible grid lines over the entire drawing area. Other advantages of having the grid lines on the paper are that (1) they may serve as guidelines in lettering notes and dimensions and (2) they may be used for measuring distances, thereby reducing the number of times the scale is used for measuring.

Reproduction Shortcuts

In the past few years, a number of reproduction techniques have been developed which, if properly used, can greatly reduce drawing preparation time. An understanding of available techniques and their limitations, supported by the close cooperation of a reproduction group familiar with drafting operations, can help the drafting supervisor to make significant cost savings.

NEW DRAWINGS MADE FROM EXISTING DRAWINGS

When a new drawing is to be made from an existing drawing with few changes, CAD makes this task easy by simply removing the unwanted material and drawing in the new. With manual drafting, reproducibles will save a great deal of preparation time. This procedure involves making a translucent or transparent print from the original drawing, removing unwanted material from this print, and adding the new information to the drawing. The main drawback to this method is that the existing drawing may not conform to the latest standard drawing practice.

SCISSORS AND PASTE-UP DRAFTING

No matter how original a design may be, a great number of part features are repetitive. With the aid of modern reproduction methods, drawings can be created by using unchanged portions of existing drawings. When manual drafting is used, transferring them from one drawing to the next is accomplished by scissors and paste-up drafting. It provides a way of using all or parts of existing drawings, notes, charts, and drawing forms to revise existing drawings and to create new drawings. Through the utilization of existing drawings much valuable drafting time is freed for creative design drafting rather than hand copying.

Finished prints can be made on paper, acetate, or vellum. They can be the same size or reduced to different sizes, depending on the reproduction equipment being used.

Another important advantage of cut-and-paste drafting is that materials copied from existing drawings do not have to be rechecked minutely, as must be done with new drawings. Rechecking time is reduced. See Fig. 13-1-13.

APPLIQUÉS

One of the most successful methods of reducing drawing time, when manual drafting is used, is the use of appliqués. When parts, shapes, symbols, or notes are used repeatedly, appliqués should be

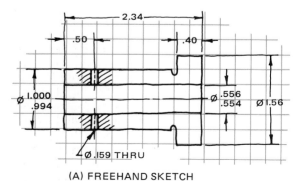

(A) FREEHAND SKETCH

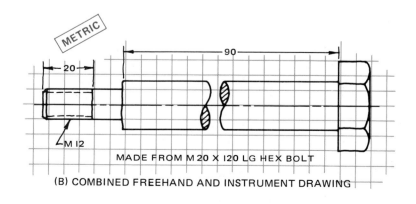

(B) COMBINED FREEHAND AND INSTRUMENT DRAWING

Fig. 13-1-12 Sketching of parts on coordinate paper.

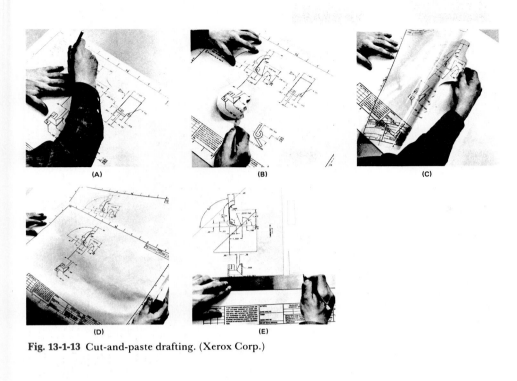

Fig. 13-1-13 Cut-and-paste drafting. (Xerox Corp.)

considered. These pressure-sensitive overlays may be printed on opaque, transparent, or translucent sheets with an adhesive backing.

Appliqués are available in a great variety of standard symbols or patterns (Fig. 13-1-14) and in blank (unprinted) sheets. A matte surface on the blank sheet will accept typewriter copy as well as pencil or ink lines. This material is often used for making corrections on drawings and for adding materials lists or detailed notes which can be typed faster than they can be lettered. Figure 13-1-15 shows a drawing which used many of the appliqués shown in Fig. 13-1-14.

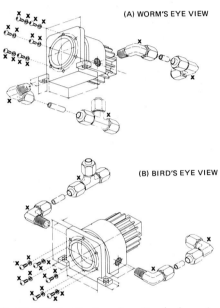

Fig. 13-1-15 Application of appliqués shown in Fig. 13-1-14. (Graphic Standard Instruments Co.)

Appliqués are available in two basic types: *cutout* and *transfer*. Cutout appliqués are applied by positioning the desired image in the correct position on the drawing, burnishing (rubbing) the image area, and cutting around it to remove the portion not wanted. The transfer type pressure-sensitive appliqué works on a somewhat different principle. The carrier is removed from the translucent image sheet, and the area to be transferred is placed in position on the drawing. The image to be transferred is then rubbed over the top surface of the transfer sheet with a burnishing stick.

The combined use of cut-and-paste drafting and appliqués for new drawings is found extensively in industry, especially in the electronics and piping fields. See Fig. 13-1-16.

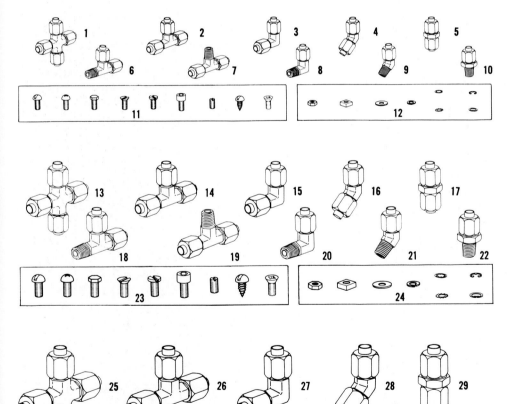

Fig. 13-1-14 A variety of shapes and sizes of appliqués. (Graphic Standard Instruments Co.)

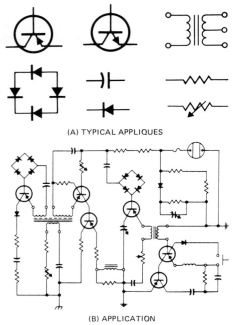

(A) TYPICAL APPLIQUES

(B) APPLICATION

Fig. 13-1-16 Electronics appliqués. (Bishop Industries Corp.)

Photodrawings

Photodrawings, that is, engineering drawings into which one photograph, or more, is incorporated, have increased in popularity because they can sometimes present a subject even more clearly than conventional drawings. Photodrawings supplement rather than replace conventional engineering drawings by eliminating much of the tedious and time-consuming effort involved when the subject is difficult to draw. They are particularly useful for assembly drawings, piping diagrams, large machine installations, switchboards, etc., provided, of course, that the subject of the drawings exists so that it may be photographed.

Photodrawings are also a comprehensive means of clearly transmitting technical information; they free the drafter from having to draw things which already exist. See Fig. 13-1-17.

Photodrawings have other advantages. They are easy to make and usually take much less time to prepare than an equivalent amount of conventional drafting.

Background Any photodrawing must begin with a photograph of an object, a part or assembly, a building, a model, or whatever else may be the subject of the drawing.

Photography The best photographic angle usually is one which shows the subject in a flat view with as little perspective as possible. (If the situation calls for a perspective, select the angle that best describes the object.) Make certain that all the parts important to the photodrawing are in view of the camera.

ASSIGNMENTS

See Assignments 1 through 10 for Unit 13-1 starting on page 277.

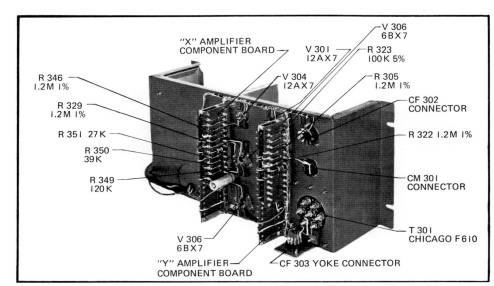

Fig. 13-1-17 Photodrawings. (Eastman Kodak Co.)

UNIT 13-2
Detail Drawings

A *working drawing* is a drawing that supplies information and instructions for the manufacture or construction of machines or structures. Generally, working drawings may be classified into two groups: detail drawings, which provide the necessary information for the manufacture of the parts, and assembly

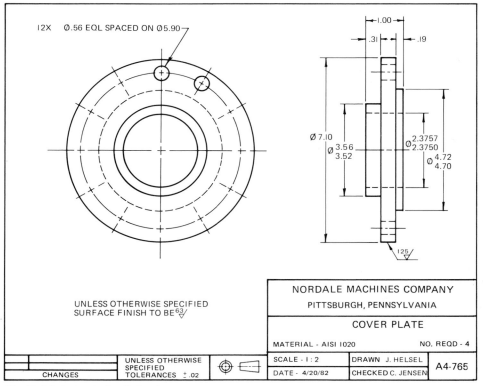

Fig. 13-2-1 A simple detail drawing.

drawings, which supply the necessary information for their assembly.

Since working drawings may be sent to other companies to make or assemble the parts, the drawings should conform with the drawing standards of that company. For this reason, most companies follow the drawing standards of their country. The drawing standards recommended by the American National Standards Institute (ANSI) has been adopted by the majority of industries in the United States.

DETAIL DRAWINGS

A detail drawing (Fig. 13-2-1) must supply the complete information for the construction of a part. This information may be classified under three headings: shape description, size description, and specifications.

Shape Description This term refers to the selection and number of views to show or describe the shape of the part. The part may be shown in either pictorial or orthographic projection, the latter being used more frequently. Sectional views, auxiliary views, and enlarged detail views may be added to the drawing in order to provide a clearer image of the part.

Size Description Dimensions which show the size and location of the shape features are then added to the drawing.

The manufacturing process will influence the selection of some dimensions, such as datum features. Tolerances are then selected for each dimension.

Specifications This term refers to general notes, material, heat treatment, finish, general tolerances, and number required. This information is located on or near the title block or strip.

Additional Drawing Information In addition to the information pertaining to the part, a detail drawing includes additional information such as drawing number, scale, method of projection, date, name of part or parts, and the drafter's name.

The selection of paper or finished plot

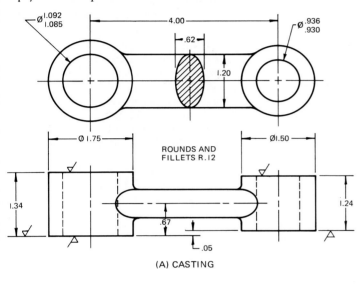

(A) CASTING

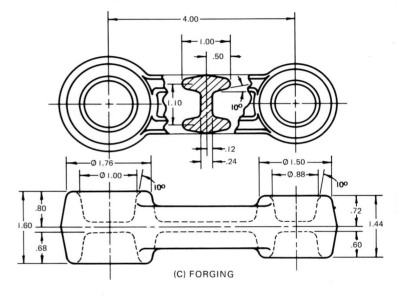

(C) FORGING

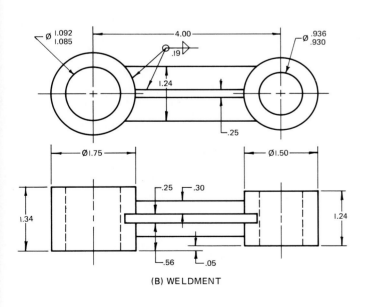

(B) WELDMENT

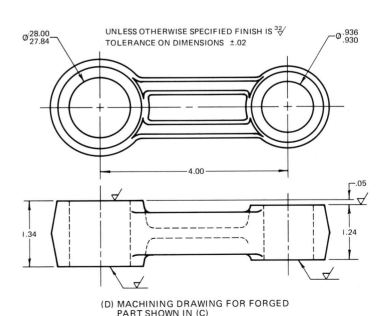

(D) MACHINING DRAWING FOR FORGED PART SHOWN IN (C)

Fig. 13-2-2 Manufacturing process influences the shape of the part.

size is determined by the number of views selected, the number of general notes required, and the drawing scale used. If the drawing is to be microformed, then the lettering size would be another factor to consider. The drawing number may carry a prefix or suffix number or letter to indicate the sheet size, such as A-571 or 4-571; the letter A indicates that it is made on an 8.50 × 11.00 in. sheet, and the number 4 indicates that the drawing is made on a 210 × 297 mm sheet.

DRAWING CHECKLIST

As an added precaution against errors occurring on a drawing, many companies have provided checklists for drafters to follow before a drawing is issued to the shop. A typical checklist may be as follows:

1. *Dimensions.* Is the part fully dimensioned, and are the dimensions clearly positioned? Is the drawing dimensioned to avoid unnecessary shop calculations?
2. *Scale.* Is the drawing to scale? Is the scale shown? What will the plot scale be?
3. *Tolerances.* Are the clearances and tolerances specified by the linear and angular dimensions and by local, general, or title block notes suitable for proper functioning? Are they realistic? Can they be liberalized?
4. *Standards.* Have standard parts, design, materials, processes, or other items been used where possible?
5. *Surface texture.* Have surface roughness values been shown where required? Are the values shown compatible with overall design requirements?
6. *Material.* Have proper material and heat treatment been specified?

Qualifications of a Detailer

The detailer should have a thorough understanding of materials, shop processes, and operations in order to properly dimension the part and call for the correct finish and material. In addition, the detailer must have a thorough knowledge of how the part functions in order to provide the correct data and tolerances for each dimension.

The detailer may be called upon to work from a complete set of instructions and drawings, or he or she may be required to make working drawings of parts which involve the design of the part. Design considerations are limited in this unit, but are covered in detail in Chap. 19.

MANUFACTURING METHODS

The type of manufacturing process will influence the selection of material and detailed feature of a part. See Fig. 13-2-2. For example, if the part is to be cast, rounds and fillets will be added. Additional material will also be required where surfaces are to be finished.

The more common manufacturing processes are machining from standard stock; prefabrication which includes welding, riveting, soldering, brazing, and gluing; forming from sheet stock; casting; and forging. The latter two processes can be justified only when large quantities are required for specially designed parts. All these processes are described in detail in other chapters.

Several drawings may be made for the same part, each one giving only the information necessary for a particular step in the manufacture of the part. A part which is to be produced by forging, for example, may have one drawing showing the original rough forged part and one detail of the finished forged part. See Figs. 13-2-2C and 13-2-2D.

ASSIGNMENTS

See Assignments 11 through 13 for Unit 13-2 starting on page 282.

UNIT 13-3
Multiple Detail Drawings

Detail drawings may be shown on separate sheets, or they may be grouped on one or more large sheets.

Often the detailing of parts is grouped according to the department in which they are made. For example, wood, fiber, and metal parts are used in the assembly of a transformer. Three separate detail sheets—one for wood parts, one for fiber parts, and the third for the metal parts—may be drawn. These parts would be made in the different shops and sent to another area for assembly. In order to facilitate assembly, each part is given an identification part number which is shown on the assembly drawing. A typical detail drawing showing multiple parts is illustrated in Fig. 13-3-1.

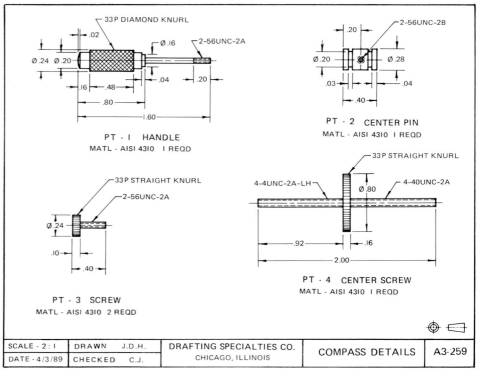

Fig. 13-3-1 Detail drawing containing many details on one drawing.

If the details are few, the assembly drawing may appear on the same sheet or sheets.

ASSIGNMENT

See Assignment 14 for Unit 13-3 on page 285.

UNIT 13-4
Drawing Revisions

Revisions are made to an existing drawing when manufacturing methods are improved, to reduce cost, to correct errors, and to improve design. A clear record of these revisions must be registered on the drawing.

All drawings must carry a change or revision table, either down the right-hand side or across the bottom of the drawing. In addition to a description of drawing changes, provision may be made for recording a revision symbol, a zone location, an issue number, a date, and the approval of the change. Should the drawing revision cause a dimension or dimensions to be other than the scale indicated on the drawing, then the dimensions that are not to scale should be indicated by the method shown in Fig. 7-1-22. Typical revision tables are shown in Fig. 13-4-1.

At times, when there are a large number of revisions to be made, it may be more economical to make a new drawing. When this is done, the words REDRAWN and REVISED should appear in the revision column of the new drawing. A new date is also shown for updating old prints.

CAD

CAD has many editing options available. By selecting the appropriate command, you can edit or revise a drawing directly on the CRT monitor. One such command is the ERASE command, see Fig. 13-4-2, which will permit you to erase all or part of an object in order for that part to be redrawn. The remainder of the drawing need not be redrawn.

Reference

1. ANSI Y14.5M, *Dimensions and Tolerancing*.

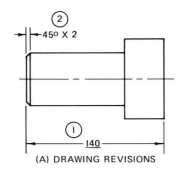

(A) DRAWING REVISIONS

REVISIONS		
SYMBOL	DESCRIPTION	DATE & APPROVAL
1	LENGTH WAS 150	J. Helsel 3-4-88
2	CHAMFER ADDED	F. Newman 2-2-89

(B) VERTICAL REVISION BLOCK

(C) HORIZONTAL REVISION BLOCK

Fig. 13-4-1 Drawing revisions.

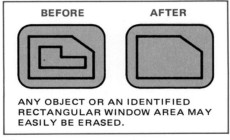

ANY OBJECT OR AN IDENTIFIED RECTANGULAR WINDOW AREA MAY EASILY BE ERASED.

Fig. 13-4-2 CAD erase command.

ASSIGNMENT

See Assignment 15 for Unit 13-4 on page 285.

UNIT 13-5
Assembly Drawings

All machines and mechanisms are composed of numerous parts. A drawing showing the product in its completed state is called an *assembly drawing*.

Assembly drawings vary greatly in the amount and type of information they give, depending on the nature of the machine or mechanism they depict. The primary functions of the assembly drawing are to show the product in its completed shape, to indicate the relationship of its various components, and to designate these components by a part or detail number. Other information that might be given includes overall dimensions, capacity dimensions, relationship dimensions between parts (necessary information for assembly), operating instructions, and data on design characteristics.

Design Assembly Drawings

When a machine is designed, an assembly drawing or a design layout is first drawn to clearly visualize the performance, shape, and clearances of the various parts. From this assembly drawing, the detail drawings are made and each part is given a part number. To assist in the assembling of the machine, part numbers of the various details are placed on the assembly drawing. The part number is attached to the corresponding part with a leader, as illustrated in Fig. 13-5-1. It is important that

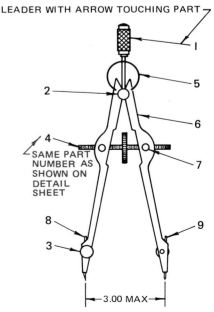

Fig. 13-5-1 Identification numbers on assembly drawings.

the detail drawings not use identical numbering schemes when several item lists are used. Circling the part number is optional.

CAD

Depending on the CAD equipment available, both color and layers can be used in the preparation of an assembly drawing. Color speeds up the recognition of specific areas, making drawing interpretation much easier.

Layering (preparing different portions of a drawing on different levels) simplifies the task of preparing an assembly drawing.

Although the completed assembly drawing may consist of several layers of information, only one or two layers may be required by one department for manufacture or assembly purposes, while another department may require information from a combination of other levels. The end result is simplification of drawing interpretation.

Installation Assembly Drawings

This type of assembly drawing is used when many unskilled people are employed to mass-assemble parts. Since these people are not normally trained to read technical drawings, simplified pictorial assembly drawings similar to the one shown in Fig. 13-5-2 are used.

Assembly Drawings for Catalogs

Special assembly drawings are prepared for company catalogs. These assembly drawings show only pertinent details and dimensions that would interest the potential buyer. Often one drawing, having letter dimensions accompanied by a chart, is used to cover a range of sizes, such as the pillow block shown in Fig. 13-5-3B.

Item List

An item list, often referred to as a *bill of material*, is an itemized list of all the components shown on an assembly drawing or a detail drawing. See Fig. 13-5-4. Often an item list is placed on a separate sheet for ease of handling and duplicat-

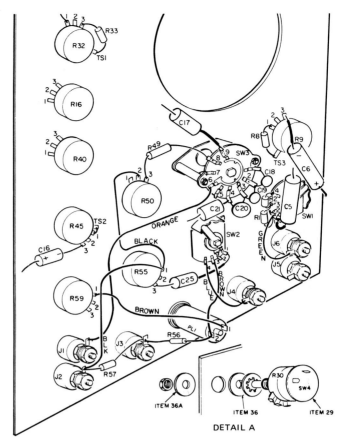

Fig. 13-5-2 Installation assembly drawing.

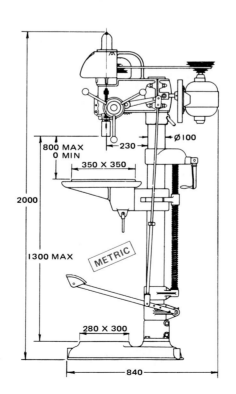

(A) DRILL PRESS

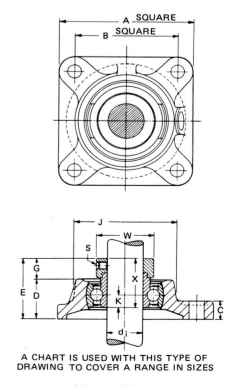

A CHART IS USED WITH THIS TYPE OF DRAWING TO COVER A RANGE IN SIZES

(B) PILLOW BLOCK

Fig. 13-5-3 Assembly drawings used in catalogs.

QTY	ITEM	MATL	DESCRIPTION	PT NO.
I	BASE	GI	PATTERN # A3154	I
I	CAP	GI	PATTERN # B7156	2
I	SUPPORT	AISI-1212	.38 X 2.00 X 4.38	3
I	BRACE	AISI-1212	.25 X 1.00 X 2.00	4
I	COVER	AISI-1035	.1345 (# 10 GA USS) X 6.00 X 7.50	5
I	SHAFT	AISI-1212	Ø1.00 X 6.50	6
2	BEARINGS	SKF	RADIAL BALL # 6200Z	7
2	RETAINING CLIP	TRUARC	N5000-725	8
I	KEY	STL	WOODRUFF # 608	9
I	SET SCREW	CUP POINT	HEX SOCKET .25UNC X 1.50	10
4	BOLT—HEX HD—REG	SEMI-FIN	.38UNC X 1.50LG	11
4	NUT—REG HEX	STL	.38UNC	12
4	LOCK WASHER—SPRING	STL	.38 - MED	13
				14

(A) TYPICAL BILL OF MATERIAL   PARTS 7 TO 13 ARE PURCHASED ITEMS

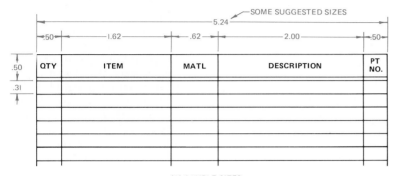

(B) SAMPLE SIZES

Fig. 13-5-4 Item list.

ing. Since the item list is used by the purchasing department to order the necessary material for the design, the item list should show the raw material size, rather than the finished size of the part.

For castings, a pattern number should appear in the size column in lieu of the physical size of the part.

Standard components, which are purchased rather than fabricated, such as bolts, nuts, and bearings, should have a part number and appear on the bill of material. Information in the descriptive column should be sufficient for the purchasing agent to order these parts.

Item lists placed on the bottom of the drawing should read from bottom to top, while item lists placed on the top of the drawings should read from top to bottom. This practice allows additions to be made at a later date.

CAD

Since automatic item list generation is complicated and so specialized, it will not be covered in this text. Use the TEXT command to key in the required items.

ASSIGNMENT

See Assignment 16 for Unit 13-5 on page 286.

UNIT 13-6
Exploded Assembly Drawings

In many instances parts must be identified or assembled by persons unskilled in the reading of engineering drawings. Examples are found in the appliance-repair industry, which relies on assembly drawings for repair work and for reordering parts. Exploded assembly drawings, like that shown in Fig. 13-6-1, are used extensively in these cases, for they are easier to read. This type of assembly drawing is also used frequently by companies that manufacture do-it-yourself assembly kits, such as model-making kits.

For this type of drawing, the parts are aligned in position. Frequently, shading techniques are used to make the drawings appear more realistic.

ASSIGNMENT

See Assignment 17 for Unit 13-6 on page 287.

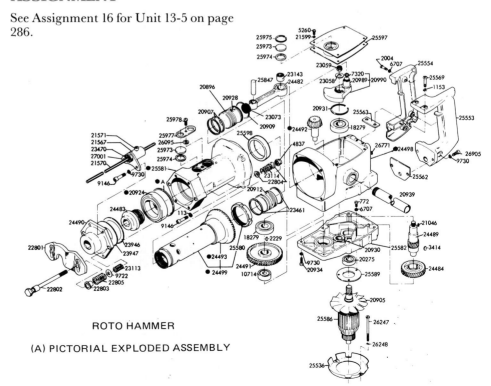

ROTO HAMMER

(A) PICTORIAL EXPLODED ASSEMBLY

Fig. 13-6-1 Exploded assembly drawings.

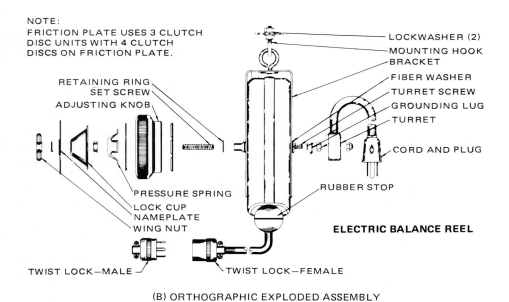

NOTE:
FRICTION PLATE USES 3 CLUTCH DISC UNITS WITH 4 CLUTCH DISCS ON FRICTION PLATE.

RETAINING RING
SET SCREW
ADJUSTING KNOB

PRESSURE SPRING
LOCK CUP
NAMEPLATE
WING NUT

TWIST LOCK—MALE

LOCKWASHER (2)
MOUNTING HOOK BRACKET
FIBER WASHER
TURRET SCREW
GROUNDING LUG
TURRET

CORD AND PLUG

RUBBER STOP

ELECTRIC BALANCE REEL

TWIST LOCK—FEMALE

(B) ORTHOGRAPHIC EXPLODED ASSEMBLY

Fig. 13-6-1 (Continued)

UNIT 13-7
Detailed Assembly Drawings

Often these are made for fairly simple objects, such as pieces of furniture, when the parts are few in number and are not intricate in shape. All the dimensions and information necessary for the construction of each part and for the assembly of the parts are given directly on the assembly drawing. Separate views of specific parts, in enlargements showing the fitting together of parts, may also be drawn in addition to the regular assembly drawing. Note that in Fig. 13-7-1 the enlarged views are drawn in picture form, not as regular orthographic views. This method is peculiar to the cabinet-making trade and is not normally used in mechanical drawing.

ASSIGNMENT

See Assignment 18 for Unit 13-7 on page 288.

UNIT 13-8
Subassembly Drawings

Many completely assembled items, such as a car and a television set, are assembled with many preassembled components as well as individual parts. These preassembled units are referred to as *subassemblies*. The assembly drawings of a transmission for an automobile and the transformer for a television set are typical examples of subassembly drawings.

Subassemblies are designed to simplify final assembly as well as permit the item to be either assembled in a more suitable area or purchased from an outside source. This type of drawing shows only those dimensions which would be required for the completed assembly. Examples are size of the mounting holes and their location, shaft locations, and overall sizes. This type of drawing is found frequently in catalogs. The pillow block shown in Fig. 13-5-3B is a typical subassembly drawing.

ASSIGNMENT

See Assignment 19 for Unit 13-8 starting on page 288.

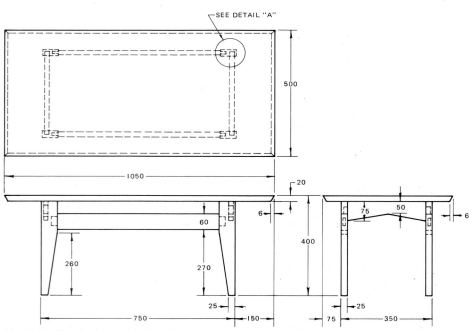

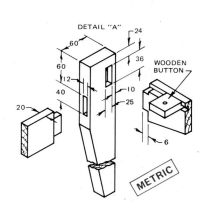

DETAIL "A"

WOODEN BUTTON

METRIC

Fig. 13-7-1 Detailed assembly drawings.

ASSIGNMENTS FOR CHAPTER 13

Note: Convert to symbolic and limit dimensioning, wherever practical, for all drawing assignments in this chapter.

Assignments for Unit 13-1, Functional Drafting

1. After the number of drawings made over the last 6 months was reviewed, it was discovered that a great number of cable straps, shown in Fig. 13-1-A, were being made which were similar in design. Prepare a standard tabulated drawing similar to Fig. 13-1-1, reducing the number of standard parts to 4. Scale 1:1.

2. The rod guide shown in Fig. 13-1-B is to be drawn twice and the drawing time for each recorded. First, on plain paper make an isometric drawing of the part, using a compass to draw the circles and arcs. Next repeat the drawing, only this time use isometric grid paper and a template for drawing the circles and arcs. From the drawing times recorded, state in percentage the time saved by the use of grid paper and templates. Scale 1:1. Do not dimension.

3. The book rack shown in Fig. 13-1-C is to be drawn twice and the drawing time for each drawing recorded. The first drawing is to show a three-view orthographic projection of the book rack assembly showing only those dimensions and instructions pertinent to the assembly. On the same drawing prepare detail drawings for the parts required. Scale to suit. On the second drawing make an isometric detailed assembly drawing of the book rack showing the dimensions and instructions necessary to completely make and assemble the parts. Scale to suit. From the drawing times recorded, state a percentage of time saved by the use of detailed assembly drawings.

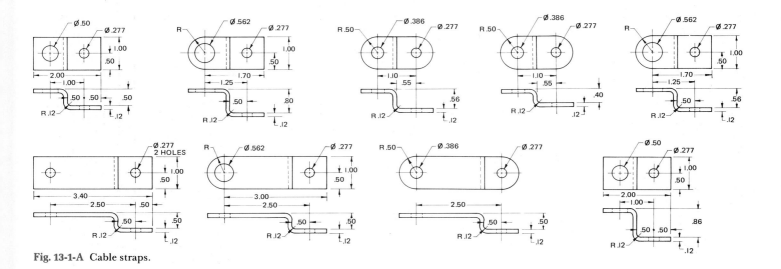

Fig. 13-1-A Cable straps.

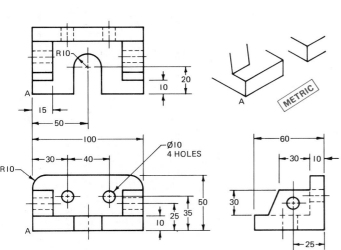

Fig. 13-1-B Rod guide.

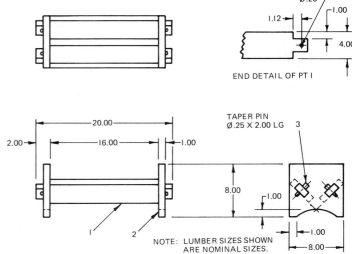

Fig. 13-1-C Book rack.

4. Redraw the two parts shown in Fig. 13-1-D using arrowless dimensioning and simplified drawing practices. Use half scale. For the cover plate use the bottom and left-hand edge for the datum surfaces. For the back plate use the bottom and the center of the part for the datum surfaces.

5. Redraw the two parts shown in Fig. 13-1-E using partial views and the symmetry symbol. Scale to suit.

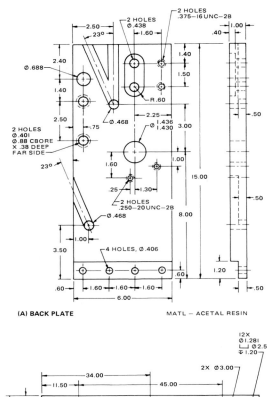

(A) BACK PLATE   MATL — ACETAL RESIN

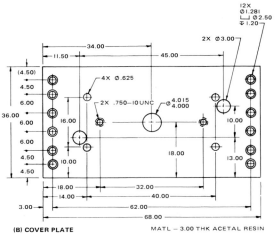

(B) COVER PLATE   MATL – 3.00 THK ACETAL RESIN

Fig. 13-1-D Arrowless dimensioning assignment.

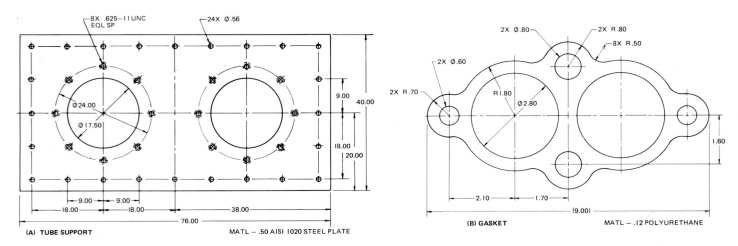

(A) TUBE SUPPORT   MATL — .50 AISI 1020 STEEL PLATE

(B) GASKET   MATL — .12 POLYURETHANE

Fig. 13-1-E Tube support and gasket.

6. Make simplified drawings of the parts shown in Fig. 13-1-F. Refer to Fig. 13-1-8. Scale to suit.

7. Redraw the terminal board shown in Fig. 13-1-G using tabular dimensioning. Scale 1:1.

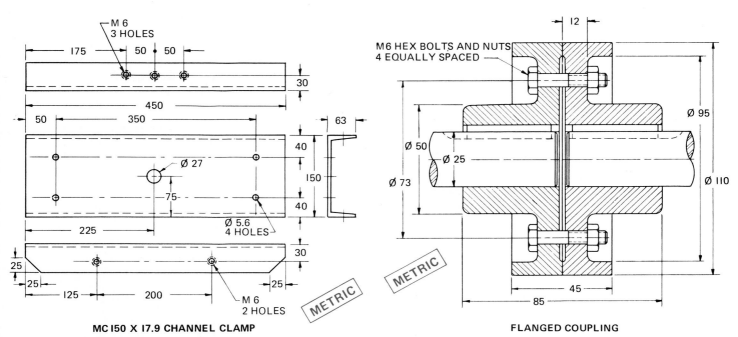

MC 150 X 17.9 CHANNEL CLAMP

FLANGED COUPLING

Fig. 13-1-F Simplification of detail assignment.

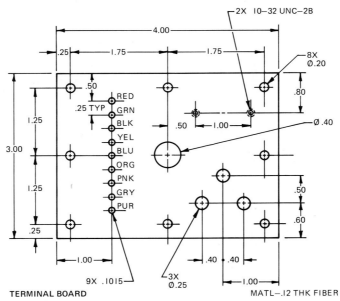

TERMINAL BOARD

MATL—.12 THK FIBER

Fig. 13-1-G Tabular dimensioning assignment.

8. An exploded assembly drawing of the wheel puller shown in Fig. 13-1-H is urgently required. Time does not permit one drafter to do the entire drawing; thus, three drafters will be required to draw the parts. Scale is 1:2. Draw all the parts. When all the parts have been completed make prints of them. Cut out the parts and assemble them in the exploded position on a B (A3) size sheet. Make a suitable print of the exploded assembly on a copying machine.

9. Draw the electronics diagram shown in Fig. 13-1-J using the CAD library (you may have to make your own) or use appliqués and templates if manual drafting is to be used. If appliqués of the electronic components are not available, make your own by photostating Fig. 13-1-6, then cut them out and glue them to your drawing. There is no scale.

10. Make a photostat of the sprocket shown in Fig. 13-1-K. The photostat, which is to serve as a photodrawing, is to replace the two-view drawing. Make a new chart listing metric sizes to replace the existing chart if directed to do so. Leaders, dimensions, and dimension lines are to be added to the photodrawing.

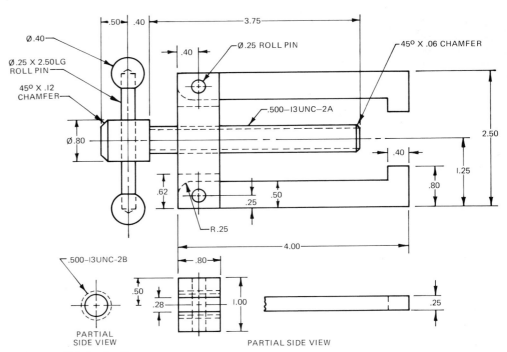

Fig. 13-1-H Wheel puller.

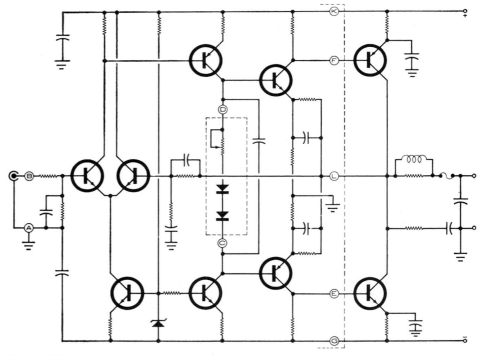

Fig. 13-1-J Electronics diagram.

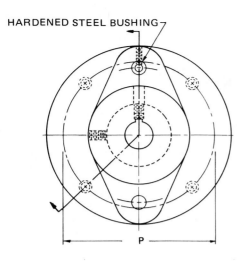

HARDENED STEEL BUSHING

P

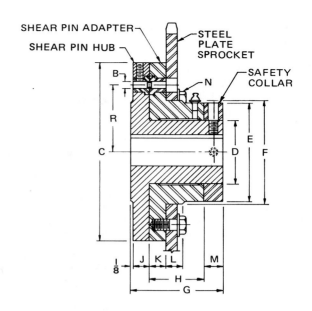

SHEAR PIN ADAPTER
SHEAR PIN HUB
STEEL PLATE SPROCKET
SAFETY COLLAR

B
R
C
D
E
F
N

1/8 J K L M
H
G

Hub Bore Range	Shear Pin Assembly Number	Shear Pin		Diameters				Length Thru			Hub Flange Thickness	Adapt Flange Thickness	Sprocket Seat Width	Bolts	
		Radius	Pin Dia.	Flange	Shear Pin Hub	Adapt Hub & Collar	Sprocket Seat	Shear Pin Hub	Adapt	Collar				Number & Size	Bolt Circle
		R	B	C	D	E	F	G	H	M	J	K	L	N	P
1.00 & under	SP-17	1.80	.25	5.25	1.75	2.50	2.62	2.44	1.38	.38	.56	.56	.44	4-.38	4.00
1.06-1.25	SP-18	2.18	.25	6.00	2.25	3.25	3.38	2.94	1.75	.50	.56	.56	.56	4-.38	4.75
1.30-1.50	SP-19	2.56	.30	6.75	2.75	4.00	4.12	3.56	2.12	.62	.68	.68	.68	4-.50	5.50
1.56-1.75	SP-20	3.00	.38	7.75	3.25	4.75	4.88	4.18	2.50	.75	.80	.80	.68	4-.50	6.25
1.80-2.00	SP-21	3.30	.45	8.75	3.75	5.25	5.38	4.80	2.88	.88	.94	.94	.94	4-.62	7.00
2.06-2.25	SP-22	3.80	.50	9.75	4.25	6.25	6.38	5.18	3.00	1.00	1.06	1.06	1.18	4-.62	8.00
2.30-2.50	SP-23	4.00	.50	10.00	4.50	6.50	6.62	5.68	3.50	1.00	1.06	1.06	1.38	4-.62	8.25
2.56-2.75	SP-24	4.40	.55	11.50	5.00	7.00	7.12	6.30	3.88	1.12	1.18	1.18	1.38	4-.62	9.25
2.80-3.00	SP-25	4.90	.62	12.50	5.50	8.00	8.12	6.94	4.25	1.25	1.30	1.30	1.38	6-.62	10.25

Fig. 13-1-K Arrowless dimensioning assignment.

Assignments for Unit 13-2, Detail Drawings

11. Prepare a detail drawing of one of the parts shown in Fig. 13-2-A or 13-2-B. Select appropriate views and dimensions and add to the drawing the information needed so that the parts can be completely manufactured. Scale is to suit.

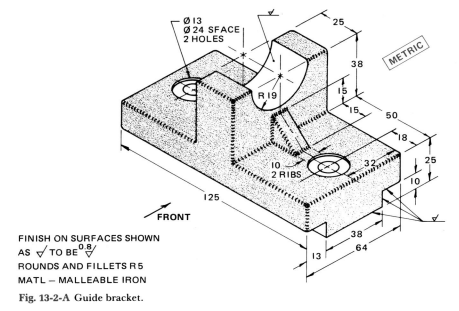

FINISH ON SURFACES SHOWN
AS ✓ TO BE $\overset{0.8}{\checkmark}$
ROUNDS AND FILLETS R 5
MATL — MALLEABLE IRON

Fig. 13-2-A Guide bracket.

FINISH ON SURFACES MARKED ✓
TO BE $\overset{32}{\checkmark}$
MATL — MALLEABLE IRON
ROUNDS AND FILLETS R .12

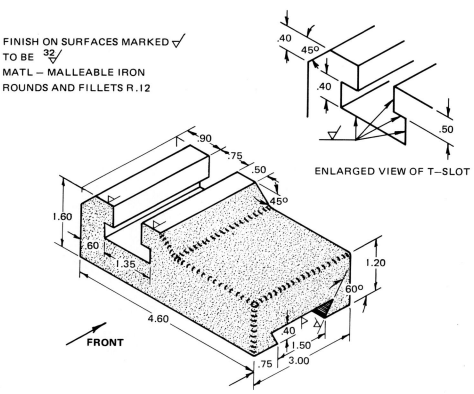

ENLARGED VIEW OF T—SLOT

Fig. 13-2-B Cross slide.

12. Select one of the parts shown in Fig. 13-2-C or 13-2-D and make a three-view working drawing. Dimensions are to be converted to millimeters. Only the dovetail and T slot dimensions are critical and must be taken to an accuracy of two points beyond the decimal point. All other dimensions are to be rounded off to whole numbers.

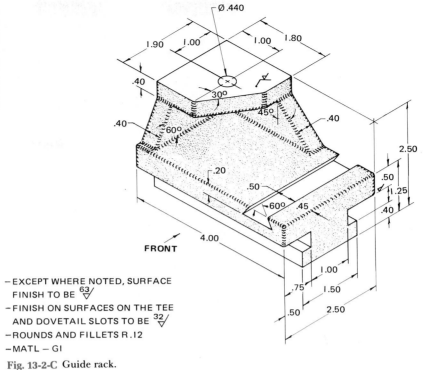

- EXCEPT WHERE NOTED, SURFACE FINISH TO BE ⁶³∇
- FINISH ON SURFACES ON THE TEE AND DOVETAIL SLOTS TO BE ³²∇
- ROUNDS AND FILLETS R.12
- MATL – GI

Fig. 13-2-C Guide rack.

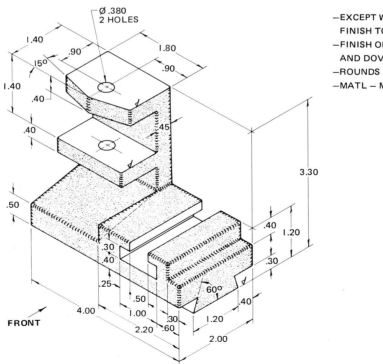

- EXCEPT WHERE NOTED, SURFACE FINISH TO BE ⁶³∇
- FINISH ON SURFACES ON THE TEE AND DOVETAIL SLOTS TO BE ³²∇
- ROUNDS AND FILLETS R.12
- MATL – MALLEABLE IRON

Fig. 13-2-D Locating stand.

13. Prepare detail drawings of any of the parts assigned by your instructor from the assembly drawings shown in Figs. 13-2-E and 13-2-F. The scale and selection of views are to be decided by the student.

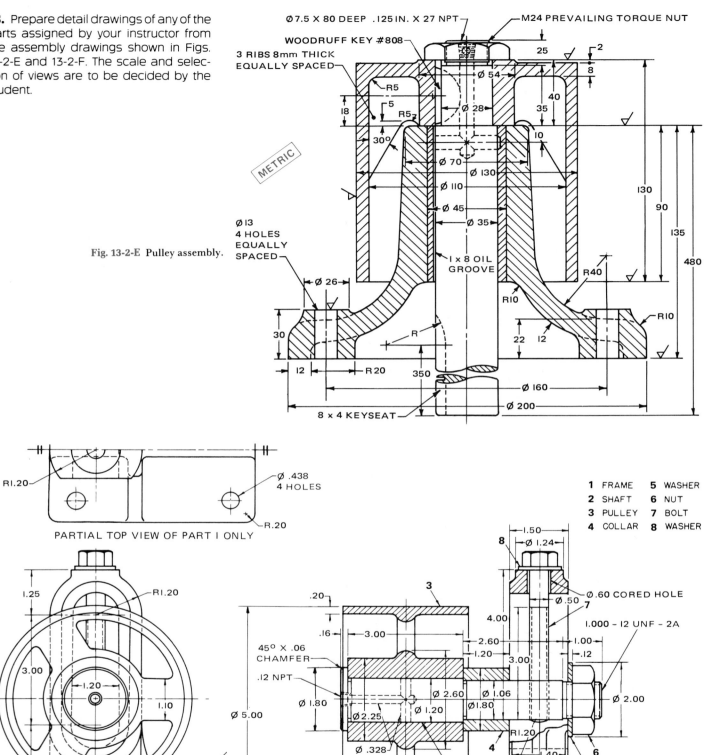

Fig. 13-2-E Pulley assembly.

1	FRAME	5	WASHER
2	SHAFT	6	NUT
3	PULLEY	7	BOLT
4	COLLAR	8	WASHER

PARTIAL TOP VIEW OF PART I ONLY

Fig. 13-2-F Adjustable pulley.

Assignment for Unit 13-3, Multiple Detail Drawings

14. Make detail drawings of all the parts shown of one of the assemblies in Fig. 13-3-A or 13-3-B. Since time is money, select only the views necessary to describe each part. Below each part show the following information: part number, name of part, material, number required. Scale 1:1.

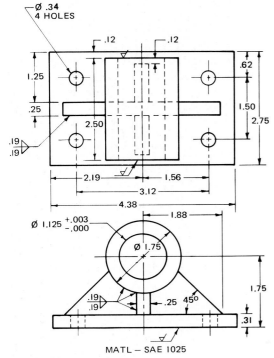

Fig. 13-3-A Shaft support.

MATL — SAE 1025

Assignment for Unit 13-4, Drawing Revisions

15. Select one of the drawings shown in Fig. 13-4-A or 13-4-B and make appropriate revisions to these drawings, recording the changes in the drawing revision column and indicate which dimensions are not to scale.

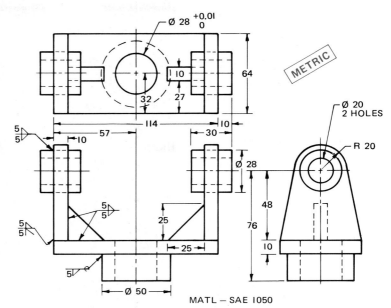

Fig. 13-3-B Shaft pivot support.

MATL — SAE 1050

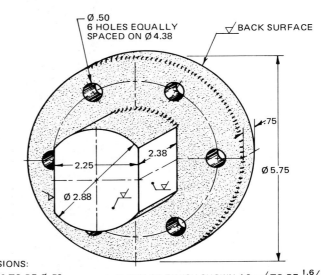

REVISIONS:
1. Ø .50 TO BE Ø .53
2. Ø 5.75 TO BE Ø 6.00
3. 2.25 TO BE 2.30
4. 2.38 TO BE 2.25

—SURFACE FINISH SHOWN AS ∨ TO BE 1.6∨
—ROUNDS AND FILLETS R .12
—MATL — GRAY IRON

Fig. 13-4-A Axle cap.

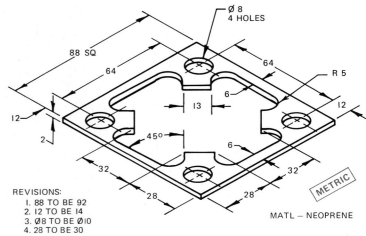

REVISIONS:
1. 88 TO BE 92
2. 12 TO BE 14
3. Ø8 TO BE Ø10
4. 28 TO BE 30

Fig. 13-4-B Gasket.

MATL — NEOPRENE

Assignment for Unit 13-5, Assembly Drawings

16. Make a one-view assembly drawing of one of the assemblies shown in Fig. 13-5-A or 13-5-B. For Fig. 13-5-A show a round bar Ø1 in.) in phantom being held in position.

Include on the drawing an item list and identification part numbers. Scale 1:1.

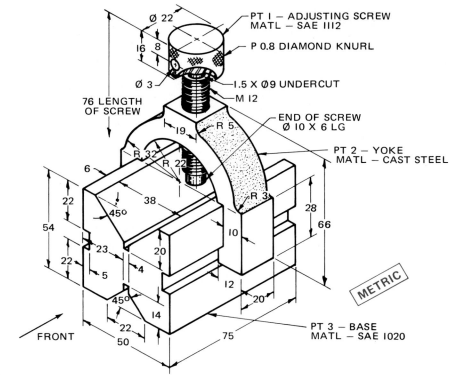

Fig. 13-5-A V-block clamp.

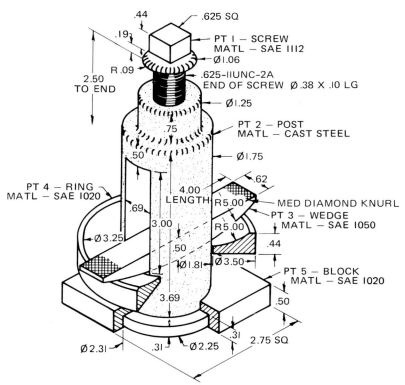

Fig. 13-5-B Tool post holder.

Assignment for Unit 13-6, Exploded Assembly Drawings

17. Make an exploded assembly drawing in orthographic projection of one of the assemblies shown in Figs. 13-6-A and 13-6-B. Use center lines to align parts and holes. To make the parts appear more realistic, shading techniques are recommended. Scale 1:1.

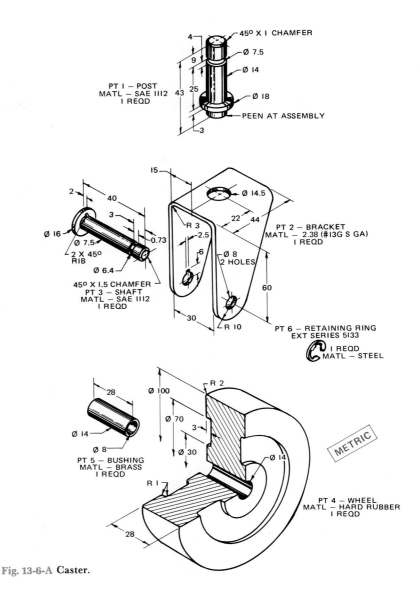

Fig. 13-6-A Caster.

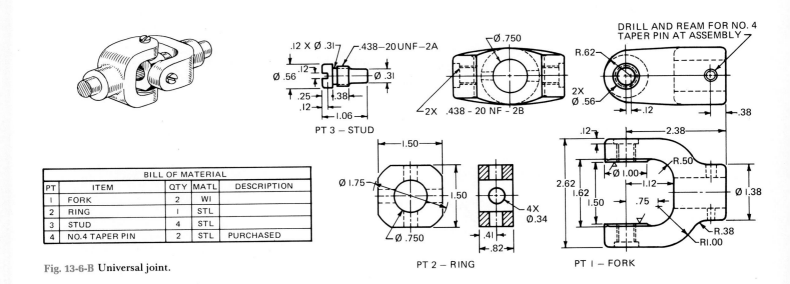

BILL OF MATERIAL			
PT	ITEM	QTY	MATL	DESCRIPTION
1	FORK	2	WI	
2	RING	1	STL	
3	STUD	4	STL	
4	NO.4 TAPER PIN	2	STL	PURCHASED

Fig. 13-6-B Universal joint.

PT 3 – STUD

PT 2 – RING

PT 1 – FORK

Assignment for Unit 13-7, Detailed Assembly Drawings

18. Make a detailed assembly drawing of one of the assemblies shown in Figs. 13-7-A and 13-7-B. Include on the drawing the method of assembly (i.e., nailing, wood screws, doweling, etc.) and an item list. Include in the item list the assembly materials. Use scale 1:5 for Fig. 13-7-A and 1.50 in. = 1 ft. for Fig. 13-7-B.

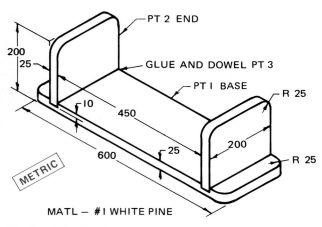

MATL — #1 WHITE PINE

Fig. 13-7-A Book rack.

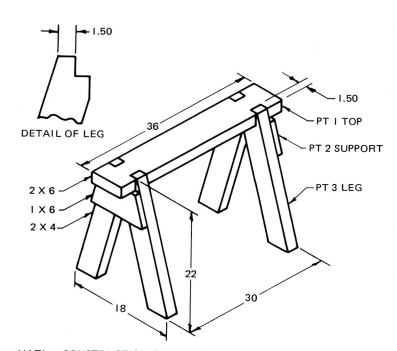

MATL — CONSTRUCTION GRADE SPRUCE

NOTE: WOOD SIZES (THICKNESS AND WIDTH) ARE NOMINAL SIZES

Fig. 13-7-B Saw horse.

Assignment for Unit 13-8, Subassembly Drawings

19. Make a one-view subassembly drawing of one of the assemblies shown in Figs. 13-8-A and 13-8-B. A broken-out or partial section view is recommended to show the interior features. Include on the drawing pertinent dimensions, identification numbers on assembly drawing, an item list, and a phantom outline of the adjoining part or features. Scale 1:1.

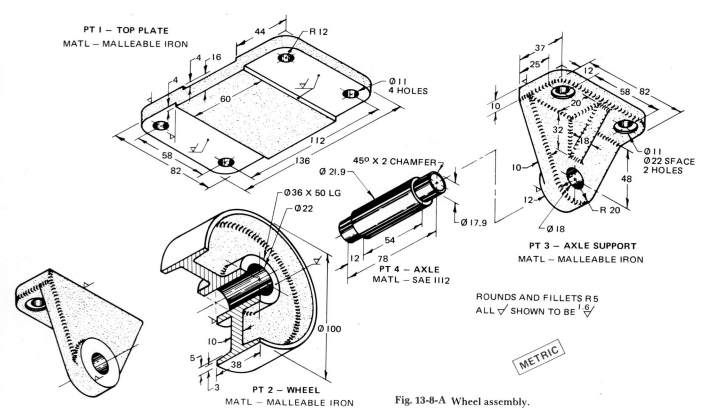

Fig. 13-8-A Wheel assembly.

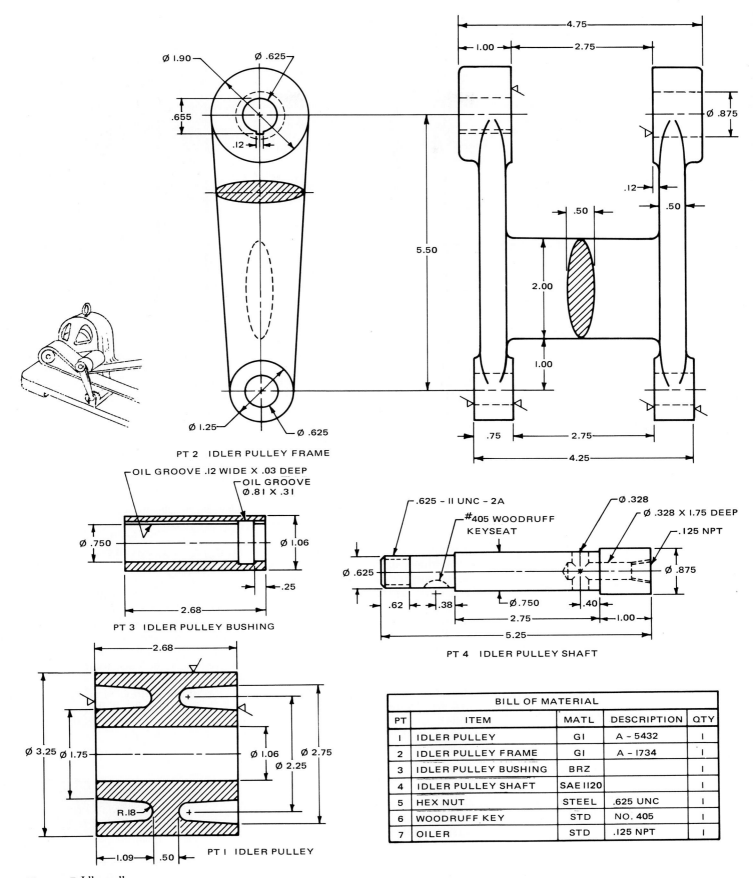

Ø 1.90   Ø .625

.655

.12

5.50

Ø 1.25   Ø .625

PT 2   IDLER PULLEY FRAME

4.75

1.00   2.75

Ø .875

.12   .50

.50

2.00

1.00

.75   2.75

4.25

OIL GROOVE .12 WIDE X .03 DEEP
OIL GROOVE Ø .81 X .31

Ø .750   Ø 1.06

.25

2.68

PT 3   IDLER PULLEY BUSHING

.625 – 11 UNC – 2A

#405 WOODRUFF
KEYSEAT

Ø .328

Ø .328 X 1.75 DEEP

.125 NPT

Ø .625

Ø .750

Ø .875

.62   .38   .40

2.75   1.00

5.25

PT 4   IDLER PULLEY SHAFT

2.68

Ø 3.25   Ø 1.75   Ø 1.06   Ø 2.75
Ø 2.25

R .18

1.09   .50

PT 1   IDLER PULLEY

BILL OF MATERIAL				
PT	ITEM	MATL	DESCRIPTION	QTY
1	IDLER PULLEY	GI	A – 5432	1
2	IDLER PULLEY FRAME	GI	A – 1734	1
3	IDLER PULLEY BUSHING	BRZ		1
4	IDLER PULLEY SHAFT	SAE 1120		1
5	HEX NUT	STEEL	.625 UNC	1
6	WOODRUFF KEY	STD	NO. 405	1
7	OILER	STD	.125 NPT	1

Fig. 13-8-B Idler pulley.

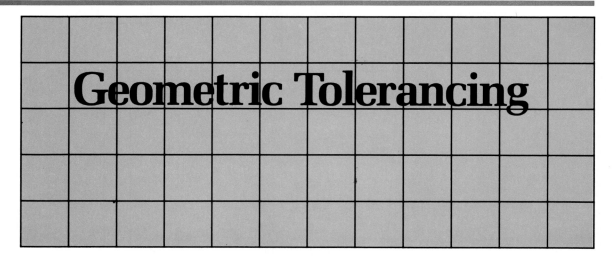

Geometric Tolerancing

UNIT 14-1

Modern Engineering Tolerancing

An engineering drawing of a manufactured part is intended to convey information from the designer to the manufacturer and inspector. It must contain all information necessary for the part to be correctly manufactured. It must also enable an inspector to make a precise determination of whether the parts are acceptable.

Therefore each drawing must convey three essential types of information:

1. The material to be used
2. The size or dimensions of the part
3. The shape or geometric characteristics

The drawing must also specify permissible variations for each of these aspects, in the form of tolerance or limits.

Materials are usually covered by separate specifications or supplementary documents, and the drawings need only make reference to these.

Size is specified by linear and angular dimensions. Tolerances may be applied directly to these dimensions or may be specified by means of a general tolerance note.

Shape and geometric characteristics, such as orientation and position, are described by views on the drawing, supplemented to some extent by dimensions.

In the past tolerances were often shown for which no precise interpretation existed, such as on dimensions which originated at nonexistent center lines. The specification of datum features was often omitted, resulting in measurements being made from actual surfaces when, in fact, datums were intended. There was confusion concerning the precise effect of various methods of expressing tolerances and of the number of decimal places used. While tolerancing of geometric characteristics was sometimes specified in the form of notes, no precise methods or interpretations were established. Straight or circular lines were drawn, without specifying how straight or round they should be. Square corners were drawn without specifying by how much the 90° angle could vary.

Modern systems of tolerancing, which include geometric and positional tolerancing, use of datum and datum targets, and more precise interpretations of linear and angular tolerances, provide designers and drafters with a means of expressing permissible variations in a very precise manner. Furthermore, the methods and symbols are international in scope and therefore help break down language barriers.

It is not necessary to use geometric tolerances for every feature on a part drawing. In most cases it is to be expected that if each feature meets all dimensional tolerances, form variations will be adequately controlled by the accuracy of the manufacturing process and equipment used.

This chapter covers the application of modern tolerancing methods on drawing.

National and International Standards References are made to technical drawing standards published by United States and ISO standardizing bodies. These bodies are generally referred to by their acronyms, as shown in Fig. 14-1-1.

Most of the symbols in all these standards are identical, but there are some variations. These are chiefly in the methods of indicating datum features and of applying the symbols to draw-

ACRONYM	STANDARDIZING BODY	STANDARD FOR DIMENSIONING AND TOLERANCING
ANSI	AMERICAN NATIONAL STANDARDS INSTITUTE	ANSI Y14.5
ISO	INTERNATIONAL ORGANIZATION FOR STANDARDIZATION	ISO R1101

Fig. 14-1-1 Standardizing bodies.

ings. In view of the exchange of drawings among the United States and other countries, it would be advantageous for drafters and designers to become acquainted with these different symbols.

For this reason whenever differences between United States and ISO standards occur, two methods are shown in some of the illustrations, and each is labeled with the acronym of the appropriate standardizing body, ANSI or ISO. However, differences in symbols or methods of application do not in any way affect the principles or interpretation of tolerances, unless specifically noted.

Illustrations

Most of the drawings in this chapter are not complete working drawings. They are intended only to illustrate a principle. Therefore, to avoid distraction from the information being presented, most of the details that are not essential to explain the principle have been omitted.

DEFINITIONS OF BASIC TERMS

Definitions of some of the basic terms used in dimensioning and tolerancing of drawings follow. While these terms are not new, their exact meanings warrant special attention in order that there be no ambiguity in the precise interpretation of tolerancing methods described in this chapter.

Dimension

A *dimension* is a geometric characteristic, of which the size is specified, such as diameter, length, angle, location, or center distance. The term is also used for convenience to indicate the magnitude or value of a dimension, as specified on a drawing. See Fig. 14-1-2.

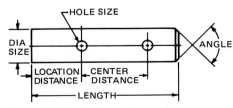

Fig. 14-1-2 Dimensions of a part.

Tolerance

The *tolerance* on a dimension is the total permissible variation in its size, which is equal to the difference between the limits of size. The plural term *tolerances* is sometimes used to denote the permissible variations from the specified size when the tolerance is expressed bilaterally.

For example, in Fig. 14-1-3A the tolerance on the center distance dimension 1.50 ± .04 is .08 in., but in common practice the values + .04 and − .04 are often referred to as the tolerances.

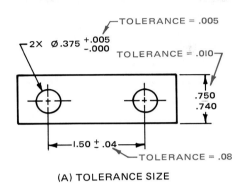

(A) TOLERANCE SIZE

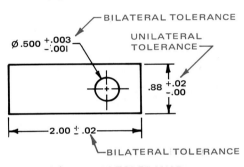

(B) TYPE OF TOLERANCE

Fig. 14-1-3 Tolerances.

Size of Dimensions

In theory, it is impossible to produce a part to an exact size, because every part, if measured with sufficient accuracy, would be found to be a slightly different size. However, for purposes of discussion and interpretation, a number of distinct sizes for each dimension have to be recognized.

Actual Size *Actual size* simply means the measured size of an individual part.

Nominal Size The *nominal size* is the designation of size used for purposes of general identification.

The nominal size is used in referring to a part in an assembly drawing stocklist, in a specification, or in other such documents. It is very often identical to the basic size but in many instances may differ widely; for example, the external diameter of a .50 in. steel pipe is 0.84 in. (21.34 mm). The nominal size is .50 in.

Specified Size This is the size specified on the drawing when the size is associated with a tolerance. The specified size is usually identical to the design size or, if no allowance is involved, to the basic size.

Figure 14-1-4 shows two mating features with the tolerance and allowance zones exaggerated, to illustrate the sizes, tolerances, and allowances. This figure also illustrates the origin of tolerance block diagrams, as shown in Fig. 14-1-5, which are commonly used to show the relationships among part limits, gage or inspection limits, and gage tolerances.

Design Size The *design size* of a dimension is the size in relation to which the tolerance for that dimension is assigned.

Theoretically, it is the size on which the design of the individual feature is based, and therefore it is the size which

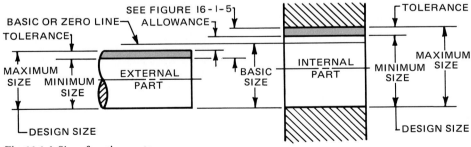

Fig. 14-1-4 Size of mating parts.

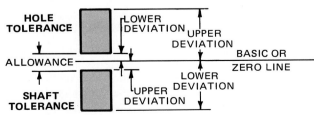

Fig. 14-1-5 Tolerance block diagram.

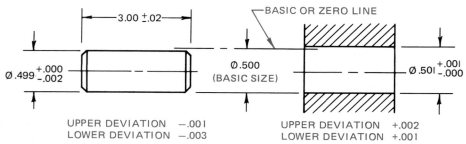

UPPER DEVIATION   −.001
LOWER DEVIATION   −.003

UPPER DEVIATION   +.002
LOWER DEVIATION   +.001

Fig. 14-1-6 Deviations.

should be specified on the drawing. For dimensions of mating features it is derived from the basic size by the application of the allowance, but when there is no allowance, it is identical to the basic size.

Deviations

The differences between the basic, or zero, line and the maximum and minimum sizes are called the *upper* and *lower deviations*, respectively.

Thus in Fig. 14-1-6 the upper deviation of the external part is − .001, and the lower deviation is − .003. For the hole diameter, the upper deviation is + .002, and the lower deviation is + .001, whereas for the length of the pin the upper and lower deviations are + .02 and − .02, respectively.

Basic (Exact) Dimensions

A *basic dimension* represents the theoretical exact size, profile, orientation, or location of a feature or datum target. It is the basis from which permissible variations are established by tolerances or other dimensions, in notes, or in feature control frames. See Fig. 14-1-7. They are shown without tolerances, and each basic dimension is enclosed in a rectangular frame to indicate that the tolerances in the general tolerance note do not apply.

Feature

A *feature* is a specific, characteristic portion of a part, such as a surface, hole, slot, screw thread, or profile.

While a feature may include one or more surfaces, the term is generally used in geometric tolerancing in a more restricted sense, to indicate a specific point, line, or surface. Some examples are the axis of a hole, the edge of a part, or a single flat or curved surface, to

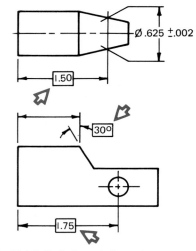

Fig. 14-1-7 Basic (exact) dimensions.

which reference is being made or which forms the basis for a datum.

Axis

An *axis* is a theoretical straight line about which a part or circular feature revolves or could be considered to revolve. See Fig. 14-1-8.

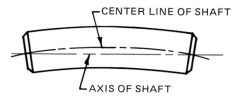

Fig. 14-1-8 Divergence of axis and center line when part is deformed.

INTERPRETATION OF DRAWINGS AND DIMENSIONS

It should not be necessary to specify the geometric shape of a feature, unless some particular precision is required. Lines which appear to be straight imply straightness; those that appear to be

round imply circularity; those that appear to be parallel imply parallelism; those that appear to be square imply perpendicularity; center lines imply symmetry; and features that appear to be concentric about a common center line imply concentricity.

Therefore it is not necessary to add angular dimensions of 90° to corners of rectangular parts nor to specify that opposite sides are parallel.

However, if a particular departure from the illustrated form is permissible, or if a certain degree of precision of form is required, these must be specified. If a slight departure from the true geometric form or position is permissible, it should be exaggerated pictorially in order to show clearly where the dimensions apply. Figure 14-1-9 shows some examples. Dimensions which are not to scale should be underlined.

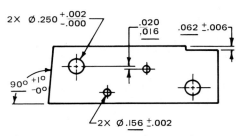

Fig. 14-1-9 Exaggeration of small dimensions.

Point-to-Point Dimensions

When datums are not specified, linear dimensions are intended to apply on a point-to-point basis, either between opposing points on the indicated surfaces or directly between the points marked on the drawing.

The examples shown in Fig. 14-1-10 should help to clarify this principle of point-to-point dimensions.

Location Dimensions with Datums

When location dimensions originate from a feature or surface specified as a datum, measurement is made from the theoretical datum, not from the actual feature or surface of the part.

There will be many cases where a curved center line, as shown in Fig. 14-1-10F, would not meet functional requirements or where the position of the hole in Fig. 14-1-10H would be required to be measured parallel to the base. This can easily be specified by referring the dimension to a datum feature, as shown in Fig. 14-1-11. This will

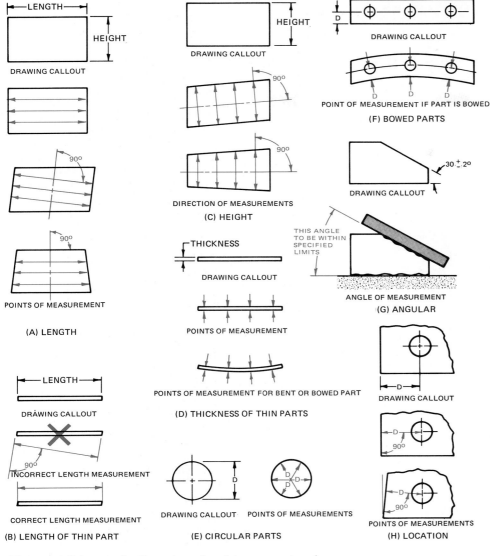

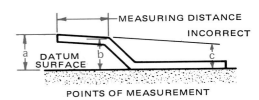

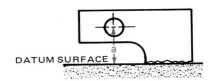

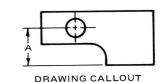

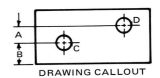

Fig. 14-1-10 Point-to-point dimensions when datums are not used.

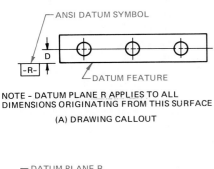

NOTE – DATUM PLANE R APPLIES TO ALL
DIMENSIONS ORIGINATING FROM THIS SURFACE

(A) DRAWING CALLOUT

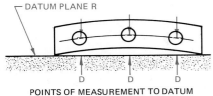

Fig. 14-1-11 Dimensions referenced to a datum.

1. If a dimension refers to two parallel edges or planes, the longer edge or larger surface, which has the greatest influence in the measurement, is assumed to be the datum feature. For example, if the surfaces of the part shown in Fig. 14-1-12A were not quite parallel, as shown in the lower view, dimension *D* would be acceptable if the top surface was within limits when measured at *a* and *b*, but need not be within limits if measured at *c*.

2. If only one of the extension lines refers to a straight edge or surface,

be more fully explained in Unit 14-7, where the interpretation of coordinate tolerances is compared with geometric and positional tolerances.

Assumed Datums

There are often cases where the basic rules for measurements on a point-to-point basis cannot be applied, because the originating points, lines, or surfaces are offset in relation to the features located by the dimensions. See Fig. 14-1-12. It is then necessary to assume a suitable datum, which is usually the theoretical extension of one of the lines or surfaces involved.

The following general rules cover three types of dimensioning procedures commonly encountered.

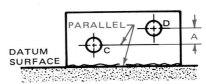

Fig. 14-1-12 Assumed datums.

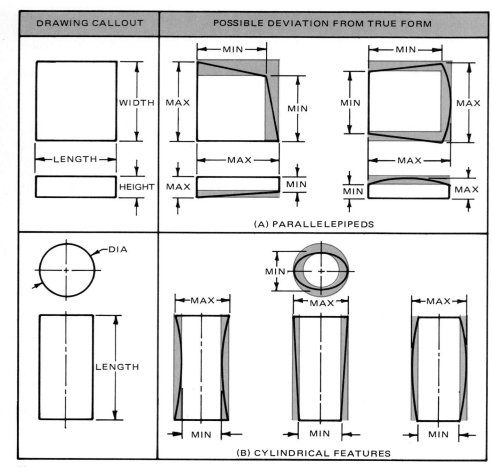

Fig. 14-1-13 Deviations permitted by toleranced dimensions.

If only size tolerances or limits of size are specified for an individual feature and no geometric tolerance is given, no element of the feature would extend beyond the maximum material boundary. Examples are shown in Fig. 14-1-15.

When geometric tolerances are applied, the boundary of perfect form may be violated. Where it is desired to permit a surface of a feature to exceed the boundary of perfect form at maximum material condition, a note such as PERFECT FORM AT MMC NOT REQD is specified, exempting the pertinent size dimension from staying within the boundary of perfect form.

Reference and Source Material

1. ANSI Y14.5M Dimensioning and Tolerancing

ASSIGNMENTS

See Assignments 1 and 2 for Unit 14-1 starting on page 339.

UNIT 14-2
Geometric Tolerancing

In principle, where only size tolerances or limits of size are specified for an individual feature and no form tolerance is given, the maximum material limit of size could be expected to define an envelope of perfect form, and no element of the feature should extend beyond the boundary. However, this principle is not guaranteed unless suitable geometric tolerances are specified on the drawing. When perfect form at MMC for an individual feature is a critical functional requirement, i.e., where no element of the feature can be allowed to cross the boundary of perfect form at the maximum material size, a geometric tolerance of zero MMC must be specified.

A geometric tolerance is the maximum permissible variation of form, profile, orientation, location and runout from that indicated or specified on the drawing. The tolerance value represents the width or diameter of the tolerance zone, within which the point, line, or surface of the feature shall lie.

From this definition it follows that a feature would be permitted to have any variation of form, or take up any position, within the specified geometric tolerance zone.

the extension of that edge or surface is assumed to be the datum. Thus in Fig. 14-1-12B measurement of dimension A is made to a datum surface as shown at a in the bottom view.

3. If both extension lines refer to offset points rather than to edges or surfaces, generally it should be assumed that the datum is a line running through one of these points and parallel to the line or surface to which it is dimensionally related. Thus in Fig. 14-1-12C dimension A is measured from the center of hole D to a line through the center of hole C which is parallel to the datum.

Permissible Form Variations

The actual size of a feature must be within the limits of size, as specified on the drawing, at all points of measurement. This means that each measurement, made at any cross section of the feature, must not be greater than the maximum limit of size nor smaller than the minimum limit of size. See Fig. 14-1-13.

By themselves, toleranced linear dimensions, or limits of size, do not give specific control over many other variations of form, orientation, and, to some extent, position, such as errors of squareness of related features or deviations caused by bending of parts, lobing, eccentricity, and the like. Therefore features may actually cross the boundaries of perfect form at the maximum material size.

In order to meet functional requirements, it is often necessary to control such deviations. This is done to ensure that parts are not only within their limits of size but also within specified limits of geometric form, orientation, and position. In the case of mating parts, such as holes and shafts, it is usually necessary to ensure that they do not deviate from perfect form at the maximum material size (envelope principle), by reason of being bent or otherwise deformed. This condition is shown in Fig. 14-1-14, where features conform to perfect form at the maximum material condition, but are permitted to deviate from perfect at the minimum material condition.

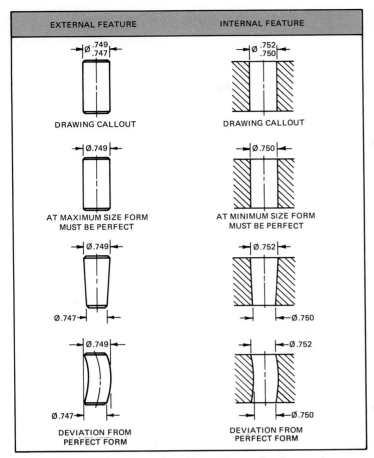

Fig. 14-1-14 Examples of deviation of form when perfect form at the maximum material condition is required.

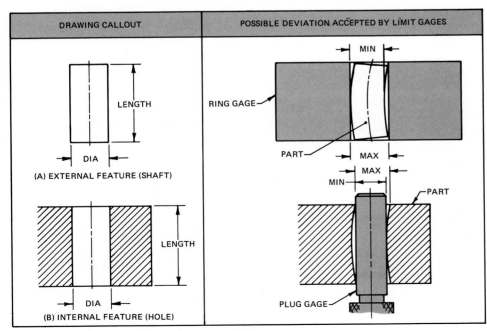

Fig. 14-1-15 Form variations accepted by gage limits.

For example, a line controlled in a single plane by a straightness tolerance of .006 in. must be contained within a tolerance zone .006 in. wide. See Fig. 14-2-1.

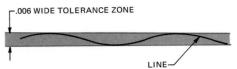

Fig. 14-2-1 Tolerance zone for straightness of a line.

Points, Lines, and Surfaces

The production and measurement of engineering parts deals, in most cases, with surfaces of objects. These surfaces may be flat, cylindrical, conical, or spherical or have some more or less irregular shape or contour.

Measurement, however, usually has to take place at specific points. A line or surface is evaluated dimensionally by making a series of measurements at various points along its length.

Therefore, geometric tolerances are chiefly concerned with points and lines, while surfaces are considered to be composed of a series of line elements running in two or more directions.

Points have position but no size, and therefore position is the only characteristic that requires control. Lines and surfaces have to be controlled for form, orientation, and location. Therefore geometric tolerances provide for control of these characteristics as shown in Fig. 14-2-2.

FEATURE CONTROL FRAME

Some geometric tolerances have been used for many years in the form of notes such as "PARALLEL WITH SURFACE *A* WITHIN .001" and "STRAIGHT WITHIN .12." While such notes are now obsolete, the reader should be prepared to recognize them on older drawings.

The current method is to specify geometric tolerances by means of the *feature control frame*. A feature control frame for an individual feature is divided into compartments containing the geometric characteristic symbol followed by the tolerance. See Fig. 14-2-3. Where applicable, the tolerance is preceded by the diameter symbol (see Unit 14-4) and may be followed by a material condition symbol (see Unit 14-3).

FEATURE	TYPE OF TOLERANCE	CHARACTERISTIC	SYMBOL	SEE UNIT
INDIVIDUAL FEATURES	FORM	STRAIGHTNESS	——	14-2, 14-4
		FLATNESS	▱	14-4
		CIRCULARITY (ROUNDNESS)	○	14-9
		CYLINDRICITY	⌭	
INDIVIDUAL OR RELATED FEATURES	PROFILE	PROFILE OF A LINE	⌒	14-10
		PROFILE OF A SURFACE	⌓	
RELATED FEATURES	ORIENTATION	ANGULARITY	∠	14-6
		PERPENDICULARITY	⊥	
		PARALLELISM	//	
	LOCATION	POSITION	⊕	14-7
		CONCENTRICITY	◎	14-11
	RUNOUT	CIRCULAR RUNOUT	*↗	14-11
		TOTAL RUNOUT	*↗↗	
SUPPLEMENTARY SYMBOLS		MAXIMUM MATERIAL CONDITION	Ⓜ	14-3
		REGARDLESS OF FEATURE SIZE	Ⓢ	
		LEAST MATERIAL CONDITION	Ⓛ	
		PROJECTED TOLERANCE ZONE	Ⓟ	14-7
		BASIC DIMENSION	☐XX☐	14-7, 14-8
		DATUM FEATURE	—A—	14-5
		DATUM TARGET	Ø.50 / A2	14-8

* MAY BE FILLED IN

Fig. 14-2-2 Geometric characteristic symbols.

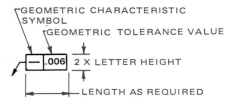

Fig. 14-2-3 Feature control frame for an individual feature.

When necessary, other compartments are added to contain datum references, as explained in Unit 14-5.

Geometric characteristic symbols relating to lines (straightness, angularity, perpendicularity, profile of a line, parallelism, position) are shown in Fig. 14-2-2. Other symbols will be introduced as required, but all are shown in the figure for reference purposes.

Application to Drawings

The feature control frame is related to the feature by one of the following methods (shown in Fig. 14-2-4):

1. Locating the frame below the size dimension to control the center line, axis, or center plane of the feature.

2. Running a leader from the frame to the feature.
3. Attaching a side or end of the frame to an extension line extending from a plane surface feature.
4. Attaching a side or end of the frame to an extension of the dimension line pertaining to a feature of size. ISO practice is to attach the dimension line to the feature control frame and place the feature of size above or below the frame.

Application to Surfaces

The arrowhead of the leader from the feature control frame should touch the surface of the feature or the extension line of the surface.

The leader from the feature control frame should be directed at the feature in its characteristic profile. Thus, in Fig. 14-2-5 the straightness tolerance is directed to the side view, and the circularity tolerance to the end view. This may not always be possible, and a tolerance connected to an alternative view, such as a circularity tolerance connected to a side view, is acceptable. When it is more convenient, or when space is limited, the arrowhead may be directed to an extension line, but not in line with the dimension line.

When two or more feature control frames apply to the same feature, they are drawn together with a single leader and arrowhead, as shown in Fig. 14-2-6.

CIRCULAR TOLERANCE ZONES

When the resulting tolerance zone is cylindrical, such as when straightness of the center line of a cylindrical feature is specified, a diameter symbol precedes the tolerance value in the feature control frame and the feature control frame is located below the dimension pertaining to the feature. See Fig. 14-2-7 and refer to Unit 14-4.

Form Tolerances

Form tolerances control straightness, flatness, circularity, and cylindricity. Orientation tolerances control angularity, parallelism, and perpendicularity.

Form tolerances are applicable to single (individual) features or elements of single features; therefore, form tolerances are not related to datums.

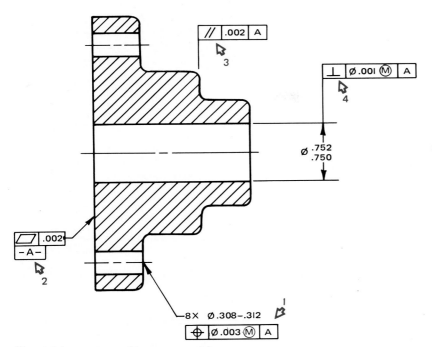

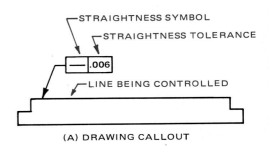

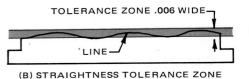

Fig. 14-2-4 Placement of feature control frame.

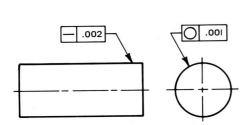

Fig. 14-2-5 Preferred location of feature control symbol.

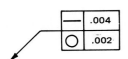

Fig. 14-2-6 Combined feature control frames directed to one surface.

Form and orientation tolerances critical to function and interchangeability are specified where the tolerances of size and location do not provide sufficient control. A tolerance of form or orientation may be specified where no tolerance of size is given, e.g., the control of flatness.

STRAIGHTNESS

Straightness is a condition where the element of a surface or a center line is a straight line. A straightness tolerance specifies a tolerance zone within which the considered element of the surface or

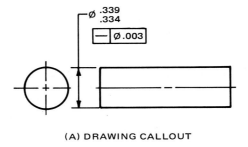

(A) DRAWING CALLOUT

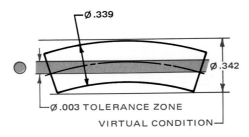

(B) INTERPRETATION

Fig. 14-2-7 Specifying straightness—RFS.

center line must lie. A straightness tolerance is applied to the view where the elements to be controlled are represented by a straight line.

Straightness Controlling Surface Elements

Lines Straightness is fundamentally a characteristic of a line, such as the edge

of a part or a line scribed on a surface. A straightness tolerance is specified on a drawing by means of a feature control frame, which is directed by a leader to the line requiring control, as shown in Fig. 14-2-8. It states in symbolic form that the line shall be straight within .006 in. This means that the line shall be contained within a tolerance zone consisting of the area between two parallel straight lines in the same plane, separated by the specified tolerance.

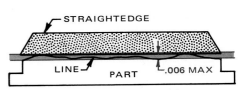

Fig. 14-2-8 Straightness symbol and application.

Theoretically, straightness could be measured by bringing a straightedge into contact with the line and determining that any space between the straightedge and the line does not exceed the specified tolerance.

Cylindrical Surfaces For cylindrical parts, or curved surfaces which are straight in one direction, the feature control frame should be directed to the side view, where line elements appear as a straight line, as shown in Figs. 14-2-9 and 14-2-10.

A straightness tolerance thus applied to the surface controls surface elements only. Therefore it would control bending or a wavy condition of the surface or a barrel-shaped part, but it would not necessarily control the straightness of

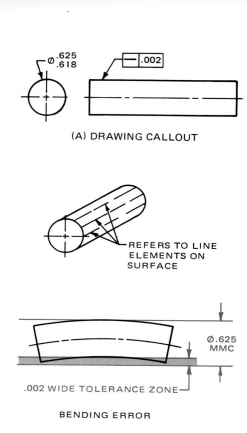

(A) DRAWING CALLOUT

REFERS TO LINE ELEMENTS ON SURFACE

Ø.625 MMC

.002 WIDE TOLERANCE ZONE

BENDING ERROR

Ø.625 MMC

.002 WIDE TOLERANCE ZONE

CONCAVE ERROR

Ø.625 MMC

.002 WIDE TOLERANCE ZONE

CONVEX ERROR

(B) INTERPRETATION

NOTE – NO PART OF THE CYLINDRICAL SURFACE MAY LIE OUTSIDE THE LIMITS OF SIZE

Fig. 14-2-9 Straightness errors in surface elements of a cylindrical part.

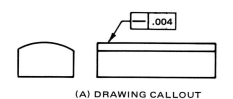

(A) DRAWING CALLOUT

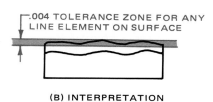

REFERS TO LINE ELEMENTS ON SURFACE

.004 TOLERANCE ZONE FOR ANY LINE ELEMENT ON SURFACE

(B) INTERPRETATION

Fig. 14-2-10 Straightness of surface line elements.

(A) DRAWING CALLOUT

DRAWING CALLOUT REFERS TO EACH LINE ON SURFACE

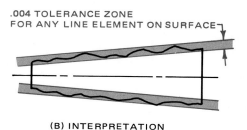

.004 TOLERANCE ZONE FOR ANY LINE ELEMENT ON SURFACE

(B) INTERPRETATION

Fig. 14-2-11 Straightness of a conical surface.

the center line or the conicity of the cylinder.

Straightness of a cylindrical surface is interpreted to mean that each line element of the surface shall be contained within a tolerance zone consisting of the space between two parallel lines, separated by the width of the specified toler-

ance, when the part is rolled along one of the planes. All circular elements of the surface must be within the specified size tolerance. When only limits of size (or a tolerance) are specified without other references to MMC, RFS, LMC, no error in straightness would be permitted if the diameter were at its maximum material size. The straightness tolerance must be less than the size tolerance. Since the limits of size must be respected, the full straightness tolerance may not be available for opposite elements in the case of waisting or barreling of the surface. See Fig. 14-2-9.

Conical Surfaces A straightness tolerance can be applied to a conical surface in the same manner as for a cylindrical surface, as shown in Fig. 14-2-11, and will ensure that the rate of taper is uniform. The actual rate of taper, or the taper angle, must be separately toleranced.

Flat Surfaces A straightness tolerance applied to a flat surface indicates straightness control in one direction only and must be directed to the line on the drawing representing the surface to be controlled and the direction in which control is required, as shown in Fig. 14-2-12A. It is then interpreted to mean that each line element on the surface in the indicated direction shall lie within a tolerance zone.

Different straightness tolerances may be specified in two or more directions when required, as shown in Fig. 14-2-12B. However, if the same straightness tolerance is required in two coordinate directions on the same surface, a flatness tolerance rather than a straightness tolerance is used.

If it is not otherwise necessary to draw all three views, the straightness tolerances may all be shown on a single view by indicating the direction with short lines terminated by arrowheads, as shown in Fig. 14-2-12C.

CAD

Any geometrical tolerance symbol may be created individually, as required, on a drawing. Use the LINE, CIRCLE, ARC, and LEADER commands to draw each one. If your drawing required multiples of the same symbol use the COPY command. It will be even faster and easier to place symbols if a standard library is available. Simply access the desired symbol from the geometrical tolerance library and place it where desired. If a geometrical tolerance library is not available one may be created.

Reference and Source Material

1. ANSI Y14.5M Dimensioning and Tolerancing.

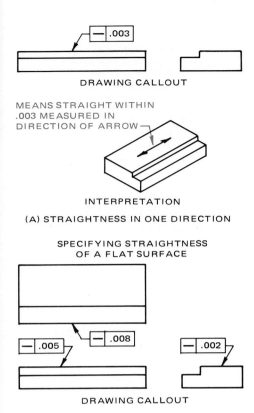

DRAWING CALLOUT

MEANS STRAIGHT WITHIN
.003 MEASURED IN
DIRECTION OF ARROW

INTERPRETATION

(A) STRAIGHTNESS IN ONE DIRECTION

SPECIFYING STRAIGHTNESS
OF A FLAT SURFACE

DRAWING CALLOUT

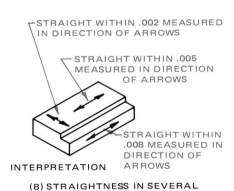

STRAIGHT WITHIN .002 MEASURED
IN DIRECTION OF ARROWS

STRAIGHT WITHIN .005
MEASURED IN DIRECTION
OF ARROWS

STRAIGHT WITHIN
.008 MEASURED IN
DIRECTION OF
ARROWS

INTERPRETATION

(B) STRAIGHTNESS IN SEVERAL
DIRECTIONS

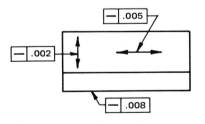

(C) THREE STRAIGHTNESS TOLERANCES
ON ONE VIEW

Fig. 14-2-12 Three straightness tolerances on
one view.

ASSIGNMENT

See Assignment 3 for Unit 14-2 on
page 342.

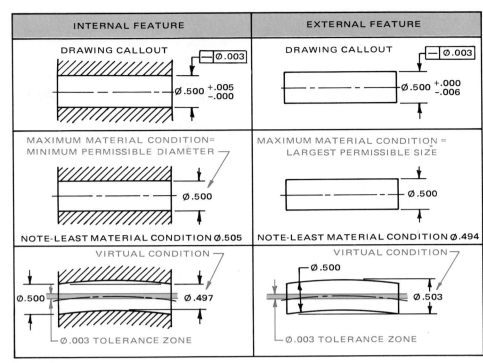

Fig. 14-3-1 Maximum material and virtual conditions.

UNIT 14-3
Relationship to Feature of Size

DEFINITIONS

Maximum Material Condition (MMC) When a feature or part is at the limit of size which results in it containing the maximum amount of material, it is said to be at MMC. Thus it is the maximum limit of size for an external feature, such as a shaft, or the minimum limit of size for an internal feature, such as a hole. See Fig. 14-3-1.

Virtual Condition (Size) *Virtual condition* refers to the overall envelope of perfect form within which the feature would just fit. For an external feature such as a shaft, it is the maximum measured size plus the effect of permissible form variations, such as straightness, flatness, roundness, cylindricity, and orientation tolerances. For an internal feature such as a hole, it is the minimum measured size minus the effect of such form variations. See Fig. 14-3-1.

Least Material Condition (LMC) This term refers to that size of a feature which results in the part containing the minimum amount of material. Thus it is the minimum limit of size for an external

feature, for example, a shaft, and the maximum limit of size for an internal feature, such as a hole. See Fig. 14-3-2.

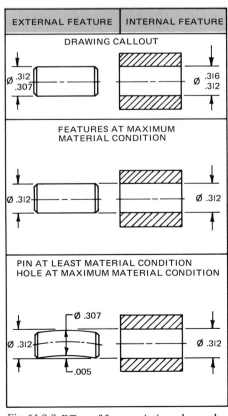

Fig. 14-3-2 Effect of form variation when only features of size are specified.

Regardless of Feature Size (RFS)

This term indicates that a geometric tolerance applies to any size of a feature which lies within its size tolerance.

MATERIAL CONDITION SYMBOLS

The symbols used to indicate "at maximum material condition," "regardless of feature size," and "at least material condition" are shown in Figure 14-3-4. The United States is the only country to adopt the latter two symbols. The use of these symbols in local or general notes is prohibited.

APPLICABILITY OF RFS, MMC, AND LMC

Applicability of RFS, MMC, and LMC is limited to features subject to variations in size. They may be datum features or other features whose axes or center planes are controlled by geometric tolerances. In such cases, the following practices apply:

1. *Tolerance of position.* RFS, MMC, or LMC must be specified on the drawing with respect to the individual tolerance, datum reference, or both, as applicable.
2. *All other geometric tolerances.* RFS applies, with respect to the individual tolerance, datum reference, or both, where no modifying symbol is specified. MMC must be specified on the drawing where it is required.

FEATURES OF SIZE

Geometric tolerances which have so far been considered concern only lines, line elements, and single surfaces. These are features having no diameter or thickness, and tolerances applied to them cannot be affected by feature size.

Features of size are features which do have diameter or thickness. These may be cylinders, such as shafts and holes. They may be slots, tabs, or rectangular or flat parts, where two parallel, flat surfaces are considered to form a single feature. With features of size, the feature control frame is associated with the size dimension. See Figs. 14-3-1 and 14-3-7.

If freedom of assembly of mating parts is the chief criterion for establishing a geometric tolerance for a feature of size, the least favorable assembly condition exists when the parts are made to the maximum material condition. Further geometric variations can then be permitted, without jeopardizing assembly, as the features approach their least material condition.

EXAMPLE 1 The effect of a form tolerance is shown in Fig. 14-3-2, where a cylindrical pin of Ø.307–.312 in. is intended to assemble into a round hole of Ø.312–.316 in. If both parts are at their maximum material condition of Ø.312 in., it is evident that both would have to be perfectly round and straight in order to assemble. However, if the pin was at its least material condition of Ø.307 in., it could be bent up to .005 in. and still assemble in the smallest permissible hole.

EXAMPLE 2 Another example, based on the location of features, is shown in Fig. 14-3-3. This shows a part with two projecting pins required to assemble into a mating part having two holes at the same center distance.

The worst assembly condition exists when the pins and holes are at their maximum material condition, which is Ø.250 in. Theoretically, these parts would just assemble if their form, orientation (squareness to the surface), and center distances were perfect. However, if the pins and holes were at their least material condition of Ø.247 and Ø.253 in., respectively, it would be evident that one center distance could be increased and the other decreased by .003 in. without jeopardizing the assembly condition.

MAXIMUM MATERIAL CONDITION (MMC)

The symbol for maximum material condition is shown in Fig. 14-3-4. The symbol dimensions are based on percentages of the recommended letter height of dimensions.

If a geometric tolerance is required to be modified on an MMC basis, it is specified on the drawing by including the symbol Ⓜ immediately after the tolerance value in the feature control frame as shown in Fig. 14-3-5.

A form tolerance modified in this way can be applied only to a feature of size; it cannot be applied to a single surface. It controls the boundary of the feature, such as a complete cylindrical surface, or two parallel surfaces of a flat feature. This permits the feature surface or surfaces to cross the maximum material

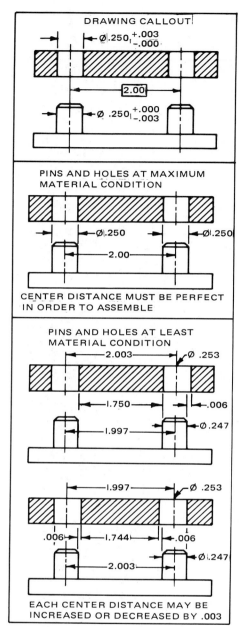

Fig. 14-3-3 Effect on location.

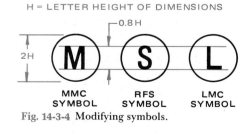

Fig. 14-3-4 Modifying symbols.

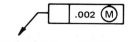

Fig. 14-3-5 Application of MMC symbol.

boundary by the amount of the form tolerance unless it is required that the virtual condition be kept within the maximum material boundary in which case the form tolerance must be specified as zero at MMC, as shown in Fig. 14-3-6.

METRIC

INCH

Fig. 14-3-6 NUMBER OF DIGITS TO CORRESPOND WITH THE SIZE DIMENSION

However, a note such as PERFECT FORM AT MMC NOT REQD must be specified on the drawing. See Figs. 14-4-1 and 14-4-2.

Application of MMC to geometric symbols is shown in Fig. 14-3-7.

Application with Maximum Value

It is sometimes necessary to ensure that the geometric tolerance does not vary over the full range permitted by the size variations. For such applications a maximum limit may be set to the geometric tolerance and this is shown in addition to that permitted at MMC, as shown in Fig. 14-3-8.

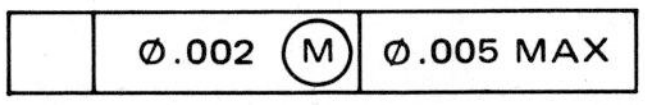

$$\boxed{\quad \varnothing.002 \; \textcircled{M} \; \varnothing.005 \; MAX \quad}$$

Fig. 14-3-8 Tolerance with a maximum specified value.

REGARDLESS OF FEATURE SIZE (RFS)

When MMC or LMC is not specified with a geometric tolerance for a feature of size, no relationship is intended to exist between the feature size and the geometric tolerance. In other words, the tolerance applies regardless of feature size.

In this case, the geometric tolerance controls the form, orientation, or location of the center line, axis, or median plane of the feature.

The regardless of feature size symbol shown in Fig. 14-3-4 is used only with a tolerance of position. See Unit 14-7 and Fig. 14-3-9.

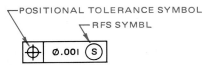

Fig. 14-3-9 Application of RFS symbol.

LEAST MATERIAL CONDITION (LMC)

The symbol for LMC is shown in Fig. 14-3-4. It is the condition in which a feature of size contains the least amount of material within the stated limits of size.

Specifying LMC is limited to positional tolerance applications where MMC does not provide the desired control and RFS is too restrictive. LMC is used to maintain a desired relationship between the surface of a feature and its true position at tolerance extremes. The LMC symbol is shown in Fig. 14-3-4. It is used only with a tolerance of position. See Unit 14-7 and Figs. 14-3-10.

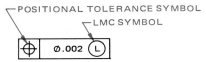

Fig. 14-3-10 Application of LMC symbol.

The symbols for RFS and LMC are used only in ANSI standards and have not been adopted internationally.

Reference and Source Material

1. ANSI Y14.5M Dimensioning and Tolerancing.

ASSIGNMENTS

See Assignments 4 through 7 for Unit 14-3 on page 342.

UNIT 14-4
Straightness of a Feature of Size and Flatness

STRAIGHTNESS OF A FEATURE OF SIZE

Figures 14-4-1 and 14-4-2 show examples of cylindrical parts where all circular elements of the surface are to be within

CHARACTERISTIC TOLERANCE		THE MAXIMUM MATERIAL CONDITION CONCEPT MAY BE APPLIED IF INDICATED BELOW, TO THE FEATURE BEING TOLERANCED AND/OR THE DATUM FEATURE ACCORDING TO THE DESIGN REQUIRED	
STRAIGHTNESS	—	**YES** FOR A FEATURE THE SIZE OF WHICH IS SPECIFIED BY A TOLERANCED DIMENSION, SUCH AS A HOLE, SHAFT OR A SLOT	**NO** FOR A PLANE SURFACE OR A LINE ON A SURFACE
PARALLELISM	//		
PERPENDICULARITY	⊥		
ANGULARITY	∠		
POSITION	⊕		
CONCENTRICITY	◎		
SYMMETRY	⊕		
FLATNESS	▱	**NO** FOR ALL FEATURES	
CIRCULARITY (ROUNDNESS)	○		
CYLINDRICITY	⌭		
PROFILE OF A LINE	⌒		
PROFILE OF A SURFACE	⌓		
CIRCULAR RUNOUT	↗		
TOTAL RUNOUT	↗↗		

Fig. 14-3-7 Application of MMC to geometric symbols.

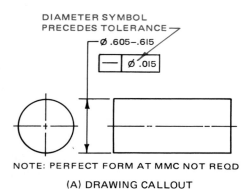

NOTE: PERFECT FORM AT MMC NOT REQD

(A) DRAWING CALLOUT

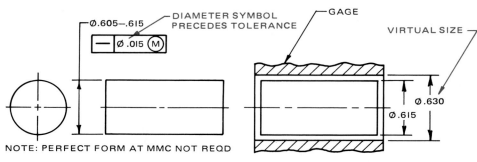

(A) DRAWING CALLOUT

THE MAXIMUM DIAMETER OF THE PIN
WITH PERFECT FORM IN A GAGE

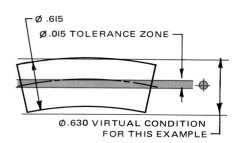

FEATURE SIZE	DIAMETER TOLERANCE ZONE ALLOWED
.615	.015
.614	.015
.613	.015
↓	↓
.606	.015
.605	.015

(B) INTERPRETATION

Fig. 14-4-1 Specifying straightness—RFS.

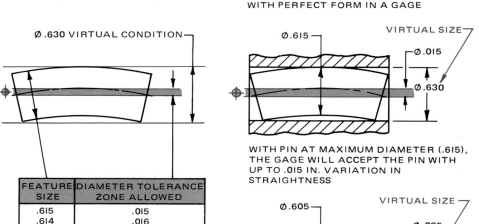

WITH PIN AT MAXIMUM DIAMETER (.615),
THE GAGE WILL ACCEPT THE PIN WITH
UP TO .015 IN. VARIATION IN
STRAIGHTNESS

WITH PIN AT MINIMUM DIAMETER (.605),
THE GAGE WILL ACCEPT THE PIN WITH
UP TO .025 IN. VARIATION IN
STRAIGHTNESS

(C) ACCEPTANCE BOUNDRY

FEATURE SIZE	DIAMETER TOLERANCE ZONE ALLOWED
.615	.015
.614	.016
.613	.017
↓	↓
.606	.024
.605	.025

(B) INTERPRETATION

Fig. 14-4-2 Specifying straightness—MMC.

the specified size tolerance; however, the boundary of perfect form at MMC may be violated. This violation is permissible when the feature control frame is associated with the size dimension, or attached to an extension of the dimension line. In the two figures a diameter symbol precedes the tolerance value and the tolerance is applied on an RFS and an MMC basis respectively. Normally the straightness tolerance is smaller than the size tolerance, but a specific design may allow the situation depicted in the figures. The collective effect of size and form variation can produce a virtual condition equal to the MMC size plus the straightness tolerance. See Fig. 14-4-2. The derived center line of the feature must lie within a cylindrical tolerance zone as specified.

Straightness—RFS

When applied on an RFS basis, as in Fig. 14-4-1, the maximum permissible deviation from straightness is .015 in. regardless of the feature size. Note the absence of a modifying symbol indicates that RFS applies. The RFS symbol ⓢ is shown only with a tolerance of position.

Straightness—MMC

If the straightness tolerance of .015 in. is required only at MMC, further straightness error can be permitted without jeopardizing assembly, as the feature approaches its least material size, Fig. 14-4-2. The maximum straightness tolerance is the specified tolerance plus the amount the feature departs from its MMC size. The center line of the actual feature must lie within the derived cylin-

drical tolerance zone such as given in the table of Fig. 14-4-2.

Straightness—Zero MMC

It is quite permissible to specify a geometric tolerance of zero MMC, which means that the virtual condition coincides with the maximum material size. See Fig. 14-4-3. Therefore, if a feature is at its maximum material limit everywhere, no errors of straightness are permitted.

Straightness on the MMC basis can be applied to any part or feature having straight-line elements in a plane which includes the diameter or thickness. This includes practically all the parts already shown on an RFS basis. However, it should not be used for features which do not have a uniform cross section.

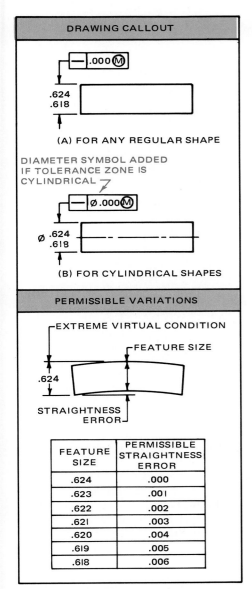

DRAWING CALLOUT

⟦— .000 Ⓜ⟧

.624
.618

(A) FOR ANY REGULAR SHAPE

DIAMETER SYMBOL ADDED
IF TOLERANCE ZONE IS
CYLINDRICAL

⟦— Ø .000 Ⓜ⟧

Ø .624
.618

(B) FOR CYLINDRICAL SHAPES

PERMISSIBLE VARIATIONS

EXTREME VIRTUAL CONDITION

FEATURE SIZE

.624

STRAIGHTNESS ERROR

FEATURE SIZE	PERMISSIBLE STRAIGHTNESS ERROR
.624	.000
.623	.001
.622	.002
.621	.003
.620	.004
.619	.005
.618	.006

Fig. 14-4-3 Straightness—zero MMC.

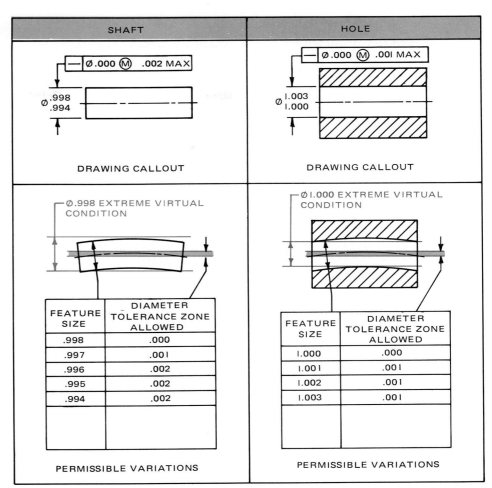

SHAFT	HOLE

DRAWING CALLOUT (SHAFT)

⟦— Ø .000 Ⓜ .002 MAX⟧

Ø .998
.994

DRAWING CALLOUT (HOLE)

⟦— Ø .000 Ⓜ .001 MAX⟧

Ø 1.003
1.000

Ø .998 EXTREME VIRTUAL CONDITION

FEATURE SIZE	DIAMETER TOLERANCE ZONE ALLOWED
.998	.000
.997	.001
.996	.002
.995	.002
.994	.002

PERMISSIBLE VARIATIONS

Ø 1.000 EXTREME VIRTUAL CONDITION

FEATURE SIZE	DIAMETER TOLERANCE ZONE ALLOWED
1.000	.000
1.001	.001
1.002	.001
1.003	.001

PERMISSIBLE VARIATIONS

Fig. 14-4-4 Straightness of a shaft and hole with a maximum value.

Straightness with a Maximum Value

If it is desired to ensure that the straightness error does not become too great when the part approaches the least material condition, a maximum value may be added, as shown in Fig. 14-4-4.

Shapes Other Than Round

A straightness tolerance, not modified by MMC, may be applied to parts or features of any size or shape, provided they have a center plane, as in Fig. 14-4-5, which is intended to be straight in the direction indicated. Examples are parts having a cross section which is hexagonal, square, or rectangular.

Tolerances directed in this manner apply to straightness of the center plane between all opposing line elements of the surfaces in the direction to which the control is directed. The width of the tolerance zone is in the direction of the arrowhead. If the cross section forms a regular polygon, such as a hexagon or square, the tolerance applies to the center plane, between each pair of sides, without its being necessary to so state on the drawing.

(A) SQUARE AND RECTANGULAR PARTS

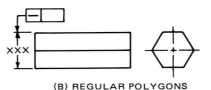

(B) REGULAR POLYGONS

Fig. 14-4-5 Straightness of the center plane— RFS.

Control in Specific Directions

As already stated, straightness of a center line applies only to center lines that run in the direction of the line or line elements to which the straightness tolerance is directed. If there could be some ambiguity, a note should be added, such as THIS DIRECTION ONLY, as shown in Fig. 14-4-6A. If the part is circular and it is intended that the tolerance apply in all directions, a diameter symbol precedes the tolerance value, as shown in Fig. 14-4-6B.

If different tolerances apply in two directions, the tolerance zone is then a parallelepiped, as shown in Fig. 14-4-7.

Straightness Per Unit Length

Straightness may be applied on a unit length basis as a means of preventing an abrupt surface variation within a relatively short length of the feature. See Fig. 14-4-8. Caution should be exercised when using unit control without specifying a maximum limit for the total length

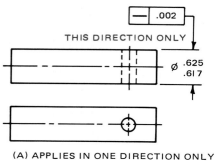

THIS DIRECTION ONLY

Ø .625
.617

(A) APPLIES IN ONE DIRECTION ONLY

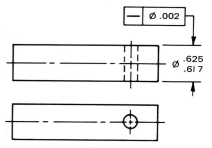

Ø .625
.617

(B) APPLIES IN ALL DIRECTIONS

Fig. 14-4-6 Direction and application of straightness.

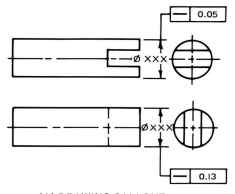

(A) DRAWING CALLOUT
Fig. 14-4-7 Straightness in two directions.

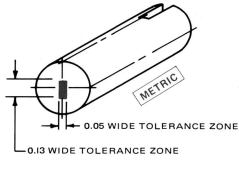

0.05 WIDE TOLERANCE ZONE

0.13 WIDE TOLERANCE ZONE

(B) TOLERANCE ZONE

because of the relatively large variations that may result if no such restriction is applied. If the feature has a uniformly continuous bow throughout its length which just conforms to the tolerance applicable to the unit length, then the overall tolerance may result in an unsatisfactory part. Figure 14-4-9 illustrates the possible condition if the straightness per unit length given in Fig.

14-4-8 is used alone, that is, if straightness for the total length is not specified.

FLATNESS

The symbol for flatness is a parallelogram, with angles of 60° as shown in Fig. 14-4-10. The length and height are based on a percentage of the height of the lettering used on the drawing.

Flatness of a Surface

Flatness of a surface is a condition in which all surface elements are in one plane.

A flatness tolerance is applied to a line representing the surface of a part by

means of a feature control frame, as shown in Fig. 14-4-11.

A flatness tolerance means that all points on the surface shall be contained within a tolerance zone consisting of the space between two parallel planes which are separated by the specified tolerance. These planes may be oriented in any manner to contain the surface, that is, they are not necessarily parallel to the base.

Where the considered surface is associated with a size dimension, the flatness tolerance must be less than the size tolerance.

If the same control is desired on two or more surfaces, a suitable note indicating the number of surfaces may be added instead of repeating the symbol, as shown in Fig. 14-4-12.

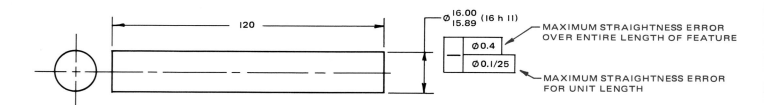

MAXIMUM STRAIGHTNESS ERROR OVER ENTIRE LENGTH OF FEATURE

MAXIMUM STRAIGHTNESS ERROR FOR UNIT LENGTH

NOTE: THE ABSENCE OF A MODIFIER INDICATES RFS APPLIES

(A) DRAWING CALLOUT

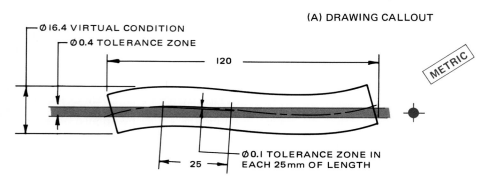

Ø16.4 VIRTUAL CONDITION
Ø0.4 TOLERANCE ZONE

Ø0.1 TOLERANCE ZONE IN EACH 25mm OF LENGTH

(B) INTERPRETATION
Fig. 14-4-8 Specifying straightness per unit length with specified total straightness, both RFS.

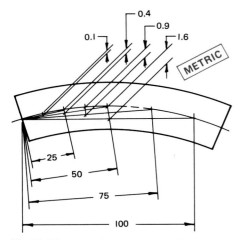

Fig. 14-4-9 Possible results of specifying straightness per unit length RFS with no maximum specified.

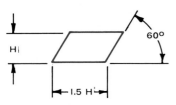

H = RECOMMENDED LETTER HEIGHT
Fig. 14-4-10 Flatness symbol.

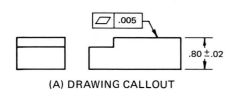

(A) DRAWING CALLOUT

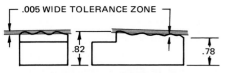

.005 WIDE TOLERANCE ZONE

THE SURFACE MUST LIE BETWEEN TWO PARALLEL PLANES .005 IN. APART. ADDITIONALLY, THE SURFACE MUST BE LOCATED WITHIN ANY SPECIFIED LIMITS OF SIZE.

(B) INTERPRETATION
Fig. 14-4-11 Specifying flatness of a surface.

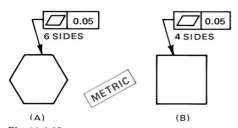

(A)   (B)
Fig. 14-4-12 Controlling flatness on two or more surfaces.

Flatness per Unit Area

Flatness may be applied, as in the case of straightness, on a unit basis as a means of preventing an abrupt surface variation within a relatively small area of the feature. The unit variation is used either in combination with a specified total variation, or alone. Caution should be exercised when using unit control alone for the same reason as was given to straightness.

Since flatness involves surface area, the size of the unit area, for example, 1.00×1.00 in., is specified to the right of the flatness tolerance, separated by a slash line. See Fig. 14-4-13.

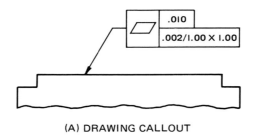

(A) DRAWING CALLOUT

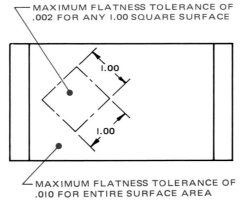

MAXIMUM FLATNESS TOLERANCE OF .002 FOR ANY 1.00 SQUARE SURFACE

MAXIMUM FLATNESS TOLERANCE OF .010 FOR ENTIRE SURFACE AREA

(B) INTERPRETATION
Fig. 14-4-13 Overall flatness tolerance combined with a flatness tolerance of a unit area.

Two or More Flat Surfaces in One Plane

Coplanarity is the condition of two or more surfaces having all elements in one plane. Coplanarity may be controlled by form, orientation, or locational tolerancing, depending on the functional requirements. See Unit 14-11.

Reference and Source Material

1. ANSI Y14.5M Dimensioning and Tolerancing.

ASSIGNMENTS

See Assignments 8 through 12 for Unit 14-4 on page 343.

UNIT 14-5
Datums and the Three-Plane Concept

DATUMS

Datum A *datum* is a point, line, plane, or other geometric surface from which dimensions are measured when so specified or to which geometric tolerances are referenced. A datum has an exact form and represents an exact or fixed location, for purposes of manufacture or measurement.

Datum Feature A *datum feature* is a feature of a part, such as an edge or a surface, which forms the basis for a datum or is used to establish its location.

DATUMS FOR GEOMETRIC TOLERANCING

As defined, datums are exact geometric points, lines, or surfaces, each based on one or more datum features of the part. Surfaces are usually either flat or cylindrical, but other shapes are used when necessary. The datum features, being physical surfaces of the part, are subject to manufacturing errors and variations. For example, a flat surface of a part, if greatly magnified, will show some irregularity. If brought into contact with a perfect plane, it will touch only at the highest points, as shown in Fig. 14-5-1. The true datums are theoretical but are considered to exist, or to be simulated, by locating surfaces of machines, fixtures, and gaging equipment on which

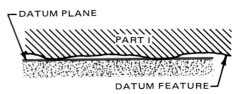

Fig. 14-5-1 Magnified section of a flat surface.

the part rests or with which it makes contact during manufacture and measurement.

THREE-PLANE SYSTEM

Geometric tolerances, such as straightness and flatness, refer to unrelated lines and surfaces and do not require the use of datums.

Orientation and locational tolerances refer to related features, that is, they control the relationship of features to one another or to a datum or datum system. Such datum features must be properly identified on the drawing.

Usually only one datum is required for orientation purposes, but positional relationships may require a datum system consisting of two or three datums. These datums are designated as *primary, secondary,* and *tertiary.* When these datums are plane surfaces that are mutually perpendicular, they are commonly referred to as a *three-plane datum system,* or a *datum reference frame.*

Primary Datum If the primary datum feature is a flat surface, it could lie on a suitable plane surface, such as the surface of a gage, which would then become a primary datum, as shown in Fig. 14-5-2. Theoretically, there will be a minimum of three high spots on the flat surface which will come in contact with the surface of the gage.

Secondary Datum If the part, while lying on this primary plane, is brought into contact with a secondary plane, it will theoretically touch at a minimum of two points.

Tertiary Datum The part can now be slid along, while maintaining contact with the primary and secondary planes, until it contacts a third plane. This plane then becomes the tertiary datum, and

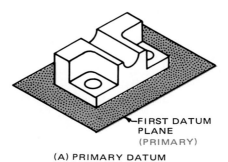

(A) PRIMARY DATUM

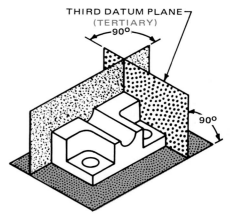

(C) TERTIARY DATUM

Fig. 14-5-2 The datum planes.

the part will theoretically touch it at only one point.

These three planes constitute a datum system from which measurements can be taken. They will appear on the drawing, as shown in Fig. 14-5-3, except that the datum features will be identified in their correct sequence by the methods described late in this unit.

It must be remembered that the majority of parts are not of the simple rectangular shape, and considerably

more ingenuity may be required to establish suitable datums for more complex shapes.

Identification of Datums

Datum symbols are required to serve two purposes:

1. To indicate which is the datum surface or feature on the drawing
2. To identify, for reference purposes, the datum feature

There are two methods of datum symbolization in general use for such purposes: one is shown and used in ANSI standards; the other, the ISO method, is used in most other countries of the world.

ANSI Datum Feature Symbol

In the ANSI system, every datum feature is identified by a capital letter, enclosed in a rectangular box.

A dash is placed before and after the letter, to identify it as applying to a datum feature, as shown in Fig. 14-5-4.

This identifying symbol may be directed to the datum feature in any one of the following ways:

1. By attaching a side or end of the

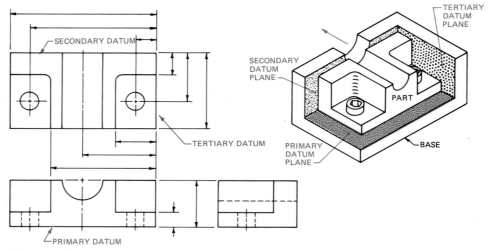

Fig. 14-5-3 Three-plane datum system.

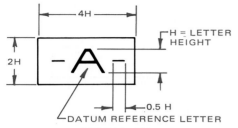

Fig. 14-5-4 ANSI datum feature symbol.

frame to an extension line from the feature, providing it is a plane surface

2. By running a leader with arrowhead from the frame to the feature
3. By adding the symbol to a note, a dimension, or a feature control frame pertaining to the feature
4. By attaching a side or end of the frame to an extension of the dimension line pertaining to a feature of size

These methods are illustrated in Fig. 14-5-5.

ISO Datum Feature Symbol

The ISO datum feature symbol is used by most other countries. The ISO datum feature symbol is a right-angle triangle, with a leader projecting from the 90° apex, as shown in Fig. 14-5-6.

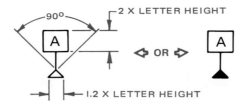

Fig. 14-5-6 ISO datum feature symbol.

The base of the triangle should be slightly greater than the height of the lettering used on the drawing. The triangle may be filled in.

The datum is identified by a capital letter placed in a square frame and connected to the leader.

The ISO datum feature symbol may be directed to the datum feature in one of the following ways:

1. Placed on the outline of the feature or an extension of the outline (but clearly separated from the dimension line) when the datum feature is the line or surface itself.
2. Shown as an extension of the dimension line when the datum feature is the axis or median plane.
3. Placed on the axis or median plane when the datum is the axis or median plane of a single feature (e.g., a cylinder) or the common axis or plane formed by two features, e.g., two holes or lugs.

4. For small features, where extension lines are not used, the symbol may be placed on the leader line.

These methods are illustrated in Fig. 14-5-7.

Association with Geometric Tolerances

The datum letter is placed in the feature control frame by adding an extra compartment for the datum reference, as shown in Fig. 14-5-8.

If two or more datum references are involved, then additional frames are added and the datum references are placed in these frames in the correct order, that is, primary, secondary, and tertiary datums, as shown in Fig. 14-5-9.

Multiple Datum Features

If a single datum is established by two datum features, such as two ends of a shaft, the features are each identified by separate letters. Both letters are then placed in the same compartment of the feature control symbol, with a dash between them, as shown in Fig. 14-5-10A. The datum, in this case, is the common line between the two datum features.

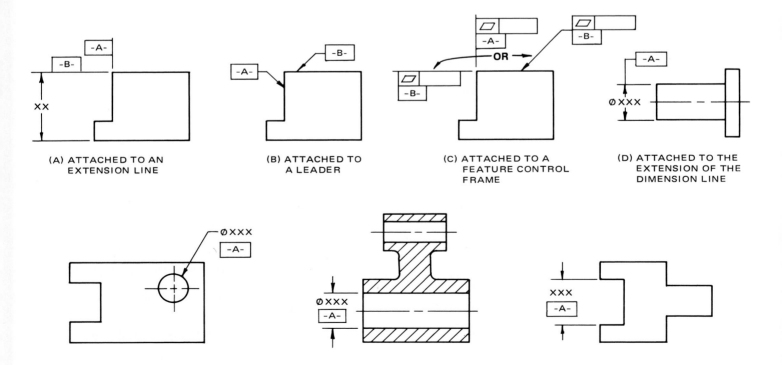

(A) ATTACHED TO AN EXTENSION LINE

(B) ATTACHED TO A LEADER

(C) ATTACHED TO A FEATURE CONTROL FRAME

(D) ATTACHED TO THE EXTENSION OF THE DIMENSION LINE

(E) ATTACHED TO A DIMENSION

Fig. 14-5-5 Placement of ANSI datum feature symbol.

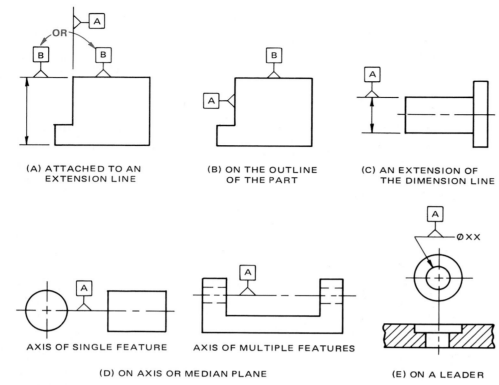

(A) ATTACHED TO AN EXTENSION LINE

(B) ON THE OUTLINE OF THE PART

(C) AN EXTENSION OF THE DIMENSION LINE

AXIS OF SINGLE FEATURE

AXIS OF MULTIPLE FEATURES

(D) ON AXIS OR MEDIAN PLANE

(E) ON A LEADER

Fig. 14-5-7 Placement of ISO datum feature symbol.

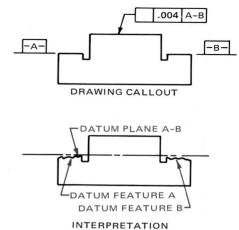

DRAWING CALLOUT

DATUM PLANE A-B

DATUM FEATURE A
DATUM FEATURE B

INTERPRETATION

(A) COPLANAR DATUM FEATURES

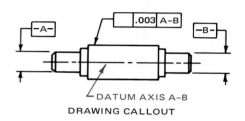

.003 A-B

DATUM AXIS A-B

DRAWING CALLOUT

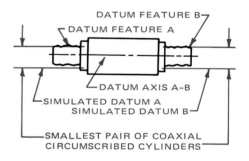

DATUM FEATURE B
DATUM FEATURE A

DATUM AXIS A-B

SIMULATED DATUM A
SIMULATED DATUM B

SMALLEST PAIR OF COAXIAL CIRCUMSCRIBED CYLINDERS

INTERPRETATION

(B) COAXIAL DATUM FEATURES

Fig. 14-5-10 Two datum features for one datum.

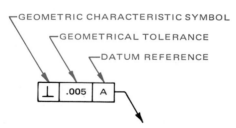

GEOMETRIC CHARACTERISTIC SYMBOL

GEOMETRICAL TOLERANCE

DATUM REFERENCE

⊥ .005 A

Fig. 14-5-8 Feature control symbol referenced to a datum.

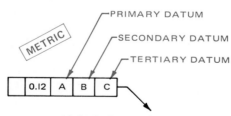

METRIC

PRIMARY DATUM

SECONDARY DATUM

TERTIARY DATUM

0.12 A B C

Fig. 14-5-9 Multiple datum references.

Datums Based on Features of Size

When a feature of size is specified as a datum feature, such as diameters and widths, the datum has to be established from the full surface of the cylindrical feature or from two opposing surfaces of other features of size. These datums are subject to variations in size as well as form. Because variations are allowed by the size dimension, it becomes necessary to determine whether RFS or MMC applies in each case. For a tolerance of position, the datum reference letter is always followed by the appropriate modifying symbol in the feature control frame. For all other geometric tolerances, RFS is implied unless otherwise specified.

Datum Features—RFS

Where a datum feature of size is applied on an RFS basis, the datum is established by physical contact between the feature surface or surfaces and surfaces of the processing or measuring equipment.

Primary Datum Feature—Diameter

For an external feature, the datum is the axis of the smallest circumscribed cylinder which contacts the feature surface. For an internal feature, the datum is the axis of the largest inscribed cylinder which contacts the feature surface. See Fig. 14-5-11.

Primary Datum Feature—Width

For an external feature, the datum is the center plane between two parallel planes which, at minimum separation, contact the corresponding surfaces of the fea-

ture. For an internal feature, the datum is the center plane between two parallel planes which, at maximum separation, contact the corresponding surfaces of the feature. See Fig. 14-5-12.

For both external and internal features, the secondary datum (axis or center plane) is established in the same manner as indicated above with an additional requirement: the contacting cylinder or parallel planes must be oriented perpendicular to the primary datum.

Datum Features—MMC

Where a datum feature of size is applied on an MMC basis, machine and gaging elements in the processing equipment,

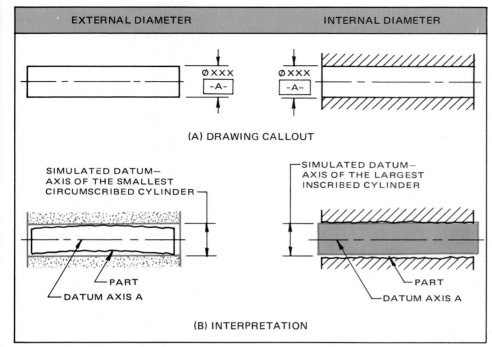

EXTERNAL DIAMETER | **INTERNAL DIAMETER**

Ø XXX
-A-

Ø XXX
-A-

(A) DRAWING CALLOUT

SIMULATED DATUM—
AXIS OF THE SMALLEST
CIRCUMSCRIBED CYLINDER

SIMULATED DATUM—
AXIS OF THE LARGEST
INSCRIBED CYLINDER

PART

DATUM AXIS A

PART

DATUM AXIS A

(B) INTERPRETATION

Fig. 14-5-11 Diameter as primary datum—RFS.

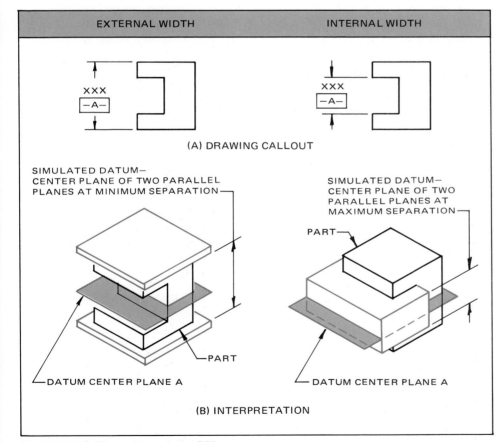

EXTERNAL WIDTH | **INTERNAL WIDTH**

XXX
-A-

XXX
-A-

(A) DRAWING CALLOUT

SIMULATED DATUM—
CENTER PLANE OF TWO PARALLEL
PLANES AT MINIMUM SEPARATION

SIMULATED DATUM—
CENTER PLANE OF TWO
PARALLEL PLANES AT
MAXIMUM SEPARATION

PART

PART

DATUM CENTER PLANE A

DATUM CENTER PLANE A

(B) INTERPRETATION

Fig. 14-5-12 Width as primary datum RFS.

which remain constant in size, may be used to simulate a true geometric counterpart of the feature and to establish the datum. In this case, the size of the simu-lated datum is established by the spec-ified MMC limit of size of the datum feature or its virtual condition, where applicable.

Where a datum feature of size is con-trolled by a specified tolerance of form, the size of the simulated datum is the MMC limit of size, with the exception that where a straightness tolerance is applied on an RFS or MMC basis, the size of the simulated datum is the virtual condition of the datum feature.

Datum features on an MMC basis always apply at their virtual condition. *Virtual condition* refers to the potential boundary of a feature, as specified on a drawing, derived from the collective effect of the maximum material limit of size and the specified form or orientation tolerance. These are added for external features, such as shafts, and subtracted for internal features, such as holes and slots.

If no form or orientation tolerance is specified, it is assumed, for datum refer-ence purposes, that the tolerance is zero MMC.

The fact that a datum applies on an MMC basis is indicated in the feature control frame by the addition of the MMC symbol Ⓜ immediately following the datum reference, as shown in Fig. 14-5-13. When there is more than one datum reference, the MMC symbol must be added for each datum where this modification is required.

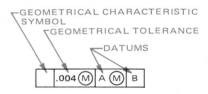

GEOMETRICAL CHARACTERISTIC
SYMBOL
GEOMETRICAL TOLERANCE
DATUMS

.004 Ⓜ | A | Ⓜ | B

Fig. 14-5-13 References to datums—MMC.

External Features with Regular Cross Sections—MMC For external features having a cross section which is circular or which comprises a regular polygon, the datum will be of the same shape as the datum feature. The width or diameter of the datum will be equal to the maximum material limit of size, plus the specified form tolerance.

In Fig. 14-5-14, because no form tol-erance is specified, the cylindrical gag-ing elements are made to the maximum material size of .565 in. This provides an exact location of the part when it is made to the maximum material condition. However, it allows a deviation of .003 in. in any direction from true position if the part is everywhere at the minimum material size of .559 in. and there are no form errors.

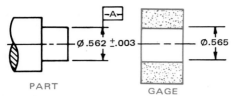

Fig. 14-5-14 Gage element for circular datum feature.

Rectangular Features—MMC If the specified datum feature consists of two flat surfaces and the cross section is not a regular polygon, then the datum consists of two parallel planes. These are separated by a distance equal to the maximum material condition, plus the specified form tolerance.

Figure 14-5-15A shows a flatness tolerance which applies separately to both datum feature surfaces. It should be noted that in such cases the form tolerance is not doubled in calculating the virtual condition. This is because if the part is everywhere at its maximum material condition, as shown in the gaging position, then a convex point on one side can only be offset by a concave point on the other side. If this were not true, the size dimension would be exceeded.

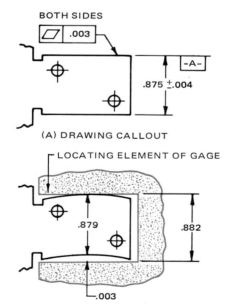

(A) DRAWING CALLOUT

(B) LOCATING ELEMENT OF GAGE

Fig. 14-5-15 Gage element for rectangular datum feature.

It should be noted that Fig. 14-5-15B shows the element of a gage which locates the datum on an MMC basis. It does not check the flatness requirement.

Reference and Source Material

1. ANSI Y14.5M Dimensioning and Tolerancing.

ASSIGNMENTS

See Assignments 13 through 16 for Unit 14-5 starting on page 343.

UNIT 14-6
Orientation

ANGULAR RELATIONSHIPS OF FLAT SURFACES

Orientation refers to the angular relationship which exists between two or more lines, surfaces, or other features. Angularity, parallelism, and perpendicularity are orientation tolerances applicable to related features.

There are three geometric symbols for these characteristics, as shown in Fig. 14-6-1. The proportions are based on the height of the lettering used on the drawing.

Angularity is the condition of a surface or axis at a specified angle (other than 90°) from a datum plane or axis. An angularity tolerance specifies one of the following:

1. A tolerance zone the width of which is defined by two parallel planes at a specified basic angle from a datum plane, or axis, within which the surface of the considered feature must lie. See Fig. 14-6-2.

2. A tolerance zone defined by two parallel planes at the specified basic angle from a datum plane, or axis, within which the axis of the considered feature must lie. See Fig. 14-6-4A.

For geometric tolerancing of angularity, the angle between the datum and the controlled feature should be stated as a basic angle. Therefore it should be enclosed in a rectangular frame, as shown in Fig. 14-6-2, to indicate that the general tolerance note does not apply. However, the angle need not be stated for either perpendicularity (90°) or parallelism (0°).

Parallelism is the condition of a surface equidistant at all points from a datum plane or an axis equidistant along its length from a datum axis or plane. A parallelism tolerance specifies: (1) a tolerance zone defined by two planes or lines parallel to a datum plane, or axis, within which the line elements of the surface (Fig. 14-6-2) or axis of the considered feature must lie (see Fig. 14-6-4); or (2) a cylindrical tolerance, the axis of which is parallel to a datum axis within which the axis of the considered feature must lie (Fig. 14-6-10).

Perpendicularity is the condition of a surface, median plane, or axis at 90° to a datum plane or axis. A perpendicularity tolerance specifies one of the following:

1. A tolerance zone defined by two parallel planes perpendicular to a datum plane or axis, within which the surface, center line or median plane of the considered feature must lie (Figs. 14-6-2 and 14-6-12).

2. A tolerance zone defined by two parallel planes perpendicular to a datum axis within which the axis of the considered feature must lie (Fig. 14-6-13).

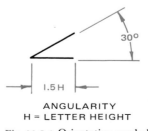

ANGULARITY
H = LETTER HEIGHT

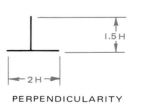

PERPENDICULARITY
(SQUARENESS)

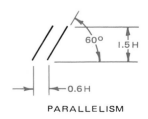

PARALLELISM

Fig. 14-6-1 Orientation symbols.

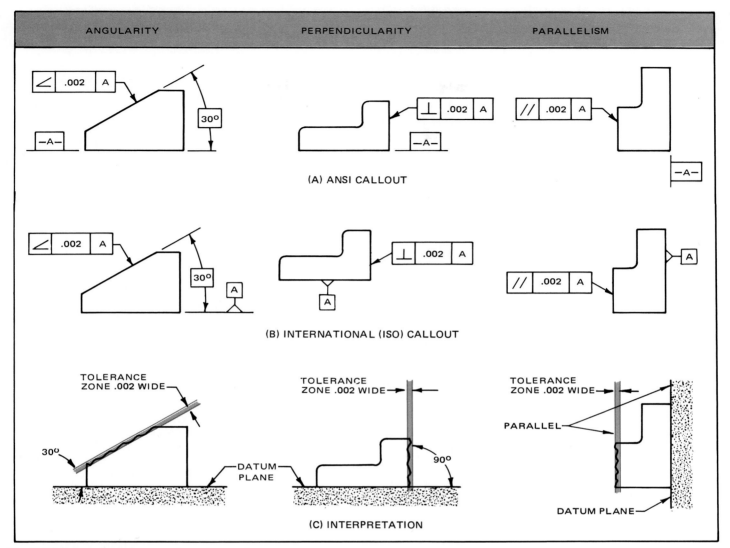

| ANGULARITY | PERPENDICULARITY | PARALLELISM |

(A) ANSI CALLOUT

(B) INTERNATIONAL (ISO) CALLOUT

(C) INTERPRETATION

Fig. 14-6-2 Orientation tolerancing for a plane surface.

3. A cylindrical tolerance zone perpendicular to a datum plane or axis within which the center line of the considered feature must lie (Figs. 14-6-14 to 14-6-17).

ORIENTATION TOLERANCING OF FLAT SURFACES

Figure 14-6-2 shows three simple parts in which one flat surface is designated as a datum feature and another flat surface is related to it by one of the orientation tolerances.

Each of these tolerances is interpreted to mean that the designated surface shall be contained within a tolerance zone consisting of the space between two parallel planes, separated by the specified tolerance (.002 in.) and related to the datum by the basic angle specified (30, 90, or 0°).

Note: Angularity, perpendicularity, and parallelism, when applied to plane surfaces, control flatness if a flatness tolerance is not specified.

Control in Two Directions

The measuring principles for angularity indicate the method of aligning the part prior to making angularity measurements. Proper alignment ensures that line elements of the surface perpendicular to the angular line elements are parallel to the datum.

For example, the part in Fig. 14-6-3 will be aligned so that line elements running horizontally in the left-hand view will be parallel to the datum. However, these line elements will bear a proper relationship with the sides, ends, and top faces only if these surfaces are true and square with the primary datum.

It may be functionally more important to measure the angle in a direction parallel to a side or perpendicular to the front or back face. In this case, one side or face must be chosen as a secondary datum, as shown in Fig. 14-6-7.

Under these circumstances, the part is aligned on the angle or sine plate so that the secondary datum *B* is exactly parallel to the side of the angle plate.

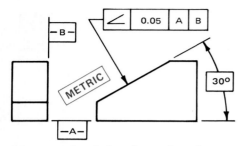

Fig. 14-6-3 Angularity referenced to a datum system.

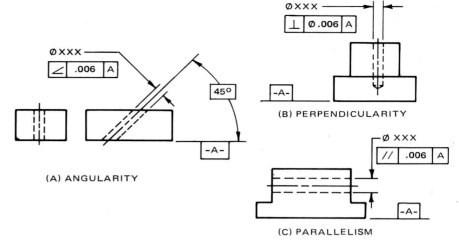

Fig. 14-6-4 Orientation of axis of holes with surfaces (feature RFS).

ORIENTATION FOR AN AXIS OF CYLINDRICAL FEATURES

Tolerances intended to control orientation of the axis of a feature are applied to drawings as shown in Fig. 14-6-4.

The axis of the cylindrical feature must be contained within a tolerance zone consisting of the space between two parallel planes separated by the specified tolerance. The parallel planes are related to the datum by the basic angle of 45, 90, or 0°.

Since the tolerance planes for perpendicularity can be revolved around the feature axis, the tolerance zone effectively becomes a cylinder. The diameter of this cylinder is equal to the specified tolerance.

Internal Cylindrical Features

Figure 14-6-4 shows some simple parts in which the axis or center line of a hole is related by an orientation tolerance to a flat surface. The flat surface is designated as the datum feature.

The axis of the hole must be contained within a tolerance zone consisting of the space between two parallel planes. These planes are separated by a specified tolerance of .006 in. Figure 14-6-5 clearly illustrates the tolerance zone for angularity.

When the tolerance is one of perpendicularity, the tolerance-zone planes can be revolved around the feature axis without the angle being affected. The tolerance zone therefore becomes a cylinder. This cylindrical zone is perpendicular to the datum and has a diameter equal to the specified tolerance, as shown in Fig. 14-6-6.

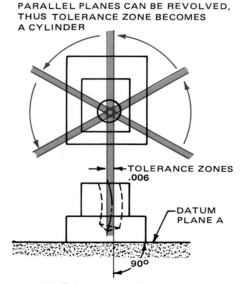

PARALLEL PLANES CAN BE REVOLVED, THUS TOLERANCE ZONE BECOMES A CYLINDER

TOLERANCE ZONES .006

DATUM PLANE A

90°

Fig. 14-6-6 Tolerance zone for perpendicularity—Fig. 14-6-4B.

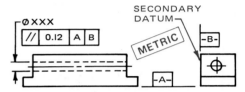

Fig. 14-6-8 Parallelism controlled in two directions.

Specifying Parallelism for an Axis

Regardless of feature size, the feature axis shown in Fig. 14-6-9 must lie between two parallel planes .005 in. apart which are parallel to datum plane A. Additionally, the feature axis must be within any specified positional tolerance zone.

Figure 14-6-10 specifies parallelism for an axis when both the feature and the datum feature are shown on an RFS basis. Regardless of feature size, the feature axis must lie within a cylindrical tolerance zone of .002 diameter whose axis is parallel to datum axis A. Additionally, the feature axis must be within any specified tolerance of location.

Figure 14-6-11 specifies parallelism for an axis when the feature is shown on an MMC basis and the datum feature is shown on an RFS basis. Where the feature is at the maximum material condition (.392 in.), the maximum parallelism tolerance is .002 in. diameter. Where the feature departs from its MMC size, an increase in the parallelism tolerance is allowed which is equal to the amount of such departure. Additionally, the feature axis must be within any specified tolerance of location.

Perpendicularity for a Median Plane

Regardless of feature size, the center plane of the feature shown in Fig. 14-6-12 must lie between two parallel planes

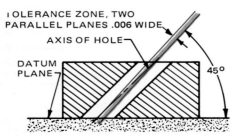

TOLERANCE ZONE, TWO PARALLEL PLANES .006 WIDE

AXIS OF HOLE

DATUM PLANE

45°

Fig. 14-6-5 Tolerance zone for angularity—Fig. 14-6-4A.

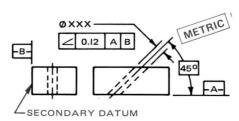

Fig. 14-6-7 Angularity referenced to two datums.

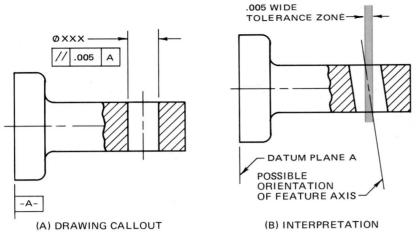

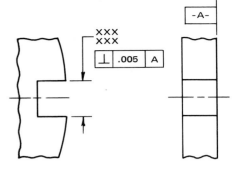

(A) DRAWING CALLOUT

Fig. 14-6-9 Specifying parallelism for an axis (feature RFS).

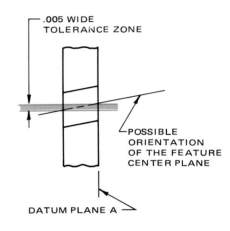

(B) INTERPRETATION

Fig. 14-6-12 Specifying perpendicularity for a median plane (feature RFS).

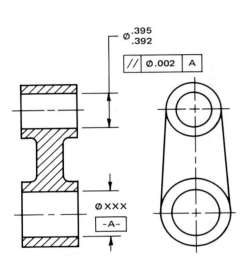

(A) DRAWING CALLOUT

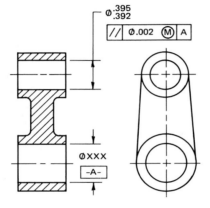

(A) DRAWING CALLOUT

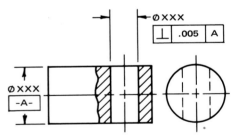

(A) DRAWING CALLOUT

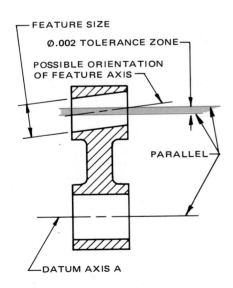

(B) INTERPRETATION

Fig. 14-6-10 Specifying parallelism for an axis (both feature and datum feature RFS).

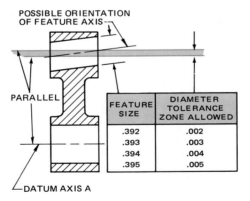

FEATURE SIZE	DIAMETER TOLERANCE ZONE ALLOWED
.392	.002
.393	.003
.394	.004
.395	.005

(B) INTERPRETATION

Fig. 14-6-11 Specifying parallelism for an axis (feature at MMC and datum feature RFS).

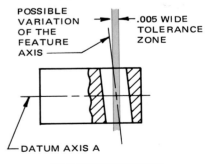

(B) INTERPRETATION

Fig. 14-6-13 Specifying perpendicularity for an axis (both feature and datum feature RFS).

.005 in. apart which are perpendicular to datum plane A. Additionally, the feature center plane must be within any specified tolerance of location.

Perpendicularity for an Axis (Both Feature and Datum RFS) Regardless of feature size, the feature axis shown in Fig. 14-6-13 must lie between two parallel planes .005 in. apart which are per-

pendicular to datum axis A. Additionally, the feature axis must be within any specified tolerance of location.

Perpendicularity for an Axis (Zero Tolerance at MMC) Where the feature shown in Fig. 14-6-14 is at the MMC (50.00 mm), its axis must be perpendicular to datum plane . Where the feature departs from MMC, a perpendicularity tolerance is allowed which is equal to the amount of such departure. Additionally, the feature axis must be within any specified tolerance of location.

Perpendicularity with a Maximum Tolerance Specified Where the feature shown in Fig. 14-6-15 is at MMC (50.00 mm), its axis must be perpendicular to datum plan *A*. Where the fea-

ture departs from MMC, a perpendicularity tolerance is allowed which is equal to the amount of such departure, up to the 0.1 mm maximum. Additionally, the feature axis must be within any specified tolerance at location.

External Cylindrical Features

Perpendicularity for an Axis (Pin or Boss RFS) Regardless of feature size, the feature axis shown in Fig. 14-6-16 must lie within a cylindrical zone 0.4 mm diameter which is perpendicular to and projects from datum plane *A* for the

feature height. Additionally, the feature axis must be within any specified tolerance of location.

Perpendicularity for an Axis (Pin or Boss at MMC) Where the feature shown in Fig. 14-6-17 is at MMC (∅15.984), the maximum perpendicularity tolerance is 0.05 diameter. Where the feature departs from its MMC size, an increase in the perpendicularity tolerance is allowed which is equal to the amount of such departure. Additionally, the feature axis must be within any specified tolerance of location.

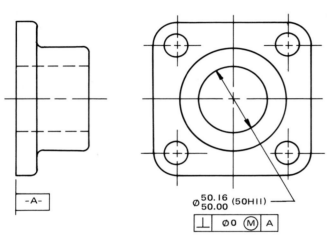

(A) DRAWING CALLOUT

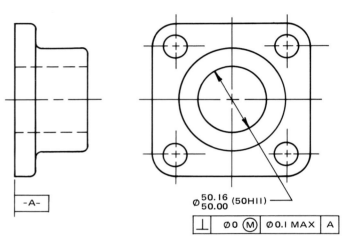

(A) DRAWING CALLOUT

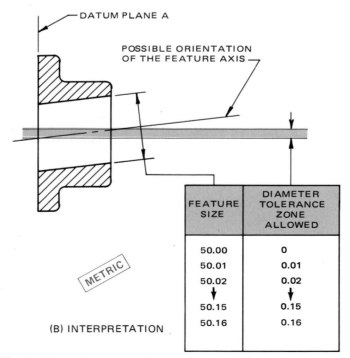

FEATURE SIZE	DIAMETER TOLERANCE ZONE ALLOWED
50.00	0
50.01	0.01
50.02	0.02
↓	↓
50.15	0.15
50.16	0.16

(B) INTERPRETATION

Fig. 14-6-14 Specifying perpendicularity for an axis (zero tolerance at MMC).

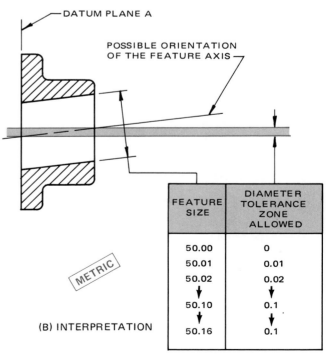

FEATURE SIZE	DIAMETER TOLERANCE ZONE ALLOWED
50.00	0
50.01	0.01
50.02	0.02
↓	↓
50.10	0.1
↓	↓
50.16	0.1

(B) INTERPRETATION

Fig. 14-6-15 Specifying perpendicularity for an axis (zero tolerance at MMC with a maximum specified).

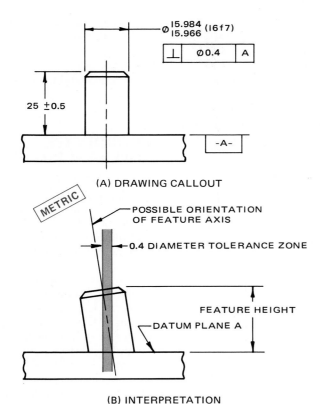

(A) DRAWING CALLOUT

METRIC

POSSIBLE ORIENTATION OF FEATURE AXIS

0.4 DIAMETER TOLERANCE ZONE

FEATURE HEIGHT

DATUM PLANE A

(B) INTERPRETATION

Fig. 14-6-16 Specifying perpendicularity for an axis (pin or boss RFS).

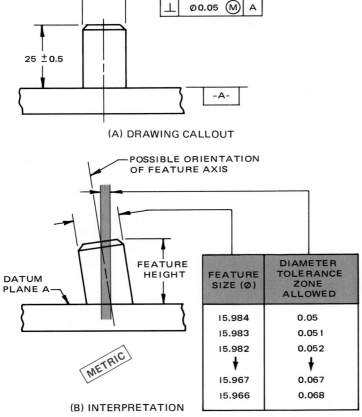

(A) DRAWING CALLOUT

POSSIBLE ORIENTATION OF FEATURE AXIS

FEATURE HEIGHT

DATUM PLANE A

METRIC

FEATURE SIZE (∅)	DIAMETER TOLERANCE ZONE ALLOWED
15.984	0.05
15.983	0.051
15.982	0.052
↓	↓
15.967	0.067
15.966	0.068

(B) INTERPRETATION

Fig. 14-6-17 Specifying perpendicularity for an axis (pin or boss at MMC).

Reference and Source Material

1. ANSI Y14.5M Dimensioning and Tolerancing.

ASSIGNMENTS

See Assignments 17 through 22 for Unit 14-6 starting on page 344.

UNIT 14-7
Positional Tolerancing

The location of features is one of the most frequently used applications of dimensions on technical drawings. Tolerancing may be accomplished either by *coordinate tolerances* applied to the dimensions or by *geometric (positional) tolerancing*.

Positional tolerancing is especially useful when applied on an MMC basis to groups or patterns of holes or other small features in the mass production of parts. This method meets functional requirements in most cases and permits assessment with simple gaging procedures.

Most examples in this unit are devoted to the principles involved in the location of small, round holes, because they represent the most commonly used applications. The same principles apply, however, to the location of other features, such as slots, tabs, bosses, and noncircular holes.

TOLERANCING METHODS

A single hole is usually located by means of rectangular coordinate dimensions, extending from suitable edges or other features of the part to the axis of the hole. Other dimensioning methods, such as polar coordinates, may be used when circumstances warrant.

There are two standard methods of tolerancing the location of holes, as illustrated in Fig. 14-7-1:

1. *Coordinate tolerancing*, which refers to tolerances applied directly to the coordinate dimensions or to applicable tolerances specified in a general tolerance note.
2. *(a) Positional tolerancing*, RFS (regardless of feature size).
(b) Positional tolerancing, MMC (maximum material condition).
(c) Positional tolerancing, LMC (least material condition).

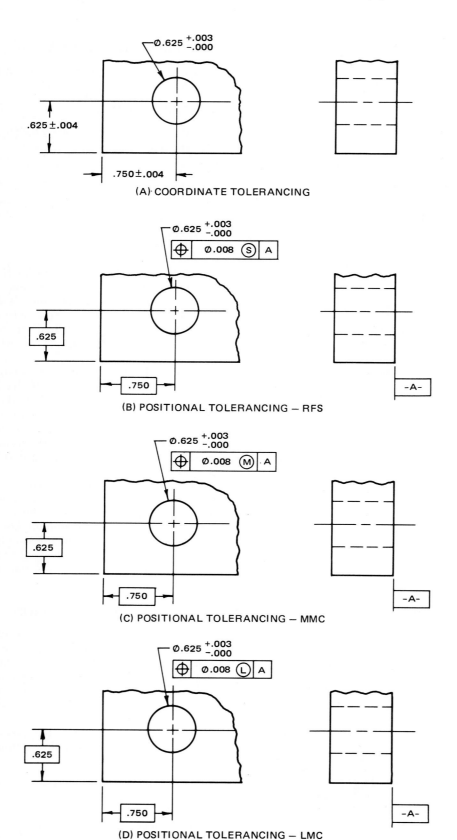

(A) COORDINATE TOLERANCING

(B) POSITIONAL TOLERANCING — RFS

(C) POSITIONAL TOLERANCING — MMC

(D) POSITIONAL TOLERANCING — LMC

Fig. 14-7-1 Comparison of tolerancing methods.

These positional tolerancing methods are part of the system of geometric tolerancing.

Any of these tolerancing methods can be substituted one for the other, although with differing results. It is necessary, however, to first analyze the widely used method of coordinate tolerancing in order to explain and understand the advantages and disadvantages of the positional tolerancing methods.

COORDINATE TOLERANCING

Coordinate dimensions and tolerances may be applied to the location of a single hole, as shown in Fig. 14-7-2. They locate the hole axis and result in either a rectangular or a wedge-shaped tolerance zone within which the axis of the hole must lie.

If the two coordinate tolerances are equal, the tolerance zone formed will be a square. Unequal tolerances result in a rectangular tolerance zone. Where one of the locating dimensions is a radius, polar dimensioning gives a circular ring section tolerance zone. For simplicity, square tolerance zones are used in the analyses of most of the examples in this section.

It should be noted that the tolerance zone extends for the full depth of the hole, that is, the whole length of the axis. This is illustrated in Fig. 14-7-3 and explained in more detail in a later unit. In most of the illustrations, tolerances will be analyzed as they apply at the surface of the part, where the axis is represented by a point.

Maximum Permissible Error

The actual position of the feature axis may be anywhere within the rectangular tolerance zone. For square tolerance zones, the maximum allowable variation from the desired position occurs in a direction of 45° from the direction of the coordinate dimensions. See Fig. 14-7-4.

For rectangular tolerance zones this maximum tolerance is the square root of the sum of the squares of the individual tolerances, or expressed mathematically

$$\sqrt{X^2 + Y^2}$$

For the examples shown in Fig. 14-7-2, the tolerance zones are shown in Fig. 14-7-5, and the maximum tolerance values are as shown in the following examples.

EXAMPLE A

$$\sqrt{.010^2 + .010^2} = .014$$

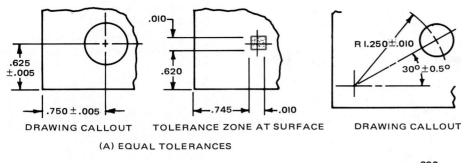

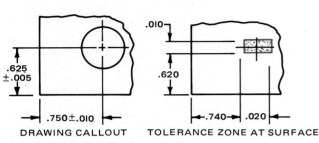

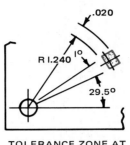

(A) EQUAL TOLERANCES

(B) UNEQUAL TOLERANCES

(C) POLAR TOLERANCES

Fig. 14-7-2 Tolerance zones for coordinate tolerancing.

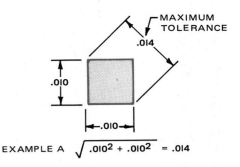

EXAMPLE A $\sqrt{.010^2 + .010^2} = .014$

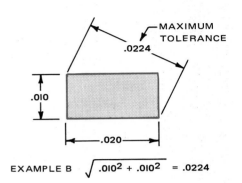

EXAMPLE B $\sqrt{.010^2 + .010^2} = .0224$

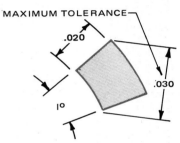

EXAMPLE C

Fig. 14-7-5 Tolerance zones for parts shown in Fig. 14-7-2.

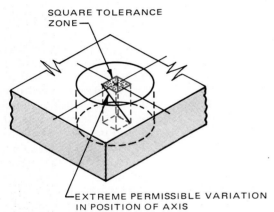

Fig. 14-7-3 Tolerance zone extending through part.

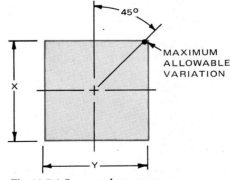

Fig. 14-7-4 Square tolerance zone.

A	tan a	A	tan a	A	tan a
0° 5'	0.00145	0°25'	0.00727	0°45'	0.01309
0°10'	0.00291	0°30'	0.00873	0°50'	0.01454
0°15'	0.00436	0°35'	0.01018	0°55'	0.01600
0°20'	0.00582	0°40'	0.01164	1° 0'	0.01745

Use of Chart

A quick and easy method of finding the maximum positional error permitted with coordinate tolerancing, without having to calculate squares and square roots, is by use of a chart like that shown in Fig. 14-7-6.

In Example A shown in Fig. 14-7-2, the tolerance in both directions is .010 in. The extensions of the horizontal and vertical lines of .010 in the chart intersect at point A, which lies between the radii of .013 and .014 in. When interpolated and rounded to three dimensional places, the maximum permissible variation of position is .014 in.

EXAMPLE B

$$\sqrt{.010^2 + .020^2} = .0224$$

EXAMPLE C For polar coordinates the extreme variation is

$$\sqrt{A^2 + T^2}$$

where $A = R \tan a$
T = tolerance on radius
R = mean radius
a = angular tolerance

Thus, the extreme variation in example C is

$$\sqrt{(1.25 \times .017\ 45)^2 + .020^2} = .03$$

Note: Mathematically, A in the above formula should be $2R \tan a/2$, instead of $R \tan a$, and T should be $T \cos A/2$; but the difference in results is quite insignificant for the tolerances normally used.

Some values of tan A for commonly used angular tolerances are as follows:

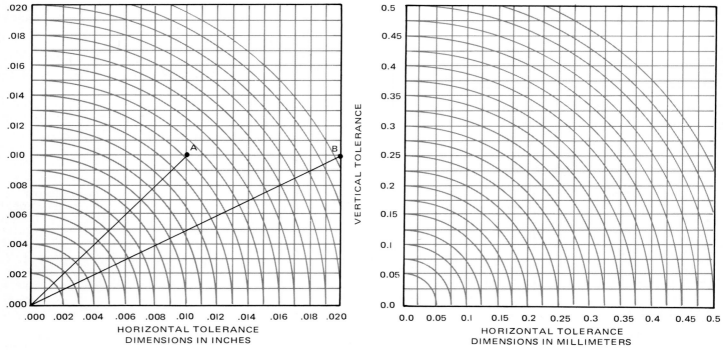

Fig. 14-7-6 Chart for calculating maximum tolerance using coordinate tolerancing.

In Example B shown in Fig. 14-7-2, the tolerances are .010 in. in one direction and .020 in. in the other. The extensions of the vertical and horizontal lines at .010 and .020 in., respectively, in the chart intersect at point *B*, which lies between the radii of .022 and .023 in. When interpolated and rounded to three decimal places, the maximum variation of position is .022 in. Figure 14-7-6 also shows a chart for use with tolerances in millimeters.

Advantages of Coordinate Tolerancing

The advantages claimed for direct coordinate tolerancing are as follows:

1. It is simple and easily understood, and, therefore, it is commonly used.
2. It permits direct measurements to be made with standard instruments and does not require the use of special-purpose functional gages or other calculations.

Disadvantages of Coordinate Tolerancing

There are a number of disadvantages of the direct tolerancing method:

1. It results in a square or rectangular tolerance zone within which the axis

must lie. For a square zone this permits a variation in a 45° direction of approximately 1.4 times the specified tolerance. This amount of variation may necessitate the specification of tolerances which are only 70 percent of those that are functionally acceptable.
2. It may result in an undesirable accumulation of tolerances when several features are involved, especially when chain dimensioning is used.
3. It is more difficult to assess clearances between mating features and components than when positional tolerancing is used, especially when a group or a pattern of features is involved.
4. It does not correspond to the control exercised by fixed functional "go" gages, often desirable in mass production of parts. This becomes particularly important in dealing with a group of holes. With direct coordinate tolerancing, the location of each hole has to be measured separately in two directions, whereas with positional tolerancing on an MMC basis one functional gage checks all holes in one operation.

POSITIONAL TOLERANCING

Positional tolerancing is part of the system of geometric tolerancing. It defines

a zone within which the center, axis, or center plane of a feature of size is permitted to vary from true (theoretically exact) position. A positional tolerance is indicated by the position symbol, a tolerance, and appropriate datum references placed in a feature control frame. Basic dimensions represent the exact values to which geometrical positional tolerances are applied elsewhere by symbols or notes on the drawing. They are enclosed in a rectangular frame (basic dimension symbol) as shown in Fig. 14-7-7. Where the dimension represents a diameter or a radius, the symbol Ø or R is included in the rectangular frame. General tolerance notes do not apply to basic dimensions. The frame size need not be any larger than that necessary to enclose the dimension. Permissible deviations from the basic dimension are then given by a positional tolerance as described in this unit.

Formerly the word BASIC or the abbreviation TP was used to indicate such dimensions.

SYMBOL FOR POSITION

The geometric characteristic symbol for position is a circle with two solid center lines, as shown in Fig. 14-7-8. This symbol is used in the feature control frame in the same manner as for other geometric tolerances.

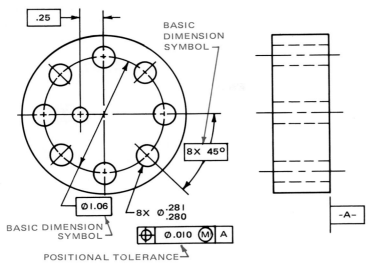

Fig. 14-7-7 Identifying basic dimensions.

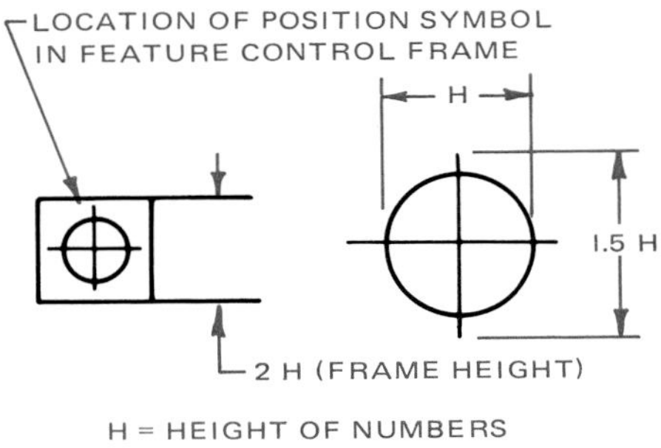

MATERIAL CONDITION BASIS

Positional tolerancing is applied on an MMC, RFS, or LMC basis. The appropriate symbol for the above follows the specified tolerance and where required the applicable datum reference in the feature control frame.

As positional tolerance controls the position of the center, axis, or center plane of a feature, the feature control frame is normally attached to the size of the feature, as shown in Fig. 14-7-9.

The positional tolerance represents the diameter of a cylindrical tolerance zone, located at true position as determined by the basic dimensions on the drawing, within which the axis or center line of the feature must lie.

Except for the fact that the tolerance zone is circular instead of square, a positional tolerance on this basis has exactly the same meaning as direct coordinate tolerancing but with equal tolerances in all directions.

It has already been shown that with rectangular coordinate tolerancing the maximum permissible error in location is not the value indicated by the horizon-

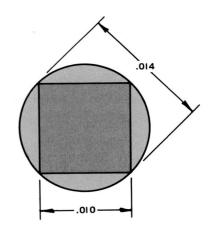

Fig. 14-7-8 Position symbol.

tal and vertical tolerances, but rather is equivalent to the length of the diagonal between the two tolerances. For square tolerance zones this is 1.4 times the specified tolerance values. The specified tolerance can therefore be increased to an amount equal to the diagonal of the coordinate tolerance zone without affecting the clearance between the hole and its mating part.

This does not affect the clearance between the hole and its mating part, yet it offers 57 percent more tolerance area, as shown in Fig. 14-7-10. Such a change would most likely result in a reduction in the number of parts rejected for positional errors.

AREA OF CIRCUMSCRIBED CIRCULAR ZONE = 157% OF SQUARE TOLERANCE ZONE

Fig. 14-7-10 Relationship of tolerance zones.

A simpler method is to make coordinate measurements and evaluate them on a chart, as shown in Fig. 14-7-11. For example, if measurements of four parts, made according to Fig. 14-7-9, are as shown in the table on page 320, only two are acceptable. These positions are shown on the chart.

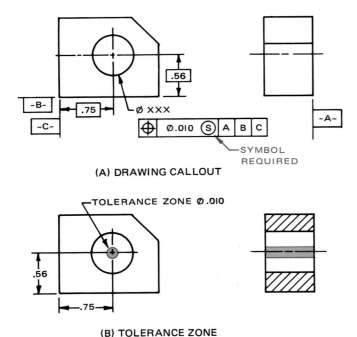

(A) DRAWING CALLOUT

(B) TOLERANCE ZONE

Fig. 14-7-9 Positional tolerancing—RFS.

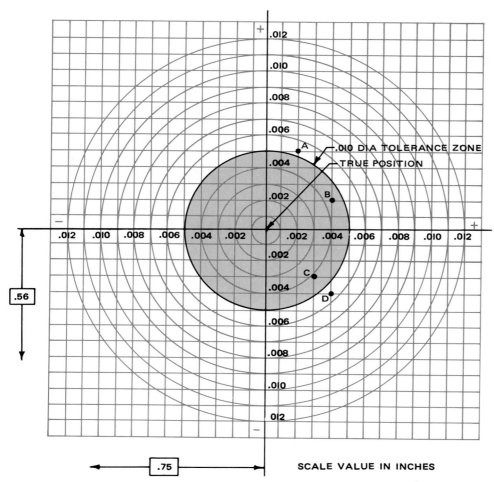

NOTE – DARK CIRCULAR AREA REPRESENTS Ø.010 TOLERANCE ZONE

Fig. 14-7-11 Chart for evaluating positional tolerancing—RFS.

Part	Measurements		Acceptability
A	.565	.752	Rejected
B	.562	.754	Accepted
C	.557	.753	Accepted
D	.556	.754	Rejected

MMC as Related to Positional Tolerancing

The positional tolerance and MMC of mating features are considered in relation to each other. MMC by itself means a feature of a finished product contains the maximum amount of material permitted by the toleranced size dimension of that feature. Thus for holes, slots, and other internal features, maximum material is the condition where these factors are at their minimum allowable sizes. For shafts, as well as for bosses, lugs, tabs, and other external features, maximum material is the condition where these are at their maximum allowable sizes.

A positional tolerance applied on an MMC basis may be explained in either of the following ways:

1. *In terms of the surface of a hole.* While maintaining the specified size limits of the hole, no element of the hole surface shall be inside a theoretical boundary having a diameter equal to the minimum limit of size minus the positional tolerance located at true position. See Fig. 14-7-12.

2. *In terms of the axis of the hole.* Where a hole is at MMC (minimum diameter), its axis must fall within a cylindrical tolerance zone whose axis is located at true position. The diameter of this zone is equal to the positional tolerance. See Figs. 14-7-13A and B. This tolerance zone also defines the limits of variation in the attitude of the axis of the hole in relation to the datum surface. See Fig. 14-7-13C. It is only when the feature is at MMC that the specified positional tolerance applies. Where the actual size of the feature is larger than

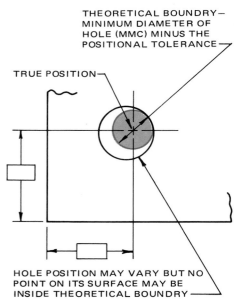

Fig. 14-7-12 Boundary for surface for a hole at MMC.

MMC, additional positional tolerance results. See Fig. 14-7-15. This increase of positional tolerance is equal to the difference between the specified maximum material limit of size (MMC) and the actual size of the feature. The specified positional tolerance for a feature may be exceeded where the actual size is larger than MMC and still satisfy function and interchangeability requirements.

The problems of tolerancing for the position of holes are simplified when positional tolerancing is applied on an MMC basis. Positional tolerancing simplifies measuring procedures of functional "go" gages. It also permits an increase in positional variations as the size departs from the maximum material size without jeopardizing free assembly of mating features.

A positional tolerance on an MMC basis is specified on a drawing, on either the front or the side view, as shown in Fig. 14-7-14. The MMC symbol Ⓜ is added in the feature control frame immediately after the tolerance.

A positional tolerance applied to a hole on an MMC basis means that the boundary of the hole must fall outside a perfect cylinder having a diameter equal to the minimum limit of size minus the positional tolerance. This cylinder is located with its axis at true position. The hole must, of course, meet its diameter limits.

The effect is illustrated in Fig. 14-7-15, where the gage cylinder is shown at true

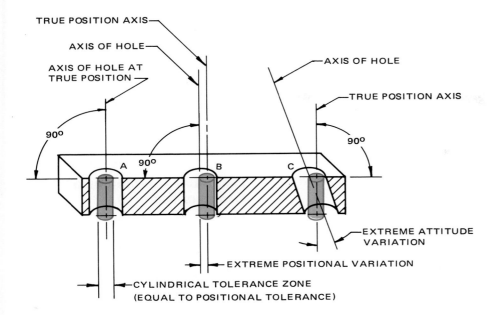

HOLE A — AXIS OF HOLE IS COINCIDENT WITH TRUE POSITION AXIS

HOLE B — AXIS OF HOLE IS LOCATED AT EXTREME POSITION TO THE LEFT OF TRUE POSITION AXIS (BUT WITHIN TOLERANCE ZONE)

HOLE C — AXIS OF HOLE IS INCLINED TO EXTREME ATTITUDE WITHIN TOLERANCE ZONE

NOTE — THE LENGTH OF THE TOLERANCE ZONE IS EQUAL TO THE LENGTH OF THE FEATURE, UNLESS OTHERWISE SPECIFIED ON THE DRAWING

Fig. 14-7-13 Hole axes in relationship to positional tolerance zones.

position and the minimum and maximum diameter holes are drawn to show the extreme permissible variations in position in one direction.

Therefore, if a hole is at its maximum material condition (minimum diameter), the position of its axis must lie within a circular tolerance zone having a diameter equal to the specified tolerance. If the hole is at its maximum diameter (least material condition), the diameter of the tolerance zone for the axis is increased by the amount of the feature tolerance. The greatest deviation of the axis in one direction from true position is therefore

$$\frac{H + P}{2} = \frac{.004 + .008}{2} = .006$$

where H = hole diameter tolerance
P = positional tolerance

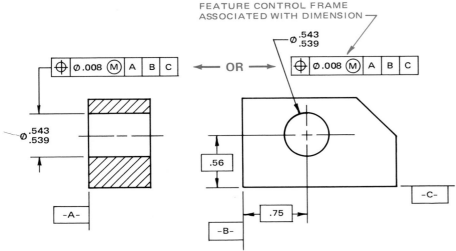

Fig. 14-7-14 Positional tolerancing—MMC.

It must be emphasized that positional tolerancing, even on an MMC basis, is not a cure-all for positional tolerancing problems; each method of tolerancing has its own area of usefulness. In each application a method must be selected which best suits that particular case.

Positional tolerancing on an MMC basis is preferred when production quantities warrant the provision of functional "go" gages, because gaging is then limited to one simple operation, even when a group of holes is involved. This method also facilitates manufacture by permitting larger variations in position when the diameter departs from the maximum material condition. It cannot be used when it is essential that variations in location of the axis be observed regardless of feature size.

Zero Positional Tolerancing at MMC. The application of MMC permits the tolerance to exceed the value specified, provided features are within size limits and parts are acceptable. This is accomplished by adjusting the minimum size limit of a hole to the absolute minimum required for the insertion of an applicable fastener located precisely at true position, and specifying a zero tolerance at MMC. See Fig. 14-7-17. In this case, the positional tolerance allowed is totally dependent on the actual size of the considered feature.

RFS as Related to Positional Tolerancing

In certain cases, the design or function of a part may require the positional tolerance or datum reference, or both, to be maintained regardless of actual feature sizes. RFS, where applied to the positional tolerance of circular features, requires the axis of each feature to be located within the specified positional tolerance regardless of the size of the feature. This requirement imposes a closer control of the features involved and introduces complexities in verification.

LMC as Related to Positional Tolerancing

Where positional tolerancing at LMC is specified, the stated positional tolerance applies when the feature contains the least amount of material permitted by its toleranced size dimension. Specification of LMC further requires perfect form at

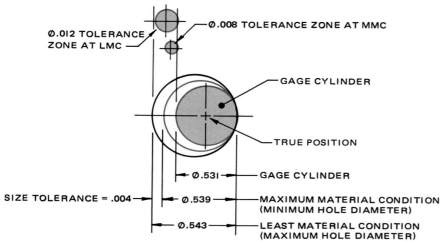

Fig. 14-7-15 Positional variations for tolerancing for Fig. 14-7-14.

LMC. Perfect form at MMC is not required. Where the feature departs from its LMC size, an increase in positional tolerance is allowed, which is equal to the amount of such departure. See Fig. 14-7-18. Specifying LMC is limited to positional tolerancing applications where MMC does not provide the desired control and RFS is too restrictive.

Positional Tolerancing a Group of Features

In many instances, a group of features (such as a group of mounting holes) must be positioned relative to a datum

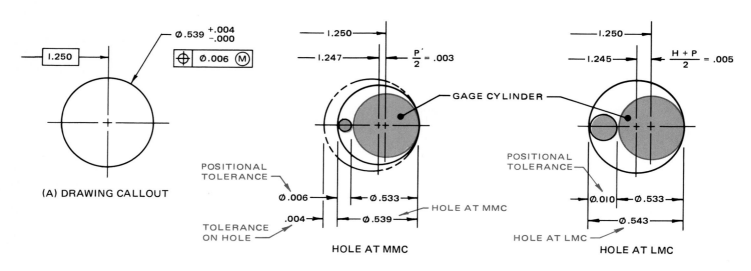

Fig. 14-7-16 Hole with an MMC positional tolerance.

feature at MMC. See Fig. 14-7-19. Where datum feature B is at MMC, its axis determines the location of the pattern of features as a group. Where the datum feature B departs from MMC, its axis may be displaced relative to the location of the datum axis (datum B at MMC) by an amount equal to one-half the difference between its actual and MMC size.

Projected Tolerance Zone

The application of this concept is recommended where the variation in perpendicularity of threaded or pressfit holes could cause fasteners, such as screws, studs, or pins, to interfere with mating parts, as shown in Fig. 14-7-20. Interference can occur where a positional tolerance is applied to the depth of a hole

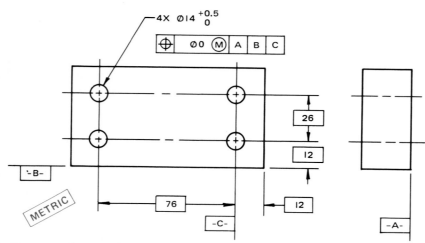

Fig. 14-7-17 Zero positional tolerance—MMC.

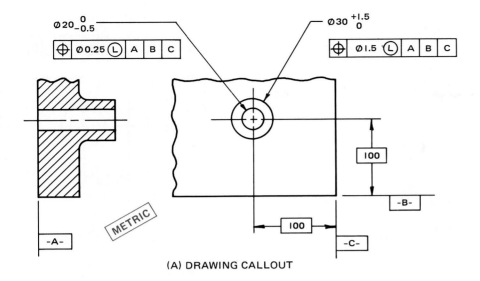

(A) DRAWING CALLOUT

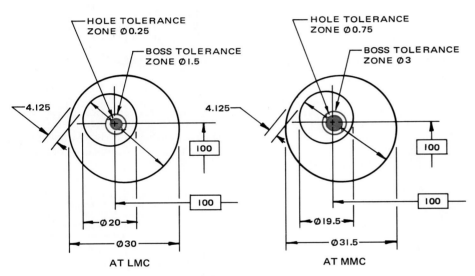

(B) INTERPRETATION

Fig. 14-7-18 LMC applied to a boss and hole.

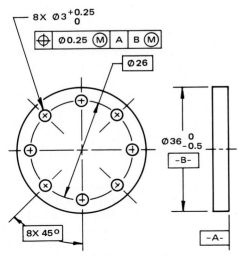

Fig. 14-7-19 Positional tolerancing a group of holes (datum feature at MMC).

and the hole axis is inclined within allowable limits. Figure 14-7-21 illustrates the application of a positional tolerance using a projected tolerance zone. The specified length for the projected tolerance zone is the minimum value and represents the maximum permissible mating part thickness or installed length or height of components such as studs or dowel pins.

For through holes, or in some situations, the direction of the projection from the datum surface may need further clarification. In such cases, the projected tolerance zone may be indicated as illustrated in Fig. 14-7-22. The minimum extent and direction of the projected tolerance zone is shown on the drawing as a dimensioned value and a heavy chain line drawn closely adjacent to an extension of the center line of the hole.

Where design considerations require a closer control in the perpendicularity of a threaded hole than that allowed by the positional tolerance, a perpendicularity tolerance applied as a projected tolerance zone may be specified.

Reference and Source Material

1. ANSI Y14.5M Dimensioning and Tolerancing

ASSIGNMENTS

See Assignments 23 through 28 for Unit 14-7 starting on page 346.

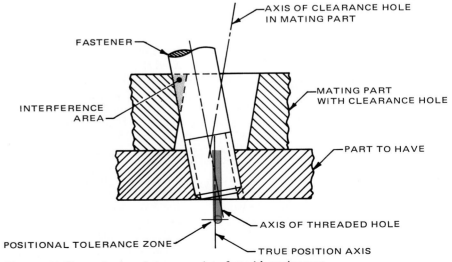

Fig. 14-7-20 Illustrating how fastener can interfere with mating part.

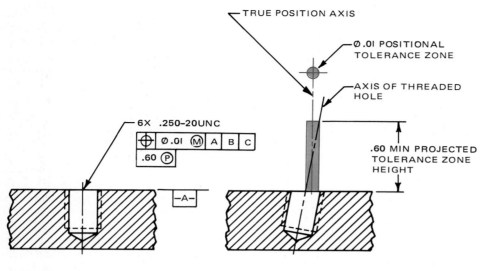

(A) DRAWING CALLOUT   (B) INTERPRETATION

Fig. 14-7-21 Specifying a projected tolerance zone.

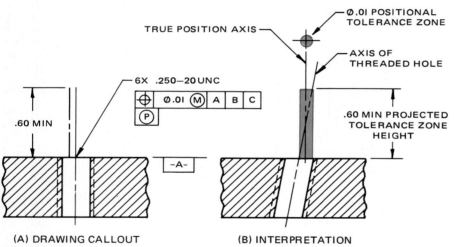

(A) DRAWING CALLOUT   (B) INTERPRETATION

Fig. 14-7-22 Projected tolerance zone indicated with a chain line.

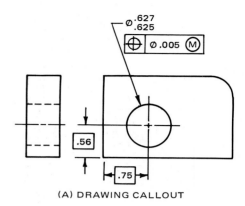

(A) DRAWING CALLOUT

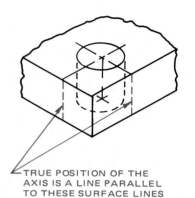

TRUE POSITION OF THE AXIS IS A LINE PARALLEL TO THESE SURFACE LINES

(B) INTERPRETATION OF TRUE POSITION

Fig. 14-8-1 Lines from which measurements are made.

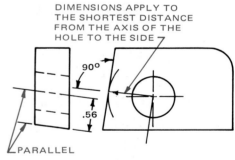

DIMENSIONS APPLY TO THE SHORTEST DISTANCE FROM THE AXIS OF THE HOLE TO THE SIDE

Fig. 14-8-2 Results when sides are off-square.

UNIT 14-8
Datums for Positional Tolerancing

In the examples given so far, the position of the axis of the hole was established by using dimensions from the actual surfaces of the part. These surfaces were not designated as datums. The true or mean position of the axis was therefore a line parallel to a surface line element on each of the surfaces from which the dimensions were drawn, as illustrated in Fig. 14-8-1.

If these sides are off-square with one another or with other surfaces of the part, the true position of the axis would be similarly off-square, as shown in a somewhat exaggerated format in Fig. 14-8-2.

In some applications this may be the desired requirement, but in most cases it is preferable to have the hole either related to other surfaces or features or related to a full side rather than a line on the surface. It is then necessary to specify the desired datum feature or features in the required order of priority.

The first consideration in such applications is to decide on the primary datum feature. The usual course of action is to specify as the primary datum the surface into which the hole is produced. This will ensure that the true position of the axis is perpendicular to this surface or at the basic angle, if it is

other than 90°. Secondary and tertiary datum features are then selected and identified, if requried.

Figure 14-8-3 shows a part similar to that shown in Fig. 14-8-1 but with the addition of a primary datum feature and the MMC modifier. Figure 14-8-4 shows the same part with three datum features specified.

Long Holes It is not always essential to have the true position of a hole perpendicular to the face into which the hole is produced. It may be functionally more important, especially with long holes, to have it parallel to one of the sides. Figure 14-8-5 is a case in point. In

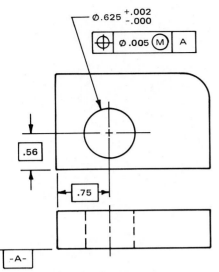

Fig. 14-8-3 Part with one datum feature specified.

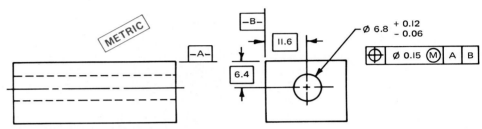

(A) DRAWING CALLOUT   (B) INTERPRETATION OF TRUE POSITION

Fig. 14-8-4 Part with three datum features specified.

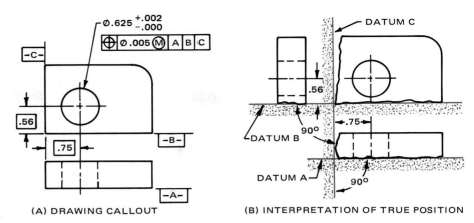

Fig. 14-8-5 Datum system for a long hole.

this example the sides are designated as primary and secondary datums. A tertiary datum is not required.

Circular Datums Circular features, such as holes or external cylindrical features, can be used as datums just as readily as flat surfaces. In the simple part shown in Fig. 14-8-6, it is quite evident that the true position of the small hole is established from the axis of the large hole. In cases like these, it may not be necessary to specify one of the holes as a datum, although it would facilitate gaging if the datum were specified on an MMC basis.

In other cases, such as that shown in Fig. 14-8-7, it is essential to specify the datum; otherwise, the origin of the true-position dimension would be left in doubt. It could be either the axis of the hole or the axis of the outside cylindrical surface.

Multiple-Hole Datums The axis of holes is sometimes specified as a datum feature with MMC being specified for the datum reference. On an MMC basis, any number of holes or similar features which form a group or pattern may be specified as a single datum. All features forming such a datum must be related with a positional tolerance on an MMC basis. See Fig. 14-8-8.

COMPOSITE POSITIONAL TOLERANCING

Where design requirements permit the location of a pattern of features, as a group, to vary within a larger tolerance

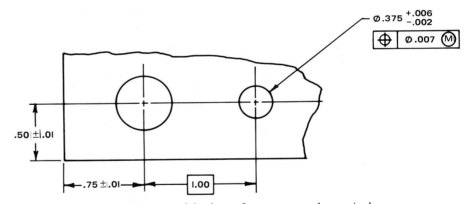

Fig. 14-8-6 Part where specification of the datum feature may not be required.

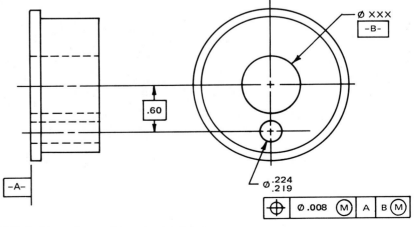

Fig. 14-8-7 Position referenced to a circular datum.

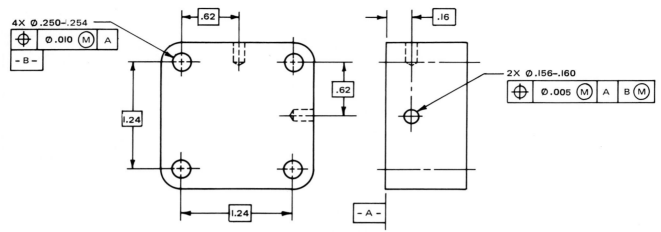

Fig. 14-8-8 Group of holes forming a single datum.

than the positional tolerance assigned to each feature within the pattern, composite positional tolerancing is used.

This provides a composite application for location of feature patterns as well as the interrelation of features within these patterns. Requirements are annotated by the use of a composite feature control frame. Each complete horizontal entry in the feature control frame of Fig. 14-8-9 constitutes a separate requirement. The position symbol is entered once and is applicable to both horizontal entries. The upper entry is referred to as the pattern-locating control. It specifies the larger positional tolerance for the location of the pattern of features as a group. Applicable datums are specified in a desired order of precedence. The lower entry is referred to as the feature-relating control. It specifies the smaller positional tolerance for each feature within the pattern and repeats the primary datum.

Each pattern of features is located from specified datums by basic dimensions. See Figs. 14-8-10 through 14-8-12. As can be seen from the sectional view of the tolerance zones in Fig. 14-8-10, the axes of both the large and small zones are parallel. The axes of the holes may vary obliquely only within the confines of the respective smaller positional tolerance zones. The axes of the holes must lie within the larger tolerance zones and also within the smaller tolerance zones.

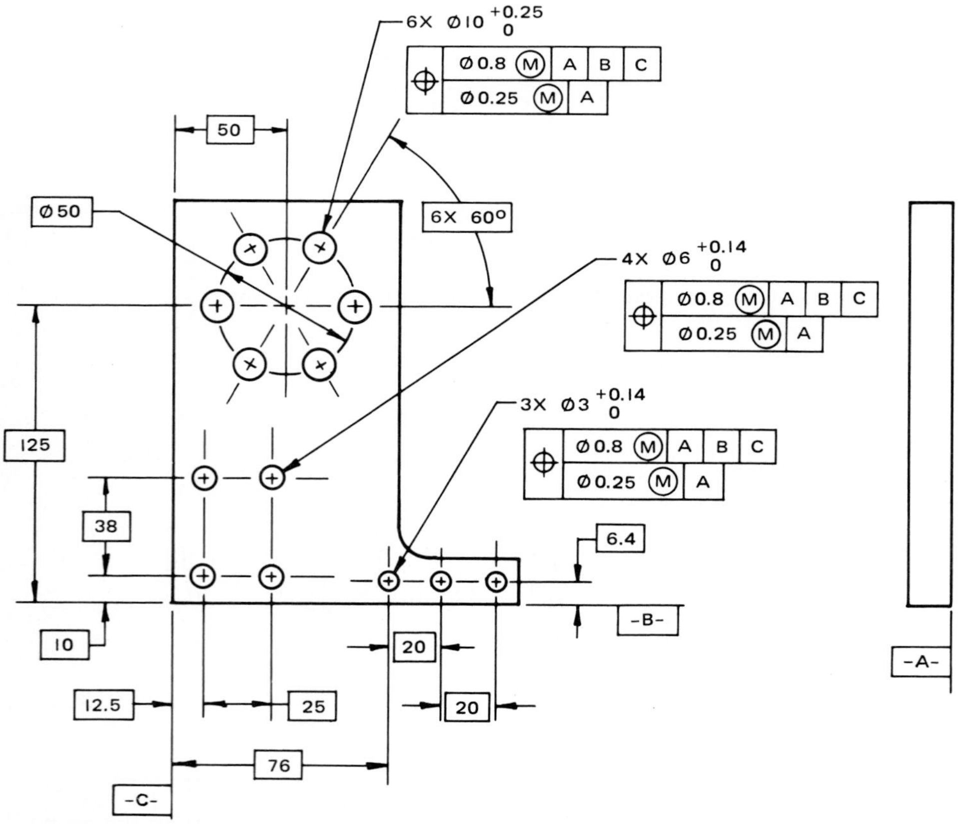

DATUM TARGETS

The full feature surface was used to establish a datum for the features so far designated as datum features. This may not always be practical for the following reasons:

1. The surface of a feature may be so large that a gage designed to make contact with the full surface may be too expensive or too cumbersome to use.
2. Functional requirements of the part may necessitate the use of only a portion of a surface as a datum feature, for example, the portion which contacts a mating part in assembly.
3. A surface selected as a datum feature may not be sufficiently true, and a flat datum feature may rock when placed on a datum plane, so that accurate and repeatable measurements from the surface would not be possible. This is particularly so for surfaces of castings, forgings, weldments, and some sheet-metal and formed parts.

A useful technique to overcome such problems is the datum target method. In this method certain points, lines, or small areas on the surfaces are selected as the bases for establishment of datums.

Fig. 14-8-9 Hole patterns located by composite positional tolerancing.

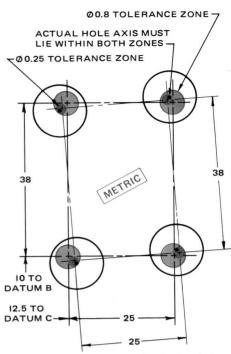

Fig. 14-8-11 Tolerance zones for the four-hole pattern shown in Fig. 14-8-9.

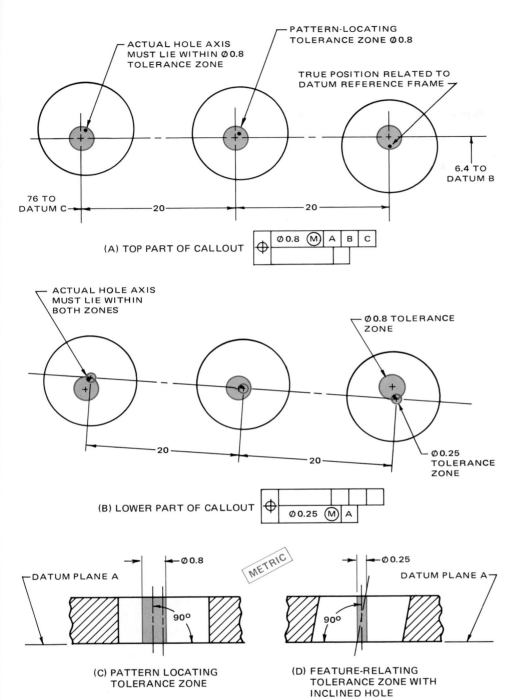

Fig. 14-8-10 Tolerance zones for the three-hole pattern shown in Fig. 14-8-9.

For flat surfaces, this usually requires three target points or areas for a primary datum, two for a secondary datum, and one for a tertiary datum.

It is not necessary to use targets for all datums. It is quite logical, for example, to use targets for the primary datum and other surfaces or features for secondary and tertiary datums if required; or to use a flat surface of a part as the primary datum and to locate fixed points or lines on the edges as secondary and tertiary datums.

Datum targets should be spaced as far apart from each other as possible to provide maximum stability for making measurements.

Datum Target Symbol

Points, lines, and areas on datum features are designated on the drawing by means of a datum target symbol. See Fig. 14-8-13. The symbol is placed outside the part outline with a radial (leader) line directed to the target point

(indicated by an "X"), target line, or target area, as applicable. See Fig. 14-8-14. The use of a solid radial (leader) line indicates that the datum target is on the near (visible) surface. The use of a dashed radial (leader) line, as in Fig. 14-8-19B, indicates that the datum target is on the far (hidden) surface. The leader, shown without an arrowhead, should not be shown in either a horizontal or vertical position. The datum feature itself is identified in the usual manner with a datum feature symbol.

The datum target symbol is a circle having a diameter approximately 3.5 times the height of the lettering used on the drawing. The circle is divided horizontally into two halves. The lower half contains a letter identifying the associated datum, followed by the target number assigned sequentially starting with 1 for each datum; for example, in a three-plane, six-point datum system, if the datums are A, B, and C, the datum target would be A_1, A_2, A_3, B_1, B_2, and C_1. See Fig. 14-8-23. Where the datum target is an area, the area size may be entered in the upper half of the symbol; otherwise, the upper half is left blank.

Identification of Targets

Datum Target Points. Each target point is shown on the surface, in its

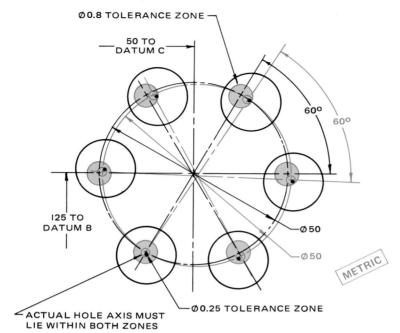

Fig. 14-8-12 Tolerance zones for the six-hole pattern shown in Fig. 14-8-9.

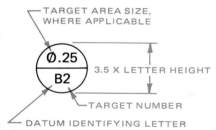

Fig. 14-8-13 Datum target symbols.

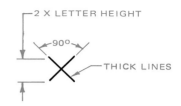

Fig. 14-8-15 Symbol for a datum target point.

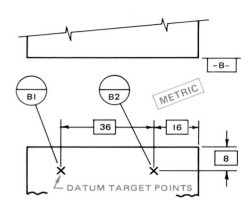

(A) DATUM POINTS SHOWN ON A SURFACE

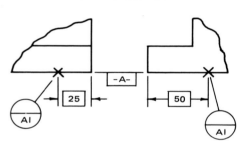

(B) DATUM POINT LOCATED BY TWO VIEWS

Fig. 14-8-16 Datum target point.

desired location, by means of a cross, drawn at approximately 45° to the coordinate dimensions. The cross is twice the height of the lettering used, as shown in Figs. 14-8-15 and 14-8-16A. Where there is no direct view, the point location is dimensioned on two adjacent views. See Fig. 14-8-16B.

Target points may be represented on tools, fixtures, and gages by spherically ended pins, as shown in Fig. 14-8-17.

Datum Target Lines. A datum target line is indicated by the symbol X on an edge view of a surface, a phantom line on the direct view, or both. See Fig. 14-8-18. Where the length of the datum target line must be controlled, its length and location are dimensioned.

Datum Target Areas. Where it is determined that an area or areas of flat contact is necessary to assure establishment of the datum (that is where spherical or pointed pins would be inadequate), a target area of the desired shape is specified. The datum target area is indicated by section lines inside a phantom outline of the desired shape, with controlling dimensions added. The diameter of circular areas is given in the upper half of the datum target symbol. See Fig. 14-8-19A. Where it becomes impractical to delineate a circular target area, the method of indication shown in Fig. 14-8-19B may be used.

Datum target areas may have any desired shape, a few of which are shown in Fig. 14-8-20. Target areas should be kept as small as possible, consistent with functional requirements, to avoid having large, cross-hatched areas on the drawing.

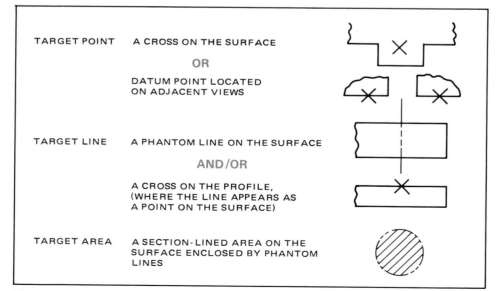

Fig. 14-8-14 Identification of datum targets.

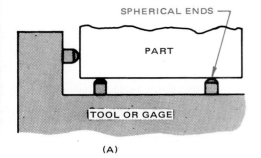

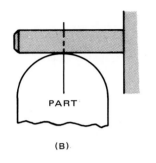

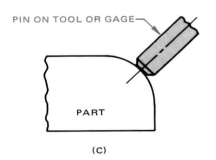

Fig. 14-8-17 Location of part on datum target points.

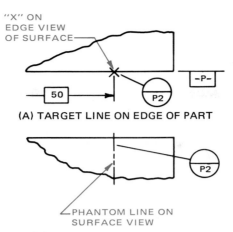

(A) TARGET LINE ON EDGE OF PART

(B) TARGET LINE ON SURFACE
Fig. 14-8-18 Datum target line.

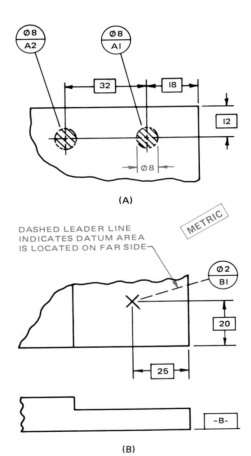

(A)

(B)
Fig. 14-8-19 Datum target areas.

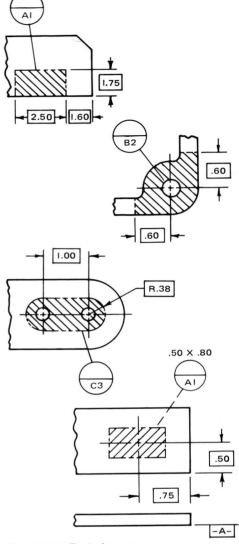

Fig. 14-8-20 Typical target areas.

Targets not in the Same Plane

In most applications datum target points which form a single datum are all located on the same surface, as shown in Fig. 14-8-16A. However, this is not essential. They may be located on different surfaces, to meet functional requirements, as shown, for example, in Fig. 14-8-21. In some cases the datum plane may be located in space, that is, not actually touching the part, as shown in Fig. 14-8-22. In such applications the controlled features must be dimensioned from the specified datum, and the position of the datum from the datum targets must be shown by means of exact datum dimensions. For example, in Fig. 14-8-22 datum *B* is positioned by means of datum dimensions .38, .50, and 1.06. The top surface is controlled from this datum by means of a toleranced dimension, and the hole is positioned by means

of the basic dimension $\boxed{1.00}$ and a positional tolerance.

Dimensioning for Target Location

The location of datum targets is shown by means of basic dimensions. Each dimension is shown, without tolerances, enclosed in a rectangular frame, indicat-

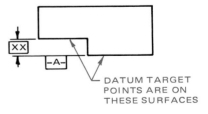

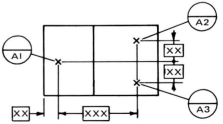

Fig. 14-8-21 Datum target points on different planes.

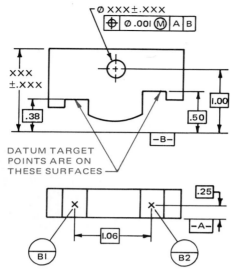

Fig. 14-8-22 Datum outside of part profile.

ing that the general tolerance does not apply. Dimensions locating a set of datum targets should be dimensionally related or have a common origin.

Application of datum targets and datum dimensioning is shown in Fig. 14-8-23.

Reference and Source Material

1. ANSI Y14.5M Dimensioning and Tolerancing.

ASSIGNMENTS

See Assignments 29 and 30 for Unit 14-8 on page 347.

UNIT 14-9
Circularity (Roundness) and Cylindricity

CIRCULARITY

Circularity refers to a condition of a circular line or the surface of a circular feature where all points on the line, or on the periphery of a plane cross section of the feature, are equidistant from a common axis or center point.

Examples of circular features would include disks, spheres, cylinders, and cones. The measurement plane for a sphere is any plane which passes through a section of maximum diameter. For a cylinder, cone, or other non-spherical feature, the measurement

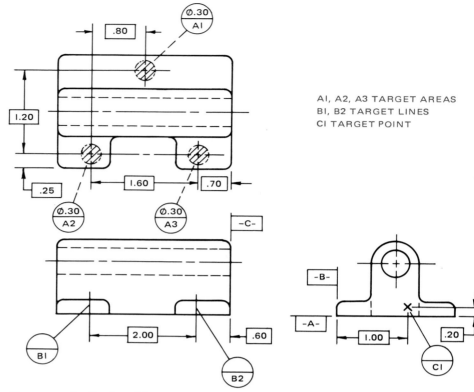

Fig. 14-8-23 Application of datum targets and dimensioning.

plane is any plane perpendicular to the axis or center line.

Errors of Circularity

Errors of circularity (out-of-roundness) of a circular line or the periphery of a cross section of a circular feature may occur as *ovality*, where differences appear between the major and minor axes; as *lobing*, where in some instances the diametral values may be constant or nearly so; or as *random irregularities* from a true circle. All these errors are illustrated in Fig. 14-9-1. The geometric characteristic symbol for circularity is a circle, having a diameter equal to 1.5 times the height of letters on the drawing, as shown in Fig. 14-9-2.

Circularity Tolerance

A circularity tolerance may be specified by using this symbol in the feature control frame. It is expressed on an RFS basis.

A circularity tolerance specifies the width of an annular tolerance zone, bounded by two concentric circles in the same plane, within which the circular line or the periphery of the feature in that plane shall lie, as shown in Fig.

AI, A2, A3 TARGET AREAS
BI, B2 TARGET LINES
CI TARGET POINT

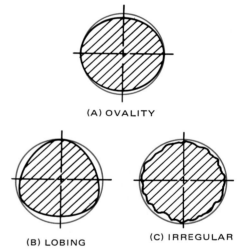

(A) OVALITY

(B) LOBING   (C) IRREGULAR

Fig. 14-9-1 Common types of circularity errors.

1.5 X LETTER HEIGHT

Fig. 14-9-2 Circularity symbol.

14-9-3. A circularity tolerance cannot be modified on an MMC basis since it controls surface elements only. The circularity tolerance must be less than the size tolerance.

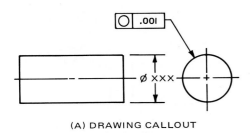

(A) DRAWING CALLOUT

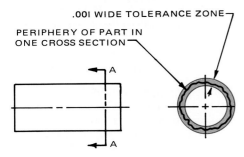

(B) TOLERANCE ZONE

Fig. 14-9-3 Circularity tolerance.

Circularity of Noncylindrical Parts

Noncylindrical parts refer to conical parts and other features which are circular in cross section but which have variable diameters, such as those shown in Fig. 14-9-4. Since many sizes of circles may be involved, it is usually best to direct the circularity tolerance to the longitudinal surfaces as shown.

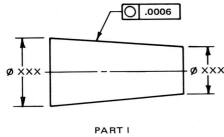

PART 1

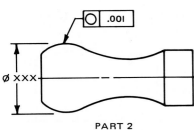

PART 2

Fig. 14-9-4 Circularity tolerance on noncylindrical parts.

Cylindricity

Cylindricity is the condition of the surface which forms a cylinder where the surface elements in cross sections parallel to the axis are straight and parallel

and in cross sections perpendicular to the axis are circular. The cylindricity tolerance is a composite control of form which includes circularity, straightness, and parallelism of the surface elements.

The geometric characteristic symbol for cylindricity consists of a circle with two tangent lines at 60°, as shown in Fig. 14-9-5.

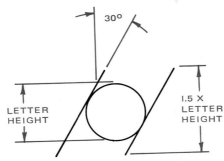

Fig. 14-9-5 Cylindricity symbol.

A cylindricity tolerance specifies a tolerance zone bounded by two concentric cylinders within which the surface must lie. In the case of cylindricity, unlike that of circularity, the tolerance applies simultaneously to both circular and longitudinal elements of the surface. See Fig. 14-9-6. The leader from the feature control symbol may be directed to either view. The cylindricity tolerance must be less than the size tolerance.

Cylindricity tolerances can be applied only to cylindrical surfaces, such as round holes and shafts. No specific geometric tolerances have been devised for other circular forms, which require the use of several geometric tolerances. A conical surface, for example, must be controlled by a combination of tolerances for circularity, straightness, and angularity.

Errors of cylindricity may be caused by out-of-roundness, like ovality or lobing, by errors of straightness caused by bending or by diametral variation, by errors of parallelism like conicity or taper, and by random irregularities from a true cylindrical form.

Since cylindricity is a form tolerance controlling surface elements only, it cannot be modified on an MMC basis.

Reference and Source Material

1. ANSI Y14.5M Dimensioning and Tolerancing.

ASSIGNMENTS

See Assignments 31 through 34 for Unit 14-9 on page 348.

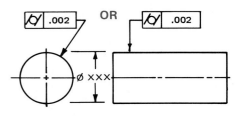

(A) DRAWING CALLOUT

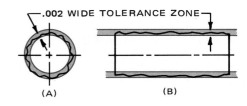

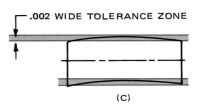

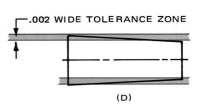

(B) PERMISSIBLE FORM ERRORS

Fig. 14-9-6 Cylindricity tolerance directed to either view.

U N I T   1 4 - 1 0
Profile Tolerancing

PROFILES

A *profile* is the outline form or shape of a line or surface. A *line profile* may be the outline of a part or feature as depicted in a view on a drawing. It may represent the edge of a part, or it may refer to line elements of a surface in a single direction, such as the outline of cross sections through the part. In contrast, a *surface profile* outlines the form or shape of a complete surface in three dimensions.

The elements of a line profile may be straight lines, arcs, or other curved lines. The elements of a surface profile may be flat surfaces, spherical surfaces, cylindrical surfaces, or surfaces composed of various line profiles in two or more directions.

A profile tolerance may be applied to a single, independent line or surface, or any part of a complex line or surface. It may be applied to a feature of size, in which case MMC is specified. It may be related to a datum, in which case it also controls orientation and, in some cases, position of the line or surface.

PROFILE SYMBOLS

There are two geometric characteristic symbols for profiles, one for lines and one for surfaces. Separate symbols are required, because it is often necessary to distinguish between line elements of a surface and the complete surface itself. The symbol for profile of a line consists of a semicircle with a diameter equal to twice the lettering size used on the drawing. The symbol for profile of a surface is identical except that the semicircle is closed by a straight line at the bottom, as shown in Fig. 14-10-1. All other geometric tolerances of form and orientation are merely special cases of profile tolerancing.

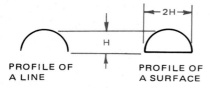

H = HEIGHT OF LETTERS

Fig. 14-10-1 Profile symbol.

Profile tolerances are used to control the position of lines and surfaces which are neither flat nor cylindrical.

PROFILE OF A LINE

A profile-of-a-line tolerance may be directed to a line of any length or shape. With profile-of-a-line tolerance, datums may be used in some circumstances but would not be used when the only requirement is the profile shape taken cross section by cross section.

A profile-of-a-line tolerance is specified in the usual manner, by including the symbol and tolerance in a feature control frame directed to the line to be controlled, as shown in Fig. 14-10-2.

If the line on the drawing to which the tolerance is directed represents a surface, the tolerance applies to all line ele-

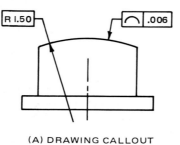

(A) DRAWING CALLOUT

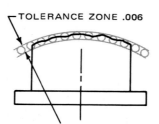

(B) BILATERAL TOLERANCE ZONE

Fig. 14-10-2 Simple profile with a bilateral profile tolerance zone.

ments of the surface parallel to the plane of the view on the drawing, unless otherwise specified.

The tolerance indicates a tolerance zone consisting of the area between two parallel lines, separated by the specified tolerance, which are themselves parallel to the basic form of the line in a plane parallel to the view on the drawing.

Bilateral and Unilateral Tolerances

The profile tolerance zone, unless otherwise specified, is equally disposed about the basic profile in a form known as a *bilateral tolerance zone*. The width of this zone is always measured perpendicular to the profile surface. The tolerance zone may be considered to be bounded by two lines enveloping a series of circles, each having a diameter equal to the specified profile tolerance, with their centers on the theoretical, basic profile, as shown in Fig. 14-10-2.

Occasionally it is desirable to have the tolerance zone wholly on one side of the basic profile instead of equally divided on both sides. Such zones are called *unilateral tolerance zones*. They are specified by showing a phantom line drawn parallel and close to the profile surface. The tolerance is directed to this line, as shown in Fig. 14-10-3. The zone line need extend only a sufficient distance to make its application clear.

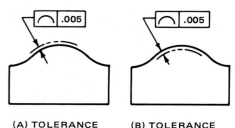

(A) TOLERANCE ZONE ON OUTSIDE OF TRUE PROFILE

(B) TOLERANCE ZONE ON INSIDE OF TRUE PROFILE

Fig. 14-10-3 Unilateral tolerance zones.

Method of Dimensioning

The true profile is established by means of basic dimensions, each of which is enclosed in a rectangular frame to indicate that the tolerance in the general tolerance note does not apply.

When the profile tolerance is not intended to control the position of the profile, there must be a clear distinction between dimensions which control the position of the profile and those which control the form or shape of the profile.

Any convenient method of dimensioning may be used to establish the basic profile. Examples are chain or common-point dimensions, dimensioning to points on a surface or to the intersection of lines, dimensioning located on tangent radii, and angles.

To illustrate, the simple part in Fig. 14-10-4 shows a dimension of .90 ± .01 controlling the height of the profile. This dimension must be separately measured. The radius of 1.500 in. is a basic dimension, and it becomes part of the profile. Therefore the profile tolerance zone has radii of 1.497 and 1.503, but is free to flat in any direction within the limits of the positional tolerance zone in order to enclose the curved profile.

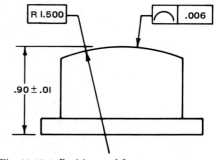

Fig. 14-10-4 Position and form as separate requirements.

If the radius were shown as a toleranced dimension, without the rectangular frame, as in Fig. 14-10-5, it would become a separate measurement.

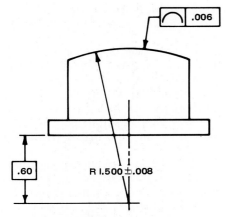

Fig. 14-10-5 Position and radius separate from form.

Figure 14-10-6 shows a more complex profile, where the profile is located by a single toleranced dimension. There are, however, five basic dimensions defining the true profile.

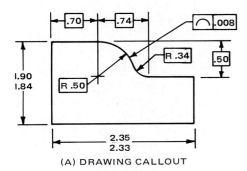

(A) DRAWING CALLOUT

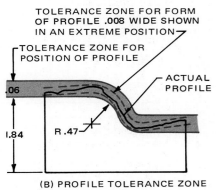

(B) PROFILE TOLERANCE ZONE

Fig. 14-10-6 Profile defined by basic dimensions.

In this case, the tolerance on the height indicates a tolerance zone .06 in. wide extending the full length of the profile. This is because the profile is established by basic dimensions. No other dimension exists to effect the orientation or height. The profile tolerance specifies a .008 in. wide tolerance zone, which may lie anywhere within the .06 in. tolerance zone.

Extent of Controlled Profile

The profile is generally intended to extend to the first abrupt change or sharp corner. For example, in Fig. 14-10-6 it extends from the upper left- to the upper right-hand corners, unless otherwise specified. If the extent of the profile is not clearly identified by sharp corners or by basic profile dimensions, it must be indicated by a note under the feature control symbol, such as FROM A TO B or BETWEEN A & B as shown in Fig. 14-10-7.

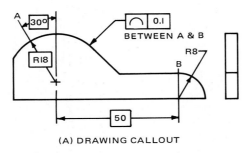

(A) DRAWING CALLOUT

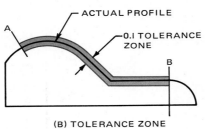

(B) TOLERANCE ZONE

Fig. 14-10-7 Specifying extent of profile.

If the controlled profile includes a sharp corner, the corner represents a discontinuity of the tolerance boundary, and the boundary is considered to extend to the intersection of the boundary lines, as shown in Fig. 14-10-8.

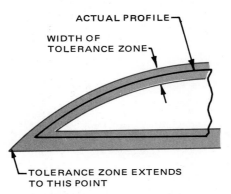

Fig. 14-10-8 Tolerance zone at a sharp corner.

If different profile tolerances are required on different segments of a surface, the extent of each profile tolerance

is indicated by the use of reference letters to identify the extremities. See Fig. 14-10-9.

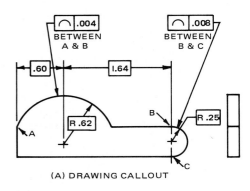

(A) DRAWING CALLOUT

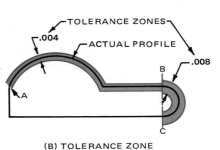

(B) TOLERANCE ZONE

Fig. 14-10-9 Dual tolerance zones.

All-Around Profile Tolerance

Where a profile tolerance applies all around the profile of a part, the symbol used to designate "all around" is placed on the leader from the feature control frame. See Figs. 14-10-10 and 14-10-11.

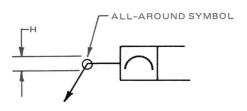

H = LETTER HEIGHT

Fig. 14-10-10 All-around symbol.

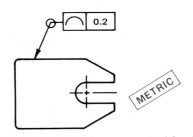

Fig. 14-10-11 Profile tolerance required for all around the surface.

PROFILE OF A SURFACE

As already stated, a profile-of-a-line tolerance, when directed to a line on a drawing which represents a surface, applies to the profile of all cross sections parallel to the view on the drawing, unless otherwise specified. This is illustrated by the cross sections *AA* and *BB* in Fig. 14-10-12.

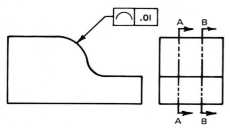

Fig. 14-10-12 Profile of a line tolerance applies to profile of cross section.

If the profile in a plane normal to the plane of the drawing requires tolerancing, a separate tolerance may be added to the side view. Frequently this profile will be straight, in which case a straightness tolerance may be substituted, as shown in Fig. 14-10-13.

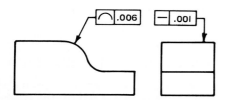

Fig. 14-10-13 Profile and straightness tolerance to same surface.

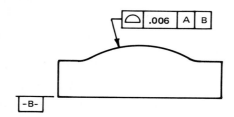

(A) DRAWING CALLOUT

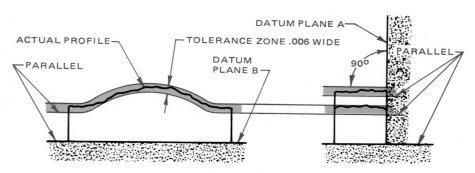

(B) TOLERANCE ZONE

Fig. 14-10-15 Profile referenced to datums.

If the same tolerance is intended to apply over the whole surface, instead of to lines or line elements in specific directions, the profile of a surface symbol is used, as shown in Fig. 14-10-14. While the profile tolerance may be directed to the surface in either view, it is usually directed to the view showing the more characteristic profile.

The profile-of-a-surface tolerance indicates a tolerance zone having the same form as the basic surface, with a

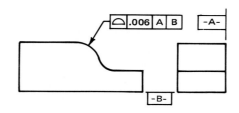

(A) DRAWING CALLOUT

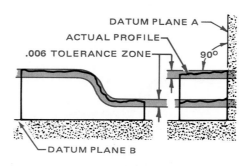

(B) TOLERANCE ZONE

Fig. 14-10-14 Profile of a surface.

uniform width equal to the specified tolerance within which the entire surface must lie.

In most cases, a profile of a surface tolerance requires reference to datums in order to provide proper orientation of the profile. This is specified simply by indicating suitable datums. Figure 14-10-15 shows a simple part where the base and one side are designated as datums.

The criterion which distinguishes a profile tolerance as applying to position or to orientation is whether the profile is related to the datum by a basic dimension or by a toleranced dimension. This is illustrated by Fig. 14-10-16.

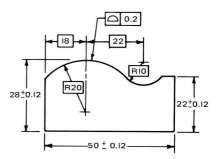

(A) PROFILE TOLERANCE CONTROLS FORM OF PROFILE ONLY

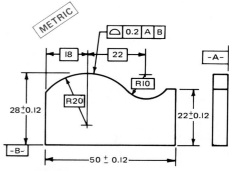

(B) PROFILE TOLERANCE CONTROLS FORM AND ORIENTATION OF PROFILE

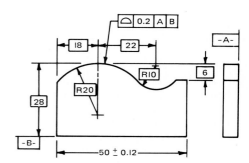

(C) PROFILE TOLERANCE CONTROLS FORM, ORIENTATION, AND POSITION OF PROFILE

Fig. 14-10-16 Comparison of profile tolerances.

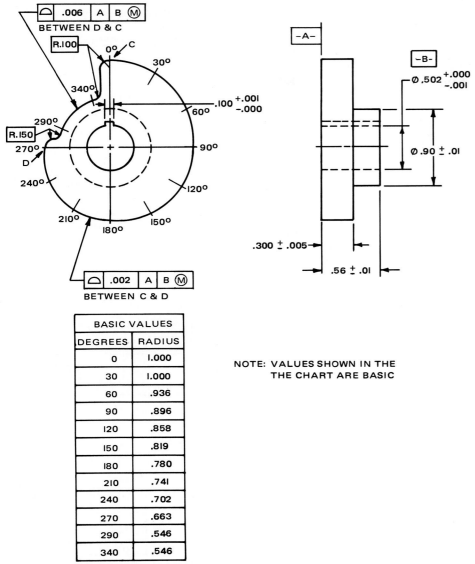

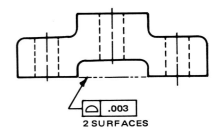

NOTE: VALUES SHOWN IN THE
THE CHART ARE BASIC

BASIC VALUES	
DEGREES	RADIUS
0	1.000
30	1.000
60	.936
90	.896
120	.858
150	.819
180	.780
210	.741
240	.702
270	.663
290	.546
340	.546

Fig. 14-10-17 Profile tolerances control form and size of cam profile.

Profile tolerances controlling position are very useful for parts which can be revolved around a cylindrical datum feature. The position or size as well as the form can easily be assessed by revolving the part on the datum axis in conjunction with a dividing head, with an indicator gage to make direct measurements on the periphery. Figure 14-10-17 shows an example.

Reference and Source Material

1. ANSI Y14.5M Dimensioning and Tolerancing.

ASSIGNMENTS

See Assignments 35 through 38 for Unit 14-10 starting on page 348.

UNIT 14-11
Correlative Tolerances

Correlative geometric tolerancing refers to tolerancing for the control of two or more features intended to be correlated in position or attitude. Examples of such correlated tolerancing include coplanarity, for control of two or more flat surfaces; positional tolerance at MMC for symmetrical relationships, such as control of features equally disposed about a center line; concentricity and coaxiality, for control of features having common axes or center lines; and runout, for control of surfaces related to an axis. These are all tolerances of location for which

special symbols have been provided for some of them to clarify and simplify drawing callout requirements.

When position is to be separately controlled, other form or orientation tolerances may be applied to control the correlation of features.

COPLANARITY

Coplanarity refers to the relative position of two or more flat surfaces, which are intended to lie in the same geometric plane. A profile of a surface tolerance may be used where it is desirable to treat two or more surfaces as a single interrupted or noncontinuous surface. See Fig. 14-11-1. Each surface must lie between two parallel planes .003 in. apart. Additionally, both surfaces must be within the specified limits of size. No datum reference is stated in Fig. 14-11-1, as in the case of flatness. Since the orientation of the tolerance zone is established from contact of the part against a reference standard, the datum is established by the surfaces themselves.

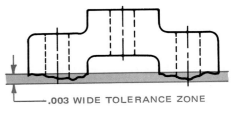

Fig. 14-11-1 Specifying profile of a surface for coplanar surfaces.

Where more than two surfaces are involved, it may be desirable to identify which specific surfaces are to be used in contacting the reference standard to establish the tolerance zone. In this case a datum-identifying symbol is applied and the appropriate surface and the datum reference letter added to the feature control frame. See Fig. 14-11-2.

Figure 14-11-3 shows a case where the coplanar surfaces are required to be perpendicular to the axis of a hole.

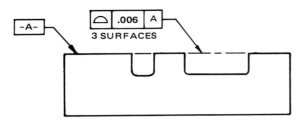

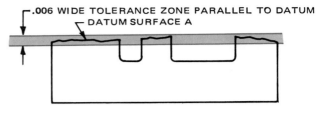

(A) DRAWING CALLOUT

(B) TOLERANCE ZONE

Fig. 14-11-2 Coplanar surfaces with one surface designated as a datum.

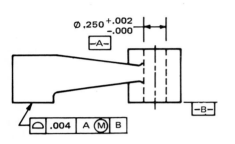

(A) DRAWING CALLOUT

Fig. 14-11-3 Surface referenced to a datum system.

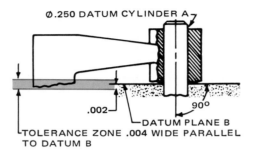

(B) TOLERANCE ZONE

TOLERANCING SYMMETRICALLY LOCATED FEATURES

Where it is required that a feature be located symmetrically with respect to the center plane of a datum feature, positional tolerancing is used. A symmetrical relationship may be controlled by specifying a positional tolerance at MMC as illustrated in Figs. 14-11-4 and 14-11-5. The datum feature may be specified either on an MMC or an RFS basis, depending upon the design requirements.

Where it is necessary to control the symmetry of related features within their limits of size, a zero positional tolerance at MMC is specified and the datum feature is normally specified on an MMC basis. Boundaries of perfect form are thereby established which are truly symmetrical when both features are at MMC. Variations in symmetry are permitted only where the features depart from their MMC size toward LMC.

Some designs may require a control of the symmetrical relationship between features regardless of their actual sizes. In such cases, both the specified positional tolerance and the datum reference apply on an RFS basis. See Fig. 14-11-6. The center plane of the slot must lie between two parallel planes, 0.8 mm apart regardless of the sizes of both datum B and the feature, which are

equally disposed about the center plane of datum B.

CONCENTRICITY

Concentricity is a condition in which two or more features, such as circles, spheres, cylinders, cones, or hexagons, share a common center or axis. An example would be a round hole through the center of a cylindrical part.

A concentricity tolerance is a particular case of a positional tolerance. It controls the permissible variation in position, or eccentricity, of the center line of the controlled feature in relation to the axis of the datum feature.

The tolerance zone for circles is a circle; for other circular features it is a cylinder, concentric with the datum axis, having a diameter equal to the specified tolerance. The center of all cross sections normal to the axis of the controlled feature must lie within this tolerance zone.

The geometric characteristic symbol used for concentricity consists of two concentric circles, having diameters equal to the actual height (1:1) and 1.5 times the height of lettering used on the drawing. See Fig. 14-11-7.

Concentricity—RFS

Concentricity tolerance and the datum reference, because of their unique char-

acteristics, are always used on an RFS basis.

Figure 14-11-8 shows a common type of part where the outer diameter is required to be concentric with the center hole, which is designated as a datum feature.

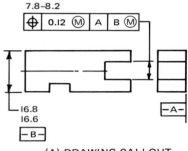

(A) DRAWING CALLOUT

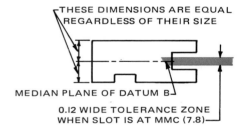

(B) TOLERANCE ZONE

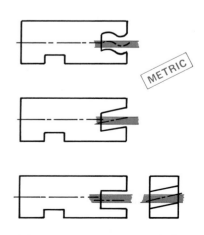

(C) PERMISSIBLE VARIATIONS

Fig. 14-11-4 Positional tolerancing at MMC for symmetry.

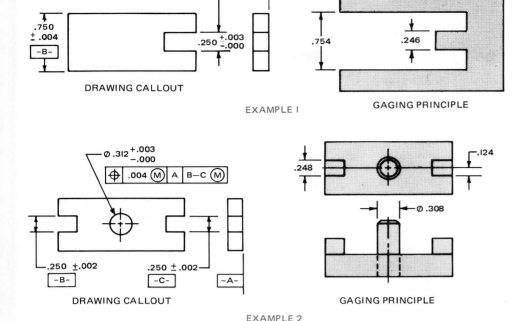

EXAMPLE I

GAGING PRINCIPLE

Fig. 14-11-5 Examples in tolerancing symmetrically-located features.

EXAMPLE 2

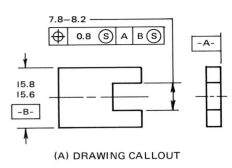

(A) DRAWING CALLOUT

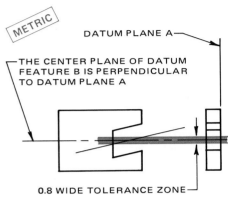

(B) INTERPRETATION

Fig. 14-11-6 Positional tolerancing RFS for symmetry.

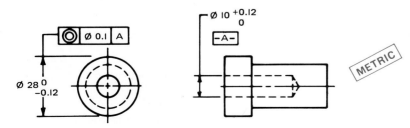

H = LETTER HEIGHT

Fig. 14-11-7 Concentricity control.

Figure 14-11-9 shows an example of a part where two cylindrical portions are intended to be coaxial. This figure also illustrates the extreme errors of eccentricity and parallelism that the concentricity tolerance would permit. The absence of a modifying symbol after the tolerance indicates that RFS applies.

A concentricity tolerance may be referenced to a datum system, instead of to a single datum, to meet certain functional requirements. Figure 14-11-10 gives an example in which the tolerance zone is perpendicular to datum *A* and also concentric with the axis of datum *B* in the plane of datum *A*.

RUNOUT

Runout is a composite tolerance used to control the functional relationship of one or more features of a part to a datum axis. The types of features controlled by runout tolerances include those surfaces constructed around a datum axis and those constructed at right angles to a datum axis. See Fig. 14-11-11.

Each feature must be within its runout tolerance when rotated about the datum axis. The tolerance specified for a controlled surface is the total tolerance or full indicator movement (FIM) in inspection and international terminology.

There are two types of runout control, circular runout and total runout. The type used is dependent upon design requirements and manufacturing considerations. The geometric characteristic symbols for runout are shown in Fig. 14-11-12.

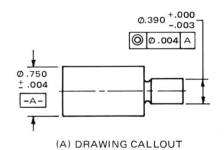

(A) DRAWING CALLOUT

(B) EXTREME ECCENTRICITY

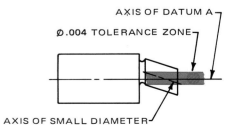

(C) EXTREME ANGULAR VARIATION

Fig. 14-11-9 Concentricity of cylindrical features.

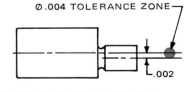

Fig. 14-11-8 Cylindrical part with concentricity tolerance.

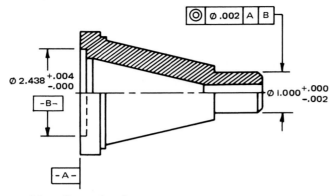

Fig. 14-11-10 Concentricity referenced to datum system.

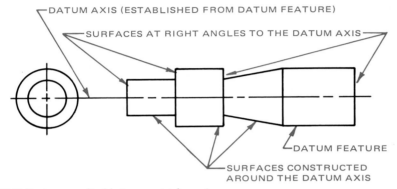

Fig. 14-11-11 Features applicable to runout tolerancing.

Circular Runout

Circular runout provides control of circular elements of a surface. The tolerance is applied independently at any usual measuring position as the part is rotated 360°. See Fig. 14-11-13. Where applied to surfaces constructed around a datum axis, circular runout controls variations such as circularity and coaxiality. Where applied to surfaces constructed at right angles to the datum axis, circular runout controls wobble at all diametral positions.

Where a runout tolerance applies to a specific portion of a surface, a thick chain line is drawn adjacent to the surface profile to show the desired length. Basic dimensions are used to define the extent of the portion so indicated. See Fig. 14-11-13.

Total Runout

Total runout concerns the runout of a complete surface, not merely the runout of each circular element. For measurement purposes the checking indicator must traverse the full length or extent of the surface while the part is revolved about its datum axis. Measurements are made

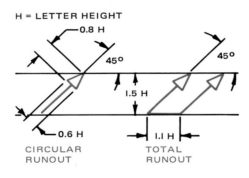

Fig. 14-11-12 Runout symbols.

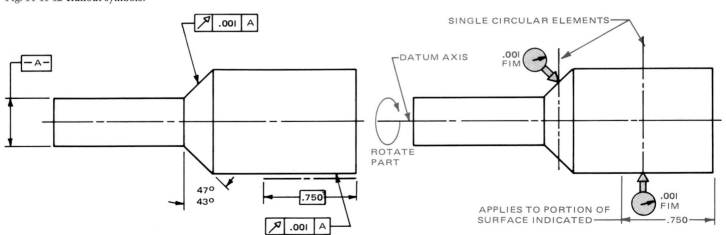

(A) DRAWING CALLOUT

(B) METHOD OF MEASURING

Fig. 14-11-13 Specifying circular runout relative to a datum diameter.

over the whole surface without resetting the indicator. Total runout is the difference between the lowest indicator reading in any position and the highest reading in that or in any other position on the same surface. Thus in Fig. 14-11-14 the tolerance zone is the space between two concentric cylinders separated by the specified tolerance and coaxial with the datum axis.

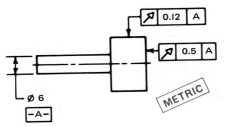

Fig. 14-11-15 Cylindrical datum feature.

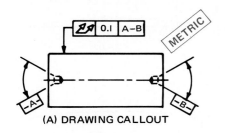

(A) DRAWING CALLOUT

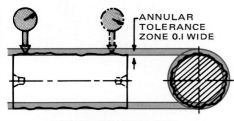

(B) TOLERANCE ZONE

Fig. 14-11-14 Tolerance zone for total runout.

Establishing Datums

In many examples the datum axis has been established from centers drilled in the two ends of the part, in which case the part is mounted between centers for measurement purposes. This is an ideal method of mounting and revolving the part. When centers are not provided, any cylindrical or conical surface may be used to establish the datum axis if it is chosen on the basis of the functional requirements of the part.

Figure 14-11-15 shows a simple, external cylindrical feature specified as the datum feature.

Figure 14-11-16 illustrates the application of runout tolerances where two datum diameters act as a single datum axis to which the features are related.

Reference and Source Material

1. ANSI Y14.5M Dimensioning and Tolerancing

See Assignments 39 through 43 for Unit 14-11 on page 350.

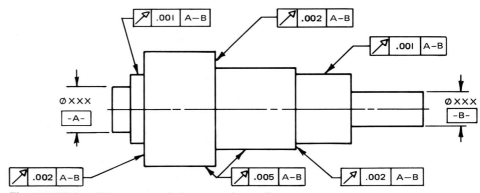

Fig. 14-11-16 Specifying runout relative to two datum diameters.

ASSIGNMENTS FOR CHAPTER 14

Assignments for Unit 14-1, Modern Engineering Tolerancing

1. Parts may deviate from true form and still be acceptable provided the measurements lie within the limits of size. Show by means of a sketch with dimensions two acceptable form variations for each part shown in Fig. 14-1-A.

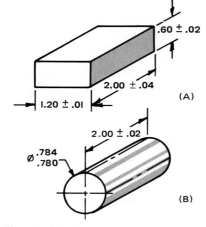

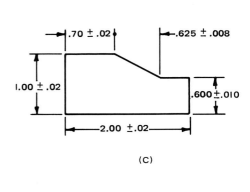

Fig. 14-1-A Assignments.

2. Prepare sketches from the drawings shown in Fig. 14-1-B or 14-1-C and the following information:

(a) Using illustration (A) make a tolerance block diagram similar to Fig. 14-1-5. Show the deviations and limits of size.

(b) Draw illustration (B) and shade in and dimension the tolerance zone.

(c) The exaggeration of sizes is used when it improves the clarity of the drawing. Draw illustration (C) and exaggerate the sizes which would improve the readability of the drawing. Dimension the exaggerated features.

(d) With reference to illustration (D), is the part acceptable? State your reason.

(e) With reference to the drawing callout shown in illustration (E), what parts would pass inspection?

(f) In the drawing callout in illustration (F), what parts in illustration (E) would pass inspection?

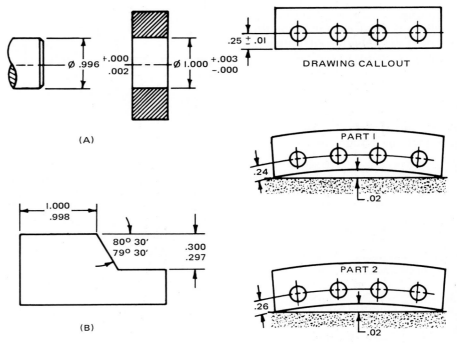

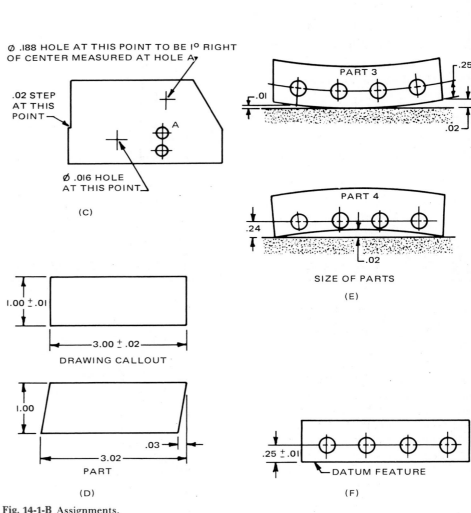

Fig. 14-1-B Assignments.

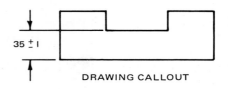

DRAWING CALLOUT

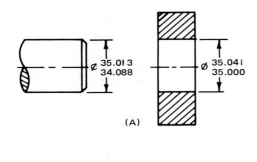

(A)

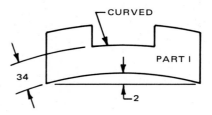

PART 1

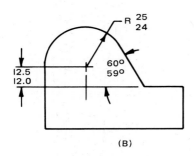

(B)

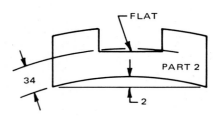

PART 2

0.5 SAWCUTS
10 mm WIDE
CENTER SECTION

4X Ø1

189°

HORIZONTAL

SLOPED

STARTING OF ANGLE

NOTE: ALL VERTICAL LINES PERPENDICULAR
TO HORIZONTAL BASE LINE

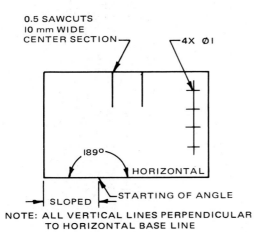

(C)

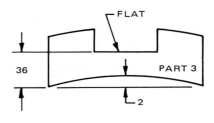

PART 3

20 ± 0.5

50 ± 0.5

DRAWING CALLOUT

METRIC

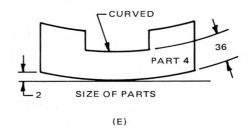

PART 4

SIZE OF PARTS

(E)

0.5

0.5

20.5

19.5

50
PART
(D)

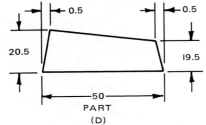

PART 5

DATUM FEATURE

(F)

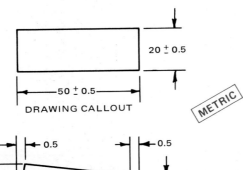

Fig. 14-1-C Assignments.

Assignment for Unit 14-2, Geometric Tolerancing

3. With reference to Fig. 14-2-A and the information given below, add the feature control frames to the following parts:

Part 1. Surface A to have a straightness tolerance of .004 in.

Part 2. Surface M to have a straightness tolerance of .006 in.
Surface N to have a straightness tolerance of .008 in.

Part 3. Surface R to be straight within 006 in. for direction A and straight within .002 in. for direction B.

Part 4. With straightness specified as shown, what is the maximum permissible deviation from straightness of the line elements if the radius is (a) .496 in., (b) .501 in., (c) .504 in.?

Part 5. Eliminate the top view and place the feature control frames on the front and side views.

Assignments for Unit 14-3, Relationship to Feature of Size

4. With reference to Fig. 14-3-A, what is the maximum deviation permitted from straightness for the surface of the diameter if the shaft was (a) at MMC, (b) at LMC, (c) Ø.621?

5. With reference to Fig. 14-3-B, calculate the MMC, LMC, and extreme virtual condition of the hole and shaft. Refer to Fig. 14-3-1.

6. With reference to Fig. 14-3-B, if the hole was straight and at its MMC, how much could the shaft be bent and still assemble if the shaft diameter was (a) at MMC, (b) at LMC, (c) at Ø17.94?

7. With reference to Fig. 14-3-C, calculate the limits for the distances between the holes when the pins and holes are at (a) MMC, and (b) LMC.

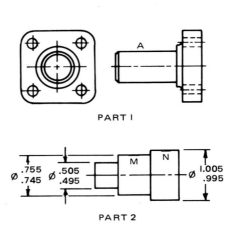

PART I

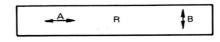

PART 2

PART 3

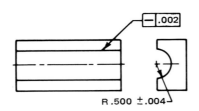

R .500 ±.004

PART 4

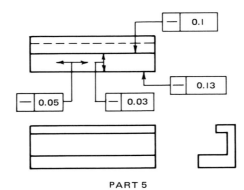

PART 5

Fig. 14-2-A Assignments.

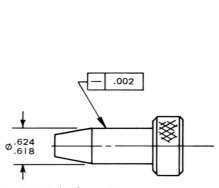

Fig. 14-3-A Asssignment.

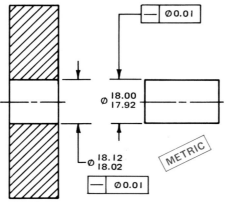

METRIC

Fig. 14-3-B Assignment.

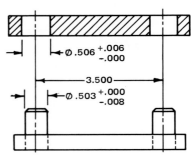

Fig. 14-3-C Assignment.

Assignments for Unit 14-4, Straightness of Features and Flatness

8. With reference to Fig. 14-4-A, are parts A to E acceptable? State your reasons if part is not acceptable.

9. If Ⓜ was added to the straightness tolerance shown in Fig. 14-4-A, what parts would be acceptable? State you reasons if part is not acceptable.

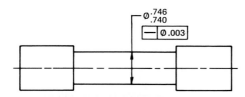

Fig. 14-4-A Assignments.

PART	FEATURE SIZE	STRAIGHTNESS DEVIATION	ACCEPTABLE
A	.747	.001	
B	.741	.004	
C	.742	.005	
D	.740	.006	
E	.740	.003	

10. Dimension the ring and snap gage shown in Fig. 14-4-B to check the pins shown. The ring gage should be of such a size as to check the entire length of pin. The two open ends of the snap gage should measure the minimum and maximum acceptable pin diameters.

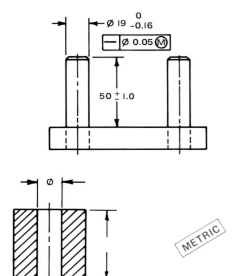

Fig. 14-4-B Assignments.

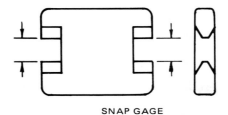

SNAP GAGE

Fig. 14-4-B (Continued)

11. In Fig. 14-4-C, part 1 is required to fit into part 2 so that there will not be any interference and the maximum clearance will never exceed .005 in. Show a flatness tolerance of .001 in. for both parts and a size with the largest size tolerance for part 2.

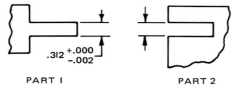

PART I   PART 2

Fig. 14-4-C Assignments.

12. Show the tolerance zones and widths, for the two parts shown in Fig. 14-4-D.

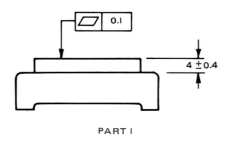

PART I

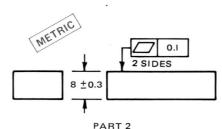

PART 2

Fig. 14-4-D Assignments.

Assignments for Unit 14-5, Datums and the Three-Plane Concept

13. Draw the top and front views of Fig. 14-5-A and show the following information:
- Surface A is datum A.
- Surface B is datum B.
- Surfaces C and D are datum features which form a single datum.
- Add the geometric tolerances from the information shown on the drawing.

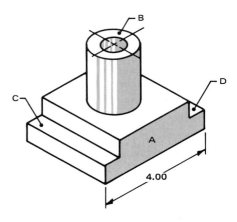

DATUM AND TOLERANCING INFORMATION

—A STRAIGHTNESS TOLERANCE OF .010 APPLIED TO THE CYLINDRICAL SURFACE

—THE BASE IS TO BE FLAT WITHIN .005

—DATUM A IS TO BE STRAIGHT WITHIN .008 FOR THE 4.00 LENGTH BUT THE STRAIGHTNESS ERROR NOT TO EXCEED .002 FOR ANY 1.00 LENGTH

—DATUM B IS TO BE FLAT WITHIN .004

—SURFACE B IS TO BE PARALLEL TO DATUM C—D WITHIN .006

Fig. 14-5-A Assignments.

14. With reference to Fig. 14-5-B:
- The surface of the base plate to be datum A.
- Pins 1, 2, and 3 are used to establish the secondary and tertiary datums for the part shown.
- Make a two-view drawing of the part shown and identify the primary, secondary, and tertiary datum planes as A, B, and C, respectively.
- Add a flatness tolerance of 0.2 to the back of the slot.
- With reference to the slot and locational tolerances are the three parts shown acceptable?

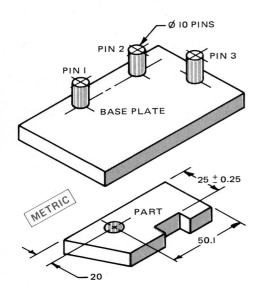

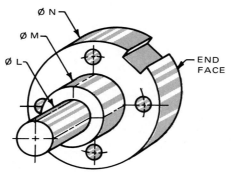

Fig. 14-5-C Assignments.

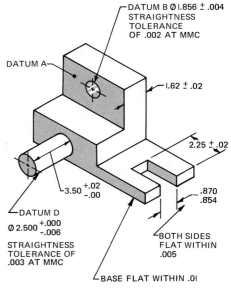

Fig. 14-5-D Assignments.

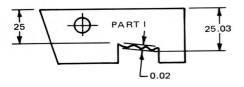

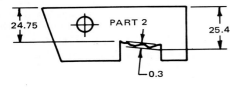

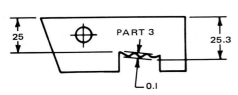

Fig. 14-5-B Assignments.

Assignments for Unit 14-6, Orientation

17. Today's drafters must be capable of interpreting and preparing drawings for use in other countries as well as in their own locality. From the information shown in Fig. 14-6-A prepare two three-view sketches, one showing the datums and geometric tolerancing symbols for use in the United States and the other sketch using ISO standards.

18. The surfaces shown in Fig. 14-6-B are required to be controlled in the following manner: Surfaces A, B, C, and E are datums, A, B, C, and E, respectively. Prepare a three-view drawing showing the datums and feature control symbols from the information supplied.

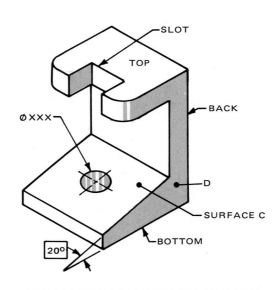

DATUM AND TOLERANCING INFORMATION

- BOTTOM TO BE DATUM A
- BACK TO BE DATUM B
- HOLE TO BE PERPENDICULAR TO BOTTOM WITHIN **.003**
- BACK TO BE PERPENDICULAR TO BOTTOM WITHIN **.004**
- TOP TO BE PARALLEL WITH BOTTOM WITHIN **.005**
- SURFACE C TO HAVE AN ANGULARITY TOLERANCE OF **.006** WITH THE BOTTOM. SURFACE D TO BE THE SECONDARY DATUM FOR THIS FEATURE.
- THE SIDES OF THE SLOT TO BE PARALLEL WITH EACH OTHER WITHIN **.002**
- BOTTOM TO HAVE A FLATNESS TOLERANCE OF **.002**
- BACK TO HAVE A FLATNESS TOLERANCE OF **.004**

Fig. 14-6-A Bracket.

15. With reference to Fig. 14-5-C: diameter *M* is to be used as datum *A*, the end face of diameter *N* is to be used as datum *B*, and the width of the slot at MMC on diameter *N* is to be used as datum *C*. Prepare two drawings, one with ANSI drawing standards the other with ISO drawing standards, which will identify these datums.

16. Prepare a three-view drawing showing the datums and feature control frames for Fig. 14-5-D.

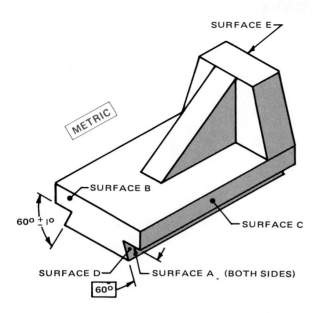

SURFACE E

METRIC

SURFACE B

$60° \pm 1°$

SURFACE C

SURFACE D — SURFACE A (BOTH SIDES)

$\boxed{60°}$

DATUM AND TOLERANCING INFORMATION

- SURFACE E TO HAVE A FLATNESS TOLERANCE OF 0.1 mm
- SURFACE D OF THE DOVETAIL MUST HAVE AN ANGULARITY TOLERANCE OF 0.05mm WITH DATUM A.
- SURFACE C SHOULD BE PERPENDICULAR TO DATUM A WITHIN 0.03mm.
- SURFACE E MUST BE PERPENDICULAR TO DATUM A WITHIN 0.02 mm.

Fig. 14-6-B Dovetail slide.

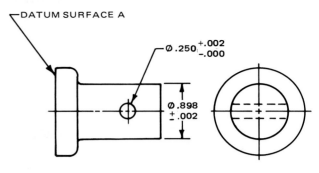

DATUM SURFACE A

$\varnothing.250 \, ^{+.002}_{-.000}$

$\varnothing.898 \pm .002$

TOLERANCING INFORMATION

— HOLE TO BE PARALLEL WITH DATUM A WITHIN $\pm 0.5°$

— THE $\varnothing.898$ SHAFT TO BE PERPENDICULAR TO DATUM A WITHIN .002 RFS

— DATUM A TO BE FLAT WITHIN .001

Fig. 14-6-C Cap.

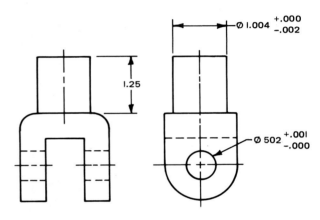

$\varnothing 1.004 \, ^{+.000}_{-.002}$

1.25

$\varnothing 502 \, ^{+.001}_{-.000}$

MAXIMUM PERPENDICULARITY TOLERANCE BETWEEN HOLES AND SHAFT .005 IN 1.00 IN.

SHAFT TO BE DESIGNATED AS DATUM A

Fig. 14-6-D Bracket.

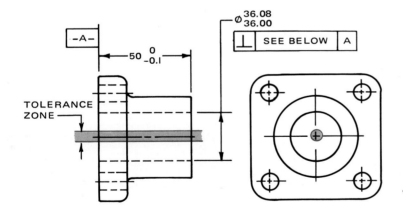

$\boxed{-A-}$

$50 \, ^{0}_{-0.1}$

$\varnothing \, ^{36.08}_{36.00}$

$\boxed{\perp}$ | SEE BELOW | A

TOLERANCE ZONE

METRIC

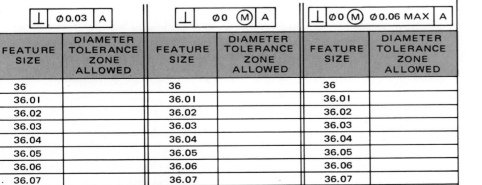

$\boxed{\perp \;\; \varnothing 0.03 \;\; A}$		$\boxed{\perp \;\; \varnothing 0 \, \text{ⓂA}}$		$\boxed{\perp \;\; \varnothing 0 \, \text{Ⓜ} \;\; \varnothing 0.06 \, \text{MAX} \;\; A}$	
FEATURE SIZE	DIAMETER TOLERANCE ZONE ALLOWED	FEATURE SIZE	DIAMETER TOLERANCE ZONE ALLOWED	FEATURE SIZE	DIAMETER TOLERANCE ZONE ALLOWED
36		36		36	
36.01		36.01		36.01	
36.02		36.02		36.02	
36.03		36.03		36.03	
36.04		36.04		36.04	
36.05		36.05		36.05	
36.06		36.06		36.06	
36.07		36.07		36.07	
36.08		36.08		36.08	

Fig. 14-6-E Shaft support.

19. Prepare a two-view sketch of Fig. 14-6-C and from the information given show the datums and geometric tolerances. Given that the tangent of 0.5° is 0.0087, sketch the tolerance zones.

20. In Fig. 14-6-D it is functionally necessary that the shaft portion of the part not depart from perpendicularity with the holes by more than the tolerance specified. Show the drawing callout for this, and indicate the shape and size of the tolerance zone.

21. Complete the tables shown in Fig. 14-6-E showing the maximum allowable perpendicularity tolerances.

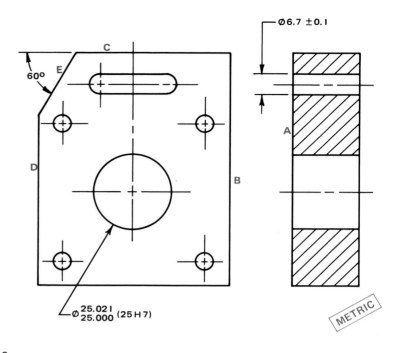

DATUM AND TOLERANCING INFORMATION

—SURFACES MARKED A, B, AND C ARE DATUMS A, B, AND C RESPECTIVELY

—SURFACE A IS PERPENDICULAR TO DATUMS B AND C WITHIN 0.2

—SURFACE D IS PARALLEL TO DATUM B WITHIN 0.1

—THE SLOT IS PARALLEL TO DATUM C WITHIN 0.2 AND PERPENDICULAR TO DATUM A WITHIN 0.3 AT MMC

—SURFACE E HAS AN ANGULARITY TOLERANCE OF 0.15 WITH DATUM C

—INDICATE THE BASIC DIMENSIONS ON THE DRAWING

Fig. 14-6-F Spacer.

22. Draw the two views shown in Fig. 14-6-F and add the information shown with the drawing.

Assignments for Unit 14-7, Tolerancing for Location of Features

23. If coordinate tolerances as shown in Fig. 14-7-A are given, what are the shapes of the tolerance zones and the distance between extreme permissible positions of the holes?

24. In Fig. 14-7-B add the largest equal tolerances so that if two such parts are assembled with the edges aligned, the distance between their hole centers could never be more than that shown.

25. If a tolerance shown in Fig. 14-7-C is specified for the vertical dimension, what tolerance should be added to the horizontal dimension to meet the same requirement as that required in Assignment 24?

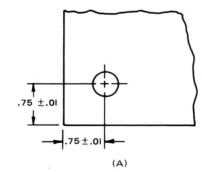

(A)

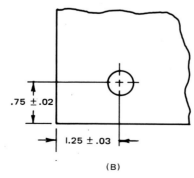

(B)

Fig. 14-7-A Assignments.

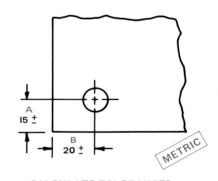

CALCULATE TOLERANCES TO NEAREST 0.02 MAXIMUM DISTANCE BETWEEN HOLE CENTERS = 0.5

Fig. 14-7-B

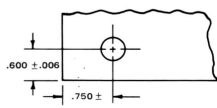

Fig. 14-7-C Assignments.

26. In order to assemble correctly, the hole in the part in Fig. 14-7-D must not vary from its true position by more than that shown on the drawing when the hole is at its smaller size.

(a) Show suitable tolerancing to achieve this.
  - By means of coordinate tolerancing
  - By positional tolerancing without MMC
  - By positional tolerancing on an MMC basis

(b) What would be the maximum permissible departure from true position if the hole were at its maximum diameter, using positional tolerancing RFS?

(c) What would be the maximum permissible departure from true position with a maximum-diameter hole and a positional tolerance on an MMC basis?

27. The part shown in Fig. 14-7-E is set on a revolving table, adjusted so that the part revolves about the true-position center of the large hole.

(a) If both indicators give identical readings and the results shown are obtained, which parts are acceptable?

(b) What is the positional error for each part?

28. With reference to Fig. 14-7-D a projected tolerance zone of .60 in. is required for the Ø.502 in. hole. Show two methods of how this could be shown.

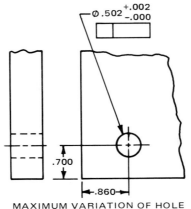

MAXIMUM VARIATION OF HOLE IN ANY DIRECTION = .0014

Fig. 14-7-D Assignments.

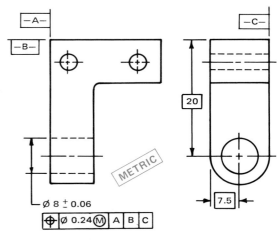

Fig. 14-7-E Assignments.

PART NO.	SIZE OF MANDREL	HIGHEST READING	LOWEST READING
1	8.00	1.54	1.32
2	8.06	0.18	-0.07
3	7.96	1.87	1.59
4	7.94	1.72	1.48
5	8.00	1.95	1.85
6	8.05	1.24	1.02

Assignments for Unit 14-8, Datums for Positional Tolerancing

29. Make a three-view drawing of the bearing housing shown in Fig. 14-8-A showing the datum features. Only the dimensions related to the datums need to be shown. Scale 1:2.

ROUNDS & FILLETS R.20

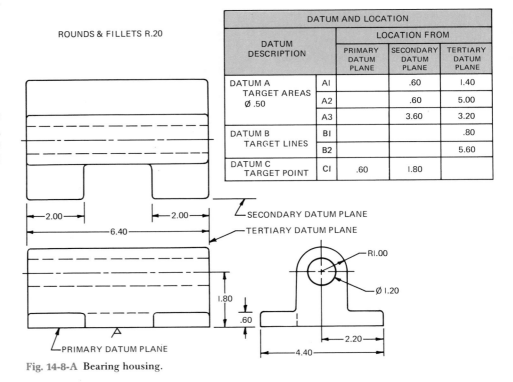

DATUM AND LOCATION				
DATUM DESCRIPTION		LOCATION FROM		
		PRIMARY DATUM PLANE	SECONDARY DATUM PLANE	TERTIARY DATUM PLANE
DATUM A TARGET AREAS Ø .50	A1		.60	1.40
	A2		.60	5.00
	A3		3.60	3.20
DATUM B TARGET LINES	B1			.80
	B2			5.60
DATUM C TARGET POINT	C1	.60	1.80	

Fig. 14-8-A Bearing housing.

30. Make a three-view drawing of the part shown in Fig. 14-8-B showing the datum features. Only the dimensions related to the datums need to be shown. Scale 1:1. Datum information is as follows:

- *Primary datum A* (three areas—Ø3). A_1 and A_2 are located on center of surface *M*, one-fifth the depth distance from the front and back, respectively. A_3 is located on the center surface *N* midway between the center of the hole and the right end.
- *Secondary datum B* is a datum line located at mid-height of surface *D*.
- *Tertiary datum C* is a datum point located on the center of surface *E*.

Assignments for Unit 14-9, Circularity (Roundness) and Cylindricity

31. Sketch the tolerance zone for the circularity tolerance in Fig. 14-9-A. If measurements made at cross sections *AA*, *BB*, and *CC* indicate that all points on the periphery fall within the annular rings shown, would you conclude that the part met the specified circularity tolerance? If not, which cross section is not acceptable? State your reason.

32. Add circularity tolerances to the diameters shown in Fig. 14-9-B. The circularity tolerances are to be one-fifth of the size tolerances for each diameter.

33. Show on each part in Fig. 14-9-C a cylindrical tolerance. The size of the cylindrical tolerance is to equal one-quarter the size tolerance for each diameter.

34. Sketch the tolerance zone for the cylindrical tolerance in Fig. 14-9-D indicating its size and shape for the part shown.

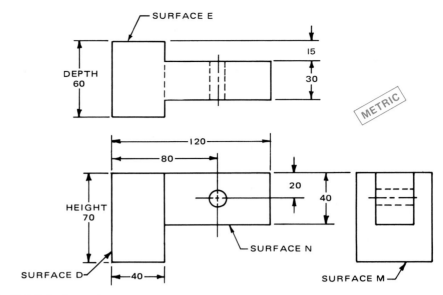

Fig. 14-8-B Assignments.

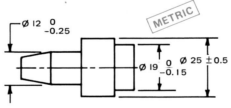

Fig. 14-9-B Assignments.

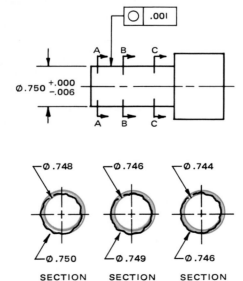

Fig. 14-9-A Assignments.

Fig. 14-9-C Assignments.

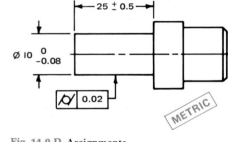

Fig. 14-9-D Assignments.

Assignments for Unit 14-10, Profile Tolerancing

35. In Fig. 14-10-A it is required to have the form of the indented portion controlled by the line (bilateral) profile tolerance of .006 in. Show the tolerance and sketch the resulting tolerance zone on the drawing and indicate which dimensions are basic.

36. It is required to control the profile in Fig. 14-10-B with the tolerance described on the drawing. Add the line profile tolerance to the drawing and sketch the resulting tolerance zone.

37. Draw the tolerance zone, showing its relationship to the datums, for the profile tolerance shown in Fig. 14-10-C.

38. A cam is dimensioned as shown in Fig. 14-10-D. If parts are measured with an indicator which was set to zero and the following readings were obtained, which parts shown in the chart would be acceptable? Of the nonacceptable parts, which could be made acceptable by regrinding?

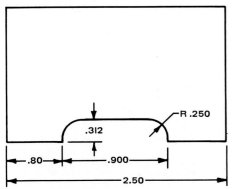

Fig. 14-10-A Assignments.

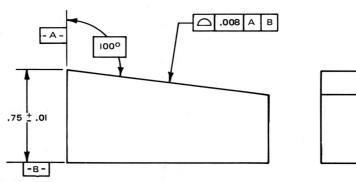

Fig. 14-10-C Assignments.

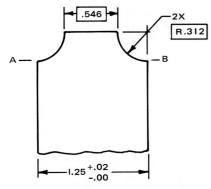

CONTROL THE PROFILE A TO B WITH
A LINE BILATERAL PROFILE TOLERANCE
OF .003 EXCEPT THAT THE .546 STRAIGHT
PORTION CAN BE PERMITTED TO VARY
VERTICALLY BY ±.01.

Fig. 14-10-B Assignments.

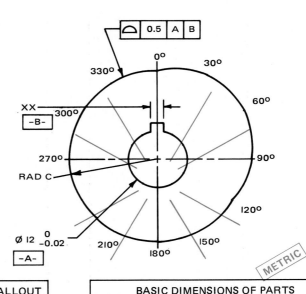

DRAWING CALLOUT		BASIC DIMENSIONS OF PARTS				
			DISTANCE FROM CENTER			
ANGLE	C RAD	ANGLE	PART 1	PART 2	PART 3	PART 4
0	24.5	0	24.6	24.7	24.5	24.4
30	27.5	30	27.8	27.7	27.4	27.4
60	27.5	60	27.6	27.6	27.25	27.5
90	25	90	25.1	25	24.6	25.1
120	22	120	22.4	21.9	21.7	22.1
150	20.5	150	20.6	20.4	20.2	20.5
180 TO 270	20	180 TO 270	20.2 TO 20.3	19.8 TO 19.75	19.6 TO 20.3	20.1 TO 20
300	20.5	300	20.4	20.3	20.2	20.4
330	22	330	21.9	22	21.7	21.9

* DIMENSIONS SHOWN ARE BASIC

Fig. 14-10-D Assignments.

Assignments for Unit 14-11, Correlative Tolerances

39. In Fig. 14-11-A show tolerances which will ensure that features are symmetrical at their maximum material conditions as shown on drawings.

40. It is required to have the three flat surfaces in Fig. 14-11-B coplanar with one another and perpendicular with the axis of the center hole within the tolerances specified. Add suitable geometric tolerances to the drawing. Show the tolerance zones for the part.

41. It is required to have the top diameter in Fig. 14-11-C concentric with the bottom diameter, when resting on surface A. Show how this should be specified on the drawing by (a) a positional tolerance, (b) a concentricity tolerance.

42. Show the size and form of the tolerance zone for the concentricity tolerance shown in Fig. 14-11-D.

43. The part shown in Fig. 14-11-E is intended to function by rotating with the two end diameters supported in bearings. The two larger diameters are required to have a total runout tolerance within the tolerance specified. Show how this would be toleranced.

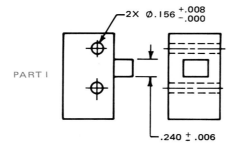

PART I

RECTANGULAR PROJECTION TO BE SYMMETRICAL WITHIN .002 AT MMC WITH THE HOLES.

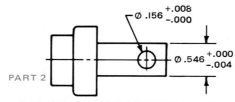

PART 2

Ø.156 HOLE TO BE SYMMETRICAL WITH Ø.546 WITHIN .001 REGARDLESS OF FEATURE SIZE.

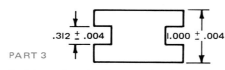

PART 3

THE 2 SLOTS TO BE SIMULTANEOUSLY SYMMETRICAL WITH THE 1.000 WIDTH WITHIN ZERO TOLERANCE WHEN BOTH THE SLOTS AND WIDTH ARE AT MMC.

Fig. 14-11-A Assignments.

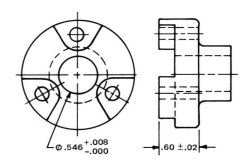

3 FLAT SURFACES COPLANAR WITH ONE ANOTHER WITHIN .001 AND PERPENDICULAR WITH AXIS OF CENTER HOLE WITHIN .002

Fig. 14-11-B Assignments.

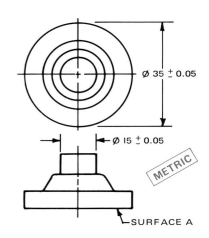

SURFACE A

Ø 15 CONCENTRIC WITH Ø 35 WITHIN 0.03 WHEN RESTING ON SURFACE A

Fig. 14-11-C Assignments.

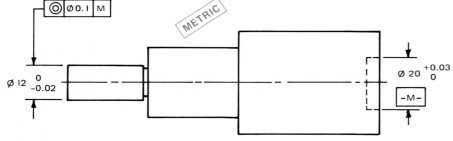

Fig. 14-11-D Assignments.

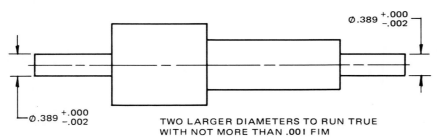

TWO LARGER DIAMETERS TO RUN TRUE WITH NOT MORE THAN .001 FIM

Fig. 14-11-E Bracket.

Review Assignments for Chapter 14

44. Draw the front and right-side views of the base plate shown in Fig. 14-12-A. Show the following information on the drawing:

(1) Hole L to be datum B.
(2) End face of diameter P to be datum A and have a flatness tolerance of .003 in. for any 2.00 in. square surface area with a maximum flatness tolerance of .008 in.
(3) Datum surface A has three target areas of Ø.50 in. equally spaced on the Ø4.25 in. hole circle. One of the targets is located diagonally opposite the slot.
(4) The center line of diameter M to be straight within .001 in. at MMC.
(5) The end face of diameter N to be parallel within .002 in. to the end face of diameter P.

(6) Diameter *N* to have a cylindrical tolerance of .0015 in. and the axis of diameter *N* to have a perpendicular tolerance of .002 in. with the end face of diameter *P*.

(7) The slot to be midway between the holes and symmetrically located on the Ø.625–.628 in. hole within .002 in. when both the slot and hole are at MMC.

(8) The Ø.625–.628 in. hole to be perpendicular to the end face of diameter *P* within .003 in.

(9) The four Ø.434–.438 in. holes to have a positional tolerance of Ø.001 in. at MMC and be referenced to the end face of diameter *P* (primary datum) and to diameter *P* (secondary datum at MMC).

45. Draw the end plate shown in Fig. 14-12-B and add the information shown below to the drawing. Only basic dimensions and dimensions for features having geometric tolerances need be shown.

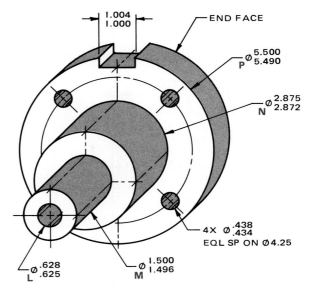

Fig. 14-12-A Base plate.

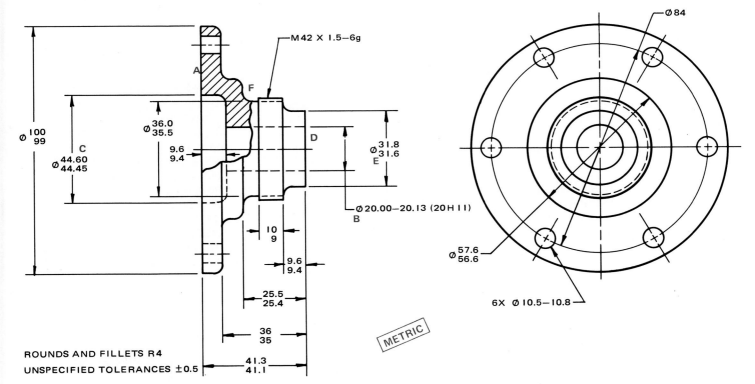

ROUNDS AND FILLETS R4
UNSPECIFIED TOLERANCES ±0.5

Fig. 14-12-B End plate.

(1) Surface *A* is datum *A*. It has a flatness tolerance of 0.02 mm.
(2) Diameter *B* is datum *B*. It has a circular runout of 0.14 mm and referenced to datums *A* and *C*.
(3) Diameter *C* is datum *C*. It has a perpendicularity tolerance of 0.08 mm at MMC and referenced to datum *A*.

(4) Surface *D* is datum *D*. It has a flatness tolerance of 0.02 mm and a parallel tolerance of 0.05 mm referenced to datum *A*.
(5) Diameter *E* has a circular runout tolerance of 0.1 mm which is referenced to datum *B*.
(6) Surface *F* is parallel to datum *A* within 0.1 mm.

(7) The Ø 10.5–10.8 mm holes have a positional tolerance of 0.2 mm at MMC and are referenced to datums *A* and *C* when datum *C* is at MMC.
(8) Datum *A* has three Ø 6 mm datum target areas equally spaced on a Ø 74 mm located midway between the Ø 10.5–10.8 mm holes.

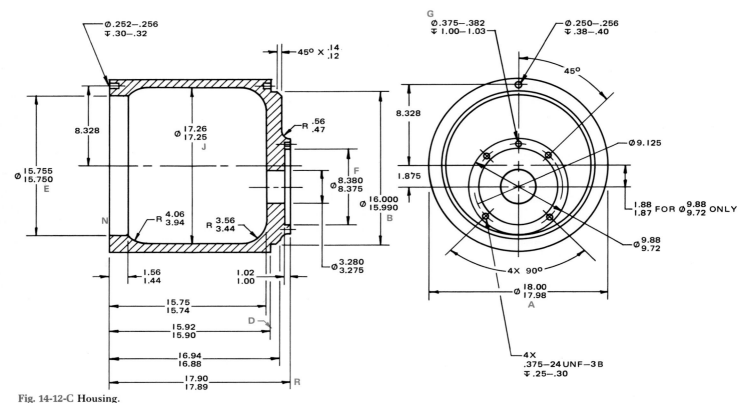

Fig. 14-12-C Housing.

46. Draw the two views of the housing shown in Fig. 14-12-C. Show the following information on the drawing. Only basic dimensions and dimensions for features having geometric tolerancing need to be shown.

(1) Diameter *A* is datum *A*. It has a perpendicularity tolerance of .020 in. related to datum *D*.

(2) Diameter *B* is datum *B*. It has a perpendicularity tolerance of .010 in. related to datum *D*, and a circular runout tolerance of .030 in. related to datum *A*.

(3) Hole *C* is datum *C*. It has a positional tolerance of .005 in. at MMC and related to datums *D* and *B* at MMC in that order.

(4) Surface *D* is datum *D*. It has a flatness tolerance of .005 in.

(5) Diameter *E* is datum *E*. It has a circular runout tolerance of .008 in. related to datums *D* and *B* in that order.

(6) Diameter *F* is datum *F*. It has a positional tolerance of .004 in. at MMC and related to datums *R*, *B* at MMC and *C* in that order.

(7) Diameter *G* is datum *G*. It has a positional tolerance of .005 in. at MMC and related to datums *R* and *F* at MMC in that order.

(8) Surface *N* is datum *N*. It has a parallel tolerance of .005 in. related to datum *D*.

(9) Surface *R* is datum *R*.

(10) The Ø3.275–3.280 in. hole has a positional tolerance of .010 in. at MMC and related to datums *R* and *F* at MMC in that order.

(11) The four .375–24 UNF tapped holes have a positional tolerance of .010 in. RFS and related to datums *R*, *F* at MMC and *G* in that order.

(12) Hole *H* has a positional tolerance of Ø.008 in. RFS and related to datums *N* and *E* at MMC in that order.

(13) Diameter *J* has a circular runout tolerance of .025 in. and is related to datum *A*.

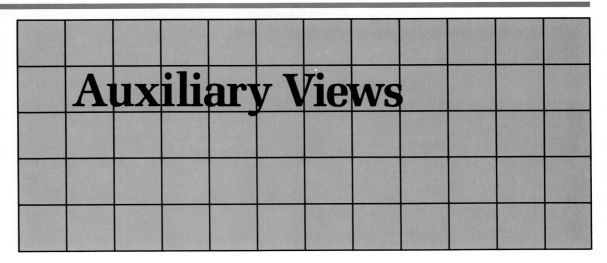

Auxiliary Views

UNIT 15-1
Primary Auxiliary Views

Many machine parts have surfaces that are not perpendicular, or at right angles, to the plane of projection. These are referred to as *sloping* or *inclined* surfaces. In the regular orthographic views, such surfaces appear to be foreshortened, and their true shape is now shown. When an inclined surface has important characteristics that should be shown clearly and without distortion, an auxiliary view is used so that the drawing completely and clearly explains the shape of the object. In many cases, the auxiliary view will replace one of the regular views on the drawing, as illustrated in Fig. 15-1-1.

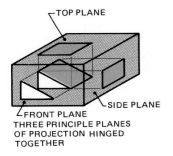

TOP PLANE

FRONT PLANE
SIDE PLANE
THREE PRINCIPLE PLANES
OF PROJECTION HINGED
TOGETHER

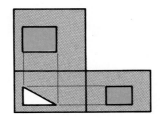

PLANES UNFOLDED

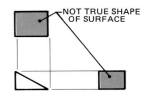

NOT TRUE SHAPE
OF SURFACE

PLANES REMOVED SHOWING
THREE REGULAR (TOP,
FRONT, SIDE) VIEWS

NOTE: IN NONE OF THESE VIEWS DOES THE SLANTED (COLORED) SURFACE APPEAR IN ITS TRUE SHAPE.

(A) WEDGED BLOCK SHOWN IN THREE REGULAR VIEWS

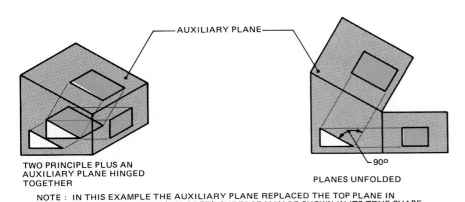

AUXILIARY PLANE

TWO PRINCIPLE PLUS AN
AUXILIARY PLANE HINGED
TOGETHER

90°
PLANES UNFOLDED

TRUE SHAPE OF
COLORED SURFACE

AUXILIARY
VIEW

90°

PLANES REMOVED SHOWING
FRONT, SIDE, AND
AUXILIARY VIEW

NOTE: IN THIS EXAMPLE THE AUXILIARY PLANE REPLACED THE TOP PLANE IN
ORDER THAT THE SLANTED (COLORED) SURFACE MAY BE SHOWN IN ITS TRUE SHAPE

(B) REPLACING THE TOP PLANE WITH AN AUXILIARY PLANE

Fig. 15-1-1 Relationship of the auxiliary plane to the three principle planes.

One of the regular orthographic views will have a line representing the edge of the inclined surface. The auxiliary view is projected from this edge line, at right angles, and is drawn parallel to the edge line.

Only the true shape features on the views need be drawn, as shown in Fig. 15-1-2. Since the auxiliary view shows only the true shape and detail of the inclined surface or features, a partial auxiliary view is all that is necessary. Likewise, the distorted features on the regular views may be omitted. Hidden lines are usually omitted unless required for clarity. This procedure is recommended for functional and production drafting where drafting costs are important consideration. However, the drafter may be called upon to draw the complete views of the part. The type of drawing is often used for catalog and standard parts drawing.

Additional examples of auxiliary view drawings are shown in Fig. 15-1-3.

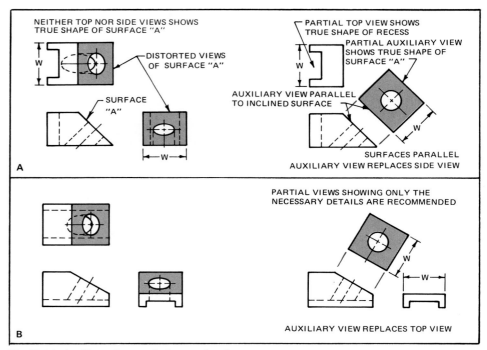

Fig. 15-1-2 Auxiliary views replacing regular views.

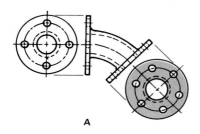

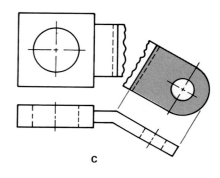

NOTE: CONVENTIONAL BREAK OR PROJECTED SURFACE ONLY NEED BE SHOWN ON PARTIAL VIEWS

Fig. 15-1-3 Examples of auxiliary view drawings.

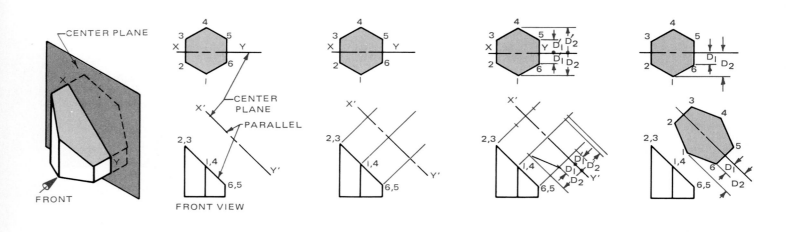

Fig. 15-1-4 To draw an auxiliary view using the center plane reference.

Assignment for Unit 15-4, Secondary Auxiliary Views

8. Make a working drawing of one of the parts shown in Fig. 15-4-A or 15-4-B. The selection and placement of views are shown beside the drawing. Only partial auxiliary views need be drawn, and hidden lines may be added to improve clarity. Scale 1:1.

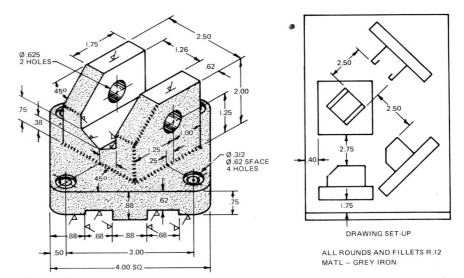

Fig. 15-3-C Angle slide.

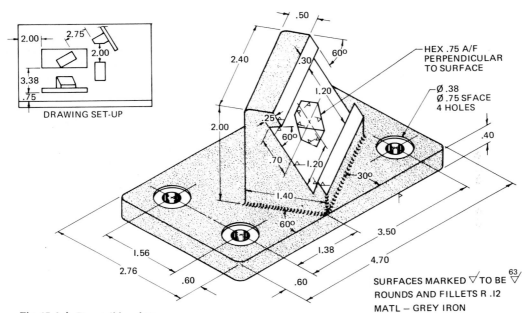

Fig. 15-4-A Dovetail bracket.

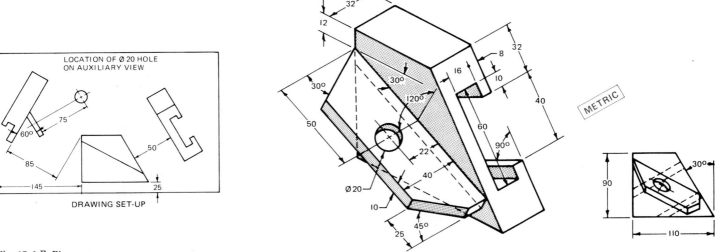

Fig. 15-4-B Pivot arm.

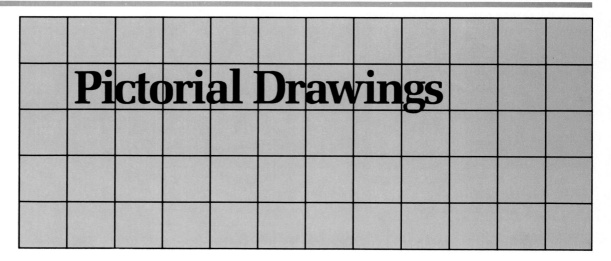

Pictorial Drawings

UNIT 16-1
Pictorial Drawings

Pictorial drawing is the oldest written method of communication known, but the character of pictorial drawing has continually changed with the advance of civilization. In this text only those kinds of pictorial drawings commonly used by the engineer, designer, and drafter are considered. Pictorial drawings are useful in design, construction or production, erection or assembly, service or repairs, and sales. They are used to explain complicated engineering drawings to people who do not have the training or ability to read the conventional multiview drawings; to help the designer work out problems in space, including clearances and interferences; to train

new employees in the shop; to speed up and clarify the assembly of a machine or the ordering of new parts; to transmit ideas from one person to another, from shop to shop, or from salesperson to purchaser; and as an aid in developing the power of visualization. The type of pictorial drawing used depends on the purpose for which it is drawn.

There are three general types into which pictorial drawings may be divided: axonometric, oblique, and perspective. These three differ from one another in the fundamental scheme of projection, as shown in Fig. 16-1-1.

AXONOMETRIC PROJECTION

A projected view in which the lines of sight are perpendicular to the plane of projection, but in which the three faces of a rectangular object are all inclined to

the plane of projection, is called an *axonometric projection*. See Fig. 16-1-2. The projections of the three principal axes may make any angle with one another except 90°. Axonometric drawings, as shown in Figs. 16-1-3, and 16-1-4, are classified into three forms: *isometric drawings*, where the three principal faces and axes of the object are equally inclined to the plane of projection; *dimetric drawings*, where two of the three principal faces and axes of the object are equally

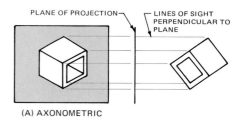

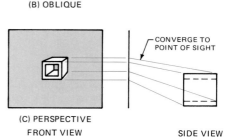

Fig. 16-1-2 Kinds of projections.

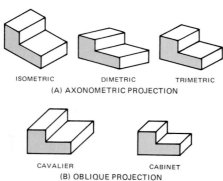

Fig. 16-1-1 Types of pictorial drawings.

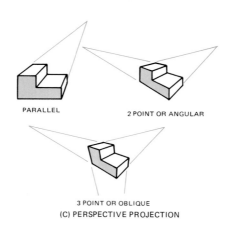

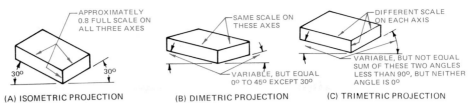

Fig. 16-1-3 Types of axonometric drawings. (Graphic Standard Instrument Co.)

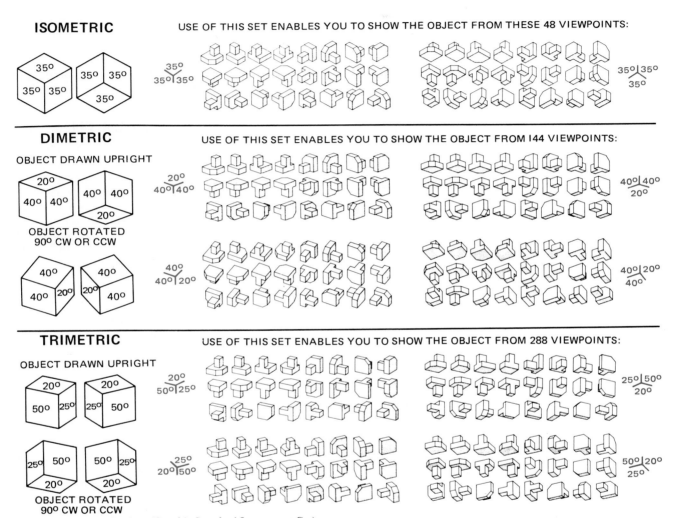

Fig. 16-1-4 Axonometric projection. (Graphic Standard Instrument Co.)

inclined to the plane of projection; and *trimetric drawings*, where all three faces and axes of the object make different angles with the plane of projection. The most popular form of axonometric projection is the isometric.

ISOMETRIC DRAWINGS

This method is based on a procedure of revolving the object at an angle of 45° to the horizontal, so that the front corner is toward the viewer, then tipping the object up or down at an angle of 35° 16′. See Fig. 16-1-5. When this is done to a

cube, the three faces visible to the viewer appear equal in shape and size, and the side faces are at an angle of 30° to the horizontal. If the isometric view were actually projected from a view of the object in the tipped position, the lines in the isometric view would be foreshortened and would, therefore, not be seen in their true length. To simplify the drawing of an isometric view, the actual measurements of the object are used. Although the object appears slightly larger without the allowance for foreshortening, the proportions are not affected. All isometric drawings are

started by constructing the isometric axes, which are a vertical line for height and isometric lines to left and right, at an angle of 30° from the horizontal, for length and width. The three faces seen in the isometric view are the same faces that would be seen in the normal orthographic views: top, front, and side. Figure 16-1-5B illustrates the selection of the front corner (A), the construction of the isometric axes, and the completed isometric view. Note that all lines are drawn to their true length, measured along the isometric axes, and that hidden lines are usually omitted. Vertical

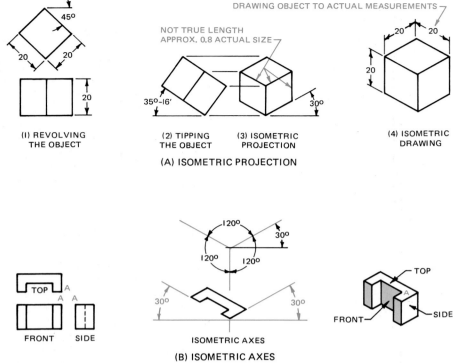

Fig. 16-1-5 Isometric axes and projection.

edges are represented by vertical lines, and horizontal edges by lines at 30° to the horizontal. Two techniques can be used for making an isometric drawing of an irregularly shaped object, as illustrated in Fig. 16-1-6. In one method, the object is divided mentally into a number of sections and the sections are created one at a time in their proper relationship to one another. In the second method, a box is created with the maximum height, width, and depth of the object; then the parts of the box that are not part of the object are removed, leaving the pieces that form the total object.

Nonisometric Lines

Many objects have sloping surfaces that are represented by sloping lines in the orthographic views. In isometric drawing, sloping surfaces appear as *nonisometric* lines. To create them, locate their endpoints, found on the ends of isometric lines, and join them with a straight line. Figures 16-1-7 and 16-1-8 illustrate examples in the construction of nonisometric lines.

Dimensioning Isometric Drawings

At times, an isometric drawing of a simple object may serve as a working drawing. In such cases, the necessary dimensions and specifications are placed on the drawing.

Dimension lines, extension lines, and the line being dimensioned are shown in the same plane. Arrowheads should be in the plane of the dimension and extension lines. See Fig. 16-1-9.

Unidirectional dimensioning is the preferred method of dimensioning isometric drawings. The letters and numbers are vertical and read from the bottom of the sheet. An example of this type of dimensioning is shown in Fig. 16-1-10.

Since the isometric is a one-view drawing, it is not usually possible to avoid placing dimensions on the view or across dimension lines. However, this practice should be avoided whenever possible.

Isometric Grid Paper

Isometric grid sheets are another time-saving device. Designers and engineers frequently use isometric grid paper on which they sketch their ideas and

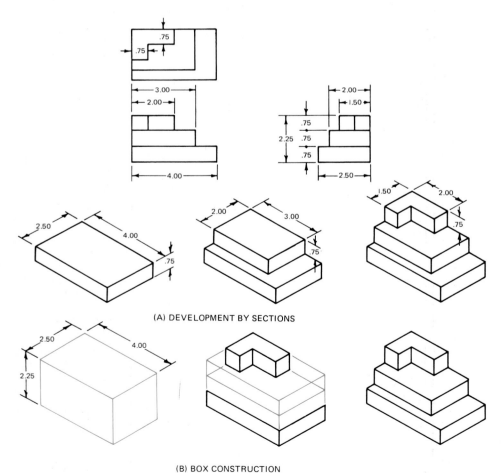

Fig. 16-1-6 Developing an isometric drawing.

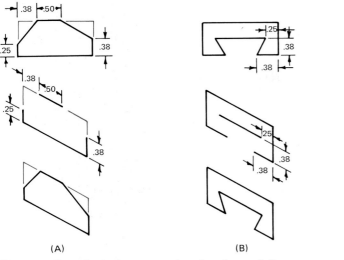

Fig. 16-1-7 Examples in the construction of nonisometric lines.

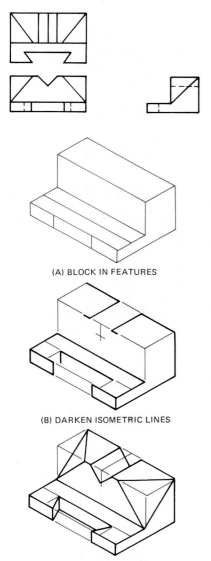

(A) BLOCK IN FEATURES

(B) DARKEN ISOMETRIC LINES

(C) COMPLETE NONISOMETRIC LINES

Fig. 16-1-8 Sequence in drawing an object having nonisometric lines.

ENDS OF ARROW PARALLEL WITH EXTENSION LINES

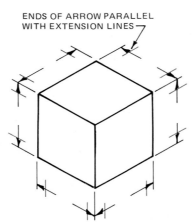

Fig. 16-1-9 Orienting the dimension line, arrowhead, and extension line.

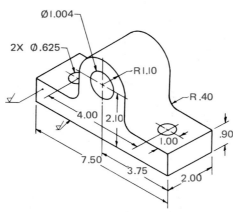

Fig. 16-1-10 Isometric dimensioning.

designs. See Fig. 16-1-11. Many companies, such as those which prepare pipe drawings, have large drawing sheets made with nonreproducible isometric grid lines.

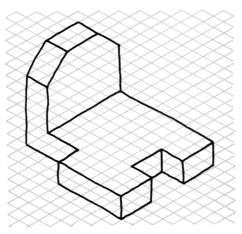

Fig. 16-1-11 Isometric grid paper.

CAD

A pictorial drawing may be created by using any CAD two-dimensional (2D) system. Generally, a grid pattern peculiar to the type of pictorial will be employed. Since isometrics are the most popular, an ISOMETRIC GRID option is found on virtually every system.

Automatic generation of a pictorial is common to the axonometric and perspective types only.

Modeling

CAD systems provide a MODELING option, often referred to as 3D MODELING. With this option the model, isometric or otherwise, is automatically generated from the multiview drawing. See Fig. 16-1-12.

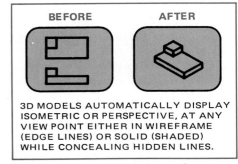

BEFORE   AFTER

3D MODELS AUTOMATICALLY DISPLAY ISOMETRIC OR PERSPECTIVE, AT ANY VIEW POINT EITHER IN WIREFRAME (EDGE LINES) OR SOLID (SHADED) WHILE CONCEALING HIDDEN LINES.

Fig. 16-1-12 CAD MODEL command.

References and Source Material

1. ANSI Y14.4 Pictorial Drawing.
2. General Motors Corporation.

ASSIGNMENTS

See Assignments 1 through 3 for Unit 16-1 on page 382.

UNIT 16-2
Curved Surfaces in Isometric

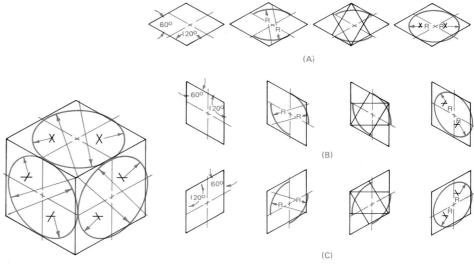

CIRCLES AND ARCS IN ISOMETRIC

A circle on any of the three faces of an object drawn in isometric has the shape of an ellipse. See Fig. 16-2-1. Figure 16-2-2 illustrates the steps in drawing circular features on isometric drawings.

Fig. 16-2-2 Sequence in drawing isometric circles.

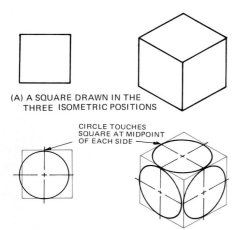

(A) A SQUARE DRAWN IN THE THREE ISOMETRIC POSITIONS

CIRCLE TOUCHES SQUARE AT MIDPOINT OF EACH SIDE

(B) A CIRCLE PLACED INSIDE A SQUARE AND DRAWN IN THE THREE ISOMETRIC POSITIONS

Fig. 16-2-1 Circles in isometric.

1. Draw the center lines and a square, with sides equal to the circle diameter, in isometric.
2. Using the obtuse-angled (120°) corners as centers, draw arcs tangent to the sides forming the obtuse-angled corners, stopping at the points where the center lines cross the sides of the square.
3. Draw construction lines from these same points to the opposite obtuse-angled corners. The points at which these construction lines intersect are the centers for arcs drawn tangent to the sides forming the acute-angled corners, meeting the first arcs.

In drawing concentric circles, each circle must have its own set of centers for the arcs, as shown in Fig. 16-2-3.

The same technique is used for drawing part-circles (arcs). See Fig. 16-2-4. Construct an isometric square with sides equal to twice the radius, and draw that portion of the ellipse necessary to join the two faces. When these faces are parallel, draw half of an ellipse (one long

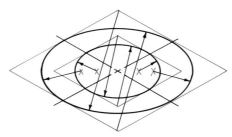

Fig. 16-2-3 Drawing concentric isometric circles.

radius and one short radius); when they are at an obtuse angle (120°), draw one long radius; and when they are at an acute angle (60°), draw one short radius.

ISOMETRIC TEMPLATES

For convenience and time saving, isometric ellipse templates should be used

whenever possible. A wide variety of elliptical templates are available. The template shown in Fig. 16-2-5 combines ellipses, scales, and angles. Markings on the ellipses coincide with the center lines of the holes which speeds up drawing circles and arcs. Figure 16-2-6 shows the same part as shown in Fig. 16-2-4 but with the arcs and circles being constructed with a template.

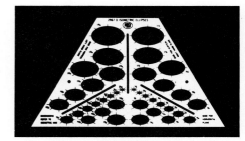

Fig. 16-2-5 Isometric ellipse template.

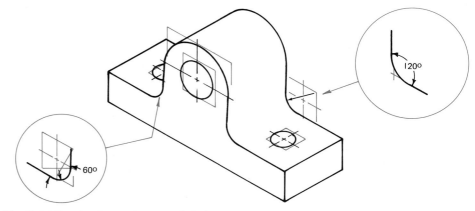

Fig. 16-2-4 Drawing isometric arcs and circles.

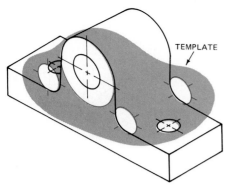

Fig. 16-2-6 Circles and arcs drawn with isometric ellipse template.

SKETCHING CIRCLES AND ARCS

In sketching circles and arcs on isometric grid paper, locate the center lines first, then lightly sketch in construction boxes (isometric squares) where the circles and arcs should be. See Fig. 16-2-7. Sketch the ellipse (isometric circle) just touching the center of each of the four sides of the square.

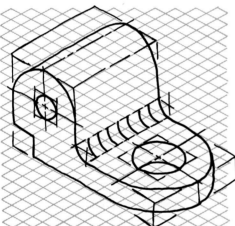

Fig. 16-2-7 Isometric sketching paper.

DRAWING IRREGULAR CURVES IN ISOMETRIC

To draw curves other than circles or arcs, the plotting method shown in Fig. 16-2-8 is used.

1. Draw an orthographic view, and divide the area enclosing the curved line into equal squares.
2. Produce an equivalent area on the isometric drawing, showing the offset squares.
3. Take positions relative to the squares

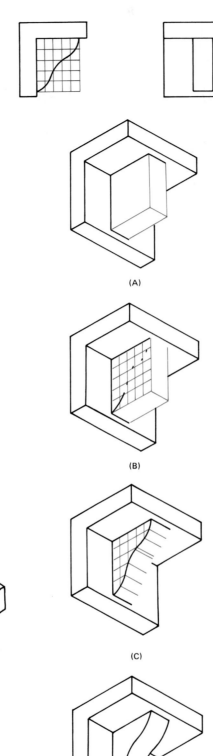

(A)

(B)

(C)

(D)

Fig. 16-2-8 Curves drawn in isometric by means of offset measurements.

from the orthographic view, and plot them on the corresponding squares on the isometric view.

4. Draw a smooth curve through the established points with the aid of an irregular curve.

CAD

The ELLIPSE command specifying the major and minor diameters is not suitable for isometric drawings. Many CAD systems are not programmed to draw an ellipse given the diameter values on the isometric axes. The four-center ellipse method outlined in Unit 6-4 is the command used to produce arcs and circles for isometric drawings.

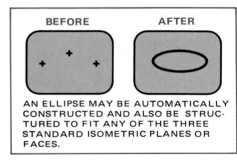

Fig. 16-2-9 CAD ELLIPSE command.

For curves other than circles or arcs a grid pattern is employed. A series of points on the curve are located on the isometric grid using the TRACE command.

ASSIGNMENTS

See Assignments 4 through 6 for Unit 16-2 on page 383.

U N I T   1 6 - 3
Common Features in Isometric

ISOMETRIC SECTIONING

Isometric drawings are generally made showing exterior views, but sometimes a sectional view is needed. The section is taken on an isometric plane, that is, on a plane parallel to one of the faces of the cube. Figure 16-3-1 shows isometric full sections taken on a different plane for each of three objects. Note the construction lines indicating the part that has been cut away. Isometric half sections are illustrated in Fig. 16-3-2.

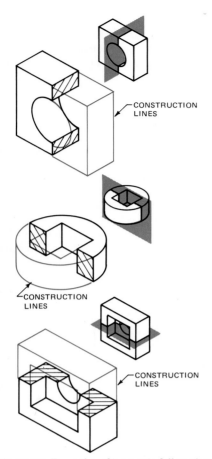

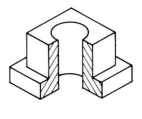
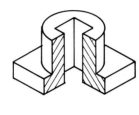

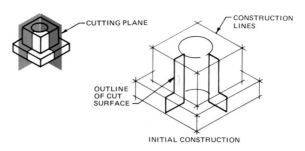

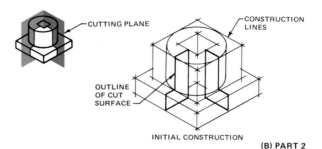

(A) PART I

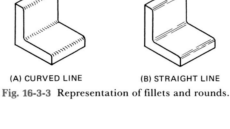

(B) PART 2

Fig. 16-3-2 Examples of isometric half sections.

Fig. 16-3-1 Examples of isometric full sections.

When an isometric drawing is sectioned, the section lines are shown at an angle of 60° with the horizontal or in a horizontal position, depending on where the cutting-plane line is located. In half sections, the section lines are sloped in opposite directions, as shown in Fig. 16-3-2.

FILLETS AND ROUNDS

For most isometric drawings of parts having small fillets and rounds, the adopted practice is to draw the corners as sharp features. However, when it is desirable to represent the part, normally a casting, as having a more realistic appearance, either of the methods shown in Fig. 16-3-3 may be used.

THREADS

The conventional method for showing threads in isometric is shown in Fig. 16-3-4. The threads are represented by a series of ellipses uniformly spaced along the center line of the thread. The spacing of the ellipses need not be the spacing of the actual pitch.

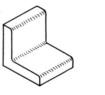

(A) CURVED LINE   **(B) STRAIGHT LINE**
Fig. 16-3-3 Representation of fillets and rounds.

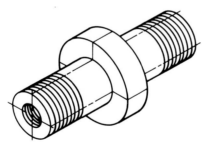

Fig. 16-3-4 Representation of threads in isometric.

BREAK LINES

For long parts, break lines should be used to shorten the length of the drawing. Freehand breaks are preferred, as shown in Fig. 16-3-5.

ISOMETRIC ASSEMBLY DRAWINGS

Regular or exploded assembly drawings are frequently used in catalogs and sales literature, as illustrated by Figs. 16-3-6 and 16-3-7.

ASSIGNMENTS

See Assignments 7 through 12 for Unit 16-3 on page 385.

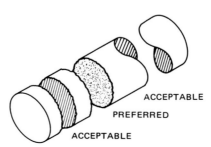
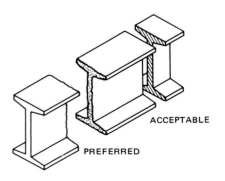

Fig. 16-3-5 Conventional breaks.

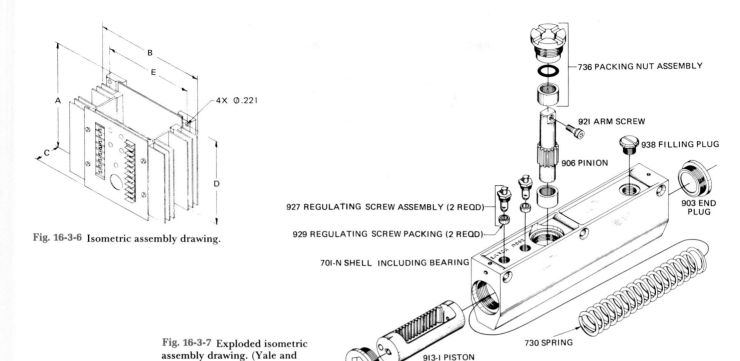

Fig. 16-3-6 Isometric assembly drawing.

736 PACKING NUT ASSEMBLY

921 ARM SCREW

938 FILLING PLUG

906 PINION

927 REGULATING SCREW ASSEMBLY (2 REQD)

929 REGULATING SCREW PACKING (2 REQD)

903 END PLUG

701-N SHELL INCLUDING BEARING

730 SPRING

913-1 PISTON

903 END PLUG

Fig. 16-3-7 Exploded isometric assembly drawing. (Yale and Towne Inc.)

UNIT 16-4
Oblique Projection

This method of pictorial drawing is based on the procedure of placing the object with one face parallel to the frontal plane and placing the other two faces on oblique (or receding) planes, to left or right, top or bottom, at a convenient angle. The three axes of projection are *vertical*, *horizontal*, and *receding*. Figure 16-4-1 illustrates a cube drawn in typical positions with the receding axis at 60°, 45° and 30°. This form of projection has the advantage of showing one face of the object without distortion. The face with the greatest irregularity of outline or contour, or the face with the greatest number of circular features, or the face with the longest dimension, faces the front. See Fig. 16-4-2.

Two types of oblique projection are used extensively. In *cavalier oblique*, all lines are made to their true length, measured on the axes of the projection. In *cabinet oblique*, the lines on the receding axis are shortened by one-half their true length to compensate for distortion and to approximate more closely what the human eye would see. For this reason, and because of the simplicity of projec-

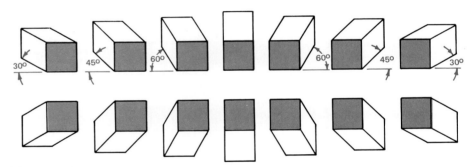

Fig. 16-4-1 Typical positions of receding axes for oblique projections.

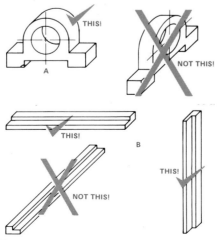

Fig. 16-4-2 Two general rules for oblique drawings.

tion, cabinet oblique is a commonly used form of pictorial representation, especially when circles and arcs are to be drawn. Figure 16-4-3 shows a comparison of cavalier and cabinet oblique. Note that hidden lines are omitted unless required for clarity. Many of the drawing techniques for isometric projection apply to oblique projection. Figure 16-4-4 illustrates the construction of an irregularly shaped object by the box method.

INCLINED SURFACES

Angles which are parallel to the picture plane are drawn as their true size. Other angles can be laid off by locating the ends of the inclined line.

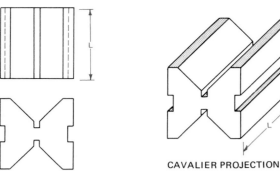

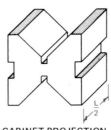

CAVALIER PROJECTION   CABINET PROJECTION

Fig. 16-4-3 Types of oblique projection.

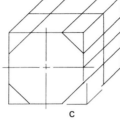 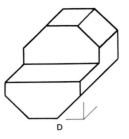

A   B   C   D

Fig. 16-4-4 Oblique construction by the box method.

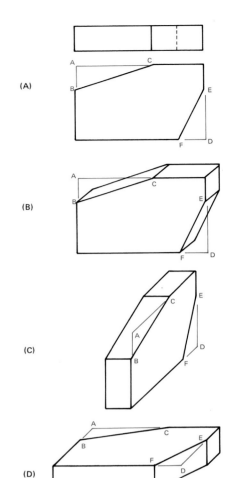

(A)

(B)

(C)

(D)

Fig. 16-4-5 Drawing inclined surfaces.

A part with notched corners is shown in Fig. 16-4-5A. An oblique drawing with the angles parallel to the picture plane is shown at Fig. 16-4-5B. In Fig. 16-4-5C the angles are parallel to the profile plane. In each case the angle is laid off by mesurement parallel to the oblique axes, as shown by the construction lines. Since the part, in each case, is drawn in cabinet oblique, the receding lines are shortened by one-half their true length.

OBLIQUE SKETCHING

Specially designed oblique sketching paper with 45° lines is available and, like isometric sketching paper, is used extensively by engineers and drafters. See Fig. 16-4-6.

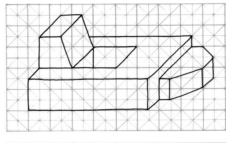

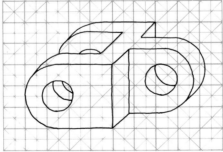

Fig. 16-4-6 Oblique sketching paper.

CAD

Specially designed oblique lines or grids with horizontal, vertical, and 45° reference lines are available on some CAD systems. If not available, a rectangular pattern may be used to develop the true-size front face. For the two oblique faces a 45 or 60° line pattern is used.

Oblique modeling is not a CAD option.

DIMENSIONING AN OBLIQUE DRAWING

Dimension lines are drawn parallel to the axes of projection. Extension lines are projected from the horizontal and vertical object lines whenever possible.

The dimensioning of an oblique drawing is similar to that of an isometric drawing. The recommended method is unidirectional dimensioning, which is shown in Fig. 16-4-7. As in isometric dimensioning, usually it is necessary to place some dimensions directly on the view.

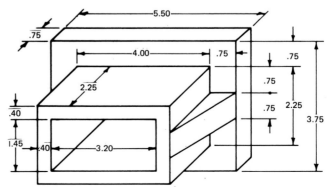

Fig. 16-4-7 Dimensioning an oblique drawing.

ASSIGNMENTS

See Assignments 13 through 15 for Unit 16-4 on page 386.

UNIT 16-5
Common Features in Oblique

CIRCLES AND ARCS

Whenever possible, the face of the object having circles or arcs should be selected as the *front* face, so that such circles or arcs can be easily drawn in their true shape. See Fig. 16-5-1. When circles or arcs must be drawn on one of the oblique faces, the offset measurement method illustrated in Fig. 16-5-2 may be used.

SCHEMATIC OF A COMPLETELY AUTOMATIC REGISTRATION CONTROL SYSTEM MAINTAINING THE LOCATION OF CUTOFF ON A CONTINUOUS PRINTED WEB.

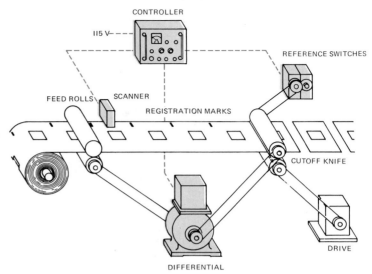

Fig. 16-5-1 Application of oblique drawing.

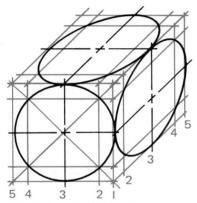

Fig. 16-5-2 Drawing oblique circles by means of offset measurements.

1. Draw an oblique square about the center lines, with sides equal to the diameter.
2. Draw a true circle within the oblique square, and establish equally spaced points about its circumference.
3. Project these point positions to the edge of the oblique square, and draw lines on the oblique axis from these positions. Similarly spaced lines are drawn on the other axis, forming off-set squares and giving intersection points for the oval shape.

Another method used when circles or arcs must be drawn on one of the oblique surfaces is the *four-center method*. In Fig. 16-5-3A a circle is shown as it would be drawn on a front plane, a side plane, and a top plane.

In Fig. 16-5-3B, the oblique drawing has some arcs in a horizontal plane. In

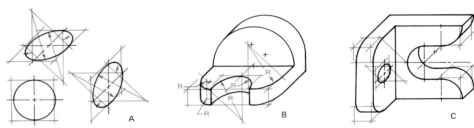

Fig. 16-5-3 Circles parallel to the picture plane are true circles; on other planes, ellipses.

Fig. 16-5-3C, the oblique drawing shown has some arcs in a profile plane.

Circles not parallel to the picture plane when drawn by the approximate method are not pleasing but are satisfactory for some purposes. Ellipse templates, when available, should be used because they reduce drawing time and give much better results. If a template is used, the oblique circle should first be blocked in as an oblique square in order to locate the proper position of the circle. Blocking in the circle first also helps the drafter select the proper size and shape of the ellipse. The construction and dimensioning of an oblique part are shown in Fig. 16-5-4.

OBLIQUE SECTIONING

Oblique drawings are generally made as outside views, but sometimes a sectional view is necessary. The section is taken on a plane parallel to one of the faces of an oblique cube. Figure 16-5-5 shows an oblique full section and an oblique half section. Construction lines show the part that has been cut away.

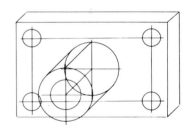

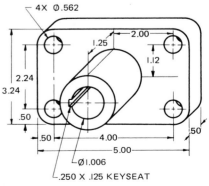

Fig. 16-5-4 Construction and dimensioning of an oblique object.

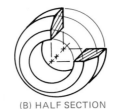

(A) FULL SECTION   (B) HALF SECTION

Fig. 16-5-5 Oblique full and half sections.

TREATMENT OF CONVENTIONAL FEATURES

Fillets and Rounds   Small fillets and rounds normally are drawn as sharp corners. When it is desirable to show the corners rounded, then either of the methods shown in Fig. 16-5-6 is recommended.

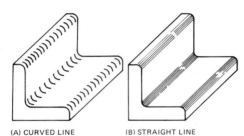

(A) CURVED LINE   (B) STRAIGHT LINE

Fig. 16-5-6 Representing rounds and fillets.

Threads   The conventional method of showing threads in oblique is shown in Fig. 16-5-7. The threads are represented by a series of circles uniformly spaced along the center line of the thread. The spacing of the circles need not be the spacing of the pitch.

Fig. 16-5-7 Representation of threads in oblique.

Breaks   Figure 16-5-8 shows the conventional method for representing breaks.

CAD

The CIRCLE, ARC, and FILLET commands are used to create circles and arcs on the front face of an oblique drawing. For the oblique faces use the

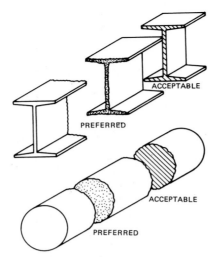

ACCEPTABLE

PREFERRED

ACCEPTABLE

PREFERRED

Fig. 16-5-8 Conventional breaks.

ELLIPSE command. If for some reason the ELLIPSE command does not produce the desired results, a four-center geometric construction command is used. See Fig. 6-4-2.

ASSIGNMENTS

See Assignments 16 and 17 for Unit 16-5 on page 387.

UNIT 16-6
Perspective Projection

Perspective is a method of drawing that depicts a three-dimensional object on a flat plane as it appears to the eye. See Fig. 16-6-1. A pictorial drawing made by the intersection of the picture plane with lines of sight converging from points on the object to the point of sight, which is located at a finite distance from the picture plane, is called a *perspective*. See Fig. 16-6-2.

Perspective drawings are more realistic than axonometric or oblique drawings because the object is shown as the eye would see it. Since they are far more difficult to draw than the other types of pictorial drawings, their use in drafting is limited mainly to production or presentation illustrations and illustrations of proposed structures by architects.

The main elements of a perspective drawing are the *picture plane* (plane of projection), the *station point* (the position

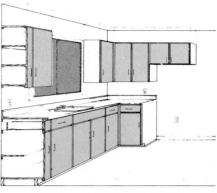

(A) PARALLEL PERSPECTIVE

(B) ANGULAR PERSPECTIVE

Fig. 16-6-1 Perspective drawings.

of the observer's eye when he or she is viewing the object), the *horizon* (an imaginary horizontal line taken at eye level), the *vanishing point or points* (a point or points on the horizon where all the receding lines converge), and the *ground line* (the base line of the picture plane and object).

To avoid undue distortion in perspective, the point of sight (station point) should be located so that the cone of rays from the observer's eye has an angle at the apex not greater than 30°. This would place the station point a distance away from the outside portion of the object of approximately 2 to 2½ times the width of the object being viewed. See Figs. 16-6-2 and 16-6-3.

TYPES OF PERSPECTIVE DRAWINGS

There are three types of perspective drawings:

1. Parallel: One vanishing point
2. Angular: Two vanishing points
3. Oblique: Three vanishing points

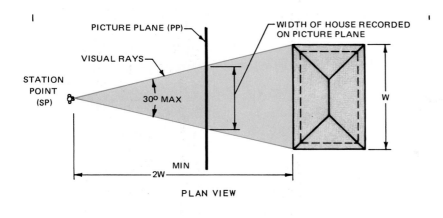

PICTURE PLANE (PP)

WIDTH OF HOUSE RECORDED ON PICTURE PLANE

VISUAL RAYS

STATION POINT (SP)

30° MAX

W

MIN

2W

PLAN VIEW

HEIGHT OF HOUSE RECORDED ON PICTURE PLANE

PICTURE PLANE (PP)

VISUAL RAYS

STATION POINT (SP)

GROUND LINE

HORIZON

ELEVATION

WIDTH FROM PLAN VIEW

PICTURE PLANE

HEIGHT FROM ELEVATION

PICTURE RECORDED ON PICTURE PLANE AS SEEN BY OBSERVER

Fig. 16-6-2 Recording the picture on the picture plane.

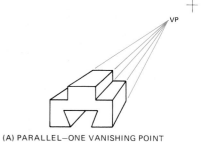

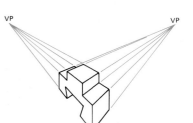
(A) PARALLEL—ONE VANISHING POINT

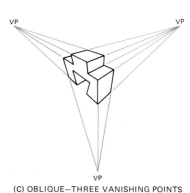
(B) ANGULAR—TWO VANISHING POINTS

(C) OBLIQUE—THREE VANISHING POINTS

Fig. 16-6-4 Types of perspective drawings.

In industry they are normally referred to as one-point, two-point, and three-point perspectives, respectively. See Fig. 16-6-4. Only parallel and angular perspectives are covered in this text.

Parallel, or One-Point, Perspective

Parallel-perspective drawings are similar to oblique drawings, except the receding lines all converge at one point on the horizon. In drawing a parallel-perspec-

tive drawing, one face of the object is placed on the picture-plane line so that it will be drawn in its true size and shape, as shown in Fig. 16-6-5. The *PP* line shown in the top view represents the picture plane line, and point *SP* (station point) is the position of the observer. The lines of the object, which are not on the picture plane, are found by projecting lines down from the top view from the point of intersection of the visual ray and the picture plane, as shown by point *N* in Fig. 16-6-5A(1).

Where the true height of a line or a point does not lie on the picture plane, such as point *P* in Fig. 16-6-5A(2), the true height may be found by extending line *PR* to point *S* on the picture plane. Since point *S* lies on the picture plane and is the same height as point *P*, it may readily be found on the perspective drawing. Point *P* will lie on the receding line joining point *S* to the line *VP*.

In drawing a one-point perspective, a

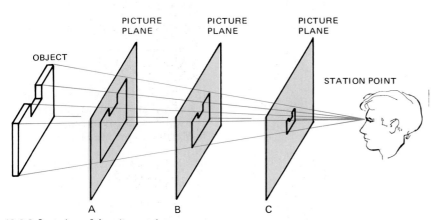

PICTURE PLANE

PICTURE PLANE

PICTURE PLANE

OBJECT

STATION POINT

A   B   C

Fig. 16-6-3 Location of the picture plane.

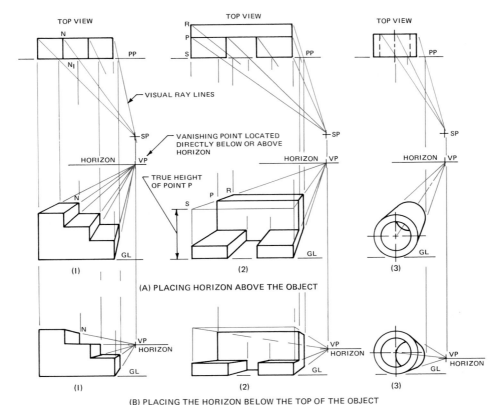

Fig. 16-6-5 Parallel or one-point perspective.

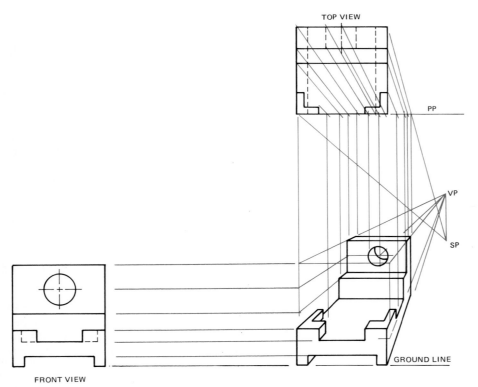

Fig. 16-6-6 Construction of a one-point perspective.

side or front view and a top view are normally drawn first—the top view to locate the part with respect to the picture plane and the side or front view to obtain the height of the various features. Figure 16-6-6 shows a simple, one-point perspective drawing with construction lines.

One of the most common uses of a parallel-perspective drawing is for representing the interior of a building. With this type of drawing, the vanishing point is located inside the room and is normally at eye level. See Figs. 16-6-7 and 16-6-10.

Parallel-Perspective Grid

A variety of perspective grid sheets is available, which enables the drafter to produce perspective drawings in less time than the conventional manner. Using a grid eliminates the tedious effort of establishing and projecting from the vanishing points for each individual feature. It also eliminates the problem of having the vanishing points located, in many instances, beyond the drawing area.

The cube grid, which is most widely used, has two basic variations: an exte-

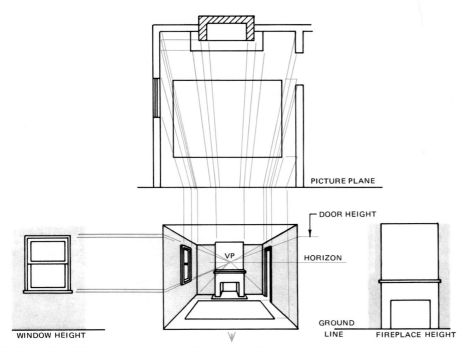

Fig. 16-6-7 Parallel-perspective drawing of an interior of a house.

rior grid and an interior grid. See Fig. 16-6-8. The grid sizes are dependent upon the desired scale of the parts to be drawn. The height and width planes are subdivided into identical increments, each increment representing any convenient size, such as 1.00 in., 1 ft or 10, 100, or 1000 mm. The plane or surface representing the depth is subdivided into increments which are proportionately foreshortened as it recedes from the picture plane and thus creates the perspective illusion. See Figs. 16-6-9 and 16-6-10.

Reference and Source Material

1. *American Drafting Standards Manual, Pictorial Drawing* (ANSI Y14.4).

ASSIGNMENTS

See Assignments 18 through 20 for Unit 16-6 on page 388.

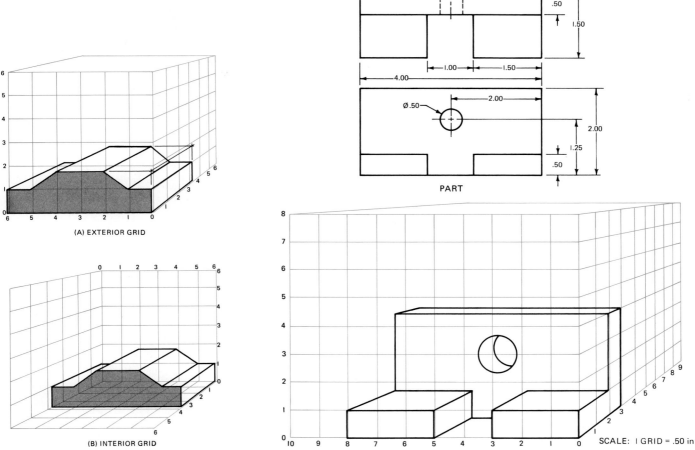

Fig. 16-6-8 Parallel-perspective grid types.

(A) EXTERIOR GRID

(B) INTERIOR GRID

PART

Fig. 16-6-9 Part drawn on parallel-perspective paper-exterior grid.

SCALE: 1 GRID = .50 in

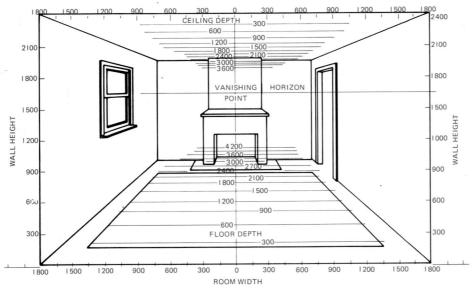

1800 1500 1200 900 600 300 0 300 600 900 1200 1500 1800

CEILING DEPTH — 300

600
2100 — 900
1200
1500
1800
2400 — 2100
3000
3600

VANISHING HORIZON

POINT

1800

WALL HEIGHT

1500

1200

900

600

300

4 200
3 600
3000 2700
2400
2100
1800
1 500
1 200
900
600
FLOOR DEPTH
300

2400

2100

1800

1 500

1 000

900

WALL HEIGHT

600

300

1800 1500 1200 900 600 300 0 300 600 900 1200 1500 1800

ROOM WIDTH

Fig. 16-6-10 Interior of a room drawn on parallel-perspective paper-interior grid.

UNIT 16-7

Angular, or Two-Point, Perspective

Two-point perspective is used quite extensively for architectural and product illustration, as shown in Fig. 16-7-1. *Angular-perspective* drawings are similar to axonometric drawings except that the receding lines converge at two vanishing points located on the horizon. Normally the height or vertical lines are parallel to the picture plane, and the length and width lines recede.

The construction for a simple prism is shown in Fig. 16-7-2. Since line 1-2 rests on the picture plane, it will appear as its true height on the perspective drawing and will be located directly below line 1-2 on the top view. The next step is to join points 1 and 2 with light receding lines to both vanishing points. These receding lines represent the width and length lines of the prism; the width lines recede to *VPL* and the length lines recede to *VPR*. Since line 3-4 on the top view does not rest on the picture plane, it will not appear in its true height nor as its true distance from line 1-2 in the perspective. To find its position on the perspective drawing, join line 3-4, which appears as a point in the top view, to *SP* with a visual ray line. Where this visual ray line intersects the picture plane at *C*, project a vertical line down to the perspective view until it intersects the

receding lines 1-*VPR* and 2-*VPR* at points 3 and 4 respectively. Line 5-6 may be found in the same manner. Next join point 3 to *VPL* and point 5 to *VPR* with light receding lines. The intersection of these lines is point 7.

Lines Not Touching on the Picture Plane Figure 16-7-3 illustrates the construction of a perspective drawing where none of the object lines touch the picture plane.

All these lines can be constructed by using the following procedure, which locates the position and size of lines 1-2 and 3-4. Extend line 1-3 in the top view to intersect the picture plane at point *C*. Project a line down from *C* to intersect horizontal lines 1-*D* and 2-*E* at *D* and *E*, respectively. Had line 1-2 been located at *C* in the top view, it would have appeared at its true height and at *D-E* on

Fig. 16-7-1 Perspective drawing.

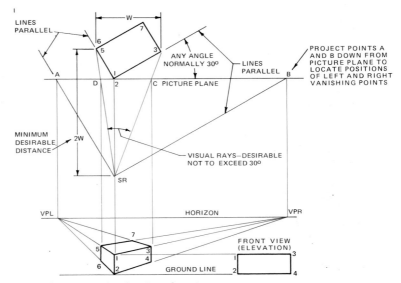

Fig. 16-7-2 Angular-perspective drawing of a prism.

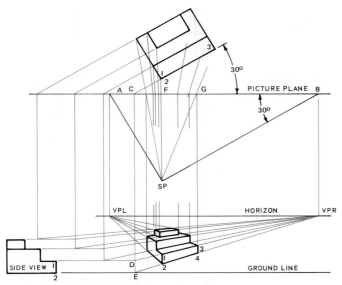

Fig. 16-7-3 Angular-perspective view of an object which does not touch the picture plane.

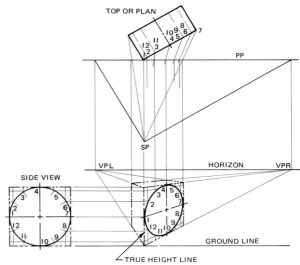

Fig. 16-7-4 Constructing a circle in angular perspective.

the perspective. Join points *D* to *VPR* and *E* to *VPR* with light receding lines. Somewhere along these lines are points 1, 2, 3, and 4. Next join lines 1-2 and 3-4 in the top view to *SP* with visual ray lines. Where these visual ray lines intersect the picture plane at *F* and *G*, respectively, project vertical lines down to the perspective view intersecting line *D-VPR* at 1 and 3 and *E-VPR* at 2 and 4.

Construction of Circles and Curves in Perspective Circles and curves

may be constructed in perspective, as illustrated in Fig. 16-7-4. Using orthographic projections, oriented with respect to the subject in the plan and side views, plot and label the desired points (using numbers) on the curved surfaces. From the plan view project these points to the picture plane, then vertically down to the perspective view. Project horizontally from the side view to the true-height line in the perspective view, the height of the plotting numbers. The position of the plotting numbers

may now be located on the perspective view. Locate the points of intersection of the lines projected down from the picture plane with the visual ray lines receding to the right vanishing point from the appropriate numbers on the true-height line.

Horizon Line Figure 16-7-5 illustrates different effects6produced by repositioning the object with respect to the horizon.

ANGULAR-PERSPECTIVE GRIDS

Exterior Grid When the three adjacent exterior planes of the cube are developed, the resultant image is referred to as an *exterior grid*. In using this grid, the points are projected from the top plane downward and from the picture planes away from the observer. See Figs. 16-7-6 an 16-7-7.

Interior Grid When the three adjacent interior planes of the cube are

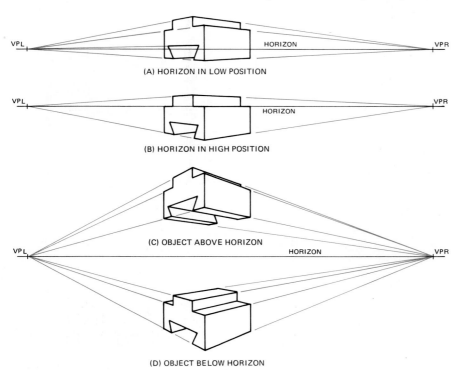

Fig. 16-7-5 Horizon lines.

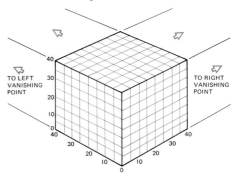

Fig. 16-7-6 Angular-perspective grid.

exposed and developed, the resultant image is referred to as an *interior grid*. In using this grid, the points are projected from the base plane upward and from the picture planes toward the observer. The choice of usage of either variation is a matter of individual preference. Each produces the same results.

Two further variations of both the exterior and interior grids are known as the *bird's-eye* and *worm's-eye grids*. These effects are achieved by rotating the vertical plane of the grid about the horizon line.

Objects drawn in the bird's-eye grid appear as if they were being viewed from above the horizon line, as seen in Fig. 16-7-8. Objects drawn in the worm's-eye grid appear is if they were being viewed from below the horizon line.

Grid Increments

The three surfaces or planes of the grid are subdivided into multiple vertical and horizontal increments. Each increment is proportionately foreshortened as it recedes from the picture plane and thus creates the perspective illusion. The grid increments can be any size desired. See Fig. 16-7-9.

Reference and Source Material

1. General Motors Corporation.

ASSIGNMENTS

See Assignments 21 and 22 for Unit 16-7 on page 389.

UNIT 16-8
Technical Illustration

Technical illustrations have an important place in all phases of engineering drawing. They form an essential part of technical manuals and catalogs, as well as illustrations appearing in technical magazines. However, this unit will not cover technical illustration techniques beyond the scope of the general drafting office.

Technical illustration drawings vary from simple sketches to rather extensive shaded drawings. They may be based upon any of the pictorial methods: isometric, oblique, or perspective, as shown in Fig. 16-8-1. The project being illustrated may be an assembly, a single

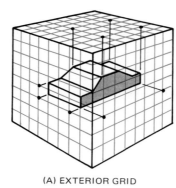

(A) EXTERIOR GRID

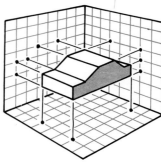

(B) INTERIOR GRID

Fig. 16-7-7 Types of angular-perspective grids.

HORIZON

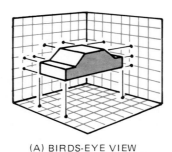

(A) BIRDS-EYE VIEW

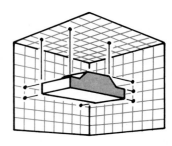

HORIZON

(B) WORMS-EYE VIEW

Fig. 16-7-8 Grid variations.

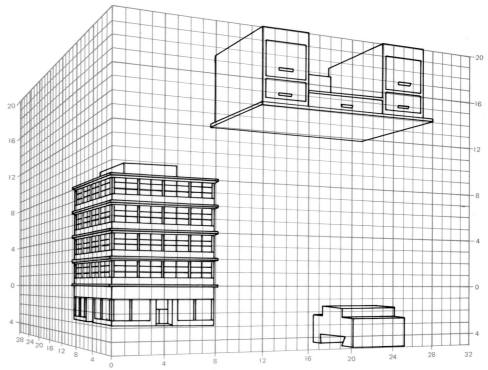

Fig. 16-7-9 Angular-perspective grid application.

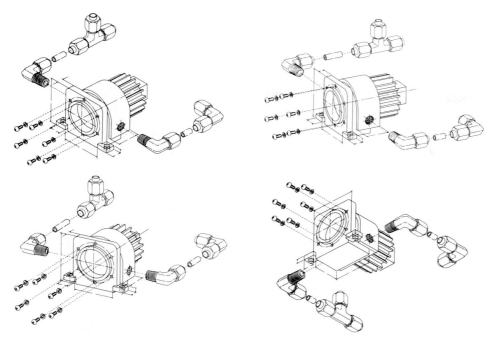

Fig. 16-8-1 A variety of pictorial methods used in technical illustrating. (Graphic Standard Instrument Co.)

part, or just a portion of a part. It may take the form of an exterior view, a sectional view, or a phantom view. The purpose in all cases is to provide a clear and easily understood drawing. See Fig. 16-8-2.

In many cases the drafting department may do the simple illustrations. However, for most purposes the special requirements of such drawings call for work by a professional technical illustrator.

PICTORIAL LINE DRAWINGS

Since nearly all technical illustrations are basically pictorial line drawings, a complete understanding of the various types and their applications is necessary.

While any type of pictorial drawing can be used as the basis for a technical illustration, some types are more suitable than others. This is especially true if the illustration is to be rendered. Figure 16-8-3 shows a V-block drawn in various types of pictorial form. Notice the differences in appearance of each. Isometric is the least natural in appearance; perspective is the most natural. This might suggest, then, that all technical illustrations should be drawn in perspective. This is not necessarily true. While perspective is the most natural in appearance, it takes more time to draw if perspective grid sheets are not used. Thus, it could be more costly.

The shape of the object and how the drawing is to be used also influence the type of pictorial drawing chosen. If the object is circular, it will be easier to draw in oblique rather than isometric if elliptical templates are not available. If the illustration is to be used in a publication such as a journal, operator's manual, technical publication, etc., dimetric, trimetric, or perspective may be the best choice.

LINE APPLICATION

Line thickness in a pictorial drawing has an illusionary meaning. A pictorial drawing created with lines only, that is, without shading, must rely on converging lines and subtle line thicknesses to appear three-dimensional. See Fig. 16-8-4A. Basically, there are three types of lines required to create a good linear pictorial drawing. They are thick, medium, and thin.

Thick Lines Thick or heavy lines are used to emphasize a part in a pictorial drawing. Also, an illusion of depth is created when certain lines are thicker than others. Similarly, a tapered line is used to illustrate depth on a curved surface.

Medium Lines Medium-thickness lines are used as secondary emphasizing lines in the main object of a pictorial. They are also used as depth lines in an unemphasized or background object, as shown in Fig. 16-8-4B.

Thin Lines Thin or fine lines are used as front-edge lines on the surface of the object nearest to the light source. This gives a feeling of light on this edge which further enhances the illusion of perspective. A second use of thin lines is in subduing or deemphasizing an object because it is of secondary importance in

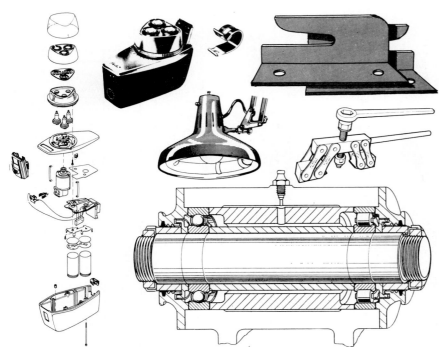

Fig. 16-8-2 Technical illustrations.

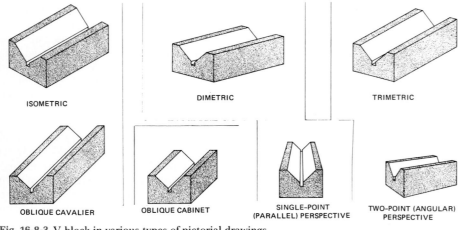

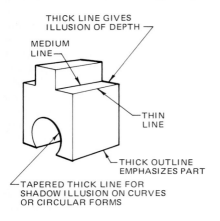

Fig. 16-8-3 V-block in various types of pictorial drawings.

THICK LINE GIVES
ILLUSION OF DEPTH

MEDIUM
LINE

THIN
LINE

THICK OUTLINE
EMPHASIZES PART

TAPERED THICK LINE FOR
SHADOW ILLUSION ON CURVES
OR CIRCULAR FORMS

(A)

Fig. 16-8-4 Line usage.

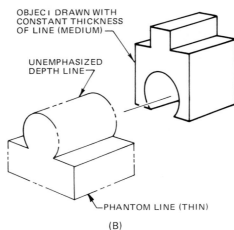

OBJECT DRAWN WITH
CONSTANT THICKNESS
OF LINE (MEDIUM)

UNEMPHASIZED
DEPTH LINE

PHANTOM LINE (THIN)

(B)

WING UPPER
SURFACE

WIRE OR HEAVY CORD. TYPICAL
METHOD OF SUPPORTING
ACTUATOR WHEN CONNECTING
OR DISCONNECTING LINKS

IDLER LINK (M)

DRIVE LINK (J)

IDLER LINK (H)

PIN (K)

ACTUATOR
SUPPORT
STRUCTURE

EXAMPLE (A)

PAD

PLATE

PEG

LOOSE PLATE

EXAMPLE (B)

Fig. 16-8-5 Line application.

RENDERING

For certain applications or where shapes are difficult to read, surface shading or rendering of some kind may be desirable. For most industrial illustrations, accurate descriptions of shapes and positions are more important than fine artistic effects. Desired results can often be

the pictorial. Phantom and broken lines are also used to illustrate background or secondary parts.

Figure 16-8-5 illustrates technical illustrations featuring line application. Notice that only the necessary detail is shown and that just enough shading is added to emphasize and give form to the parts.

IDENTIFICATION ILLUSTRATIONS

Pictorial drawings are very useful in identifying parts. They help save time when the parts are manufactured or assembled in place and are useful for illustrating operating instruction manuals and parts catalogs.

Identification illustrations often take the form of exploded views. If parts are few, they can be identified by names and leaders. The identification illustration in Fig. 16-8-6 is an example showing numbers for the parts and a tabulated parts list. This method is especially recom-

mended where identification of parts is desirable and the viewer is not trained to read technical drawings.

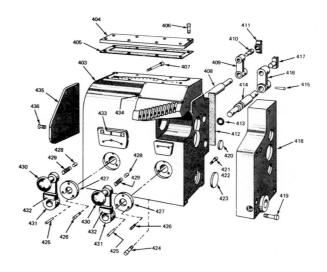

Fig. 16-8-6 Exploded perspective view.

PART NO.	PART NAME	QTY.
403	QUICK CHANGE BOX	1
404	COVER, TOP	1
405	GASKET, COVER	1
406	SCREW, SOCKET HEAD CAP	8
407	SCREW	2
408	SHAFT, SHIFTER	1
409	LINK, SHIFTER	1
410	PIN	1
411	SHOE, SHIFTER	1
412	GASKET (MAKE IN PATTERN SHOP-BOX TO BED)	2
413	"O" RING	2
414	SHAFT, SHIFTER	1
415	PIN, TAPER	2
416	LINK, SHIFTER	1
417	SHOE, SHIFTER	1
418	COVER, SLIP GEAR	1
419	SCREW	4
420	PLUG	1
421	SCREW	3
422	SCREW	1
423	PLUG (NOT USED WITH SCREW REVERSE)	1
424	SCREW	3
425	PIN	2
426	SCREW	6
427	COLLAR	2
428	PLUNGER	2
429	SPRING	2
430	KNOB	2
431	LEVER	2
432	PLATE, FEED-THD.	1
433	PLATE, COMPOUND	1
434	PLATE, ENGLISH INDEX	1
435	COVER	1
436	SCREW	7

obtained without any shading. In general, surface shading should be limited to the least amount necessary to define the shapes illustrated. Different ways of rendering technical illustrations include line shading, screen tints, and special appliqués and pencil shading (smudge).

Some shaded surfaces are indicated in Fig. 16-8-7. An unshaded view is shown at A for comparison. Ruled-surface or crosshatch shading is shown at B, freehand shading at C, and appliqué shading at D.

Line Shading Line shading is a simple, fast, and effective method of defining form. The technique may use straight or curved lines, as shown in Figs. 16-8-8 and 16-8-9.

With the light rays coming in the usual conventional direction, as in Fig. 16-8-10A, the top and front surfaces would be lighted and the right-hand surface would be shaded, as in Fig. 16-8-10B. The front surface can have light shading with heavy shading on the right-hand side, as in Fig. 16-8-10C.

Appliqué Shading Commercial products of varied patterns are used to render areas which require distinction. See Fig. 16-8-11. These products are self-adhering and are easily cut out to match the areas to be covered. Shadow effects are achieved by the addition of another layer of the same material or of a contrast pattern. Both methods are illustrated in Fig. 16-8-12.

Stippling consists of dots, short crooked lines, or similar treatment to produce a shaded effect. It is a good method when it is well done.

Screens (measured in the number of dots per area) are available in a great variety of patterns and are very effective. A combination of line and appliqué rendering is shown in Fig. 16-8-13.

Tonal Pencil Rendering Tonal pencil rendering, while in limited use, produces a finished drawing of more professional quality. See Fig. 16-8-14.

SPECIAL FEATURES

Screw Threads Both internal and external threads are conveniently shown by a series of ellipses uniformly spaced along the axis of the thread. The spacing of these lines may be greater than the actual thread pitch to allow room for the effective line shading. See Fig. 16-8-15.

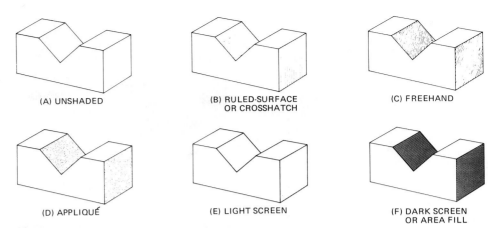

(A) UNSHADED (B) RULED-SURFACE OR CROSSHATCH (C) FREEHAND

(D) APPLIQUÉ (E) LIGHT SCREEN (F) DARK SCREEN OR AREA FILL

Fig. 16-8-7 Examples of various kinds of rendering.

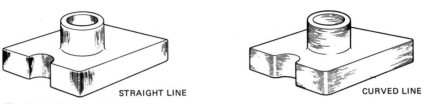

STRAIGHT LINE CURVED LINE

Fig. 16-8-8 Line shading.

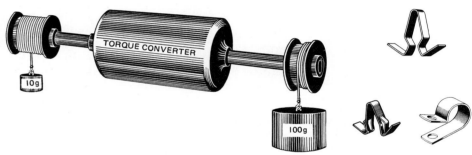

Fig. 16-8-9 Application of line shading.

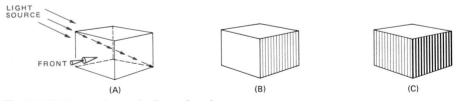

(A) (B) (C)

Fig. 16-8-10 Line rendering the faces of a cube.

SCREENS

LINES

PATTERNS

SHADOW SCREEN

Fig. 16-8-11 Appliqués.

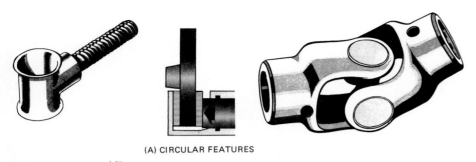

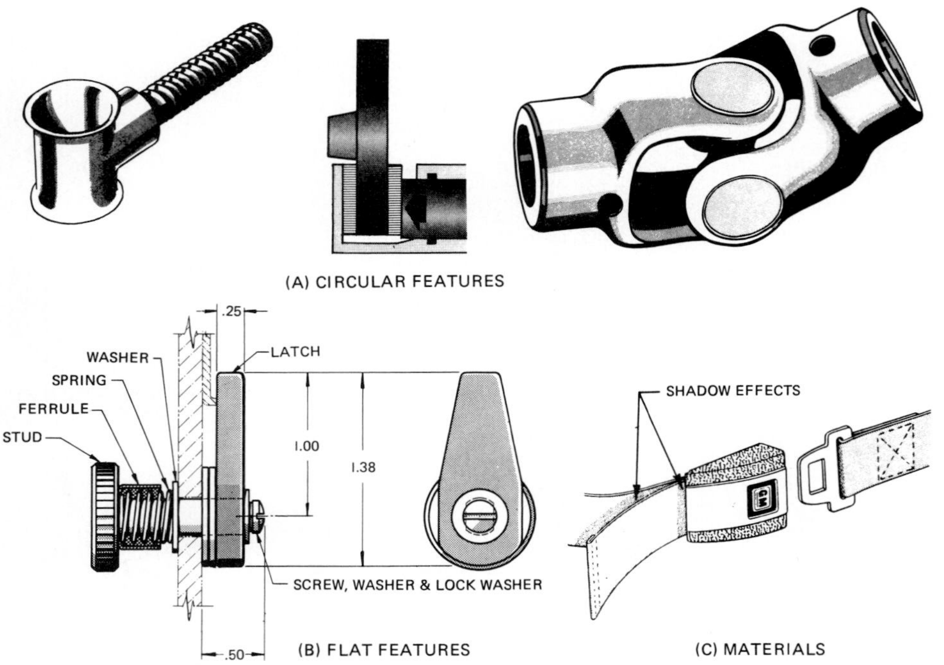

(A) CIRCULAR FEATURES

WASHER
SPRING
FERRULE
STUD
LATCH
.25
1.00
1.38
SCREW, WASHER & LOCK WASHER
.50

(B) FLAT FEATURES

SHADOW EFFECTS

(C) MATERIALS

Fig. 16-8-12 Examples of appliqué rendering.

Fig. 16-8-14 Tonal pencil rendering. (General Motors Corp.)

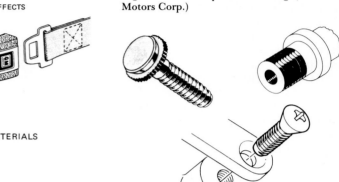

Fig. 16-8-15 Shaded screw threads.

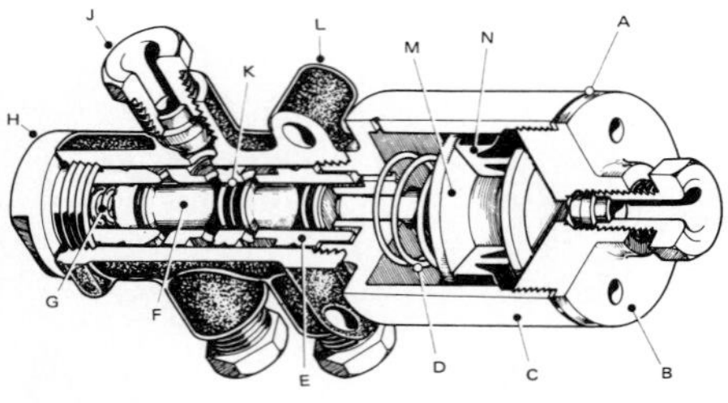

J   L   M   N   A
K
H
G   F   E   D   C   B

Fig. 16-8-13 Combination of line and appliqué rendering. (Holman Bros.)

CAD

The drawing outline and area fill of an illustration may be prepared by CAD. It is then transferred to the illustrator to be finished. He or she will manually apply the necessary shading or rendering. This is a good example of combining the talents of CAD (outline) and manual drafting (rendering) techniques.

Reference and Source Material

1. General Motors Corporation.

ASSIGNMENTS

See Assignments 23 and 24 for Unit 16-8 on page 390.

ASSIGNMENTS FOR CHAPTER 16

Assignments for Unit 16-1, Pictorial Drawings

1. On isometric grid paper or using the CAD isometric grid, draw the parts shown in Fig. 16-1-A. Do not show hidden lines. Each square shown on the drawing represents one isometric square on the grid.
2. On isometric grid paper or using the CAD isometric grid, draw the parts shown in Fig. 16-1-B. Do not show hidden lines. Each square shown on the drawing represents one isometric square on the grid.
3. Make an isometric drawing, complete with dimensions, of one of the parts shown in Figs. 16-1-C to 16-1-F. Scale 1:1.

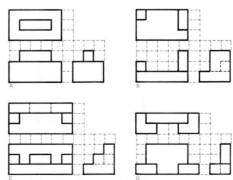

Fig. 16-1-A Isometric flat surface problems.

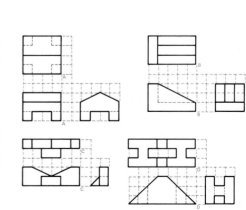

Fig. 16-1-B Isometric flat surface problems.

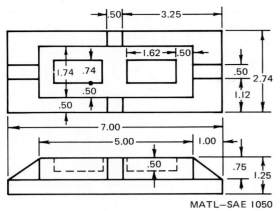

Fig. 16-1-C Base plate.

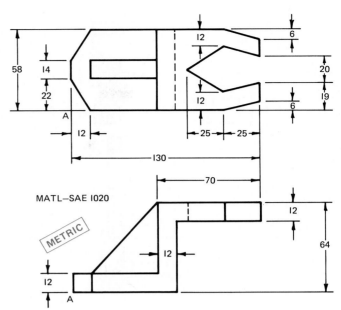

MATL—SAE 1020

METRIC

Fig. 16-1-D Support bracket.

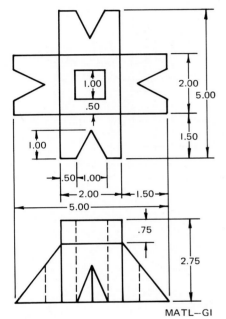

Fig. 16-1-E Base plate.

METRIC

MATL—SAE 1050

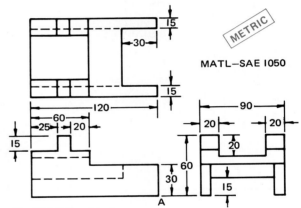

Fig. 16-1-F Step block.

Assignments for Unit 16-2, Curved Surfaces in Isometric

4. On isometric grid paper or using the CAD isometric grid, draw the parts shown in Fig. 16-2-A. Each square shown on the figure represents one square on the isometric grid. Hidden lines may be omitted for clarity.

5. On isometric grid paper or using the CAD isometric grid, draw the parts shown in Fig. 16-2-B. Each square shown on the figure represents one square on the isometric grid. Hidden lines may be omitted for clarity.

6. Make an isometric drawing complete with dimensions of one of the parts shown in Figs. 16-2-C to 16-2-G. Scale 1:2 for Fig. 16-2-G. For all others the scale is 1:1.

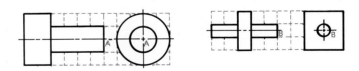

NOTE: LETTERS INDICATE FRONT ON THE ISOMETRIC DRAWING

Fig. 16-2-A Isometric curved surface problems.

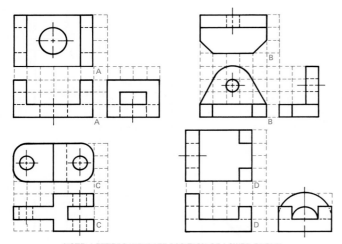

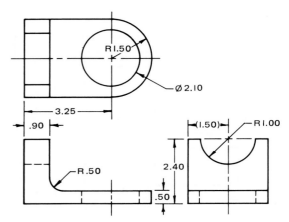

Fig. 16-2-C Cradle bracket.

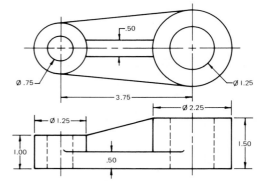

NOTE: LETTERS INDICATE POSITION OF LOWER FRONT
CORNER ON THE ISOMETRIC DRAWING

Fig. 16-2-B Isometric curved surface problems.

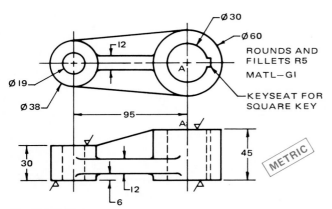

Ø30
Ø60
ROUNDS AND
FILLETS R5
MATL—GI
KEYSEAT FOR
SQUARE KEY
Ø19
Ø38

Fig. 16-2-E Link.

Fig. 16-2-D Link.

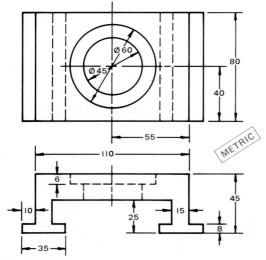

Fig. 16-2-F T-guide.

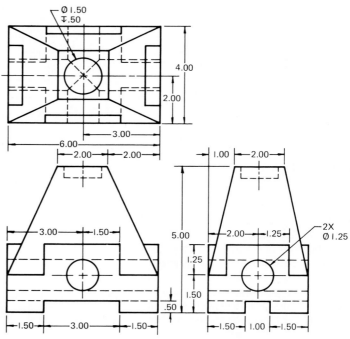

Fig. 16-2-G Base.

Assignments for Unit 16-3, Common Features in Isometric

7. Make an isometric half-section view of one of the parts shown in Fig. 16-3-A or 16-3-B. Scale 1:1.

8. Make an isometric full section drawing of one of the parts shown in Fig. 16-3-C or 16-3-D. Scale 1:1.

9. Make an isometric drawing of the shaft shown in Fig. 16-3-E. Use a conventional break to shorten the length. Scale 1:1.

10. Make an isometric assembly drawing of the two-post die set, model 302, shown in Fig. 16-3-F. Allow 2 in. between the top and base. Scale 1:2. Do not dimension. Include on the drawing an item list. Using part numbers, identify the parts on the assembly.

11. Make an isometric exploded assembly drawing of the book rack shown in Fig. 16-3-G. Use B (A3) sheet. Scale 1:2. Do not dimension. Include on the drawing an item list. Using part numbers, identify the parts on the assembly.

12. Make an isometric exploded assembly drawing of the universal joint shown in Fig. 16-3-H. Scale 1:1. Do not dimension. Include on the drawing an item list. Using part numbers, identify the parts on the assembly.

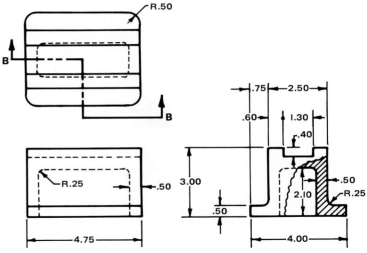

Fig. 16-3-B Base.

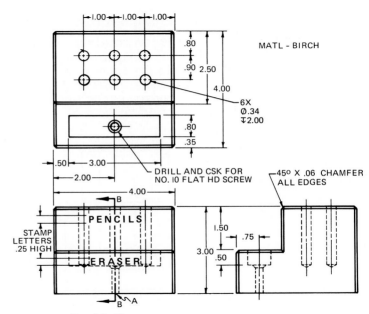

Fig. 16-3-C Pencil holder.

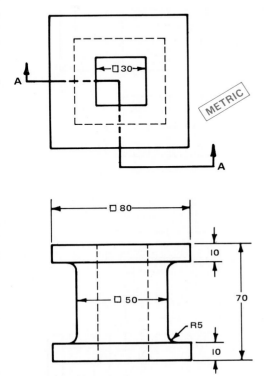

Fig. 16-3-A Guide block.

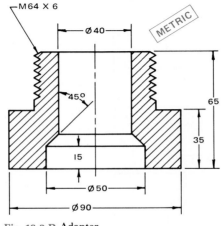

Fig. 16-3-D Adapter.

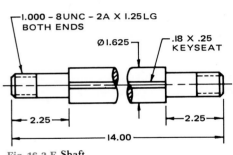

Fig. 16-3-E Shaft.

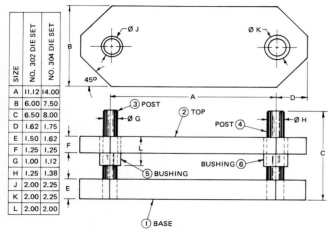

SIZE	NO. 302 DIE SET	NO. 304 DIE SET
A	11.12	14.00
B	6.00	7.50
C	6.50	8.00
D	1.62	1.75
E	1.50	1.62
F	1.25	1.25
G	1.00	1.12
H	1.25	1.38
J	2.00	2.25
K	2.00	2.25
L	2.00	2.00

Fig. 16-3-F Two-post die set.

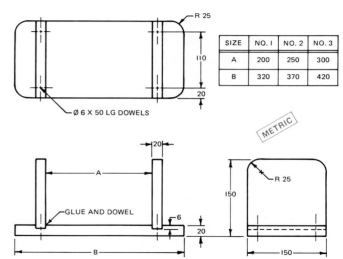

SIZE	NO. I	NO. 2	NO. 3
A	200	250	300
B	320	370	420

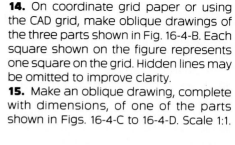

Fig. 16-3-G Book rack.

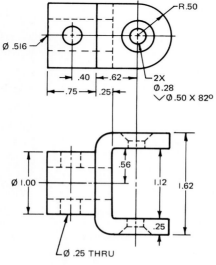

PT I – FORK – 2 REQD

4X
.250 - 20 UNC - 2B X ⤓ .3I

PT 2 – RING – I REQD
PT 3 – Ø .25 SPRING PIN – 2 REQD
PT 4 – .250 - 20 FHMS – .62 LG – 4 REQD

Fig. 16-3-H Universal joint.

Assignments for Unit 16-4, Oblique Projection

13. On coordinate grid paper or using the CAD grid, make oblique drawings of the three parts shown in Fig. 16-4-A. Each square shown on the figure represents one square on the grid. Hidden lines may be omitted to improve clarity.

14. On coordinate grid paper or using the CAD grid, make oblique drawings of the three parts shown in Fig. 16-4-B. Each square shown on the figure represents one square on the grid. Hidden lines may be omitted to improve clarity.

15. Make an oblique drawing, complete with dimensions, of one of the parts shown in Figs. 16-4-C to 16-4-D. Scale 1:1.

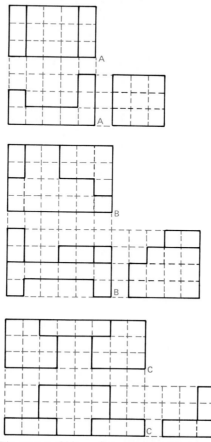

Fig. 16-4-A Oblique flat surface problems.

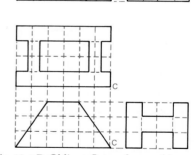

Fig. 16-4-B Oblique flat surface problems.

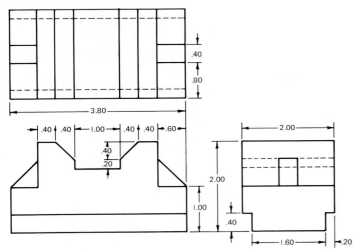

Fig. 16-4-C V-block rest.

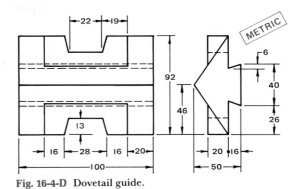

Fig. 16-4-D Dovetail guide.

Assignments for Unit 16-5, Common Features in Oblique

16. Make an oblique drawing, complete with dimensions, of one of the parts shown in Fig. 16-5-A or 16-5-B. Scale 1:2.

17. Make an oblique drawing, complete with dimensions, of one of the parts shown in Fig. 16-5-C or 16-5-D. Use the straight-line method of showing the rounds and fillets. Scale 1:2 for Fig. 16-5-C and 1:1 for Fig. 16-5-D.

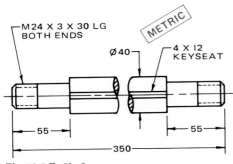

Fig. 16-5-B Shaft.

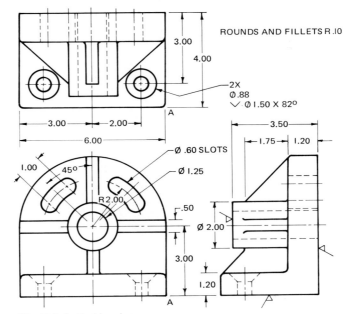

ROUNDS AND FILLETS R .10

2X
Ø .88
⌵ Ø 1.50 X 82°

Ø .60 SLOTS

Ø 1.25

R 2.00

Fig. 16-5-A End bracket.

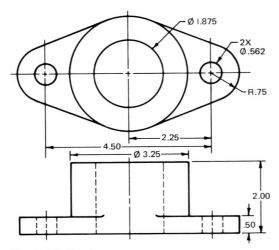

Fig. 16-5-C Bearing support.

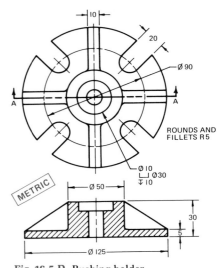

ROUNDS AND FILLETS R5

Fig. 16-5-D Bushing holder.

Assignments for Unit 16-6, Perspective Projection

18. Make a parallel-perspective drawing of the vise base shown in Fig. 16-6-A. Scale 1:1.

19. Make a parallel-perspective drawing of one of the parts shown in Fig. 16-6-B or 16-6-C. Scale 1:1.

20. With the aid of a parallel-perspective grid, make a perspective drawing of the triple bookcase shown in Fig. 16-6-D. Scale is to suit. Each unit is 36 × 12 in. × 6 ft or 900 mm W × 300 mm D × 1800 mm H. Use your judgment for sizes not shown.

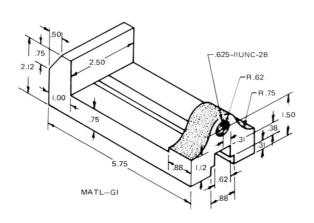

Fig. 16-6-A Vise base.

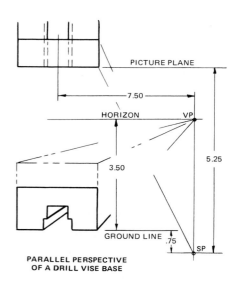

PARALLEL PERSPECTIVE OF A DRILL VISE BASE

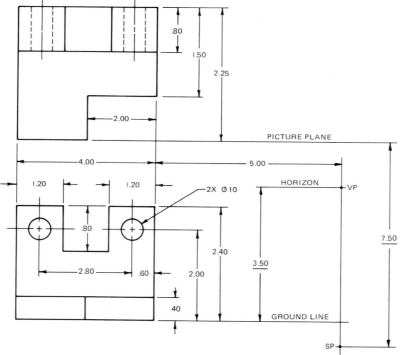

Fig. 16-6-B Spacer block.

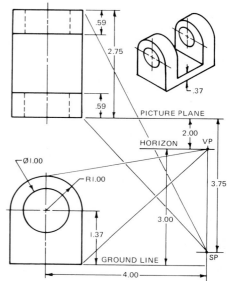

Fig. 16-6-C Shaft support.

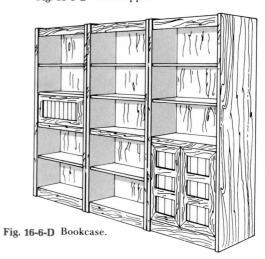

Fig. 16-6-D Bookcase.

Assignments for Unit 16-7, Angular, or Two-Point, Perspective

21. Make an angular-perspective drawing of the planter box or monument shown in Fig. 16-7-A or 16-7-B. Scale 1:5 or 1:4.

22. On angular-perspective grid paper make a bird's-eye perspective view of the tool holder or cross slide shown in Fig. 16-7-C or 16-7-D. Scale is to suit.

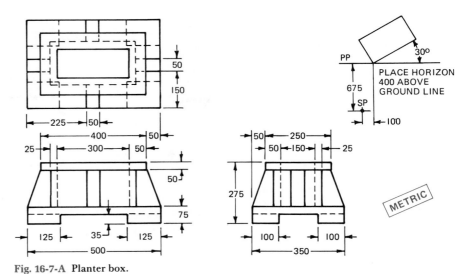

Fig. 16-7-A Planter box.

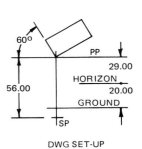

DWG SET-UP

Fig. 16-7-B Monument.

Fig. 16-7-C Tool holder.

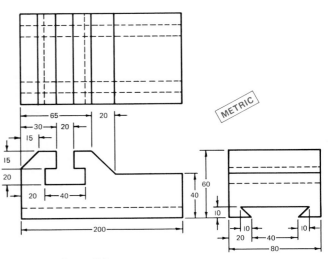

Fig. 16-7-D Cross slide.

Assignments for Unit 16-8, Technical Illustration

23. Prepare two identical pictorial drawings of one of the parts shown in Fig. 16-8-A or 16-8-B. On the first drawing use line shading; on the second, use appliqués. Do not dimension. Select any pictorial method you wish. Scale 1:1.

24. Make an exploded isometric assembly drawing of one of the assemblies shown in 16-8-C or 16-8-D. Scale 1:1. Do not dimension. Use a rendering technique of your choice to improve the appearance of the drawing.

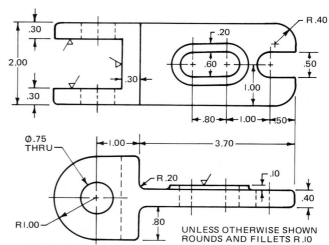

Fig. 16-8-A Swing bracket.

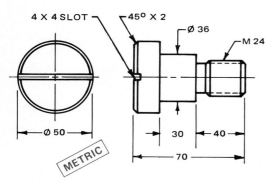

METRIC

Fig. 16-8-B Stop button.

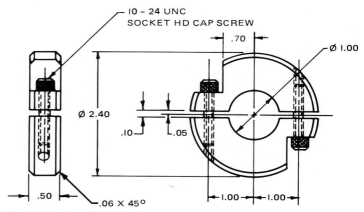

Fig. 16-8-C Gear clamp.

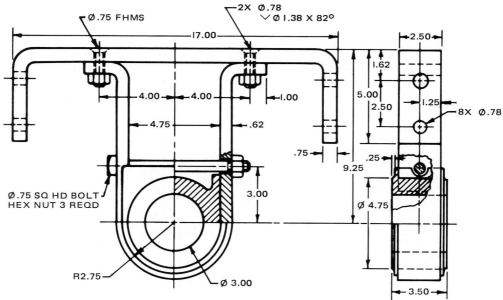

Fig. 16-8-D Bearing bracket.

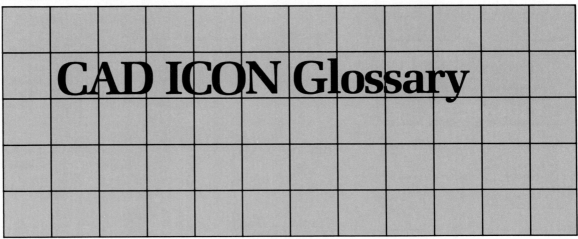

Graphical representation and definition for each major command used in CAD.

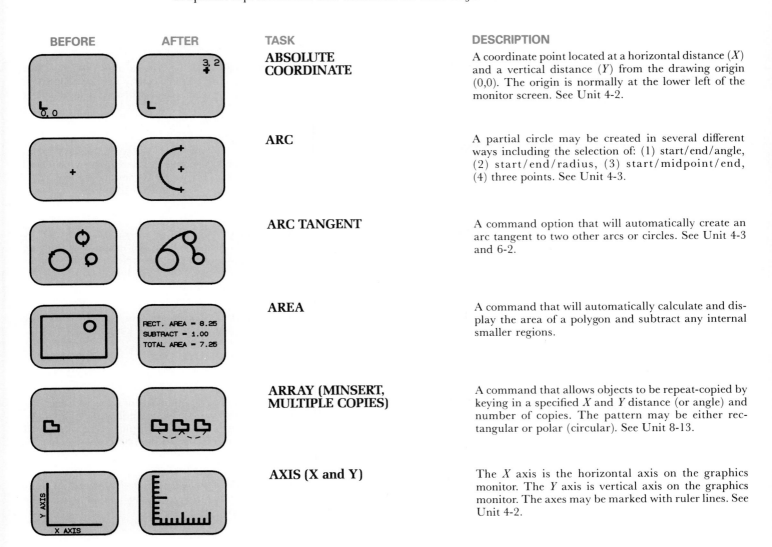

BEFORE	AFTER	TASK	DESCRIPTION
		ABSOLUTE COORDINATE	A coordinate point located at a horizontal distance (X) and a vertical distance (Y) from the drawing origin (0,0). The origin is normally at the lower left of the monitor screen. See Unit 4-2.
		ARC	A partial circle may be created in several different ways including the selection of: (1) start/end/angle, (2) start/end/radius, (3) start/midpoint/end, (4) three points. See Unit 4-3.
		ARC TANGENT	A command option that will automatically create an arc tangent to two other arcs or circles. See Unit 4-3 and 6-2.
		AREA	A command that will automatically calculate and display the area of a polygon and subtract any internal smaller regions.
		ARRAY (MINSERT, MULTIPLE COPIES)	A command that allows objects to be repeat-copied by keying in a specified X and Y distance (or angle) and number of copies. The pattern may be either rectangular or polar (circular). See Unit 8-13.
		AXIS (X and Y)	The X axis is the horizontal axis on the graphics monitor. The Y axis is vertical axis on the graphics monitor. The axes may be marked with ruler lines. See Unit 4-2.

BEFORE	AFTER	TASK	DESCRIPTION
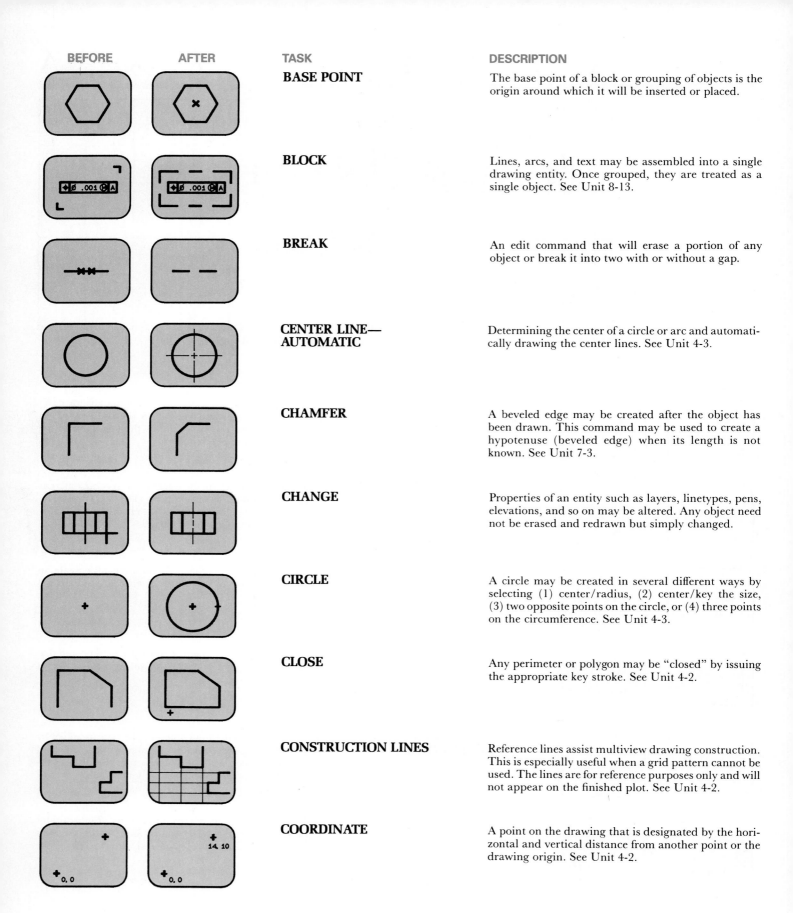		**BASE POINT**	The base point of a block or grouping of objects is the origin around which it will be inserted or placed.
		BLOCK	Lines, arcs, and text may be assembled into a single drawing entity. Once grouped, they are treated as a single object. See Unit 8-13.
		BREAK	An edit command that will erase a portion of any object or break it into two with or without a gap.
		CENTER LINE— AUTOMATIC	Determining the center of a circle or arc and automatically drawing the center lines. See Unit 4-3.
		CHAMFER	A beveled edge may be created after the object has been drawn. This command may be used to create a hypotenuse (beveled edge) when its length is not known. See Unit 7-3.
		CHANGE	Properties of an entity such as layers, linetypes, pens, elevations, and so on may be altered. Any object need not be erased and redrawn but simply changed.
		CIRCLE	A circle may be created in several different ways by selecting (1) center/radius, (2) center/key the size, (3) two opposite points on the circle, or (4) three points on the circumference. See Unit 4-3.
		CLOSE	Any perimeter or polygon may be "closed" by issuing the appropriate key stroke. See Unit 4-2.
		CONSTRUCTION LINES	Reference lines assist multiview drawing construction. This is especially useful when a grid pattern cannot be used. The lines are for reference purposes only and will not appear on the finished plot. See Unit 4-2.
		COORDINATE	A point on the drawing that is designated by the horizontal and vertical distance from another point or the drawing origin. See Unit 4-2.

BEFORE	AFTER	TASK	DESCRIPTION
		COPY	Entities or blocks of entities may be copied at any location. See Unit 7-7.
		CURSOR	An indicating mark on the monitor which is controlled by any of several types of input pointing devices such as, a joystick, puck, or stylus. See Unit 2-3.
		DIMENSION—ANGULAR	The automatic generation of a dimension indicating the angle between two nonparallel lines. See Unit 7-1.
		DIMENSION— ASSOCIATIVE (STRETCH)	Automatic revision of a dimension and text when an object is stretched, scaled, or rotated. The new dimension is created in its entirety according to the new size and orientation. See Unit 14-4.
		DIMENSION— BASE LINE	Linear dimensions continued from a common base line. The first extension line is common to each dimension. See Unit 7-4.
		DIMENSION—LINEAR	An automatic command that will calculate and place a dimension of an entity after it has been identified. See Unit 7-1.
		DIMENSION— RADIAL	The size of a selected circle or arc is calculated and the dimension is automatically placed. See Unit 7-2.
		DIMENSION VARIABLES	The values which determine the manner in which system dimensions are placed on the drawing may be varied. See Unit 7-1.
		DISTANCE/ANGLE (LIST, ID)	Describes spatial characteristics of a segment or between two points. The characteristics may include distance between end points, angle, the change in horizontal distance, the change in vertical distance, and location with respect to the origin. See Unit 4-3.
		DIVIDE (MEASURE)	An entity may be divided into any specified number of equal-length intervals. See Unit 6-1.

DONUT

A solid (filled) circle, with or without a hole, may be automatically created. Elimination of the fill results in two circles concentric about a common center. See Unit 4-3.

DRAG

Dragging provides visual movement of an object. Each time the cursor is moved (dragged), the object will be redisplayed at the new position.

EDIT-LINE (STRETCH, EXTEND, MODIFY, TRIM, GAP)

Any line or vertex may be altered. Modifications include adding or deleting a vertex and joining, moving, stretching, extending, gapping, or shortening a line.

ELLIPSE (ISOPLANE)

An ellipse may be automatically constructed and also be structured to fit any of the three standard isometric planes or faces. See Unit 6-4.

END (QUIT)

When finishing work, deactivate and return to the main menu prior to logging off. Be careful to first save the drawing if it is to be used again. See Unit 2-3.

ERASE

Any object or an identified rectangular window area may easily be erased. See Unit 4-2.

ERASE LAST

Mistakes often are immediately recognized as such. This handy command will cancel the last command entered. See Unit 4-2.

EXPLODE

Multiple entities blocked into a single object, as well as dimensions, may later be broken apart. Individual entities may then be altered without affecting the others.

FILING (FILES IN/OUT)

After a drawing is complete, it may be permanently stored. A common filing procedure involves "off-loading" the contents on to a floppy disk prior to storage. See Unit 3-2.

FILLET

A command that will draw an arc tangent to two selected lines at a specified radius. See Unit 4-3 and 6-2.

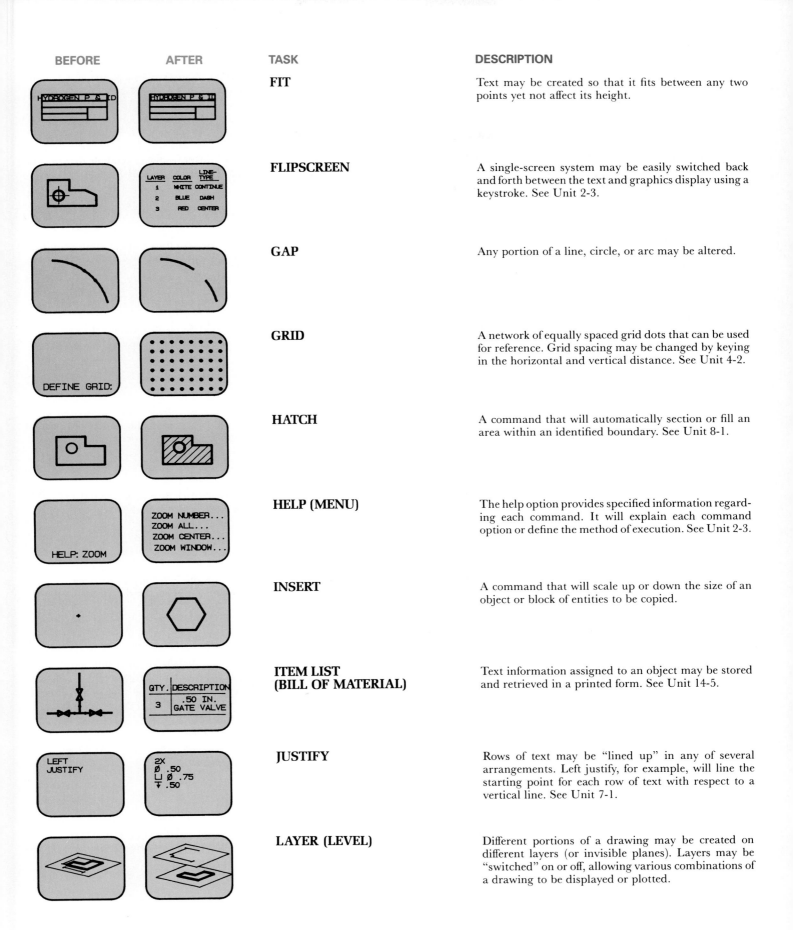

FIT

Text may be created so that it fits between any two points yet not affect its height.

FLIPSCREEN

A single-screen system may be easily switched back and forth between the text and graphics display using a keystroke. See Unit 2-3.

GAP

Any portion of a line, circle, or arc may be altered.

GRID

A network of equally spaced grid dots that can be used for reference. Grid spacing may be changed by keying in the horizontal and vertical distance. See Unit 4-2.

HATCH

A command that will automatically section or fill an area within an identified boundary. See Unit 8-1.

HELP (MENU)

The help option provides specified information regarding each command. It will explain each command option or define the method of execution. See Unit 2-3.

INSERT

A command that will scale up or down the size of an object or block of entities to be copied.

ITEM LIST (BILL OF MATERIAL)

Text information assigned to an object may be stored and retrieved in a printed form. See Unit 14-5.

JUSTIFY

Rows of text may be "lined up" in any of several arrangements. Left justify, for example, will line the starting point for each row of text with respect to a vertical line. See Unit 7-1.

LAYER (LEVEL)

Different portions of a drawing may be created on different layers (or invisible planes). Layers may be "switched" on or off, allowing various combinations of a drawing to be displayed or plotted.

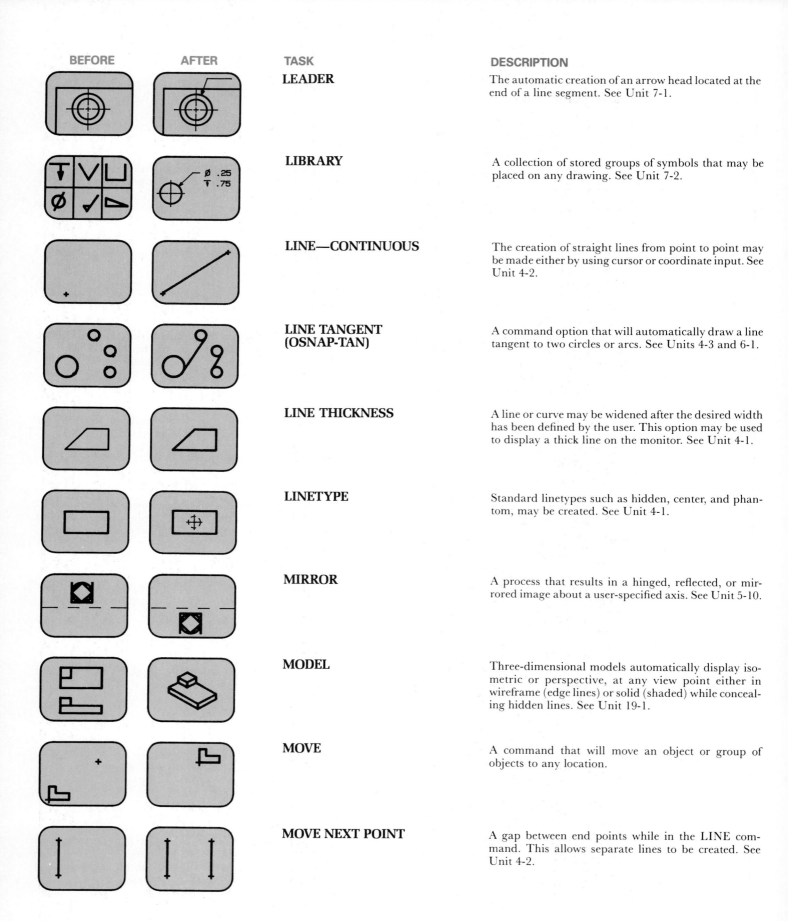

BEFORE	AFTER	TASK	DESCRIPTION
		LEADER	The automatic creation of an arrow head located at the end of a line segment. See Unit 7-1.
		LIBRARY	A collection of stored groups of symbols that may be placed on any drawing. See Unit 7-2.
		LINE—CONTINUOUS	The creation of straight lines from point to point may be made either by using cursor or coordinate input. See Unit 4-2.
		LINE TANGENT (OSNAP-TAN)	A command option that will automatically draw a line tangent to two circles or arcs. See Units 4-3 and 6-1.
		LINE THICKNESS	A line or curve may be widened after the desired width has been defined by the user. This option may be used to display a thick line on the monitor. See Unit 4-1.
		LINETYPE	Standard linetypes such as hidden, center, and phantom, may be created. See Unit 4-1.
		MIRROR	A process that results in a hinged, reflected, or mirrored image about a user-specified axis. See Unit 5-10.
		MODEL	Three-dimensional models automatically display isometric or perspective, at any view point either in wireframe (edge lines) or solid (shaded) while concealing hidden lines. See Unit 19-1.
		MOVE	A command that will move an object or group of objects to any location.
		MOVE NEXT POINT	A gap between end points while in the LINE command. This allows separate lines to be created. See Unit 4-2.

BEFORE	AFTER	TASK	DESCRIPTION
		MULTISEGMENT LINE	Continuous line segments may be created from point to point. See Unit 4-2.
		OBJECT SNAP (OSNAP)	Entities may be "locked" to various positions of an object. Common positions include: endpoints, midpoints, tangent, center, nearest, quadrant, and perpendicular. See Unit 4-2.
		ORTHO-ON	This option automatically forces a line to be drawn on a horizontal or vertical axis only. See Unit 4-2.
		ORTHO-OFF	Allows the creation of diagonal lines at any angle. See Unit 4-2.
		PAN	A scan to a different portion of a drawing while zoomed in. See Unit 4-2.
		PARALLEL (OFFSET)	Identical lines or curves may be created from an existing line or curve at any offset distance. See Unit 6-1.
		PEN	Pen numbers refer to the plotting of a drawing to produce different line weights. A number 2 pen selection will be plotted with the pen that has been placed in the number 2 plotter position. See Unit 4-2.
		PERIMETER	A command that will calculate and display the perimeter distance around an object.
		PERPENDICULAR	An option that will create a line at 90° to another line.
		PLINE (SPLINE)	An irregular curve may be constructed after points along that curve have been identified. See Units 4-4 and 6-4.

BEFORE	AFTER	TASK	DESCRIPTION
		PLOT	A plotter will create a finished drawing. Linework and lettering of professional quality will be produced. See Unit 2-3.
		POINT	Points may be located any place on the screen by either keying in the coordinates or picking a cursor position. See Unit 4-2.
		POLAR COORDINATES	A line or second point is located by keying its distance and angle from an existing point. See Unit 4-2.
		POLYGON	A polygon of any number of sides (greater than 2) may be inscribed or circumscribed about a circle of any radius. See Unit 6-3.
		POLYLINE	A series of connecting entities between cursor points and is considered a single object. See Unit 4-4.
		RECTANGLE	Creation of a rectangle by picking diagonal corners with the cursor or keying in the coordinates.
		REDRAW	After deleting, the display often will appear messy. Redrawing will redisplay the objects on the screen, "cleaning up" the display. See Unit 4-2.
		REGENERATE (PACK)	Accumulated deletions and revision work utilize a significant amount of useless drawing storage (disk) space. Data packing will discard this from drawing memory, leaving more room for useful work.
		RELATIVE COORDINATE	A coordinate position located at a horizontal and vertical distance from (relative to) the last point. See Unit 4-2.
		ROTATE	An object or group of objects may rotate about an origin at any specified angle.

BEFORE	AFTER	TASK	DESCRIPTION

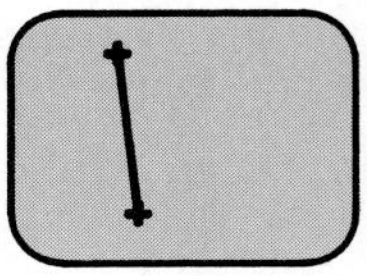

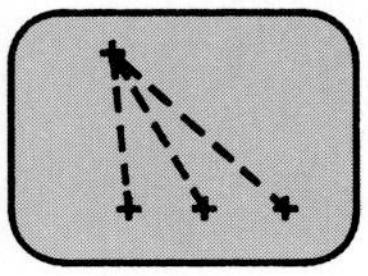

RUBBERBAND

As the cursor moves across the screen, the entity (line, circle, arc) will be redisplayed. This provides an excellent visual guide for optimizing the entity size and position. See Unit 4-2.

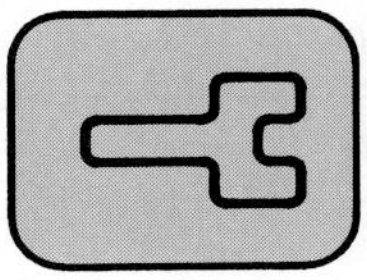

SAVE

COMMAND: SAVE
NAME: YOKE

It is desirable to periodically update the drawing file to disk. This way in the event of a power failure or lock up, only work since the last save will be lost. See Unit 3-2.

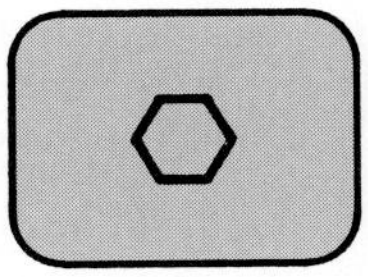

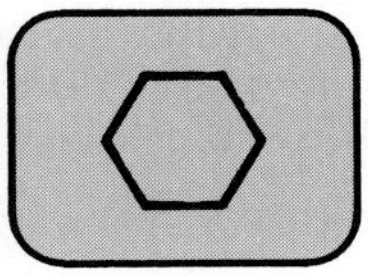

SCALE

A command that will scale up or down the size of an object or group of objects when specifying any scale factor.

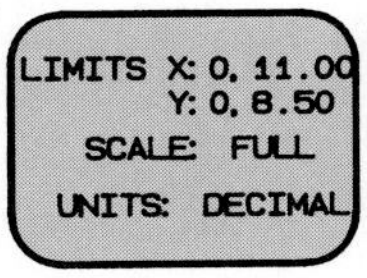

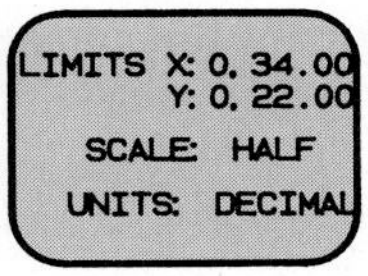

SETUP

LIMITS X: 0, 11.00
Y: 0, 8.50
SCALE: FULL
UNITS: DECIMAL

LIMITS X: 0, 34.00
Y: 0, 22.00
SCALE: HALF
UNITS: DECIMAL

When beginning a new drawing, the parameters such as drawing size, scale, and units may be selected. See Unit 3-1.

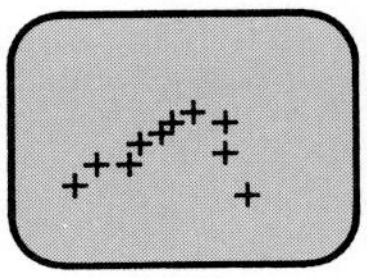

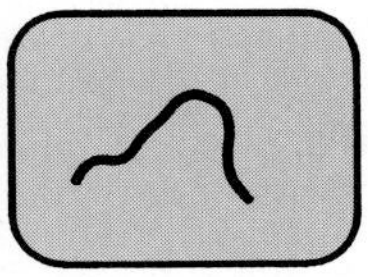

SKETCH

This command permits freehand drawings to be created. It is used for irregular shape applications. See Unit 4-5.

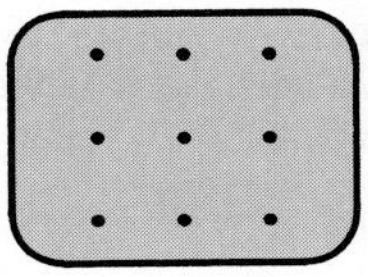

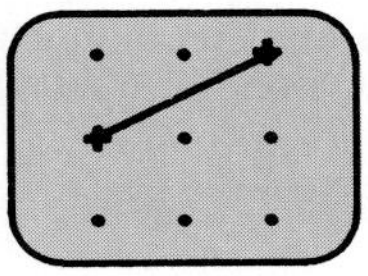

SNAP-ON

A mode that will automatically snap the cursor to the nearest snap increment. The snap increment is often set to the grid spacing or half the grid spacing. See Unit 4-2.

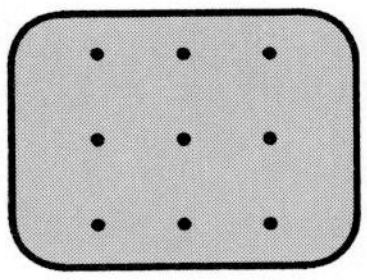

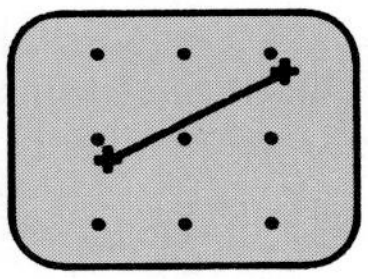

SNAP-OFF

A mode that will allow any point to be located at any position on the screen regardless of the grid pattern. See Unit 4-2.

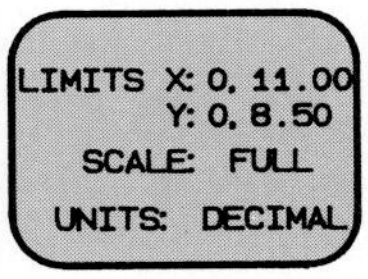

STATUS (SETVAR)

LIMITS X: 0, 11.00
Y: 0, 8.50
SCALE: FULL
UNITS: DECIMAL

The established parameters for a particular drawing may be displayed on the monitor. See Unit 3-1.

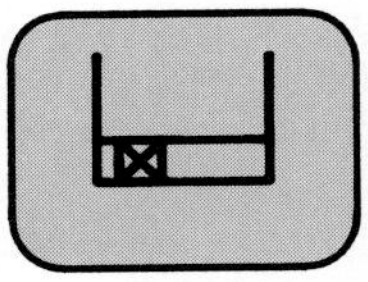

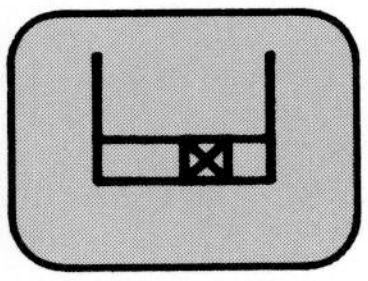

STRETCH

Moves a selected portion of a drawing to a new position while retaining the original shape.

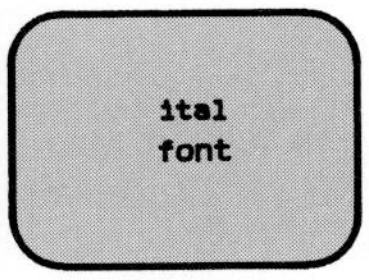

TEXT STYLE (FONT, SCRIPT)

STANDARD
FONT

ital
font

Various text-style options are available and may be selected for use on any drawing. See Unit 4-2.

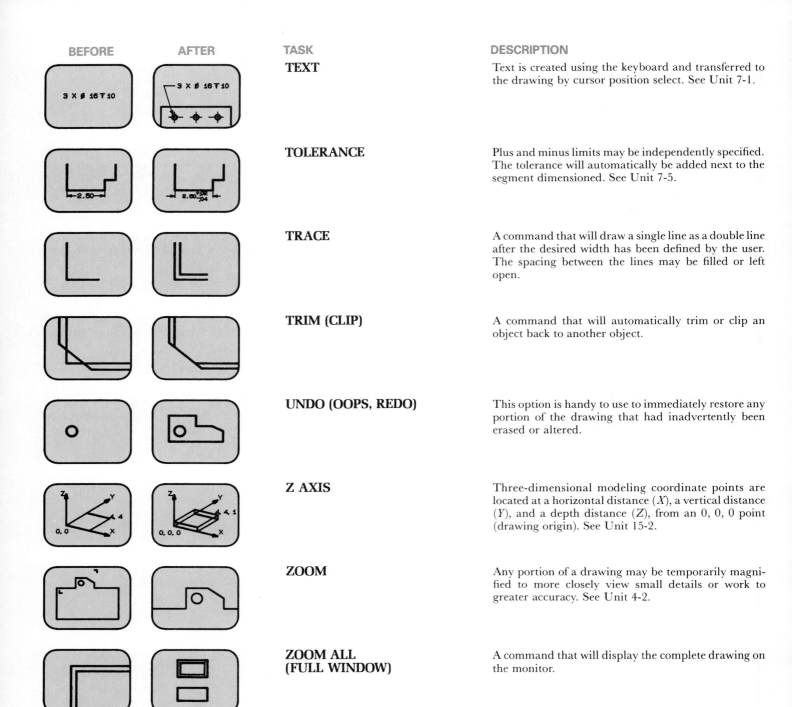

BEFORE	AFTER	TASK	DESCRIPTION
		TEXT	Text is created using the keyboard and transferred to the drawing by cursor position select. See Unit 7-1.
		TOLERANCE	Plus and minus limits may be independently specified. The tolerance will automatically be added next to the segment dimensioned. See Unit 7-5.
		TRACE	A command that will draw a single line as a double line after the desired width has been defined by the user. The spacing between the lines may be filled or left open.
		TRIM (CLIP)	A command that will automatically trim or clip an object back to another object.
		UNDO (OOPS, REDO)	This option is handy to use to immediately restore any portion of the drawing that had inadvertently been erased or altered.
		Z AXIS	Three-dimensional modeling coordinate points are located at a horizontal distance (X), a vertical distance (Y), and a depth distance (Z), from an 0, 0, 0 point (drawing origin). See Unit 15-2.
		ZOOM	Any portion of a drawing may be temporarily magnified to more closely view small details or work to greater accuracy. See Unit 4-2.
		ZOOM ALL (FULL WINDOW)	A command that will display the complete drawing on the monitor.

APPENDIX B

Standard Parts and Technical Data

ANSI	Y1.1	**Abbreviations for Use on Drawings and in Text**				
ANSI	Y14.1	Drawing Sheet Size and Format	ANSI	Y32.9	Graphic Electrical Wiring Symbols for Architectural and Electrical Layout Drawings
ANSI	Y14.2M	Line Conventions and Lettering			
ANSI	Y14.3	Projections	ANSI	B1.1	Unified Screw Threads
ANSI	Y14.4	Pictorial Drawing	ANSI	B4.2	Preferred Metric Limits and Fits
ANSI	Y14.5M	Dimensioning and Tolerancing for Engineering Drawings	ANSI	B17.1	Keys and Keyseats
			ANSI	B17.2	Woodruff Key and Keyslot Dimensions
ANSI	Y14.6	Screw Thread Representation	ANSI	B18.2.1	Square and Hex Bolts and Screws
ANSI	Y14.7	Gears, Splines, and Serrations	ANSI	B18.2.2	Square and Hex Nuts
ANSI	Y14.7.	Gear Drawing Standards—Part 1, for Spur, Helical, Double Helical and Rack	ANSI	B18.3	Socket Cap, Shoulder, and Setscrews
			ANSI	B18.6.2	Slotted-Head Cap Screws, Square-Head Setscrews, Slotted-Headless Setscrews
ANSI	Y14.9	Forgings			
ANSI	Y14.10	Metal Stampings	ANSI	B18.6.3	Machine Screws and Machine Screw Nuts
ANSI	Y14.11	Plastics	ANSI	B18.21.1	Lock Washers
ANSI	Y14.14	Mechanical Assemblies	ANSI	B27.2	Plain Washers
ANSI	Y14.15	Electrical and Electronics Diagrams	ANSI	B46.1	Surface Texture
ANSI	Y14.15A	Interconnection Diagrams	ANSI	B94.6	Knurling
ANSI	Y14.17	Fluid Power Diagrams	ANSI	B94.11M	Twist Drills
ANSI	Y14.36	Surface Texture Symbols	ANSI .	Z210.1	Metric Practice
ANSI	Y32.2	Graphic Symbols for Electrical and Electronics Diagrams			

Table 1   ANSI publications.

Quantity	Metric Unit	Symbol	Metric to Inch-Pound Unit	Inch-Pound to Metric Unit
Length	millimeter	mm	1 mm = 0.0394 in.	1 in. = 25.4 mm
	centimeter	cm	1 cm = 0.394 in.	1 ft. = 30.5 cm
	meter	m	1 m = 39.37 in. = 3.28 ft	1 yd. = 0.914 m = 914 mm
	kilometer	km	1 km = 0.62 mile	1 mile = 1.61 km
Area	square millimeter	mm²	1 mm² = 0.001 55 sq. in.	1 sq. in. = 6 452 mm²
	square centimeter	cm²	1 cm² = 0.155 sq. in.	1 sq. ft. = 0.093 m²
	square meter	m²	1 m² = 10.8 sq. ft. = 1.2 sq. yd.	1 sq. yd. = 0.836 m²
Mass	milligram	mg	1 g = 0.035 oz.	1 oz. = 28.3 g
	gram	g	1 kg = 2.205 lb.	1 lb. = 0.454 kg
	kilogram	kg	1 tonne = 1.102 tons	1 ton = 907.2 kg
	tonne	t		= 0.907 tonnes
Volume	cubic centimeter	cm³	1 mm³ = 0.000 061 cu. in.	1 fl. oz. = 28.4 cm³
	cubic meter	m³	1 cm³ = 0.061 cu. in.	1 cu. in. = 16.387 cm³
	milliliter	m	1 m³ = 35.3 cu ft. = 1.308 cu. yd.	1 cu. ft. = 0.028 m³
			1 mℓ = 0.035 fl. oz.	1 cu. yd. = 0.756 m³
Capacity	liter	L	U.S. Measure 1 pt. = 0.473 L 1 qt. = 0.946 L 1 gal = 3.785 L Imperial Measure 1 pt. = 0.568 L 1 qt. = 1.137 L 1 gal = 4.546 L	U.S. Measure 1 L = 2.113 pt. = 1.057 qt. = 0.264 gal. Imperial Measure 1 L = 1.76 pt. = 0.88 qt. = 0.22 gal.
Temperature	Celsius degree	°C	$°C = \frac{5}{9}(°F-32)$	$°F = \frac{9}{5} \times °C + 32$
Force	newton	N	1 N = 0.225 lb (f)	1 lb (f) = 4.45N
	kilonewton	kN	1 kN = 0.225 kip (f) = 0.112 ton (f)	= 0.004 448 kN
Energy/Work	joule	J	1 J = 0.737 ft · lb	1 ft · lb = 1.355 J
	kilojoule	kJ	1 J = 0.948 Btu	1 Btu = 1.055 J
	megajoule	MJ	1 MJ = 0.278 kWh	1 kWh = 3.6 MJ
Power	kilowatt	kW	1 kW = 1.34 hp 1 W = 0.0226 ft · lb/min.	1 hp (550 ft · lb/s) = 0.746 kW 1 ft · lb/min = 44.2537 W
Pressure	kilopascal	kPa	1 kPa = 0.145 psi = 20.885 psf = 0.01 ton-force per sq. ft.	1 psi = 6.895 kPa 1 lb-force/sq. ft. = 47.88 Pa 1 ton-force/sq. ft. = 95.76 kPa
	*kilogram per square centimeter	kg/cm²	1 kg/cm² = 13.780 psi	
Torque	newton meter	N · m	1 N · m = 0.74 lb · ft	1 lb · ft = 1.36 N · m
	*kilogram meter	kg/m	1 kg/m = 7.24 lb · ft	1 lb · ft = 0.14 kg/m
	*kilogram per centimeter	kg/cm	1 kg/cm = 0.86 lb · in	1 lb · in = 1.2 kg/cm
Speed/Velocity	meters per second	m/s	1 m/s = 3.28 ft/s	1 ft/s = 0.305 m/s
	kilometers per hour	km/h	1 km/h = 0.62 mph	1 mph = 1.61 km/h

*Not SI units, but included here because they are employed on some of the gages and indicators currently in use in industry.

Table 2   Metric conversion tables.

ONE HUNDREDTH OF AN INCH INCREMENTS TO ONE INCH

Inch	.00	.01	.02	.03	.04	.05	.06	.07	.08	.09
.00	0.00	0.25	0.51	0.76	1.02	1.27	1.52	1.78	2.03	2.29
.10	2.54	2.79	3.05	3.30	3.56	3.81	4.06	4.32	4.57	4.83
.20	5.08	5.33	5.59	5.84	6.10	6.35	6.60	6.86	7.11	7.37
.30	7.62	7.87	8.13	8.38	8.64	8.89	9.14	9.40	9.65	9.91
.40	10.16	10.41	10.67	10.92	11.18	11.43	11.68	11.94	12.19	12.45
.50	12.70	12.95	13.21	13.46	13.72	13.97	14.22	14.48	14.73	14.99
.60	15.24	15.49	15.75	16.00	16.26	16.51	16.76	17.02	17.27	17.53
.70	17.78	18.03	18.29	18.54	18.80	19.05	19.30	19.56	19.81	20.07
.80	20.32	20.57	20.83	21.08	21.34	21.59	21.84	22.10	22.35	22.61
.90	22.86	23.11	23.37	23.62	23.88	24.13	24.38	24.64	24.89	25.15

ONE TENTH OF AN INCH INCREMENTS TO TWENTY INCHES

Inches	0	.10	.20	.30	.40	.50	.60	.70	.80	.90
0	0.0	2.5	5.1	7.6	10.2	12.7	15.2	17.8	20.3	22.9
1	25.4	27.9	30.5	33.0	35.6	38.1	40.6	43.2	45.7	48.3
2	50.8	53.3	55.9	58.4	61.0	63.5	66.0	68.6	71.1	73.7
3	76.2	78.7	81.3	83.8	86.4	88.9	91.4	94.0	96.5	99.1
4	101.6	104.1	106.7	109.2	111.8	114.3	116.8	119.4	121.9	124.5
5	127.0	129.5	132.1	134.6	137.2	139.7	142.2	144.8	147.3	149.9
6	152.4	154.9	157.5	160.0	162.6	165.1	167.6	170.2	172.7	175.3
7	177.8	180.3	182.9	185.4	188.0	190.5	193.0	195.6	198.1	200.7
8	203.2	205.7	208.3	210.8	213.4	215.9	218.4	221.0	223.5	226.1
9	228.6	231.1	233.7	236.2	238.8	241.3	243.8	246.4	248.9	251.5
10	254.0	256.5	259.1	261.6	264.2	266.7	269.2	271.8	274.3	276.9
11	279.4	281.9	284.5	287.0	289.6	292.1	294.6	297.2	299.7	302.3
12	304.8	307.3	309.9	312.4	315.0	317.5	320.0	322.6	325.1	327.7
13	330.2	332.7	335.3	337.8	340.4	342.9	345.4	348.0	350.5	353.1
14	355.6	358.1	360.7	363.2	365.8	368.3	370.8	373.4	375.9	378.5
15	381.0	383.5	386.1	388.6	391.2	393.7	396.2	398.8	401.3	403.9
16	406.4	408.9	411.5	414.0	416.6	419.1	421.6	424.2	426.7	429.3
17	431.8	434.3	436.9	439.4	442.0	444.5	447.0	449.6	452.1	454.7
18	457.2	459.7	462.3	464.8	467.4	469.9	472.4	475.0	477.5	480.1
19	482.6	485.1	487.7	490.2	492.8	495.3	497.8	500.4	502.9	505.5
20	508.0	510.5	513.1	515.6	518.2	520.7	523.2	525.8	528.3	530.9

Table 3   Conversion of decimals of an inch to millimeters.

IN.	0	1/16	1/8	3/16	1/4	5/16	3/8	7/16	1/2	9/16	5/8	11/16	3/4	13/16	7/8	15/16
0	.0	1.6	3.2	4.8	6.4	7.9	9.5	11.1	12.7	14.3	15.9	17.5	19.1	20.6	22.2	23.8
1	25.4	27.0	28.6	30.2	31.8	33.3	34.9	36.5	38.1	39.7	41.3	42.9	44.5	46.0	47.6	49.2
2	50.8	52.4	54.0	55.6	57.2	58.7	60.3	61.9	63.5	65.1	66.7	68.3	69.9	71.4	73.0	74.6
3	76.2	77.8	79.4	81.0	82.6	84.1	85.7	87.3	88.9	90.5	92.1	93.7	95.3	96.8	98.4	100.0
4	101.6	103.2	104.8	106.4	108.0	109.5	111.1	112.7	114.3	115.9	117.5	119.1	120.7	122.2	123.8	125.4
5	127.0	128.6	130.2	131.8	133.4	134.9	136.5	138.1	139.7	141.3	142.9	144.5	146.1	147.6	149.2	150.8
6	152.4	154.0	155.6	157.2	158.8	160.3	161.9	163.5	165.1	166.7	168.3	169.9	171.5	173.0	174.6	176.2
7	177.8	179.4	181.0	182.6	184.2	185.7	187.3	188.9	190.5	192.1	193.7	195.3	196.9	198.4	200.0	201.6
8	203.2	204.8	206.4	208.0	209.6	211.1	212.7	214.3	215.9	217.5	219.1	220.7	222.3	223.8	225.4	227.0
9	228.6	230.2	231.8	233.4	235.0	236.5	238.1	239.7	241.3	242.9	244.5	246.1	247.7	249.2	250.8	252.4
10	254.0	255.6	257.2	258.8	260.4	261.9	263.5	265.1	266.7	268.3	269.9	271.5	273.1	274.6	276.2	277.8
11	279.4	281.0	282.6	284.2	285.8	287.3	288.9	290.5	292.1	293.7	295.3	296.9	298.5	300.0	301.6	303.2
12	304.8	306.4	308.0	309.6	311.2	312.7	314.3	315.9	317.5	319.1	320.7	322.3	323.9	325.4	327.0	328.6
13	330.2	331.8	333.4	335.0	336.6	338.1	339.7	341.3	342.9	344.5	346.1	347.7	349.3	350.8	352.4	354.0
14	355.6	357.2	358.8	360.4	362.0	363.5	365.1	366.7	368.3	369.9	371.5	373.1	374.7	376.2	377.8	379.4

Table 4   Conversion of fractions of an inch to millimeters.

Fraction	Decimal	Fraction	Decimal
$\frac{1}{64}$	0.015625	$\frac{33}{64}$	0.515625
$\frac{1}{32}$	0.03125	$\frac{17}{32}$	0.53125
$\frac{3}{64}$	0.046875	$\frac{35}{64}$	0.546875
$\frac{1}{16}$	0.0625	$\frac{9}{16}$	0.5625
$\frac{5}{64}$	0.078125	$\frac{37}{64}$	0.578125
$\frac{3}{32}$	0.09375	$\frac{19}{32}$	0.59375
$\frac{7}{64}$	0.109375	$\frac{39}{64}$	0.609375
$\frac{1}{8}$	0.1250	$\frac{5}{8}$	0.6250
$\frac{9}{64}$	0.140625	$\frac{41}{64}$	0.640625
$\frac{5}{32}$	0.15625	$\frac{21}{32}$	0.65625
$\frac{11}{64}$	0.171875	$\frac{43}{64}$	0.671875
$\frac{3}{16}$	0.1875	$\frac{11}{16}$	0.6875
$\frac{13}{64}$	0.203125	$\frac{45}{64}$	0.703125
$\frac{7}{32}$	0.21875	$\frac{23}{32}$	0.71875
$\frac{15}{64}$	0.234375	$\frac{47}{64}$	0.734375
$\frac{1}{4}$	0.2500	$\frac{3}{4}$	0.7500
$\frac{17}{64}$	0.265625	$\frac{49}{64}$	0.765625
$\frac{9}{32}$	0.28125	$\frac{25}{32}$	0.78125
$\frac{19}{64}$	0.296875	$\frac{51}{64}$	0.796875
$\frac{5}{16}$	0.3125	$\frac{13}{16}$	0.8125
$\frac{21}{64}$	0.328125	$\frac{53}{64}$	0.828125
$\frac{11}{32}$	0.34375	$\frac{27}{32}$	0.84375
$\frac{23}{64}$	0.359375	$\frac{55}{64}$	0.859375
$\frac{3}{8}$	0.3750	$\frac{7}{8}$	0.8750
$\frac{25}{64}$	0.390625	$\frac{57}{64}$	0.890625
$\frac{13}{32}$	0.40625	$\frac{29}{32}$	0.90625
$\frac{27}{64}$	0.421875	$\frac{59}{64}$	0.921875
$\frac{7}{16}$	0.4375	$\frac{15}{16}$	0.9375
$\frac{29}{64}$	0.453125	$\frac{61}{64}$	0.953125
$\frac{15}{32}$	0.46875	$\frac{31}{32}$	0.96875
$\frac{31}{64}$	0.484375	$\frac{63}{64}$	0.984375
$\frac{1}{2}$	0.5000	1	1.0000

Table 5 Decimal equivalents of common inch fractions.

ANGLE	SINE	COSINE	TAN	COTAN	ANGLE
0°	.0000	1.0000	.0000	θ	90°
1°	0.0175	0.9998	0.0175	57.290	89°
2°	0.0349	0.9994	0.0349	28.636	88°
3°	0.0523	0.9986	0.0524	19.081	87°
4°	0.0698	0.9976	0.0699	14.301	86°
5°	0.0872	0.9962	0.0875	11.430	85°
6°	0.1045	0.9945	0.1051	9.5144	84°
7°	0.1219	0.9925	0.1228	8.1443	83°
8°	0.1392	0.9903	0.1405	7.1154	82°
9°	0.1564	0.9877	0.1584	6.3138	81°
10°	0.1736	0.9848	0.1763	5.6713	80°
11°	0.1908	0.9816	0.1944	5.1446	79°
12°	0.2079	0.9781	0.2126	4.7046	78°
13°	0.2250	0.9744	0.2309	4.3315	77°
14°	0.2419	0.9703	0.2493	4.0108	76°
15°	0.2588	0.9659	0.2679	3.7321	75°
16°	0.2756	0.9613	0.2867	3.4874	74°
17°	0.2924	0.9563	0.3057	3.2709	73°
18°	0.3090	0.9511	0.3249	3.0777	72°
19°	0.3256	0.9455	0.3443	2.9042	71°
20°	0.3420	0.9397	0.3640	2.7475	70°
21°	0.3584	0.9336	0.3839	2.6051	69°
22°	0.3746	0.9272	0.4040	2.4751	68°
23°	0.3907	0.9205	0.4245	2.3559	67°
24°	0.4067	0.9135	0.4452	2.2460	66°
25°	0.4226	0.9063	0.4663	2.1445	65°
26°	0.4384	0.8988	0.4877	2.0503	64°
27°	0.4540	0.8910	0.5095	1.9626	63°
28°	0.4695	0.8829	0.5317	1.8807	62°
29°	0.4848	0.8746	0.5543	1.8040	61°
30°	0.5000	0.8660	0.5774	1.7321	60°
31°	0.5150	0.8572	0.6009	1.6643	59°
32°	0.5299	0.8480	0.6249	1.6003	58°
33°	0.5446	0.8387	0.6494	1.5399	57°
34°	0.5592	0.8290	0.6745	1.4826	56°
35°	0.5736	0.8192	0.7002	1.4281	55°
36°	0.5878	0.8090	0.7265	1.3764	54°
37°	0.6018	0.7986	0.7536	1.3270	53°
38°	0.6157	0.7880	0.7813	1.2799	52°
39°	0.6293	0.7771	0.8098	1.2349	51°
40°	0.6428	0.7660	0.8391	1.1918	50°
41°	0.6561	0.7547	0.8693	1.1504	49°
42°	0.6691	0.7431	0.9004	1.1106	48°
43°	0.6820	0.7314	0.9325	1.0724	47°
44°	0.6947	0.7193	0.9657	1.0355	46°
45°	0.7071	0.7071	0.0000	1.0000	45°
ANGLE	COSINE	SINE	COTAN	TAN	ANGLE

Table 6 Trigonometric functions.

NUM-BER	SQUARE	SQUARE ROOT	CIRCUM-FERENCE OF CIRCLE	AREA OF CIRCLE	NUM-BER	SQUARE	SQUARE ROOT	CIRCUM-FERENCE OF CIRCLE	AREA OF CIRCLE	NUM-BER	SQUARE	SQUARE ROOT	CIRCUM-FERENCE OF CIRCLE	AREA OF CIRCLE
1	1	1	3.14	0.78	36	1296	6.0000	113.10	1017.88	71	5041	8.4261	223.05	3959.19
2	4	1.41	6.28	3.14	37	1369	6.0828	116.24	1075.21	72	5184	8.4853	226.19	4071.50
3	9	1.73	9.43	7.07	38	1444	6.1644	119.38	1134.11	73	5329	8.5440	229.34	4185.39
4	16	2.00	12.57	12.57	39	1521	6.2450	122.52	1194.59	74	5476	8.6023	232.48	4300.84
5	25	2.34	15.71	19.64	40	1600	6.3246	125.66	1256.64	75	5625	8.6603	235.62	4417.88
6	36	2.4495	18.85	28.27	41	1681	6.4031	128.81	1320.25	76	5776	8.7178	238.76	4536.47
7	49	2.6458	21.99	38.48	42	1764	6.4807	131.95	1385.44	77	5929	8.7750	241.90	4656.64
8	64	2.8284	25.13	50.27	43	1849	6.5574	135.09	1452.20	78	6084	8.8318	245.04	4778.37
9	81	3.0000	28.27	63.62	44	1936	6.6332	138.23	1520.53	79	6241	8.8882	248.19	4901.68
10	100	3.1623	31.46	78.54	45	2025	6.7082	141.37	1590.43	80	6400	8.9443	251.33	5026.56
11	121	3.3166	34.56	95.03	46	2116	6.7823	144.51	1661.90	81	6561	9.0000	254.47	5183.01
12	144	3.4641	37.70	113.09	47	2209	6.8557	147.65	1734.94	82	6724	9.0554	257.61	5281.03
13	169	3.6056	40.84	132.73	48	2304	6.9282	150.80	1809.56	83	6889	9.1104	260.75	5410.62
14	196	3.7417	43.98	153.94	49	2401	7.0000	153.94	1885.74	84	7056	9.1652	263.89	5541.78
15	225	3.8730	47.12	176.72	50	2500	7.0711	157.08	1963.50	85	7225	9.2200	267.04	5674.52
16	256	4.0000	50.27	201.06	51	2601	7.1414	160.22	2042.82	86	7396	9.2736	270.18	5808.82
17	289	4.1231	53.41	226.98	52	2704	7.2111	163.36	2123.72	87	7569	9.3274	273.32	5944.69
18	324	4.2426	56.55	254.47	53	2809	7.2801	166.50	2206.18	88	7744	9.3808	276.46	6082.14
19	361	4.3589	59.69	283.53	54	2916	7.3485	169.65	2290.22	89	7921	9.4340	279.60	6221.15
20	400	4.4721	62.83	314.16	55	3025	7.4162	172.79	2375.83	90	8100	9.4868	282.74	6361.74
21	441	4.5826	65.97	346.36	56	3136	7.4833	175.93	2463.01	91	8281	9.5393	285.89	6503.90
22	484	4.6904	69.12	380.13	57	3249	7.5498	179.07	2551.76	92	8464	9.5917	289.03	6647.63
23	529	4.7958	72.26	415.48	58	3364	7.6158	182.21	2642.08	93	8649	9.6437	292.17	6792.92
24	576	4.8990	75.39	452.39	59	3481	7.6811	185.35	2733.97	94	8836	9.6954	295.31	6939.79
25	625	5.0000	78.54	490.87	60	3600	7.7460	188.50	2827.43	95	9025	9.7468	298.45	7088.24
26	676	5.0990	81.68	530.93	61	3721	7.8102	191.64	3922.47	96	9216	9.7979	301.59	7238.25
27	729	5.1962	84.82	572.56	62	3844	7.8740	194.78	3019.07	97	9409	9.8489	304.74	7389.83
28	784	5.2915	87.97	615.75	63	3969	7.9373	197.92	3117.25	98	9604	9.8995	307.88	7542.98
29	841	5.3852	91.11	660.52	64	4096	8.0000	201.06	3216.99	99	9801	9.9509	311.02	7697.71
30	900	5.4772	94.25	706.86	65	4225	8.0623	204.20	3318.31	100	10 000	10.000	314.16	7854.00
31	961	5.5678	97.39	754.77	66	4356	8.1240	207.35	3421.19	101	10 201	10.0499	317.30	8011.87
32	1024	5.6569	100.53	804.25	67	4489	8.1854	210.49	3525.65	102	10 404	10.0995	320.44	8171.30
33	1089	5.7446	103.67	855.30	68	4624	8.2462	213.63	3631.68	103	10 609	10.1489	323.58	8332.31
34	1156	5.8310	106.81	907.92	69	4761	8.3066	216.77	3739.28	104	10 816	10.1980	326.73	8494.89
35	1225	5.9161	109.96	962.113	70	4900	8.3666	219.91	3848.50	105	11 025	10.2470	329.87	8659.04

Table 7   Function of numbers.

Across Flats	ACRFLT	Machine Steel	MST
American National Standards Institute	ANSI	Machined	✓
And	&	Malleable Iron	MI
Angular	ANLR	Material	MATL
Approximate	APPROX	Maximum	MAX
Assembly	ASSY	Maximum Material Condition	Ⓜ or MMC
Basic	BSC	Meter	m
Bill of Material	B/M	Metric Thread	M
Bolt Circle	BC	Micrometer	μm
Brass	BR	Millimeter	mm
Brown and Sharpe Gage	B&S GA	Minimum	MIN
Bushing	BUSH	Module	MDL
Canada Standards Institute	CSI	Newton	N
Carbon Steel	CS	Nominal	NOM
Casting	CSTG	Not to Scale	x̲x̲
Cast Iron	CI	Number	NO
Center Line	CL or ℄	On Center	OC
Center to Center	C to C	Outside Diameter	OD
Centimeter	cm	Parallel	PAR
Chamfer	CHAM	Pascal	Pa
Circularity	CIR	Perpendicular	PERP
Cold-Rolled Steel	CRS	Pitch	P
Concentric	CONC	Pitch Circle	PC
Counterbore	⌴ or CBORE	Pitch Diameter	PD
Counterdrill	CDRILL	Plate	PL
Countersink	⌵ or CSK	Radius	R
Cubic Centimeter	cm³	Reference or Reference Dimension	( ) or REF
Cubic Meter	m³	Regardless of Feature Size	Ⓢ
Datum	DAT	Revolutions per Minute	rev/min
Degree (Angle)	° or DEG	Right Hand	RH
Depth	DP or ▼	Root Diameter	RD
Diameter	⌀ or DIA	Second (Arc)	(")
Diametral Pitch	DP	Second (Time)	SEC
Dimension	DIM	Section	SECT
Drawing	DWG	Slotted	SLOT
Eccentric	ECC	Socket	SOCK
Equally Spaced	EQL SP	Spherical	SPHER
Figure	FIG	Spotface	⌴ or SFACE
Finish All Over	FAO	Square	☐ or SQ
Flat	FL	Square Centimeter	cm²
Gage	GA	Square Meter	m²
Gray Iron	GI	Steel	STL
Head	HD	Straight	STR
Heat Treat	HT TR	Symmetrical	⫲ or SYM
Heavy	HVY	Taper—Flat	
Hexagon	HEX	—Round	
Hydraulic	HYDR	Taper Pipe Thread	NPT
Inside Diameter	ID	Thread	THD
International Organization for Standardization	ISO	Through	THRU
International Pipe Standard	IPS	Tolerance	TOL
Kilogram	kg	True Profile	TP
Kilometer	km	U.S. Gage	USG
Left Hand	LH	Watt	W
Length	LG	Wrought Iron	WI
Liter	L	Wrought Steel	WS

ISO [A]   ANSI [-A-] (Datum)

Table 8 Abbreviations and symbols used on technical drawings.

NUMBER OR LETTER SIZE DRILL	SIZE		NUMBER OR LETTER SIZE DRILL	SIZE		NUMBER OR LETTER SIZE DRILL	SIZE		NUMBER OR LETTER SIZE DRILL	SIZE	
	mm	INCHES		mm	INCHES		mm	INCHES		mm	INCHES
80	0.343	.014	50	1.778	.070	20	4.089	.161	K	7.137	.281
79	0.368	.015	49	1.854	.073	19	4.216	.166	L	7.366	.290
78	0.406	.016	48	1.930	.076	18	4.305	.170	M	7.493	.295
77	0.457	.018	47	1.994	.079	17	4.394	.173	N	7.671	.302
76	0.508	.020	46	2.057	.081	16	4.496	.177	O	8.026	.316
75	0.533	.021	45	2.083	.082	15	4.572	.180	P	8.204	.323
74	0.572	.023	44	2.184	.086	14	4.623	.182	Q	8.433	.332
73	0.610	.024	43	2.261	.089	13	4.700	.185	R	8.611	.339
72	0.635	.025	42	2.375	.094	12	4.800	.189	S	8.839	.348
71	0.660	.026	41	2.438	.096	11	4.851	.191	T	9.093	.358
70	0.711	.028	40	2.489	.098	10	4.915	.194	U	9.347	.368
69	0.742	.029	39	2.527	.100	9	4.978	.196	V	9.576	.377
68	0.787	.031	38	2.578	.102	8	5.080	.199	W	9.804	.386
67	0.813	.032	37	2.642	.104	7	5.105	.201	X	10.084	.397
66	0.838	.033	36	2.705	.107	6	5.182	.204	Y	10.262	.404
65	0.889	.035	35	2.794	.110	5	5.220	.206	Z	10.490	.413
64	0.914	.036	34	2.819	.111	4	5.309	.209			
63	0.940	.037	33	2.870	.113	3	5.410	.213			
62	0.965	.038	32	2.946	.116	2	5.613	.221			
61	0.991	.039	31	3.048	.120	1	5.791	.228			
60	1.016	.040	30	3.264	.129	A	5.944	.234			
59	1.041	.041	29	3.354	.136	B	6.045	.238			
58	1.069	.042	28	3.569	.141	C	6.147	.242			
57	1.092	.043	27	3.658	.144	D	6.248	.246			
56	1.181	.047	26	3.734	.147	E	6.350	.250			
55	1.321	.052	25	3.797	.150	F	6.528	.257			
54	1.397	.055	24	3.861	.152	G	6.629	.261			
53	1.511	.060	23	3.912	.154	H	6.756	.266			
52	1.613	.064	22	3.988	.157	I	6.909	.272			
51	1.702	.067	21	4.039	.159	J	7.036	.277			

Table 9 Number and letter-size drills.

METRIC DRILL SIZES		Reference Decimal Equivalent (Inches)	METRIC DRILL SIZES		Reference Decimal Equivalent (Inches)	METRIC DRILL SIZES		Reference Decimal Equivalent (Inches)
Preferred	Available		Preferred	Available		Preferred	Available	
—	0.40	.0157	—	2.7	.1063	14	—	.5512
—	0.42	.0165	2.8	—	.1102	—	14.5	.5709
—	0.45	.0177	—	2.9	.1142	15	—	.5906
—	0.48	.0189	3.0	—	.1181	—	15.5	.6102
0.50	—	.0197	—	3.1	.1220	16	—	.6299
—	0.52	.0205	3.2	—	.1260	—	16.5	.6496
0.55	—	.0217	—	3.3	.1299	17	—	.6693
—	0.58	.0228	3.4	—	.1339	—	17.5	.6890
0.60	—	.0236	—	3.5	.1378	18	—	.7087
—	0.62	.0244	3.6	—	.1417	—	18.5	.7283
0.65	—	.0256	—	3.7	.1457	19	—	.7480
—	0.68	.0268	3.8	—	.1496	—	19.5	.7677
0.70	—	.0276	—	3.9	.1535	20	—	.7874
—	0.72	.0283	4.0	—	.1575	—	20.5	.8071
0.75	—	.0295	—	4.1	.1614	21	—	.8268
—	0.78	.0307	4.2	—	.1654	—	21.5	.8465
0.80	—	.0315	—	4.4	.1732	22	—	.8661
—	0.82	.0323	4.5	—	.1772	—	23	.9055
0.85	—	.0335	—	4.6	.1811	24	—	.9449
—	0.88	.0346	4.8	—	.1890	25	—	.9843
0.90	—	.0354	5.0	—	.1969	26	—	1.0236
—	0.92	.0362	—	5.2	.2047	—	27	1.0630
0.95	—	.0374	5.3	—	.2087	28	—	1.1024
—	0.98	.0386	—	5.4	.2126	—	29	1.1417
1.00	—	.0394	5.6	—	.2205	30	—	1.1811
—	1.03	.0406	—	5.8	.2283	—	31	1.2205
1.05	—	.0413	6.0	—	.2362	32	—	1.2598
—	1.08	.0425	—	6.2	.2441	—	33	1.2992
1.10	—	.0433	6.3	—	.2480	34	—	1.3386
—	1.15	.0453	—	6.5	.2559	—	35	1.3780
1.20	—	.0472	6.7	—	.2638	36	—	1.4173
1.25	—	.0492	—	6.8	.2677	—	37	1.4567
1.3	—	.0512	—	6.9	.2717	38	—	1.4361
—	1.35	.0531	7.1	—	.2795	—	39	1.5354
1.4	—	.0551	—	7.3	.2874	40	—	1.5748
—	1.45	.0571	7.5	—	.2953	—	41	1.6142
1.5	—	.0591	—	7.8	.3071	42	—	1.6535
—	1.55	.0610	8.0	—	.3150	—	43.5	1.7126
1.6	—	.0630	—	8.2	.3228	45	—	1.7717
—	1.65	.0650	8.5	—	.3346	—	46.5	1.8307
1.7	—	.0669	—	8.8	.3465	48	—	1.8898
—	1.75	.0689	9.0	—	.3543	50	—	1.9685
1.8	—	.0709	—	9.2	.3622	—	51.5	2.0276
—	1.85	.0728	9.5	—	.3740	53	—	2.0866
1.9	—	.0748	—	9.8	.3858	—	54	2.1260
—	1.95	.0768	10	—	.3937	56	—	2.2047
2.0	—	.0787	—	10.3	.4055	—	58	2.2835
—	2.05	.0807	10.5	—	.4134	60	—	2.3622
2.1	—	.0827	—	10.8	.4252			
—	2.15	.0846	11	—	.4331			
2.2	—	.0866	—	11.5	.4528			
—	2.3	.0906	12	—	.4724			
2.4	—	.0945	12.5	—	.4921			
2.5	—	.0984	13	—	.5118			
2.6	—	.1024	—	13.5	.5315			

Table 10   Metric twist drill sizes.

THREADS PER INCH AND TAP DRILL SIZES

| SIZE INCHES | | Graded Pitch Series | | | | | | Constant Pitch Series | | | | | |
| | | Coarse UNC | | Fine UNF | | Extra Fine UNEF | | 8 UN | | 12 UN | | 16 UN | |
Number or Fraction	Deci-mal	Threads per Inch	Tap Drill Dia.	Threads per Inch	Tap Drill Dia.	Threads per Inch	Tap Drill Dia.	Threads per Inch	Tap Drill Dia.	Threads per Inch	Tap Drill Dia.	Threads per Inch	Tap Drill Dia.
0	.060	—	—	80	3/64	—	—	—	—	—	—	—	—
2	.086	56	No. 50	64	No. 49	—	—	—	—	—	—	—	—
4	.112	40	No. 43	48	No. 42	—	—	—	—	—	—	—	—
5	.125	40	No. 38	44	No. 37	—	—	—	—	—	—	—	—
6	.138	32	No. 36	40	No. 33	—	—	—	—	—	—	—	—
8	.164	32	No. 29	36	No. 29	—	—	—	—	—	—	—	—
10	.190	24	No. 25	32	No. 21	—	—	—	—	—	—	—	—
1/4	.250	20	7	28	3	32	.219	—	—	—	—	—	—
5/16	.312	18	F	24	1	32	.281	—	—	—	—	—	—
3/8	.375	16	.312	24	Q	32	.344	—	—	—	—	UNC	—
7/16	.438	14	U	20	.391	28	Y	—	—	—	—	16	V
1/2	.500	13	.422	20	.453	28	.469	—	—	—	—	16	.438
9/16	.562	12	.484	18	.516	24	.516	—	—	UNC	—	16	.500
5/8	.625	11	.531	18	.578	24	.578	—	—	12	.547	16	.562
3/4	.750	10	.656	16	.688	20	.703	—	—	12	.672	UNF	—
7/8	.875	9	.766	14	.812	20	.828	—	—	12	.797	16	.812
1	1.000	8	.875	12	.922	20	.953	UNC	—	UNF	—	16	.938
1 1/8	1.125	7	.984	12	1.047	18	1.078	8	1.000	UNF	—	16	1.062
1 1/4	1.250	7	1.109	12	1.172	18	1.188	8	1.125	UNF	—	16	1.188
1 3/8	1.375	6	1.219	12	1.297	18	1.312	8	1.250	UNF	—	16	1.312
1 1/2	1.500	6	1.344	12	1.422	18	1.438	8	1.375	UNF	—	16	1.438
1 5/8	1.625	—	—	—	—	18	—	8	1.500	12	1.547	16	1.562
1 3/4	1.750	5	1.562	—	—	—	—	8	1.625	12	1.672	16	1.688
1 7/8	1.875	—	—	—	—	—	—	8	1.750	12	1.797	16	1.812
2	2.000	4.5	1.781	—	—	—	—	8	1.875	12	1.922	16	1.938
2 1/4	2.250	4.5	2.031	—	—	—	—	8	2.125	12	2.172	16	2.188
2 1/2	2.500	4	2.250	—	—	—	—	8	2.375	12	2.422	16	2.438
2 3/4	2.750	4	2.500	—	—	—	—	8	2.625	12	2.672	16	2.688
3	3.000	4	2.750	—	—	—	—	8	2.875	12	2.922	16	2.938
3 1/4	3.250	4	3.000	—	—	—	—	8	3.125	12	3.172	16	3.188
3 1/2	3.500	4	3.250	—	—	—	—	8	3.375	12	3.422	16	3.438
3 3/4	3.750	4	3.500	—	—	—	—	8	3.625	12	3.668	16	3.688
4	4.000	4	3.750	—	—	—	—	8	3.875	12	3.922	16	3.938

Note: The tap diameter sizes shown are nominal. The class and length of thread will govern the limits on the tapped hole size.

Table 11   Inch screw threads.

Table 12 — Isometric screw threads.

Nominal Size Dia (mm) Preferred	Series with Graded Pitches — Coarse		Series with Graded Pitches — Fine		Series with Constant Pitches — 4		3		2		1.5		1.25		1		0.75		0.5		0.35	
	Thread Pitch	Tap Drill Size	Thread Pitch	Tap Drill Size	Thread Pitch	Tap Drill Size	Thread Pitch	Tap Drill Size	Thread Pitch	Tap Drill Size	Thread Pitch	Tap Drill Size	Thread Pitch	Tap Drill Size	Thread Pitch	Tap Drill Size	Thread Pitch	Tap Drill Size	Thread Pitch	Tap Drill Size	Thread Pitch	Tap Drill Size
1.6	0.35	1.25																				
1.8	0.35	1.45																				
2	0.4	1.6																				
2.2	0.45	1.75																				
2.5	0.45	2.05																			0.35	2.15
3	0.5	2.5																			0.35	2.65
3.5	0.6	2.9																			0.35	3.15
4	0.7	3.3																	0.5	3.5		
4.5	0.75	3.7																	0.5	4.0		
5	0.8	4.2																	0.5	4.5		
6	1	5.0															0.75	5.2				
8	1.25	6.7	1	7.0											1	7.0	0.75	7.2				
10	1.5	8.5	1.25	8.7									1.25	8.7	1	9.0	0.75	9.2				
12	1.75	10.2	1.25	10.8							1.5	10.5	1.25	10.7	1	11						
14	2	12	1.5	12.5							1.5	12.5	1.25	12.7	1	13						
16	2	14	1.5	14.5							1.5	14.5			1	15						
18	2.5	15.5	1.5	16.5					2	16	1.5	16.5			1	17						
20	2.5	17.5	1.5	18.5					2	18	1.5	18.5			1	19						
22	2.5	19.5	1.5	20.5					2	20	1.5	20.5			1	21						
24	3	21	2	22					2	22	1.5	22.5			1	23						
27	3	24	2	25					2	25	1.5	25.5			1	26						
30	3.5	26.5	2	28					2	28	1.5	28.5			1	29						
33	3.5	29.5	2	31					2	31	1.5	31.5										
36	4	32	3	33			3	33	2	34	1.5	34.5										
39	4	35	3	36			3	36	2	37	1.5	37.5										
42	4.5	37.5	3	39	4	38	3	39	2	40	1.5	40.5										
45	4.5	39	3	42	4	41	3	42	2	43	1.5	43.5										
48	5	43	3	45	4	44	3	45	2	46	1.5	46.5										

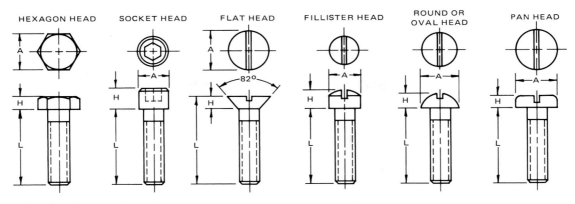

	U.S. CUSTOMARY (INCHES)										METRIC (MILLIMETERS)										
Nominal Size	Hexagon Head		Socket Head		Flat Head		Fillister Head		Round or Oval Head		Nominal Size	Hexagon Head		Socket Head		Flat Head		Fillister Head		Pan Head	
	A	H	A	H	A	H	A	H	A	H		A	H	A	H	A	H	A	H	A	H
.250	.44	.17	.38	.25	.50	.14	.38	.17	.44	.19	M3	5.5	2	5.5	3	5.6	1.6	6	2.4	5.6	1.9
.312	.50	.22	.47	.31	.62	.18	.44	.20	.56	.25	4	7	2.8	7	4	7.5	2.2	8	3.1	7.5	2.5
.375	.56	.25	.56	.38	.75	.21	.56	.25	.62	.27	5	8.5	3.5	9	5	9.2	2.5	10	3.8	9.2	3.1
.438	.62	.30	.66	.44	.81	.21	.62	.30	.75	.33	6	10	4	10	6	11	3	12	4.6	11	3.8
.500	.75	.34	.75	.50	.88	.21	.75	.33	.81	.35	8	13	5.5	13	8	14.5	4	16	6	14.5	5
.625	.94	.42	.94	.62	1.12	.28	.88	.42	1.00	.44	10	17	7	16	10	18	5	20	7.5	18	6.2
.750	1.12	.50	1.12	.75	1.38	.35	1.00	.50	1.25	.55	12	19	8	18	12						
											14	22	9	22	14						
											16	24	10	24	16						

Table 13   Common cap screws.

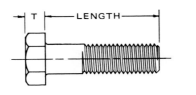

U.S. CUSTOMARY (INCHES)			METRIC (MILLIMETERS)		
Nominal Bolt Size	Width Across Flats F	Thickness T	Nominal Bolt Size and Thread Pitch	Width Across Flats F	Thickness T
.250	.438	.172	M5 x 0.8	8	3.9
.312	.500	.219	M6 x 1	10	4.7
.375	.562	.250	M8 x 1.25	13	5.7
.438	.625	.297			
.500	.750	.344	M10 x 1.5	15	6.8
.625	.938	.422	M12 x 1.75	18	8
.750	1.125	.500	M14 x 2	21	9.3
.875	1.312	.578	M16 x 2	24	10.5
1.000	1.500	.672	M20 x 2.5	30	13.1
1.125	1.688	.750	M24 x 3	36	15.6
1.250	1.875	.844	M30 x 3.5	46	19.5
1.375	2.062	.906	M36 x 4	55	23.4
1.500	2.250	1.000			

Table 14   Hexagon-head bolts and cap screws.

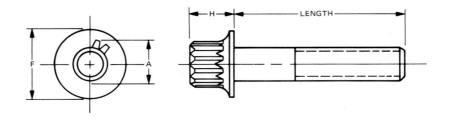

NOMINAL BOLT SIZE AND THREAD PITCH	HEAD SIZES		
	F	A	H
M5 x 0.8	9.4	5.9	5
M6 x 1	11.8	7.4	6.3
M8 x 1.25	15	9.4	8
M10 x 1.5	18.6	11.7	10
M12 x 1.75	22.8	14	12
M14 x 2	26.4	16.3	14
M16 x 2	30.3	18.7	16
M20 x 2.5	37.4	23.4	20

Table 15   Twelve-spline flange screws.

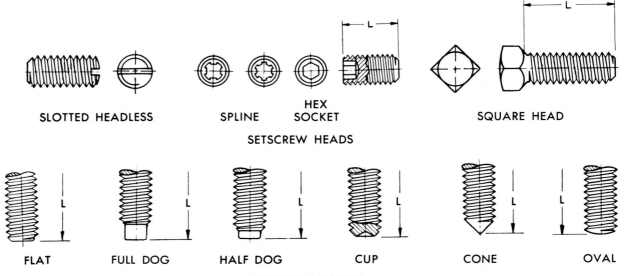

SLOTTED HEADLESS   SPLINE   HEX SOCKET   SQUARE HEAD

SETSCREW HEADS

FLAT   FULL DOG   HALF DOG   CUP   CONE   OVAL

SETSCREW POINTS

U.S. CUSTOMARY (INCHES)		METRIC (MILLIMETERS)	
Nominal Size	Key Size	Nominal Size	Key Size
.125	.06	M1.4	0.7
.138	.06	2	0.9
.164	.08	3	1.5
.190	.09	4	2
.250	.12	5	2
.312	.16	6	3
.375	.19	8	4
.500	.25	10	5
.625	.31	12	6
.750	.38	16	8

Table 16   Setscrews.

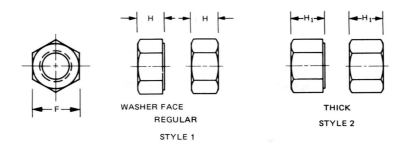

WASHER FACE
REGULAR
STYLE 1

THICK
STYLE 2

U.S. CUSTOMARY (INCHES)			
	Distance Across Flats F	Thickness Max.	
Nominal Nut Size		Style 1 H	Style 2 H₁
.250	.438	.218	.281
.312	.500	.266	.328
.375	.562	.328	.406
.438	.625	.375	.453
.500	.750	.438	.562
.562	.875	.484	.609
.625	.938	.547	.719
.750	1.125	.641	.812
.875	1.312	.750	.906
1.000	1.500	.859	1.000
1.125	1.688	.969	1.156
1.250	1.875	1.062	1.250
1.375	2.062	1.172	1.375
1.500	2.250	1.281	1.500

METRIC (MILLIMETERS)			
	Distance Across Flats F	Thickness Max.	
Nominal Nut Size and Thread Pitch		Style 1 H	Style 2 H₁
M4 x 0.7	7	—	3.2
M5 x 0.8	8	4.5	5.3
M6 x 1	10	5.6	6.5
M8 x 1.25	13	6.6	7.8
M10 x 1.5	15	9	10.7
M12 x 1.75	18	10.7	12.8
M14 x 2	21	12.5	14.9
M16 x 2	24	14.5	17.4
M20 x 2.5	30	18.4	21.2
M24 x 3	36	22	25.4
M30 x 3.5	46	26.7	31
M36 x 4	55	32	37.6

Table 17   Hexagon-head nuts.

METRIC (MILLIMETERS)							
Nominal Nut Size and Thread Pitch	Width Across Flats F	Style 1				Style 2	
		H	J	K	M	H	J
M6 x 1	10	5.8	3	1	14.2	6.7	3.7
M8 x 1.25	13	6.8	3.7	1.3	17.6	8	4.5
M10 x 1.5	15	9.6	5.5	1.5	21.5	11.2	6.7
M12 x 1.75	18	11.6	6.7	2	25.6	13.5	8.2
M14 x 2	21	13.4	7.8	2.3	29.6	15.7	9.6
M16 x 2	24	15.9	9.5	2.5	34.2	18.4	11.7
M20 x 2.5	30	19.2	11.1	2.8	42.3	22	12.6

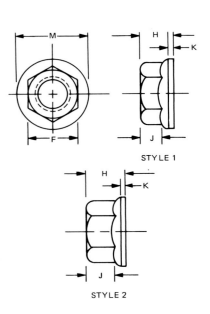

STYLE 1

STYLE 2

Table 18   Hex flange nuts.

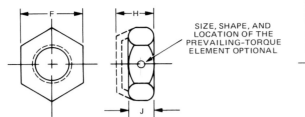

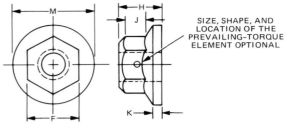

HEX NUTS

HEX FLANGE NUTS

NOMINAL NUT SIZE AND THREAD PITCH	WIDTH ACROSS FLATS F	HEX NUTS				HEX FLANGE NUTS					
		Style 1		Style 2		Style 1				Style 2	
		H max.	J max.	H max.	J max.	H	J	K	M	H	J
M5 × 0.8	8.0	6.1	2.3	7.6	2.9						
M6 × 1	10	7.6	3	8.8	3.7	7.6	3	1	14.2	8.8	3.7
M8 × 1.25	13	9.1	3.7	10.3	4.5	9.1	3.7	1.3	17.6	10.3	4.5
M10 × 1.5	15	12	5.5	14	6.7	12	5.5	1.5	21.5	14	6.7
M12 × 1.75	18	14.2	6.7	16.8	8.2	14.4	6.7	2	25.6	16.8	8.2
M14 × 2	21	16.5	7.8	18.9	9.6	16.6	7.8	2.3	29.6	18.9	9.6
M16 × 2	24	18.5	9.5	21.4	11.7	18.9	9.5	2.5	34.2	21.4	11.7
M20 × 2.5	30	23.4	11.1	26.5	12.6	23.4	11.1	2.8	42.3	26.5	12.6
M24 × 3	36	28	13.3	31.4	15.1						
M30 × 3.5	46	33.7	16.4	38	18.5						
M36 × 4	55	40	20.1	45.6	22.8						

Table 19   Prevailing-torque insert-type nuts.

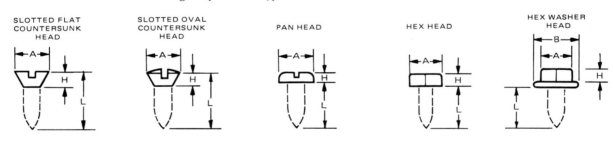

U.S. CUSTOMARY (INCHES)												METRIC (MILLIMETERS)											
NOMINAL SIZE		SLOTTED FLAT COUNTER-SUNK HEAD		SLOTTED OVAL COUNTER-SUNK HEAD		PAN HEAD		HEX HEAD		HEX WASHER HEAD			NOMINAL SIZE	SLOTTED FLAT COUNTER-SUNK HEAD		SLOTTED OVAL COUNTER-SUNK HEAD		PAN HEAD			HEX HEAD		HEX WASHER HEAD		
No.	DIA.	A	H	A	H	A	H	A	H	A	B	H		A	H	A	H	A	H Slot	H Recess	A	H	A	B	H
2	.086	.17	.05	.17	.05	.17	.05	.12	.05	.12	.17	.05	2	3.6	1.2	3.6	1.2	3.9	1.4	1.6	3.2	1.3	3.2	4.2	1.3
4	.112	.23	.07	.23	.07	.22	.07	.19	.06	.19	.24	.06	2.5	4.6	1.5	4.6	1.5	4.9	1.7	2	4	1.4	4	5.3	1.4
6	.138	.28	.08	.28	.08	.27	.08	.25	.09	.25	.33	.09	3	5.5	1.8	5.5	1.8	5.8	1.9	1.3	5	1.5	5	6.2	1.5
8	.164	.33	.10	.33	.10	.32	.10	.25	.11	.25	.35	.11	3.5	6.5	2.1	6.5	2.1	6.8	2.3	2.5	5.5	2.4	5.5	7.5	2.4
10	.190	.39	.17	.39	.17	.37	.11	.31	.12	.31	.41	.12	4	7.5	2.3	7.5	2.3	7.8	2.6	2.8	7	2.8	7	9.2	2.8
													5	9.5	2.9	9.5	2.9	9.8	3.1	3.5	8	3	8	10.5	3
													6	11.9	3.6	11.9	3.6	12	3.9	4.3	10	4.8	10	13.2	4.8
													8	15.2	4.4	15.2	4.4	15.6	5	5.6	13	5.8	13	17.2	5.8
													10	1.9	5.4	19	5.4	19.5	6.2	7	15	7.5	15	19.8	7.5
													12	22.9	6.4	22.9	6.4	23.4	7.5	8.3	18	9.5	18	23.8	9.5

Table 20   Tapping screws.

KIND OF MATERIAL	THREAD-FORMING								THREAD-CUTTING			SELF DRILLING	
	Type A	Type B	Type AB	HEX HEAD B	SWAGE FORM*	SWAGE FORM* B	Type U	Type 21	Type F*	Type L	Type B-F*	DRIL-KWICK	TAPITS*
SHEET METAL .0156 to .0469in. thick (0.4 to 1.2mm) (Steel, Brass, Aluminum, Monel, etc.)	✔	✔	✔	✔	✔	✔		✔				✔	✔
SHEET STAINLESS STEEL .0156 to .0469in thick (0.4 to 1.2mm)	✔	✔	✔	✔	✔	✔		✔	✔			✔	✔
SHEET METAL .20 to .50in. thick (1.2 to 5mm) (Steel, Brass, Aluminum, etc.)		✔	✔	✔	✔	✔	✔	✔	✔			✔	
STRUCTURAL STEEL .20 to .50in. thick (1.2 to .5mm)				✔	✔	✔	✔		✔				
CASTINGS (Aluminum, Magnesium, Zinc, Brass, Bronze, etc.)		✔	✔	✔	✔	✔	✔		✔				
CASTINGS (Grey Iron, Malleable Iron, Steel, etc.)					✔		✔		✔				
FORGINGS (Steel, Brass, Bronze, etc.)					✔		✔		✔				
PLYWOOD, Resin Impregnated: Compreg, Pregwood, etc. NATURAL WOODS	✔	✔	✔	✔			✔		✔		✔	✔	✔
ASBESTOS and other compositions: Ebony, Asbestos, Transite, Fiberglas, Insurok, etc.	✔	✔	✔	✔		✔						✔	✔
PHENOL FORMALDEHYDE: Molded: Bakelite, Durez, etc. Cast: Catalin, Marblette, etc. Laminated: Formica, Textolite, etc.		✔	✔	✔		✔	✔		✔		✔		
UREA FORMALDEHYDE: Molded: Plaskon, Beetle, etc. MELAMINE FORMALDEHYDE: Melantite, Melamac						✔	✔				✔		
CELLULOSE ACETATES and NITRATES: Tenite, Lumarith, Plastacele Pyralin, Celanese, etc. ACRYLATE & STYRENE RESINS: Lucite, Plexiglas, Styron, etc.		✔	✔	✔	✔		✔		✔		✔		
NYLON PLASTICS: Nylon, Zytel					✔	✔	✔			✔			

Table 21   Selector guide to thread cutting screws.

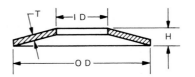

FLAT WASHER **LOCKWASHER** **SPRING LOCKWASHER**

Table 22 — Common washer sizes.

U.S. CUSTOMARY (INCHES)

Bolt Size	Flat Washers Type A–N			Lockwashers Regular		
	ID	OD	Thick	ID	OD	Thick
#6	.156	.375	.049	.141	.250	.031
#8	.188	.438	.049	.168	.293	.040
#10	.219	.500	.049	.194	.334	.047
#12	.250	.562	.065	.221	.377	.056
.250	.281	.625	.065	.255	.489	.062
.312	.344	.688	.065	.318	.586	.078
.375	.406	.812	.065	.382	.683	.094
.438	.469	.922	.065	.446	.779	.109
.500	.531	1.062	.095	.509	.873	.125
.562	.594	1.156	.095	.572	.971	.141
.625	.656	1.312	.095	.636	1.079	.156
.750	.812	1.469	.134	.766	1.271	.188
.875	.938	1.750	.134	.890	1.464	.219
1.000	1.062	2.000	.134	1.017	1.661	.250
1.125	1.250	2.250	.134	1.144	1.853	.281
1.250	1.375	2.500	.165	1.271	2.045	.312
1.375	1.500	2.750	.165	1.398	2.239	.344
1.500	1.625	3.000	.165	1.525	2.430	.375

METRIC (MILLIMETERS)

Bolt Size	Flat Washers			Lockwashers			Spring Lockwashers		
	Id	Od	Thick	Id	Od	Thick	Id	Od	Thick
2	2.2	5.5	0.5	2.1	3.3	0.5			
3	3.2	7	0.5	3.1	5.7	0.8			
4	4.3	9	0.8	4.1	7.1	0.9	4.2	8	0.3 / 0.4
5	5.3	11	1	5.1	8.7	1.2	5.2	10	0.4 / 0.5 / 0.5
6	6.4	12	1.5	6.1	11.1	1.6	6.2	12.5	0.7
7	7.4	14	1.5	7.1	12.1	1.6	7.2	14	0.5 / 0.8
8	8.4	17	2	8.2	14.2	2	8.2	16	0.6 / 0.9
10	10.5	21	2.5	10.2	17.2	2.2	10.2	20	0.8 / 1.1
12	13	24	2.5	12.3	20.2	2.5	12.2	25	0.9 / 1.5
14	15	28	2.5	14.2	23.2	3	14.2	28	1.0 / 1.5
16	17	30	3	16.2	26.2	3.5	16.3	31.5	1.2 / 1.7
18	19	34	3	18.2	28.2	3.5	18.3	35.5	1.2 / 2.0
20	21	36	3	20.2	32.2	4	20.4	40	1.5 / 2.25 / 1.75
22	23	39	4	22.5	34.5	4	22.4	45	2.5
24	25	44	4	24.5	38.5	5			
27	28	50	4	27.5	41.5	5			
30	31	56	4	30.5	46.5	6			

Table 23 — Belleville washers. (Wallace Barnes Co. Ltd.)

U.S. CUSTOMARY (INCHES)

Outside Diameter Max.	Inside Diameter Min.	Stock Thickness T	H Approx.
.187	.093	.010	.015
.250	.125	.009	.017
		.013	.020
.281	.138	.010	.020
		.015	.023
.312	.156	.011	.022
		.017	.025
.343	.164	.013	.024
		.019	.028
.375	.190	.015	.027
		.020	.030
.500	.255	.018	.034
		.025	.038
.625	.317	.022	.042
		.032	.048
.750	.380	.028	.051
		.040	.059
.875	.442	.031	.059
		.045	.067
1.000	.505	.035	.067
		.050	.075
1.125	.567	.038	.073
		.056	.084
1.250	.630	.040	.082
		.062	.092
1.375	.692	.044	.088
		.067	.101

METRIC (MILLIMETERS)

Outside Diameter Max.	Inside Diameter Min.	Stock Thickness T	H Approx.
4.8	2.4	0.16	0.33
		0.25	0.38
6.4	3.2	0.22	0.44
		0.34	0.51
7.9	4	0.27	0.55
		0.42	0.64
9.5	4.8	0.38	0.69
		0.51	0.76
12.7	6.5	0.46	0.86
		0.64	0.97
		0.97	1.20
15.9	8.1	0.56	1.07
		0.81	1.22
19.1	9.7	0.71	1.3
		1.02	1.5
		1.42	1.8
22.2	11.2	0.79	1.5
		1.14	1.7
25.4	12.8	0.89	1.7
		1.27	1.9
		1.85	2.3
28.6	14.4	0.97	1.9
		1.42	2.1
31.8	16	1.02	2.1
		1.58	2.3
34.9	17.6	1.12	2.2
		1.70	2.6
38.1	19.2	1.14	2.4
		1.83	2.7
44.5	22.4	1.45	2.9
		2.16	3.3
50.8	25.4	1.65	3.3
		2.46	3.7
63.5	31.8	2.03	4.1
		3.05	4.6

| U.S. CUSTOMARY (INCHES) | | | | | | METRIC (MILLIMETERS) | | | | | |
| Diameter of Shaft | | Square Key Nominal Size | | Flat Key Nominal Size | | Diameter of Shaft | | Square Key Nominal Size | | Flat Key Nominal Size | |
From	To	W	H	W	H	Over	Up To	W	H	W	H
.500	.562	.125	.125	.125	.094	6	8	2	2		
.625	.875	.188	.188	.188	.125	8	10	3	3		
.938	1.250	.250	.250	.250	.188	10	12	4	4		
1.312	1.375	.312	.312	.312	.250	12	17	5	5		
1.438	1.750	.375	.375	.375	.250	17	22	6	6		
1.812	2.250	.500	.500	.500	.375	22	30	7	7	8	7
2.375	2.750	.625	.625			30	38	8	8	10	8
2.875	3.250	.750	.750			38	44	9	9	12	8
3.375	3.750	.875	.875			44	50	10	10	14	9
3.875	4.500	1.000	1.000			50	58	12	12	16	10

Table 24   Square and flat stock keys.

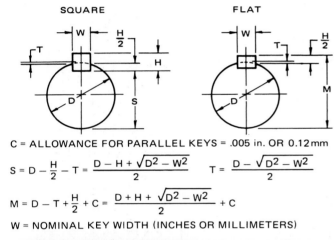

SQUARE   FLAT

C = ALLOWANCE FOR PARALLEL KEYS = .005 in. OR 0.12 mm

$$S = D - \frac{H}{2} - T = \frac{D - H + \sqrt{D^2 - W^2}}{2} \qquad T = \frac{D - \sqrt{D^2 - W^2}}{2}$$

$$M = D - T + \frac{H}{2} + C = \frac{D + H + \sqrt{D^2 - W^2}}{2} + C$$

W = NOMINAL KEY WIDTH (INCHES OR MILLIMETERS)

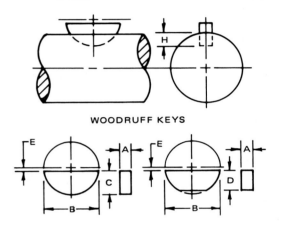

WOODRUFF KEYS

| U.S. CUSTOMARY (INCHES) | | | | | Key No. | METRIC (MILLIMETERS) | | | | |
| Nominal Size | Key | | Keyseat | | | Nominal Size | Key | | Keyseat | |
A × B	E	C	D	H		A × B	E	C	D	H
.062 × .500	.047	.203	.194	.172	204	1.6 × 12.7	1.5	5.1	4.8	4.2
.094 × .500	.047	.203	.194	.156	304	2.4 × 12.7	1.3	5.1	4.8	3.8
.094 × .625	.062	.250	.240	.203	305	2.4 × 15.9	1.5	6.4	6.1	5.1
.125 × .500	.049	.203	.194	.141	404	3.2 × 12.7	1.3	5.1	4.8	3.6
.125 × .625	.062	.250	.240	.188	405	3.2 × 15.9	1.5	6.4	6.1	4.6
.125 × .750	.062	.313	.303	.251	406	3.2 × 19.1	1.5	7.9	7.6	6.4
.156 × .625	.062	.250	.240	.172	505	4.0 × 15.9	1.5	6.4	6.1	4.3
.156 × .750	.062	.313	.303	.235	506	4.0 × 19.1	1.5	7.9	7.6	5.8
.156 × .875	.062	.375	.365	.297	507	4.0 × 22.2	1.5	9.7	9.1	7.4
.188 × .750	.062	.313	.303	.219	606	4.8 × 19.1	1.5	7.9	7.6	5.3
.188 × .875	.062	.375	.365	.281	607	4.8 × 22.2	1.5	9.7	9.1	7.1
.188 × 1.000	.062	.438	.428	.344	608	4.8 × 25.4	1.5	11.2	10.9	8.6
.188 × 1.125	.078	.484	.475	.390	609	4.8 × 28.6	2.0	12.2	11.9	9.9
.250 × .875	.062	.375	.365	.250	807	6.4 × 22.2	1.5	9.7	9.1	6.4
.250 × 1.000	.062	.438	.428	.313	808	6.4 × 25.4	1.5	11.2	10.9	7.9

NOTE: METRIC KEY SIZES WERE NOT AVAILABLE AT THE TIME OF PUBLICATION. SIZES SHOWN ARE INCH-DESIGNED KEY-SIZES SOFT CONVERTED TO MILLIMETERS. CONVERSION WAS NECESSARY TO ALLOW THE STUDENT TO COMPARE KEYS WITH SLOT SIZES GIVEN IN MILLIMETERS.

Table 25   Woodruff keys.

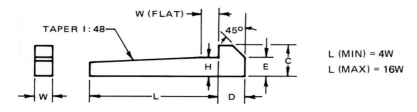

L (MIN) = 4W
L (MAX) = 16W

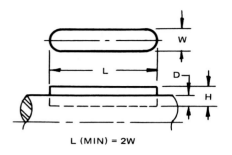

L (MIN) = 2W

U.S. CUSTOMARY (INCHES)										
Shaft Diameter	Square Type					Flat Type				
	W	H	C	D	E	W	H	C	D	E
.500–.562	.125	.125	.250	.219	.156	.125	.094	.188	.125	.125
.625–.875	.188	.188	.312	.281	.219	.188	.125	.250	.188	.156
.938–1.250	.250	.250	.438	.344	.344	.250	.188	.312	.250	.188
1.312–1.375	.312	.312	.562	.406	.406	.312	.250	.375	.312	.250
1.438–1.750	.375	.375	.688	.469	.469	.375	.250	.438	.375	.312
1.812–2.250	.500	.500	.875	.594	.625	.500	.375	.625	.500	.438
2.312–2.750	.625	.625	1.062	.719	.750	.625	.438	.750	.625	.500
2.875–3.250	.750	.750	1.250	.875	.875	.750	.500	.875	.750	.625

METRIC (MILLIMETERS)										
Shaft Diameter	Square Type					Flat Type				
	W	H	C	D	E	W	H	C	D	E
12–14	3.2	3.2	6.4	5.4	4	3.2	2.4	5	3.2	3.2
16–22	4.8	4.8	10	7	5.4	4.8	3.2	6.4	5	4
24–32	6.4	6.4	11	8.6	8.6	6.4	5	8	6.4	5
34–35	8	8	14	10	10	8	6.4	10	8	6.4
36–44	10	10	18	12	12	10	6.4	11	10	8
46–58	13	13	22	15	16	13	10	16	13	11
60–70	16	16	27	19	20	16	11	20	16	13
72–82	20	20	32	22	22	20	13	22	20	16

Note: Metric standards governing key sizes were not available at the time of publication. The sizes given in the above chart are "soft conversion" from current standards and are not representative of the precise metric key sizes which may be available in the future. Metric sizes are given only to allow the student to complete the drawing assignment.

Table 26 Square and flat gib-head keys.

U.S. CUSTOMARY (INCHES)				
Key No.	L	W	H	D
2	.500	.094	.141	.094
4	.625	.094	.141	.094
6	.625	.156	.234	.156
8	.750	.156	.234	.156
10	.875	.156	.234	.156
12	.875	.234	.328	.219
14	1.00	.234	.328	.234
16	1.125	.188	.281	.188
18	1.125	.250	.375	.250
20	1.250	.219	.328	.219
22	1.375	.250	.375	.250
24	1.50	.250	.375	.250
26	2.00	.188	.281	.188
28	2.00	.312	.469	.312
30	3.00	.375	.562	.375
32	3.00	.500	.750	.500
34	3.00	.625	.938	.625

METRIC (MILLIMETERS)				
Key No.	L	W	H	D
2	12	2.4	3.6	2.4
4	16	2.4	3.6	2.4
6	16	4	6	4
8	20	4	6	4
10	22	4	6	4
12	22	6	8.4	7
14	25	6	8.4	6
16	28	5	7	5
18	28	6.4	10	6.4
20	32	7	8	5
22	35	6.4	10	6.4
24	38	6.4	10	6.4
26	50	5	7	5
28	50	8	12	8
30	75	10	14	10
32	75	12	20	12
34	75	16	24	16

Table 27 Pratt and Whitney keys.

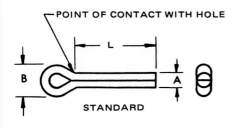

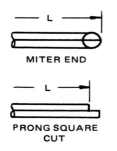

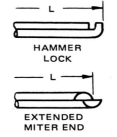

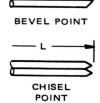

U.S. CUSTOMARY (INCHES)				METRIC (MILLIMETERS)			
Nominal Bolt or Thread Size Range	Nominal Cotter-Pin Size (A)	Cotter-Pin Hole	Min. End Clearance*	Nominal Bolt or Thread-Size Range	Nominal Cotter-Pin Size (A)	Cotter-Pin Hole	Min. End Clearance*
.125	.031	.047	.06	−2.5	0.6	0.8	1.5
.188	.047	.062	.08	2.5−3.5	0.8	1.0	2.0
.250	.062	.078	.11	3.5−4.5	1.0	1.2	2.0
.312	.078	.094	.11	4.5−5.5	1.2	1.4	2.5
.375	.094	.109	.14	5.5−7.0	1.6	1.8	2.5
.438	.109	.125	.14	7.0−9.0	2.0	2.2	3.0
.500	.125	.141	.18	9.0−11	2.5	2.8	3.5
.562	.141	.156	.25	11−14	3.2	3.6	5
.625	.156	.172	.40	14−20	4	4.5	6
1.000−1.125	.188	.203	.40	20−27	5	5.6	7
1.250−1.375	.219	.234	.46	27−39	6.3	6.7	10
1.500−1.625	.250	.266	.46	39−56	8.0	8.5	15
				56−80	10	10.5	20

*End of bolt to center of hole

Table 28 Cotter pins.

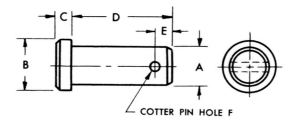

COTTER PIN HOLE F

U.S. CUSTOMARY (INCHES)						METRIC (MILLIMETERS)					
Pin Dia A	B	C	Min. D	E	Drill Size F	Pin Dia. A	B	C	Min. D	E	Drill Size F
.188	.31	.06	.59	.11	.078	4	6	1	16	2.2	1
.250	.38	.09	.80	.12	.078	6	10	2	20	3.2	1.6
.312	.44	.09	.97	.16	.109	8	14	3	24	3.5	2
.375	.50	.12	1.09	.16	.109	10	18	4	28	4.5	3.2
.500	.62	.16	1.42	.22	.141	12	20	4	36	5.5	3.2
.625	.81	.20	1.72	.25	.141	16	25	4.5	44	6	4
.750	.94	.25	2.05	.30	.172	20	30	5	52	8	5
1.000	1.19	.34	2.62	.36	.172	24	36	6	66	9	6.3

Table 29 Clevis pins.

Table 30 Taper pins.

U.S. CUSTOMARY (INCHES)

NUMBER	7/0	6/0	5/0	4/0	3/0	2/0	0	1	2	3	4	5	6	7	8	9
SIZE (LARGE END)	.062	.078	.094	.109	.125	.141	.152	.172	.193	.219	.250	.289	.314	.409	.492	.591
LENGTH																
.375	X	X														
.500	X	X	X	X	X	X	X									
.625	X	X	X	X	X	X	X									
.750		X	X	X	X	X	X	X	X	X						
.875						X	X	X	X	X						
1.000			X	X	X	X	X	X	X	X	X	X				
1.250						X	X	X	X	X	X	X	X			
1.500							X	X	X	X	X	X	X			
1.750								X	X	X	X	X	X			
2.000								X	X	X	X	X	X	X	X	
2.250									X	X	X	X	X	X	X	
2.500										X	X	X	X	X	X	
2.750										X	X	X	X	X	X	X

METRIC (MILLIMETERS)

NUMBER	7/0	6/0	5/0	4/0	3/0	2/0	0	1	2	3	4	5	6	7	8	9
SIZE (LARGE END)	1.6	2	2.4	2.8	3.2	3.6	4	4.4	4.9	5.6	6.4	7.4	8	10.4	12.5	15
LENGTH																
10	X	X														
12	X	X	X	X	X	X	X									
16	X	X	X	X	X	X	X									
20		X	X	X	X	X	X	X	X	X						
22				X	X	X	X	X	X	X						
25			X	X	X	X	X	X	X	X	X	X				
30						X	X	X	X	X	X	X	X			
40							X	X	X	X	X	X	X			
45								X	X	X	X	X	X			
50								X	X	X	X	X	X	X	X	
55									X	X	X	X	X	X	X	
65									X	X	X	X	X	X	X	
70										X	X	X	X	X	X	X

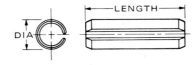

Table 31 Spring pins.

PIN DIAMETER (INCHES)								PIN DIAMETER (MILLIMETERS)										
Length	.062	.094	.125	.156	.188	.250	.312	Length	1.5	2	2.5	3	4	5	6	8	10	12
.250	X	X						5	X	X								
.375	X	X	X					10	X	X	X	X						
.500	X	X	X	X	X			15	X	X	X	X	X	X				
.625	X	X	X	X	X	X		20	X	X	X	X	X	X	X			
.750	X	X	X	X	X	X	X	25	X	X	X	X	X	X	X	X		
.875	X	X	X	X	X	X	X	30		X	X	X	X	X	X	X	X	X
1.00	X	X	X	X	X	X	X	35		X	X	X	X	X	X	X	X	X
1.250		X	X	X	X	X	X	40		X	X	X	X	X	X	X	X	X
1.500		X	X	X	X	X	X	45				X	X	X	X	X	X	X
1.750			X	X	X	X	X	50				X	X	X	X	X	X	X
2.000		X	X	X	X	X	X	55					X	X	X	X	X	X
2.225			X	X	X	X	X	60					X	X	X	X	X	X
2.500				X	X	X	X	70							X	X	X	X
3.000						X	X	75							X	X	X	X
3.500						X	X	80								X	X	X

	U.S. CUSTOMARY (INCHES)								METRIC (MILLIMETERS)							
	PIN DIAMETER									PIN DIAMETER						
Length	.09	.125	.188	.250	.312	.375	.500	Length	2	3	4	5	6	8	10	12
.250	x	x						5	x	x	x					
.375	x	x	x					10	x	x	x	x	x			
.500	x	x	x	x				15	x	x	x	x	x	x		
.625	x	x	x	x	x			20	x	x	x	x	x	x	x	
.750	x	x	x	x	x	x		25	x	x	x	x	x	x	x	x
.875	x	x	x	x	x	x		30	x	x	x	x	x	x	x	x
1.000	x	x	x	x	x	x	x	35		x	x	x	x	x	x	x
1.250	x	x	x	x	x	x	x	40			x	x	x	x	x	x
1.500		x	x	x	x	x	x	45				x	x	x	x	x
1.750			x	x	x	x	x	50					x	x	x	x
2.000			x	x	x	x	x	55					x	x	x	x
2.250			x	x	x	x	x	60					x	x	x	x
2.275				x	x	x	x	65					x	x	x	x
3.000				x	x	x	x	70						x	x	x
								75						x	x	x

Note: Metric size pins were not available at the time of publication. Sizes were soft converted to allow students to complete drawing assignments.

Table 32 Groove pins.

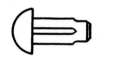

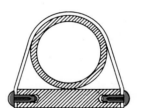

WIDELY USED FOR FASTENING BRACKETS

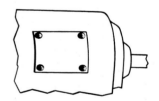

ATTACHING NAMEPLATES, INSTRUCTION PANELS

U.S. CUSTOMARY (INCHES)											METRIC (MILLIMETERS)									
STUD NUMBER	SHANK DIA.	DRILL SIZE	HEAD DIA.	STANDARD LENGTHS						STUD NUMBER	SHANK DIA.	DRILL SIZE	HEAD DIA.	STANDARD LENGTHS					
				.125	.188	.250	.312	.375	.500					4	6	8	10	12	14
0	.067	51	.130	•	•	•				0	1.7	1.7	3.3	•	•	•			
2	.086	44	.162	•	•	•				2	2.2	2.2	4.1	•	•	•			
4	.104	37	.211		•	•	•			4	2.6	2.6	5.4		•	•	•		
6	.120	31	.260			•	•	•		6	3.0	3.0	6.6			•	•	•	
7	.136	29	.309				•	•	•	7	3.4	3.4	7.8				•	•	•
8	.144	27	.309					•	•	8	3.8	3.8	7.8					•	•
10	.161	20	.359					•	•	10	4.1	4.1	9.1					•	•
12	.196	9	.408						•	12	5.0	5.0	10.4						•
14	.221	2	.457						•	14	5.6	5.6	11.6						•
16	.250	1/4	.472						•	16	6.3	6.3	12						•

Note: Metric size studs were not available at the time of publication. Sizes were soft converted to allow students to complete drawing assignments.

Table 33 Grooved studs. (Drive-Lok)

Use these columns first to locate your correct GRIP LENGTH

Grip = Total thickness of all sheets fastened together

Minimum Grip	Nominal Grip	Maximum Grip
1	2	3
2	3	4
4	5	6
5	6	7
7	8	9
9	10	11
11	12	13
13	14	15
15	16	17
19	20	21
23	24	25
27	28	29

.125in. (3mm) Dia — Part Numbers

Length Under Head L	Universal Head	100° Csk Head	Full Brazier Head
5	✓	✓	✓
6	✓	✓	✓
8	✓	✓	✓
10	✓	✓	✓
12	✓	✓	✓
14	✓	✓	✓
16	✓		
18			
20			
24			
28			
32			

.156in. (4mm) Dia — Part Numbers

Length Under Head L	Universal Head	100° Csk Head
5	✓	
6	✓	✓
8	✓	✓
10	✓	✓
12	✓	✓
14	✓	✓
16	✓	✓
18	✓	✓
20	✓	✓
24		
28		
32		

.188in. (5mm) Dia — Part Numbers

Length Under Head L	Universal Head	100° Csk Head	Full Brazier Head	All Purpose Liner Head
5	✓			
6	✓		✓	✓
8	✓	✓	✓	✓
10	✓	✓	✓	✓
12	✓	✓	✓	✓
14	✓	✓	✓	✓
16	✓	✓	✓	✓
18	✓	✓	✓	✓
20	✓	✓	✓	✓
24	✓			
28	✓			
32	✓			

.250in. (6mm) Dia — Part Numbers

Length Under Head L	Universal Head	100° Csk Head	Full Brazier Head
5	✓	✓	✓
6	✓	✓	✓
8	✓	✓	✓
10	✓	✓	✓
12	✓	✓	✓
14	✓	✓	✓
16	✓	✓	✓
18	✓	✓	✓
20	✓	✓	✓
24	✓	✓	✓
28	✓	✓	✓
32	✓	✓	✓

Note: Metric drive rivets were not available at the time of publication. Sizes were soft converted to allow students to complete drawing assignments.

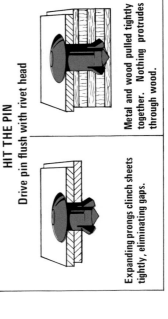

HIT THE PIN
Drive pin flush with rivet head

Expanding prongs clinch sheets tightly, eliminating gaps.

Metal and wood pulled tightly together. Nothing protrudes through wood.

IN METAL — GRIP

Grip Length = Total thickness of sheets to be fastened.

IN WOOD

Use "L" Dimension (length under head) instead of grip length. L = M (thickness of metal) + D (hole depth in wood).

Table 34 Aluminum drive rivets.

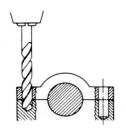

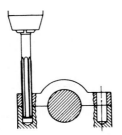

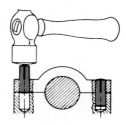

| DRILL HOLE SLIGHTLY UNDERSIZE | REAM FULL SIZE | DRIVE OR PRESS LOK DOWELS INTO PLACE | LOK DOWELS LOCK SECURELY AND PARTS SEPARATE EASILY |

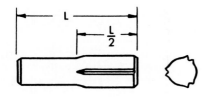

U.S. CUSTOMARY (INCHES)							METRIC (MILLIMETERS)						
	DIAMETER							**DIAMETER**					
LENGTH	.125	.188	.250	.312	.375	.500	LENGTH	4	6	8	10	12	14
.375	•						10	•					
.500	•	•	•	•			12	•	•	•	•		
.625	•	•	•	•			16	•	•	•	•		
.750	•	•	•	•	•		20	•	•	•	•	•	
.875	•	•	•	•	•		22	•	•	•	•	•	
1.000	•	•	•	•	•	•	26	•	•	•	•	•	•
1.250		•	•	•	•	•	32		•	•	•	•	•
1.500			•	•	•	•	38			•	•	•	•
1.750				•	•	•	44				•	•	•
2.000					•	•	50					•	•

Note: Metric size dowels were not available at the time of publication. Sizes were soft converted to allow students to complete drawing assignments.

Table 35 Lok dowels. (Drive-Lok)

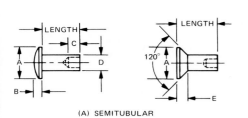

(A) SEMITUBULAR

(B) SPLIT

	D	A	B	C	E	MIN. LENGTH
U.S. CUSTOMARY (INCHES)	.062	.125	.031	.062	—	.062
	.094	.156	.031	.062	.031	.078
	.109	.188	.031	.078	—	.078
	.125	.218	.047	.109	.049	.109
	.141	.250	.047	.125	.049	.125
	.188	.312	.062	.141	.062	.156
	.219	.438	.062	.188	.062	.188
	.250	.500	.078	.219	.094	.219
	.312	.562	.109	.250		.250
METRIC (MILLIMETERS)	1.5	2.8	0.4	1.2		1.6
	2.2	3.7	0.6	1.6	0.8	2.0
	2.5	4.7	0.7	2.0		2.0
	3.1	5.5	0.9	2.4	1.0	2.4
	3.6	5.9	1.0	3.2	1.2	3.2
	4.7	7.9	1.5	3.9	1.6	4.0
	5.4	11.1	1.7	4.8	1.8	4.8
	6.3	12.7	2.0	5.6	2.2	5.6
	7.7	14.3	2.4	6.2		6.4

Table 36 Semi-tubular and split rivets.

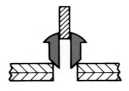

U.S. CUSTOMARY (INCHES)				METRIC (MILLIMETERS)			
HOLE DIA.	PANEL RANGE	**HEAD**		HOLE DIA.	PANEL RANGE	**HEAD**	
		Dia.	Height			Dia.	Height
.125	.031–.140	.188	.047	3.18	0.8– 3.6	4.8	1.2
	.031–.125	.218	.062		0.8– 3.2	5.5	1.5
.156	.250–.375	.218	.047	4.01	5.9– 9.4	5.5	1.3
.188	.062–.156	.375	.125	4.75	1.6– 4.0	9.5	3.2
	.156–.281	.438	.094		4.0– 7.1	11.1	1.9
.219	.062–.125	.375	.094	5.54	1.6– 3.2	9.5	2.4
	.094–.312		.078		2.4– 8.0		2.0
.250	.094–.219	.625	.125	6.35	2.3– 5.6	16	3.2
	.125–.375	.750	.062		3.2– 9.5	19	1.3
.297	.140–.328	.500	.078	7.14	3.4– 8.1	12.3	1.9
.375	.250–.500	.438	.109	9.53	6.4–12.7	11.1	2.6
.500	.312–.375	.750	.109	12.7	8.1– 9.4	19	2.5

Table 37 Plastic rivets.

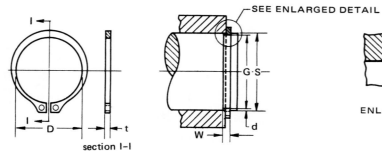
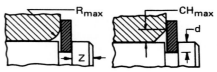

ENLARGED DETAIL OF GROOVE PROFILE
AND EDGE MARGIN (Z)

| | SHAFT DIA. | EXTERNAL SERIES | RETAINING RING DIMENSIONS | | GROOVE DIMENSIONS | | | | MAX. CORNER RADII AND CHAMFER OF RETAINED PARTS | | EDGE MARGIN | NOMINAL GROOVE DEPTH (REF) |
| | | | | | DIAMETER | | WIDTH | | | | | |
	S	Size—No.	D	t	G	Tol.	W	Tol.	R Max.	Ch. Max.	z	d
U.S. CUSTOMARY (INCHES)	.188	5100-18	.168	.015	.175	±.0015	.018	+.002	.014	.008	.018	.006
	.250	5100-25	.225	.025	.230	±.0015	.029	+.003	.018	.011	.030	.010
	.312	5100-31	.281	.025	.290	±.002	.029	+.003	.020	.012	.033	.011
	.375	5100-37	.338	.025	.352	±.002	.029	+.003	.026	.015	.036	.012
	.500	5100-50	.461	.035	.468	±.002	.039	+.003	.034	.020	.048	.016
	.625	5100-62	.579	.035	.588	±.003	.039	+.003	.041	.025	.055	.018
	.750	5100-75	.693	.042	.704	±.003	.046	+.003	.046	.027	.069	.023
	.875	5100-87	.810	.042	.821	±.003	.046	+.003	.051	.031	.081	.027
	1.000	5100-100	.925	.042	.940	±.003	.046	+.003	.057	.034	.090	.030
	1.125	5100-112	1.041	.050	1.059	±.004	.056	+.004	.063	.038	.099	.033
	1.250	5100-125	1.156	.050	1.176	±.004	.056	+.004	.068	.041	.111	.037
	1.375	5100-137	1.272	.050	1.291	±.004	.056	+.004	.072	.043	.126	.042
	1.500	5100-150	1.387	.050	1.406	±.004	.056	+.004	.079	.047	.141	.047
METRIC (MILLIMETERS)	4	M5100-4	3.6	0.25	3.80	−0.08	0.32	+0.05	0.35	0.25	0.3	0.10
	6	M5100-6	5.5	0.4	5.70	−0.08	0.5	+0.1	0.35	0.25	0.5	0.15
	8	M5100-8	7.2	0.6	7.50	−0.1	0.7	+0.15	0.5	0.35	0.8	0.25
	10	M5100-10	9.0	0.6	9.40	−0.1	0.7	+0.15	0.7	0.4	0.9	0.30
	12	M5100-12	10.9	0.6	11.35	−0.12	0.7	+0.15	0.8	0.45	1.0	0.33
	14	M5100-14	12.9	0.9	13.25	−0.12	1.0	+0.15	0.9	0.5	1.2	0.38
	16	M5100-16	14.7	0.9	15.10	−0.15	1.0	+0.15	1.1	0.6	1.4	0.45
	18	M5100-18	16.7	1.1	17.00	−0.15	1.2	+0.15	1.2	0.7	1.5	0.50
	20	M5100-20	18.4	1.1	18.85	−0.15	1.2	+0.15	1.2	0.7	1.7	0.58
	22	M5100-22	20.3	1.1	20.70	−0.15	1.2	+0.15	1.3	0.8	1.9	0.65
	24	M5100-24	22.2	1.1	22.60	−0.15	1.2	+0.15	1.4	0.8	2.1	0.70
	25	M5100-25	23.1	1.1	23.50	−0.15	1.2	+0.15	1.4	0.8	2.3	0.75
	30	M5100-30	27.9	1.3	28.35	−0.2	1.4	+0.15	1.6	1.0	2.5	0.83
	35	M5100-35	32.3	1.3	32.9	−0.2	1.4	+0.15	1.8	1.1	3.1	1.05
	40	M5100-40	36.8	1.6	37.7	−0.3	1.75	+0.2	2.1	1.2	3.4	1.15
	45	M5100-45	41.6	1.6	42.4	−0.3	1.75	+0.2	2.3	1.4	3.9	1.3
	50	M5100-50	46.2	1.6	47.2	−0.3	1.75	+0.2	2.4	1.4	4.2	1.4

Table 38 **Retaining rings—external.** (© 1965, 1958 Waldes Koh-i-noor, Inc. Reprinted with permission.)

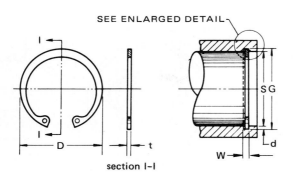

SEE ENLARGED DETAIL

section I-I

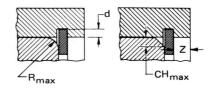

ENLARGED DETAIL OF GROOVE PROFILE
AND EDGE MARGIN (Z)

| | HOUSING DIA. | INTERNAL SERIES | RETAINING RING DIMENSIONS | | GROOVE DIMENSIONS | | | | MAX. CORNER RADII AND CHAMFER OF RETAINED PARTS | | EDGE MARGIN | NOMINAL GROOVE DEPTH |
| | | | | | DIAMETER | | WIDTH | | | | | |
	S	Size—No.	D	t	G	Tol.	W	Tol.	R Max.	Ch. Max.	z	d
U.S. CUSTOMARY (INCHES)	.250	N5000-25	.280	.015	.268	±.001	.018	+.002	.011	.008	.027	.009
	.312	N5000-31	.346	.015	.330	±.001	.018	+.002	.016	.013	.027	.009
	.375	N5000-37	.415	.025	.397	±.002	.029	+.003	.023	.018	.033	.011
	.500	N5000-50	.548	.035	.530	±.002	.039	+.003	.027	.021	.045	.015
	.625	N5000-62	.694	.035	.665	±.002	.039	+.003	.027	.021	.060	.020
	.750	N5000-75	.831	.035	.796	±.002	.039	+.003	.032	.025	.069	.023
	.875	N5000-87	.971	.042	.931	±.003	.046	+.003	.035	.028	.084	.028
	1.000	N5000-100	1.111	.042	1.066	±.003	.046	+.003	.042	.034	.099	.033
	1.125	N5000-112	1.249	.050	1.197	±.004	.056	+.004	.047	.036	.108	.036
	1.250	N5000-125	1.388	.050	1.330	±.004	.056	+.004	.048	.038	.120	.040
	1.375	N5000-137	1.526	.050	1.461	±.004	.056	+.004	.048	.038	.129	.043
	1.500	N5000-150	1.660	.050	1.594	±.004	.056	+.004	.048	.038	.141	.047
METRIC (MILLIMETERS)	8	MN5000-8	8.80	0.4	8.40	+0.6	0.5	+0.1	0.4	0.3	0.6	0.2
	10	MN5000-10	11.10	0.6	10.50	+0.1	0.7	+0.15	0.5	0.35	0.8	0.25
	12	MN5000-12	13.30	0.6	12.65	+0.1	0.7	+0.15	0.6	0.4	1.0	0.33
	14	MN5000-14	15.45	0.9	14.80	+0.1	1.0	+0.15	0.7	0.5	1.2	0.40
	16	MN5000-16	17.70	0.9	16.90	+0.1	1.0	+0.15	0.7	0.5	1.4	0.45
	18	MN5000-18	20.05	0.9	19.05	+0.1	1.0	+0.15	0.75	0.6	1.6	0.53
	20	MN5000-20	22.25	0.9	21.15	+0.15	1.0	+0.15	0.9	0.7	1.7	0.57
	22	MN5000-22	24.40	1.1	23.30	+0.15	1.2	+0.15	0.9	0.7	1.9	0.65
	24	MN5000-24	26.55	1.1	25.4	+0.15	1.2	+0.15	1.0	0.8	2.1	0.70
	25	MN5000-25	27.75	1.1	26.6	+0.15	1.2	+0.15	1.0	0.8	2.4	0.80
	30	MN5000-30	33.40	1.3	31.9	+0.2	1.4	+0.15	1.2	1.0	2.9	0.95
	35	MN5000-35	38.75	1.3	37.2	+0.2	1.4	+0.15	1.2	1.0	3.3	1.10
	40	MN5000-40	44.25	1.6	42.4	+0.2	1.75	+0.2	1.7	1.3	3.6	1.20
	45	MN5000-45	49.95	1.6	47.6	+0.2	1.75	+0.2	1.7	1.3	3.9	1.30
	50	MN5000-50	55.35	1.6	53.1	+0.2	1.75	+0.2	1.7	1.3	4.6	1.55

Table 39   **Retaining rings—internal.** (© 1965, 1958 Waldes Koh-i-noor, Inc. Reprinted with permission.)

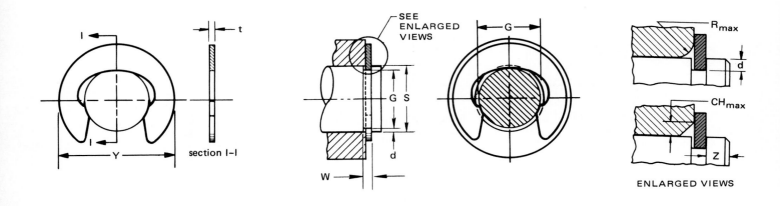

| | SHAFT DIA. | EXTERNAL SERIES 11-410 | RETAINING RING DIMENSIONS | | GROOVE DIMENSIONS | | | | MAXIMUM ALLOWABLE CORNER RADII AND CHAMFER OF RETAINED PARTS | | EDGE MARGIN | NOMINAL GROOVE DEPTH (REF) |
| | | | | | Diameter | | Width | | | | | |
	S	Size—No.	Y	t	G	Tol.	W	Tol.	R Max.	Ch. Max.	z	d
U.S. CUSTOMARY (INCHES)	.250	11-410-25	.311	.025	.222	−.004	.029	+.003	.023	.018	.030	.015
	.312	11-410-31	.376	.025	.278	−.004	.029	+.003	.024	.018	.036	.018
	.375	11-410-37	.448	.025	.337	−.004	.029	+.003	.026	.020	.040	.020
	.500	11-410-50	.581	.025	.453	−.006	.039	+.003	.030	.023	.050	.025
	.625	11-410-62	.715	.035	.566	−.006	.039	+.003	.033	.025	.062	.031
	.750	11-410-75	.845	.042	.679	−.006	.046	+.003	.036	.027	.074	.037
	.875	11-410-87	.987	.042	.792	−.006	.046	+.003	.040	.031	.086	.043
	1.000	11-410-100	1.127	.042	.903	−.006	.046	+.003	.046	.035	.100	.050
	1.125	11-410-112	1.267	.050	1.017	−.008	.056	+.004	.052	.040	.112	.056
	1.250	11-410-125	1.410	.050	1.130	−.008	.056	+.004	.057	.044	.124	.062
	1.375	11-410-137	1.550	.050	1.241	−.008	.056	+.004	.062	.048	.138	.069
	1.500	11-410-150	1.691	.050	1.354	−.008	.056	+.004	.069	.053	.150	.075
	1.750	11-410-175	1.975	.062	1.581	−.010	.068	+.004	.081	.062	.174	.087
	2.000	11-410-200	2.257	.062	1.805	−.010	.068	+.004	.091	.070	.200	.100
METRIC (MILLIMETERS)	8	11-410-080	10	0.6	7	−0.1	0.7	+0.15	0.6	0.45	1.5	0.5
	10	11-410-100	12.2	0.6	9	−0.1	0.7	+0.15	0.6	0.45	1.5	0.5
	12	11-410-120	14.4	0.6	10.9	−0.1	0.7	+0.15	0.6	0.45	1.7	0.5
	14	11-410-140	16.3	1	12.7	−0.1	1.1	+0.15	1	0.8	2	0.65
	16	11-410-160	18.5	1	14.5	−0.1	1.1	+0.15	1	0.8	2.3	0.75
	18	11-410-180	20.4	1.2	16.3	−0.1	1.3	+0.15	1.2	0.9	2.6	0.85
	20	11-410-200	22.6	1.2	18.1	−0.2	1.3	+0.15	1.2	0.9	2.9	0.95
	22	11-410-220	25	1.2	19.9	−0.2	1.3	+0.15	1.2	0.9	3.2	1.05
	24	11-410-240	27.1	1.2	21.7	−0.2	1.3	+0.15	1.2	0.9	3.5	1.15
	25	11-410-250	28.3	1.2	22.6	−0.2	1.3	+0.15	1.2	0.9	3.6	1.2
	30	11-410-300	33.7	1.5	27	−0.2	1.3	+0.2	1.5	1.15	4.5	1.5
	35	11-410-350	39.4	1.5	31.5	−0.25	1.6	+0.2	1.5	1.15	5.3	1.75
	40	11-410-400	45	1.5	36	−0.25	1.6	+0.2	1.5	1.15	6	2
	45	11-410-450	50.6	1.5	40.5	−0.25	1.6	+0.2	1.5	1.15	6.8	2.25
	50	11-410-500	56.4	2	45	−0.25	2.2	+0.2	2	1.5	7.5	2.5

Table 40 **Retaining rings—radial assembly.** (© 1965, 1958 Waldes Koh-i-noor, Inc. Reprinted with permission.)

MORSE TAPERS		
	TAPER	
No. of Taper	inches per Foot	mm per 100 mm
0	.625	5.21
1	.599	4.99
2	.599	4.99
3	.602	5.02
4	.623	5.19
5	.631	5.26
6	.626	5.22
7	.624	5.20

BROWN AND SHARPE TAPERS		
	TAPER	
No. of Taper	inches per Foot	mm per 100 mm
1	.502	4.18
2	.502	4.18
3	.502	4.18
4	.502	4.18
5	.502	4.18
6	.503	4.19
7	.502	4.18
8	.501	4.18
9	.501	4.18
10	.516	4.3
11	.501	4.18
12	.500	4.17
13	.500	4.17
14	.500	4.17
15	.500	4.17
16	.500	4.17

Table 41 Machine tapers.

PRODUCT	OUTSTANDING FEATURES	APPLICATION METHOD	COLOR
1357	A high performance adhesive with long bonding range, excellent initial strength. Meets specification requirements of MMM-A-121 (supersedes MIL-A-1154 C), MIL-A-5092 B, Type II, and MIL-A-21366. Bonds rubber, cloth, wood, foamed glass, paper honeycomb, decorative plastic laminates. Also used with metal-to-metal for bonds of moderate strength.	Spray or Brush	Gray/ Green or Olive
2210	Fast drying, exhibits aggressive tack that allows coated surfaces to knit easily under hand roller pressure. Excellent water and oil resistance. Meets specification requirements of MMM-A-121 (supersedes MIL-A-1154 C), MIL-A-21366, and MMM-A-00130a. Bonds a wide range of materials including rubber, leather, cloth, aluminum, wood, hardboard. Used extensively for bonding decorative plastic laminates.	Brush, Roller, or Trowel	Yellow
2215	Fast drying and has a rapid rate of strength build-up. Its aggressive tack permits adhesive coated surfaces to bond easily with moderate pressure. Bonds decorative plastic laminates to metal, wood, and particle board. Also used for general bonding of rubber, leather, cloth, aluminum, wood, hardboard, etc.	Spray	Light Yellow
2218	Offers rapid strength build-up, high-ultimate strength. Has a high softening point and excellent resistance to plastic flow. Adhesion to steel is especially good. Meets specification requirements of MMM-A-121 (supersedes MIL-A-1154 C). Bonds high density decorative plastic laminates to metal or wood. Widely used to fabricate honeycomb and sandwich-type building panels with various face sheets, including porcelain enamel steel.	Spray	Green
2226	Water dispersed, has high immediate bond strength, long bonding range. Changes color from blue to green while drying. Used to bond foamed plastics, plastic laminate, wood, rubber, plywood, wallboard, wood veneer, plaster, and canvas to themselves and to each other.	Spray or Brush	Wet: Lt. Blue Dry: Green
4420	High performance, fast-drying adhesive designed for application by pressure curtain coating and mechanical roll coating. Bonds decorative plastic laminates to plywood or particle board and is suitable for conveyor line production of laminated panels of various types such as aluminum to wood or hardboard.	Roll Coating	Yellow
4488	Lower viscosity version of Cement 4420 for use specifically with flow-over or Weir-type curtain coaters. Bonds decorative plastic laminates to plywood or particle board and is suitable for conveyor line production of laminated panels of various types, such as aluminum to wood or hardboard.	Curtain and Roll Coating	Yellow
4518	Designed for spray application with automatic or production line equipment. Dries very fast, requires pressure from a niproll (rotary press) or platen press to assure proper bonding. Bonds decorative plastic laminates to plywood or particle board on both flat work and postforming. Also used for conveyor line production of laminated panels of various types such as aluminum to wood or hardboard. Meets requirements of MIL-A-5092 B, Type II.	Spray	Green
4729	Similar to Cement 2218 except that it is formulated with a nonflammable solvent. Requires force drying to prevent blushing.	Spray	Red
5034	Water dispersed, has high immediate bond strength and long bonding range. Bonds foamed plastics, plastic laminate, wood, rubber, plywood, wallboard, wood veneer, plaster, and canvas to themselves and to each other.	Spray or Brush	Neutral

Table 42 Physical properties and application data of adhesives. (3M Company.)

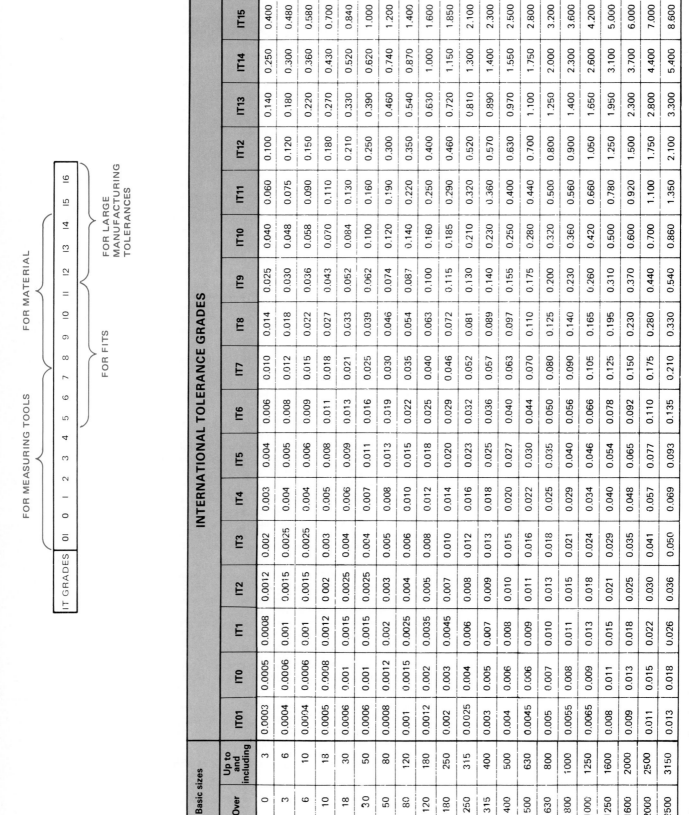

FOR MEASURING TOOLS | FOR MATERIAL

| IT GRADES | 01 | 0 | 1 | 2 | 3 | 4 | 5 | 6 | 7 | 8 | 9 | 10 | 11 | 12 | 13 | 14 | 15 | 16 |

FOR FITS | FOR LARGE MANUFACTURING TOLERANCES

INTERNATIONAL TOLERANCE GRADES

Basic sizes Over	Up to and including	IT01	IT0	IT1	IT2	IT3	IT4	IT5	IT6	IT7	IT8	IT9	IT10	IT11	IT12	IT13	IT14	IT15	IT16
0	3	0.0003	0.0005	0.0008	0.0012	0.002	0.003	0.004	0.006	0.010	0.014	0.025	0.040	0.060	0.100	0.140	0.250	0.400	0.600
3	6	0.0004	0.0006	0.001	0.0015	0.0025	0.004	0.005	0.008	0.012	0.018	0.030	0.048	0.075	0.120	0.180	0.300	0.480	0.750
6	10	0.0004	0.0006	0.001	0.0015	0.0025	0.004	0.006	0.009	0.015	0.022	0.036	0.058	0.090	0.150	0.220	0.360	0.580	0.900
10	18	0.0005	0.0008	0.0012	0.002	0.003	0.005	0.008	0.011	0.018	0.027	0.043	0.070	0.110	0.180	0.270	0.430	0.700	1.100
18	30	0.0006	0.001	0.0015	0.0025	0.004	0.006	0.009	0.013	0.021	0.033	0.052	0.084	0.130	0.210	0.330	0.520	0.840	1.300
30	50	0.0006	0.001	0.0015	0.0025	0.004	0.007	0.011	0.016	0.025	0.039	0.062	0.100	0.160	0.250	0.390	0.620	1.000	1.600
50	80	0.0008	0.0012	0.002	0.003	0.005	0.008	0.013	0.019	0.030	0.046	0.074	0.120	0.190	0.300	0.460	0.740	1.200	1.900
80	120	0.001	0.0015	0.0025	0.004	0.006	0.010	0.015	0.022	0.035	0.054	0.087	0.140	0.220	0.350	0.540	0.870	1.400	2.200
120	180	0.0012	0.002	0.0035	0.005	0.008	0.012	0.018	0.025	0.040	0.063	0.100	0.160	0.250	0.400	0.630	1.000	1.600	2.500
180	250	0.002	0.003	0.0045	0.007	0.010	0.014	0.020	0.029	0.046	0.072	0.115	0.185	0.290	0.460	0.720	1.150	1.850	2.900
250	315	0.0025	0.004	0.006	0.008	0.012	0.016	0.023	0.032	0.052	0.081	0.130	0.210	0.320	0.520	0.810	1.300	2.100	3.200
315	400	0.003	0.005	0.007	0.009	0.013	0.018	0.025	0.036	0.057	0.089	0.140	0.230	0.360	0.570	0.890	1.400	2.300	3.600
400	500	0.004	0.006	0.008	0.010	0.015	0.020	0.027	0.040	0.063	0.097	0.155	0.250	0.400	0.630	0.970	1.550	2.500	4.000
500	630	0.0045	0.006	0.009	0.011	0.016	0.022	0.030	0.044	0.070	0.110	0.175	0.280	0.440	0.700	1.100	1.750	2.800	4.400
630	800	0.005	0.007	0.010	0.013	0.018	0.025	0.035	0.050	0.080	0.125	0.200	0.320	0.500	0.800	1.250	2.000	3.200	5.000
800	1000	0.0055	0.008	0.011	0.015	0.021	0.029	0.040	0.056	0.090	0.140	0.230	0.360	0.560	0.900	1.400	2.300	3.600	5.600
1000	1250	0.0065	0.009	0.013	0.018	0.024	0.034	0.046	0.066	0.105	0.165	0.260	0.420	0.660	1.050	1.650	2.600	4.200	6.600
1250	1600	0.008	0.011	0.015	0.021	0.029	0.040	0.054	0.078	0.125	0.195	0.310	0.500	0.780	1.250	1.950	3.100	5.000	7.800
1600	2000	0.009	0.013	0.018	0.025	0.035	0.048	0.065	0.092	0.150	0.230	0.370	0.600	0.920	1.500	2.300	3.700	6.000	9.200
2000	2500	0.011	0.015	0.022	0.030	0.041	0.057	0.077	0.110	0.175	0.280	0.440	0.700	1.100	1.750	2.800	4.400	7.000	11.000
2500	3150	0.013	0.018	0.026	0.036	0.050	0.069	0.093	0.135	0.210	0.330	0.540	0.860	1.350	2.100	3.300	5.400	8.600	13.500

Table 43   International Tolerance Grades. (Values in millimeters.)

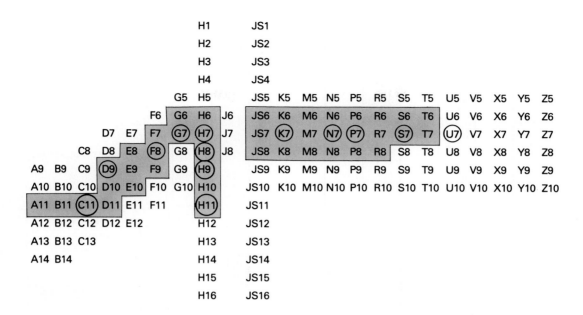

Legend: First choice tolerance zones encircled (ANSI B4.2 preferred)
Second choice tolerance zones framed (ISO 1829 selected)
Third choice tolerance zones open

TOLERANCE ZONES FOR INTERNAL DIMENSIONS (HOLES)

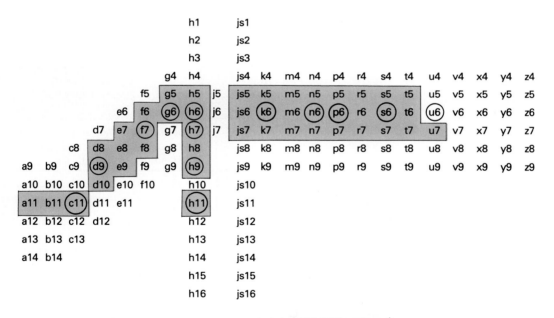

Legend: First choice tolerance zones encircled (ANSI B4.2 preferred)
Second choice tolerance zones framed (ISO 1829 selected)
Third choice tolerance zones open

TOLERANCE ZONES FOR EXTERNAL DIMENSIONS (SHAFTS)

Table 43 (cont'd.)   International Tolerance Grades.

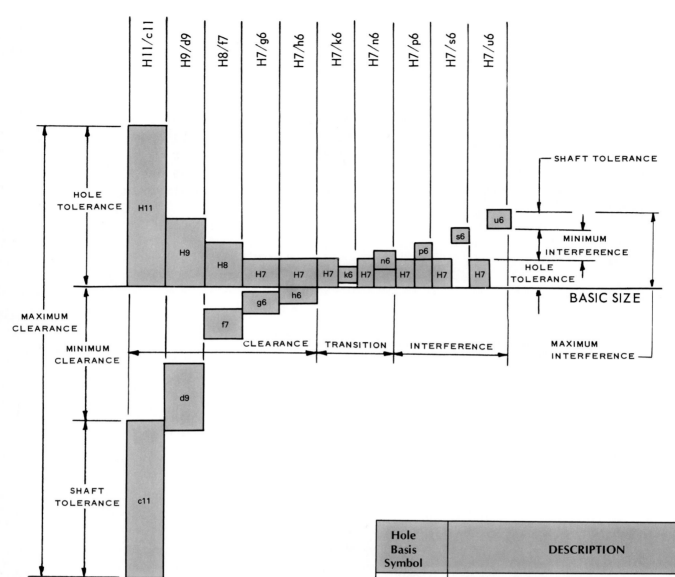

Hole Basis Symbol	DESCRIPTION
H7/h6	*Locational clearance* fit provides snug fit for locating stationary parts; but can be freely assembled and disassembled.
H7/k6	*Locational transition* fit for accurate location, a compromise between clearance and interference.
H7/n6	*Locational transition* fit for more accurate location where greater interference is permissible.
H7/p6	*Locational interference* fit for parts requiring rigidity and alignment with prime accuracy of location but without special bore pressure requirements.
H7/s6	*Medium drive* fit for ordinary steel parts or shrink fits on light sections, the tightest fit usable with cast iron.
H7/u6	*Force* fit suitable for parts which can be highly stressed or for shrink fits where the heavy pressing forces required are impractical.

Hole Basis Symbol	DESCRIPTION
H11/c11	*Loose running* fit for wide commercial tolerances or allowances on external members.
H9/d9	*Free running* fit not for use where accuracy is essential, but good for large temperature variations, high running speeds, or heavy journal pressures.
H8/f7	*Close running* fit for running on accurate machines and for accurate location at moderate speeds and journal pressures.
H7/g6	*Sliding* fit not intended to run freely, but to move and turn freely and locate accurately.

Table 44 Preferred hole basis fits description.

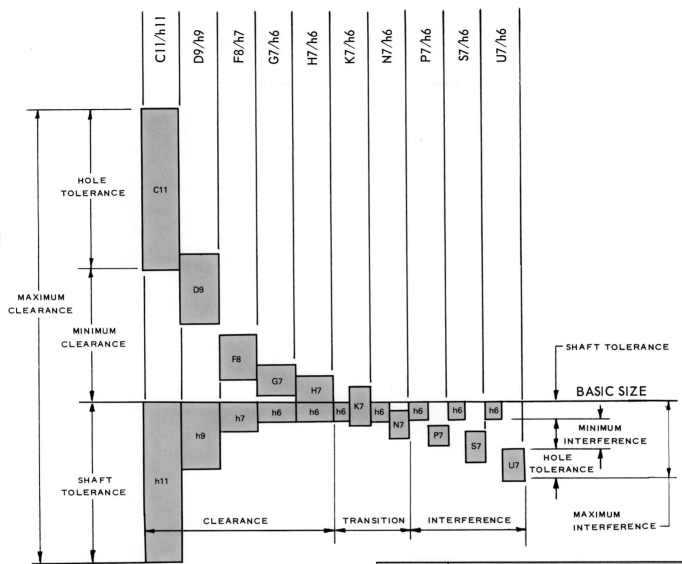

Shaft Basis Symbol	DESCRIPTION
C11/h11	*Loose running* fit for wide commercial tolerances or allowances on external members.
D9/h9	*Free running* fit not for use where accuracy is essential, but good for large temperature variations, high running speeds, or heavy journal pressures.
F8/h7	*Close running* fit for running on accurate machines and for accurate location at moderate speeds and journal pressures.
G7/h6	*Sliding* fit not intended to run freely, but to move and turn freely and locate accurately.
H7/h6	*Locational clearance* fit provides snug fit for locating stationary parts; but can be freely assembled and disassembled.

Shaft Basis Symbol	DESCRIPTION
K7/h6	*Locational transition* fit for accurate location, a compromise between clearance and interference.
N7/h6	*Locational transition* fit for more accurate location where greater interference is permissible.
P7/h6	*Locational interference* fit for parts requiring rigidity and alignment with prime accuracy of location but without special bore pressure requirements.
S7/h6	*Medium drive* fit for ordinary steel parts or shrink fits on light sections, the tightest fit usable with cast iron.
U7/h6	*Force* fit suitable for parts which can be highly stressed or for shrink fits where the heavy pressing forces required are impractical.

Table 45 **Preferred shaft basis fits description.**

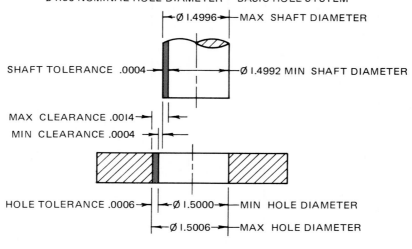

EXAMPLE: RC2 SLIDING FIT FOR A
Ø 1.50 NOMINAL HOLE DIAMETER — BASIC HOLE SYSTEM

Ø 1.4996 — MAX SHAFT DIAMETER

SHAFT TOLERANCE .0004 → Ø 1.4992 MIN SHAFT DIAMETER

MAX CLEARANCE .0014

MIN CLEARANCE .0004

HOLE TOLERANCE .0006 — Ø 1.5000 — MIN HOLE DIAMETER

Ø 1.5006 — MAX HOLE DIAMETER

Nominal Size Range Inches		Class RC1 Precision Sliding			Class RC2 Sliding Fit			Class RC3 Precision Running			Class RC4 Close Running		
		Hole Tol. GR5	Minimum Clearance	Shaft Tol. GR4	Hole Tol. GR6	Minimum Clearance	Shaft Tol. GR5	Hole Tol. GR7	Minimum Clearance	Shaft Tol. GR6	Hole Tol. GR8	Minimum Clearance	Shaft Tol. GR7
Over	To	−0		+0	−0		+0	−0		+0	−0		+0
0	.12	+0.15	0.1	−0.12	+0.25	0.1	−0.15	+0.4	0.3	−0.25	+0.6	0.3	−0.4
.12	.24	+0.2	0.15	−0.15	+0.3	0.15	−0.2	+0.5	0.4	−0.3	+0.7	0.4	−0.5
.24	.40	+0.25	0.2	−0.15	+0.4	0.2	−0.25	+0.6	0.4	−0.4	+0.9	0.5	−0.6
.40	.71	+0.3	0.25	−0.2	+0.4	0.25	−0.3	+0.7	0.6	−0.4	+1.0	0.6	−0.7
.71	1.19	+0.4	0.3	−0.25	+0.5	0.3	−0.4	+0.8	0.8	−0.5	+1.2	0.8	−0.8
1.19	1.97	+0.4	0.4	−0.3	+0.6	0.4	−0.4	+1.0	1.0	−0.6	+1.6	1.0	−1.0
1.97	3.15	+0.5	0.4	−0.3	+0.7	0.4	−0.5	+1.2	1.2	−0.7	+1.8	1.2	−1.2
3.15	4.73	+0.6	0.5	−0.4	+0.9	0.5	−0.6	+1.4	1.4	−0.9	+2.2	1.4	−1.4
4.73	7.09	+0.7	0.6	−0.5	+1.0	0.6	−0.7	+1.6	1.6	−1.0	+2.5	1.6	−1.6
7.09	9.85	+0.8	0.6	−0.6	+1.2	0.6	−0.8	+1.8	2.0	−1.2	+2.8	2.0	−1.8
9.85	12.41	+0.9	0.8	−0.6	+1.2	0.8	−0.9	+2.0	2.5	−1.2	+3.0	2.5	−2.0
12.41	15.75	+1.0	1.0	−0.7	+1.4	1.0	−1.0	+2.2	3.0	−1.4	+3.5	3.0	−2.2

Table 46 **Running and sliding fits.** (Values in thousandths of an inch.)

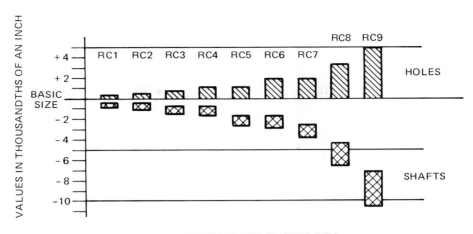

RUNNING AND SLIDING FITS
BASIC HOLE SYSTEM

Class RC5 Medium Running			Class RC6 Medium Running			Class RC7 Free Running			Class RC8 Loose Running			Class RC9 Loose Running		
Hole Tol. GR8	Minimum Clearance	Shaft Tol. GR7	Hole Tol. GR9	Minimum Clearance	Shaft Tol. GR8	Hole Tol. GR9	Minimum Clearance	Shaft Tol. GR8	Hole Tol. GR10	Minimum Clearance	Shaft Tol. GR9	Hole Tol. GR11	Minimum Clearance	Shaft Tol. GR10
−0		+0	−0		+0	−0		+0	−0		+0	−0		+0
+0.6	0.6	−0.4	+1.0	0.6	−0.6	+1.0	1.0	−0.6	+1.6	2.5	−1.0	+2.5	4.0	−1.6
+0.7	0.8	−0.5	+1.2	0.8	−0.7	+1.2	1.2	−0.7	+1.8	2.8	−1.2	+3.0	4.5	−1.8
+0.9	1.0	−0.6	+1.4	1.0	−0.9	+1.4	1.6	−0.9	+2.2	3.0	−1.4	+3.5	5.0	−2.2
+1.0	1.2	−0.7	+1.6	1.2	−1.0	+1.6	2.0	−1.0	+2.8	3.5	−1.6	+4.0	6.0	−2.8
+1.2	1.6	−0.8	+2.0	1.6	−1.2	+2.0	2.5	−1.2	+3.5	4.5	−2.0	+5.0	7.0	−3.5
+1.6	2.0	−1.0	+2.5	2.0	−1.6	+2.5	3.0	−1.6	+4.0	5.0	−2.5	+6.0	8.0	−4.0
+1.8	2.5	−1.2	+3.0	2.5	−1.8	+3.0	4.0	−1.8	+4.5	6.0	−3.0	+7.0	9.0	−4.5
+2.2	3.0	−1.4	+3.5	3.0	−2.2	+3.5	5.0	−2.2	+5.0	7.0	−3.5	+9.0	10.0	−5.0
+2.5	3.5	−1.6	+4.0	3.5	−2.5	+4.0	6.0	−2.5	+6.0	8.0	−4.0	+10.0	12.0	−6.0
+2.8	4.5	−1.8	+4.5	4.0	−2.8	+4.5	7.0	−2.8	+7.0	10.0	−4.5	+12.0	15.0	−7.0
+3.0	5.0	−2.0	+5.0	5.0	−3.0	+5.0	8.0	−3.0	+8.0	12.0	−5.0	+12.0	18.0	−8.0
+3.5	6.0	−2.2	+6.0	6.0	−3.5	+6.0	10.0	−3.5	+9.0	14.0	−6.0	+14.0	22.0	−9.0

Table 46 (cont'd.) **Running and sliding fits.** (Values in thousandths of an inch.)

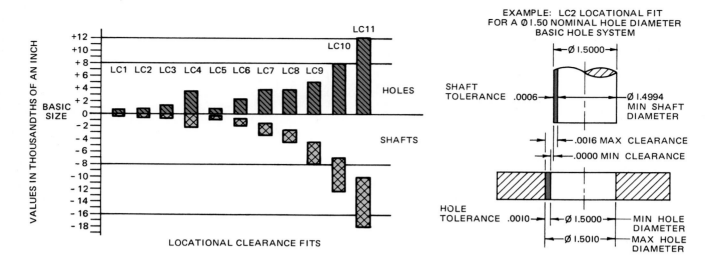

EXAMPLE: LC2 LOCATIONAL FIT
FOR A Ø1.50 NOMINAL HOLE DIAMETER
BASIC HOLE SYSTEM

LOCATIONAL CLEARANCE FITS

Nominal Size Range Inches		Class LC1			Class LC2			Class LC3			Class LC4			Class LC5		
		Hole Tol. GR6	Minimum Clearance	Shaft Tol. GR5	Hole Tol. GR7	Minimum Clearance	Shaft Tol. GR6	Hole Tol. GR8	Minimum Clearance	Shaft Tol. GR7	Hole Tol. GR10	Minimum Clearance	Shaft Tol. GR9	Hole Tol. GR7	Minimum Clearance	Shaft Tol. GR6
Over	To	−0		+0	−0		+0	−0		+0	−0		+0	−0		+0
0	.12	+0.25	0	−0.15	+0.4	0	−0.25	+0.6	0	−0.4	+1.6	0	−1.0	+0.4	0.1	−0.25
.12	.24	+0.3	0	−0.2	+0.5	0	−0.3	+0.7	0	−0.5	+1.8	0	−1.2	+0.5	0.15	−0.3
.24	.40	+0.4	0	−0.25	+0.6	0	−0.4	+0.9	0	−0.6	+2.2	0	−1.4	+0.6	0.2	−0.4
.40	.71	+0.4	0	−0.3	+0.7	0	−0.4	+1.0	0	−0.7	+2.8	0	−1.6	+0.7	0.25	−0.4
.71	1.19	+0.5	0	−0.4	+0.8	0	−0.5	+1.2	0	−0.8	+3.5	0	−2.0	+0.8	0.3	−0.5
1.19	1.97	+0.6	0	−0.4	+1.0	0	−0.6	+1.6	0	−1.0	+4.0	0	−2.5	+1.0	0.4	−0.6
1.97	3.15	+0.7	0	−0.5	+1.2	0	−0.7	+1.8	0	−1.2	+4.5	0	−3.0	+1.2	0.4	−0.7
3.15	4.73	+0.9	0	−0.6	+1.4	0	−0.9	+2.2	0	−1.4	+5.0	0	−3.5	+1.4	0.5	−0.9
4.73	7.09	+1.0	0	−0.7	+1.6	0	−1.0	+2.5	0	−1.6	+6.0	0	−4.0	+1.6	0.6	−1.0
7.09	9.85	+1.2	0	−0.8	+1.8	0	−1.2	+2.8	0	−1.8	+7.0	0	−4.5	+1.8	0.6	−1.2
9.85	12.41	+1.2	0	−0.9	+2.0	0	−1.2	+3.0	0	−2.0	+8.0	0	−5.0	+2.0	0.7	−1.2
12.41	15.75	+1.4	0	−1.0	+2.2	0	−1.4	+3.5	0	−2.2	+9.0	0	−6.0	+2.2	0.7	−1.4

Table 47 **Locational clearance fits.** (Values in thousandths of an inch.)

Nominal Size Range Inches		Class LT1			Class LT2		
		Hole Tol. GR7	Maximum Interference	Shaft Tol. GR6	Hole Tol. GR8	Maximum Interference	Shaft Tol. GR7
Over	To	−0		+0	−0		+0
0	.12	+0.4	0.1	−0.25	+0.6	0.2	−0.4
.12	.24	+0.5	0.15	−0.3	+0.7	0.25	−0.5
.24	.40	+0.6	0.2	−0.4	+0.9	0.3	−0.6
.40	.71	+0.7	0.2	−0.4	+1.0	0.3	−0.7
.71	1.19	+0.8	0.25	−0.5	+1.2	0.4	−0.8
1.19	1.97	+1.0	0.3	−0.6	+1.6	0.5	−1.0
1.97	3.15	+1.2	0.3	−0.7	+1.8	0.6	−1.2
3.15	4.73	+1.4	0.4	−0.9	+2.2	0.7	−1.4
4.73	7.09	+1.6	0.5	−1.0	+2.5	0.8	−1.6
7.09	9.85	+1.8	0.6	−1.2	+2.8	0.9	−1.8
9.85	12.41	+2.0	0.6	−1.2	+3.0	1.0	−2.0
12.41	15.75	+2.2	0.7	−1.4	+3.5	1.0	−2.2

Table 48 **Transition fits.** (Values in thousandths of an inch.)

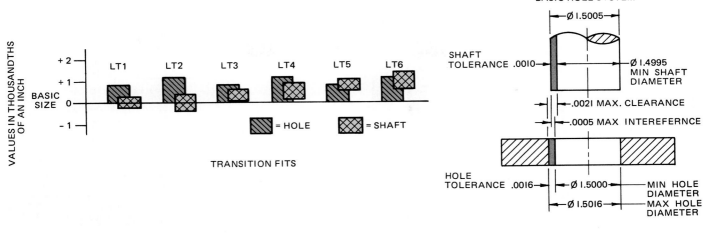

Class LC6			Class LC7			Class LC8			Class LC9			Class LC10			Class LC11		
Hole Tol. GR9	Minimum Clearance	Shaft Tol. GR8	Hole Tol. GR10	Minimum Clearance	Shaft Tol. GR9	Hole Tol. GR10	Minimum Clearance	Shaft Tol. GR9	Hole Tol. GR11	Minimum Clearance	Shaft Tol. GR10	Hole Tol. GR12	Minimum Clearance	Shaft Tol. GR11	Hole Tol. GR13	Minimum Clearance	Shaft Tol. GR12
−0		+0	−0		+0	−0		+0	−0		+0	−0		+0	−0		+0
+1.0	0.3	−0.6	+1.6	0.6	−1.0	+1.6	1.0	−1.0	+2.5	2.5	−1.6	+4.0	4.0	−2.5	+6.0	5.0	−4.0
+1.2	0.4	−0.7	+1.8	0.8	−1.2	+1.8	1.2	−1.2	+3.0	2.8	−1.8	+5.0	4.5	−3.0	+7.0	6.0	−5.0
+1.4	0.5	−0.9	+2.2	1.0	−1.4	+2.2	1.6	−1.4	+3.5	3.0	−2.2	+6.0	5.0	−3.5	+9.0	7.0	−6.0
+1.6	0.6	−1.0	+2.8	1.2	−1.6	+2.8	2.0	−1.6	+4.0	3.5	−2.8	+7.0	6.0	−4.0	+10.0	8.0	−7.0
+2.0	0.8	−1.2	+3.5	1.6	−2.0	+3.5	2.5	−2.0	+5.0	4.5	−3.5	+8.0	7.0	−5.0	+12.0	10.0	−8.0
+2.5	1.0	−1.6	+4.0	2.0	−2.5	+4.0	3.6	−2.5	+6.0	5.0	−4.0	+10.0	8.0	−6.0	+16.0	12.0	−10.0
+3.0	1.2	−1.8	+4.5	2.5	−3.0	+4.5	4.0	−3.0	+7.0	6.0	−4.5	+12.0	10.0	−7.0	+18.0	14.0	−12.0
+3.5	1.4	−2.2	+5.0	3.0	−3.5	+5.0	5.0	−3.5	+9.0	7.0	−5.0	+14.0	11.0	−9.0	+22.0	16.0	−14.0
+4.0	1.6	−2.5	+6.0	3.5	−4.0	+6.0	6.0	−4.0	+10.0	8.0	−6.0	+16.0	12.0	−10.0	+25.0	18.0	−16.0
+4.5	2.0	−2.8	+7.0	4.0	−4.5	+7.0	7.0	−4.5	+12.0	10.0	−7.0	+18.0	16.0	−12.0	+28.0	22.0	−18.0
+5.0	2.2	−3.0	+8.0	4.5	−5.0	+8.0	7.0	−5.0	+12.0	12.0	−8.0	+20.0	20.0	−12.0	+30.0	28.0	−20.0
+6.0	2.5	−3.5	+9.0	5.0	−6.0	+9.0	8.0	−6.0	+14.0	14.0	−9.0	+22.0	22.0	−14.0	+35.0	30.0	−22.0

Table 47 (cont'd.) **Locational clearance fits.** (Values in thousandths of an inch.)

Class LT3			Class LT4			Class LT5			Class LT6		
Hole Tol. GR7	Maximum Interference	Shaft Tol. GR6	Hole Tol. GR8	Maximum Interference	Shaft Tol. GR7	Hole Tol. GR7	Maximum Interference	Shaft Tol. GR6	Hole Tol. GR8	Maximum Interference	Shaft Tol. GR7
−0		+0	−0		+0	−0		+0	−0		+0
+0.4	0.25	−0.25	+0.6	0.4	−0.4	+0.4	0.5	−0.25	+0.6	0.65	−0.4
+0.5	0.4	−0.3	+0.7	0.6	−0.5	+0.5	0.6	−0.3	+0.7	0.8	−0.5
+0.6	0.5	−0.4	+0.9	0.7	−0.6	+0.6	0.8	−0.4	+0.9	1.0	−0.6
+0.7	0.5	−0.4	+1.0	0.8	−0.7	+0.7	0.9	−0.4	+1.0	1.2	−0.7
+0.8	0.6	−0.5	+1.2	0.9	−0.8	+0.8	1.1	−0.5	+1.2	1.4	−0.8
+1.0	0.7	−0.6	+1.6	1.1	−1.0	+1.0	1.3	−0.6	+1.6	1.7	−1.0
+1.2	0.8	−0.7	+1.8	1.3	−1.2	+1.2	1.5	−0.7	+1.8	2.0	−1.2
+1.4	1.0	−0.9	+2.2	1.5	−1.4	+1.4	1.9	−0.9	+2.2	2.4	−1.4
+1.6	1.1	−1.0	+2.5	1.7	−1.6	+1.6	2.2	−1.0	+2.5	2.8	−1.6
+1.8	1.4	−1.2	+2.8	2.0	−1.8	+1.8	2.6	−1.2	+2.8	3.2	−1.8
+2.0	1.4	−1.2	+3.0	2.2	−2.0	+2.0	2.6	−1.2	+3.0	3.4	−2.0
+2.2	1.6	−1.4	+3.5	2.4	−2.2	+2.2	3.0	−1.4	+3.5	3.8	−2.2

Table 48 (cont'd.) **Transition fits.** (Values in thousandths of an inch.)

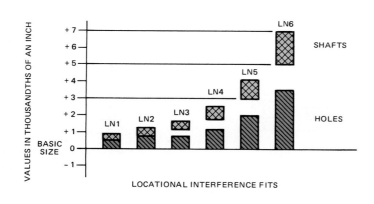

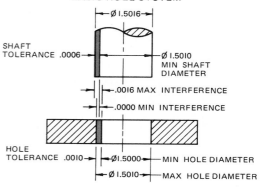

EXAMPLE: LN2 LOCATIONAL INTERFERENCE
FIT FOR A Ø1.50 NOMINAL HOLE DIAMETER
BASIC HOLE SYSTEM

Nominal Size Range Inches		Class LN1 Light Press Fit			Class LN2 Medium Press Fit		
		Hole Tol. GR6	Maximum Interference	Shaft Tol. GR5	Hole Tol. GR7	Maximum Interference	Shaft Tol. GR6
Over	To	−0		+0	−0		+0
0	.12	+0.25	0.4	−0.15	+0.4	0.65	−0.25
.12	.24	+0.3	0.5	−0.2	+0.5	0.8	−0.3
.24	.40	+0.4	0.65	−0.25	+0.6	1.0	−0.4
.40	.71	+0.4	0.7	−0.3	+0.7	1.1	−0.4
.71	1.19	+0.5	0.9	−0.4	+0.8	1.3	−0.5
1.19	1.97	+0.6	1.0	−0.4	+1.0	1.6	−0.6
1.97	3.15	+0.7	1.3	−0.5	+1.2	2.1	−0.7
3.15	4.73	+0.9	1.6	−0.6	+1.4	2.5	−0.9
4.73	7.09	+1.0	1.9	−0.7	+1.6	2.8	−1.0
7.09	9.85	+1.2	2.2	−0.8	+1.8	3.2	−1.2
9.85	12.41	+1.2	2.3	−0.9	+2.0	3.4	−1.2
12.41	15.75	+1.4	2.6	−1.0	+2.2	3.9	−1.4

Table 49 **Locational interference fits.** (Values in thousandths of an inch.)

Nominal Size Range Inches		Class FN1 Light Drive Fit			Class FN2 Medium Drive Fit		
		Hole Tol. GR6	Maximum Interference	Shaft Tol. GR5	Hole Tol. GR7	Maximum Interference	Shaft Tol. GR6
Over	To	−0		+0	−0		+0
0	.12	+0.25	0.5	−0.15	+0.4	0.85	−0.25
.12	.24	+0.3	0.6	−0.2	+0.5	1.0	−0.3
.24	.40	+0.4	0.75	−0.25	+0.6	1.4	−0.4
.40	.56	+0.4	0.8	−0.3	+0.7	1.6	−0.4
.56	.71	+0.4	0.9	−0.3	+0.7	1.6	−0.4
.71	.95	+0.5	1.1	−0.4	+0.8	1.9	−0.5
.95	1.19	+0.5	1.2	−0.4	+0.8	1.9	−0.5
1.19	1.58	+0.6	1.3	−0.4	+1.0	2.4	−0.6
1.58	1.97	+0.6	1.4	−0.4	+1.0	2.4	−0.6
1.97	2.56	+0.7	1.8	−0.5	+1.2	2.7	−0.7
2.56	3.15	+0.7	1.9	−0.5	+1.2	2.9	−0.7
3.15	3.94	+0.9	2.4	−0.6	+1.4	3.7	−0.9

Table 50 **Force and shrink fits.** (Values in thousandths of an inch.)

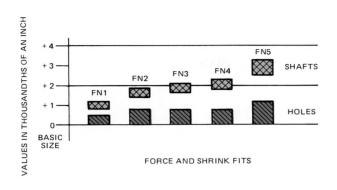

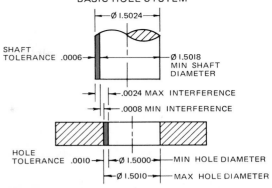

EXAMPLE: FN2 MEDIUM DRIVE
FIT FOR A Ø 1.50 NOMINAL HOLE DIAMETER
BASIC HOLE SYSTEM

	Class LN3 Heavy Press Fit			Class LN4			Class LN5			Class LN6		
Hole Tol. GR7	Maximum Interference	Shaft Tol. GR6	Hole Tol. GR8	Maximum Interference	Shaft Tol. GR7	Hole Tol. GR9	Maximum Interference	Shaft Tol. GR8	Hole Tol. GR10	Maximum Interference	Shaft Tol. GR9
−0		+0	−0		+0	−0		+0	−0		+0
+0.4	0.75	−0.25	+0.6	1.2	−0.4	+1.0	1.8	−0.6	+1.6	3.0	−1.0
+0.5	0.9	−0.3	+0.7	1.5	−0.5	+1.2	2.3	−0.7	+1.8	3.6	−1.2
+0.6	1.2	−0.4	+0.9	1.8	−0.6	+1.4	2.8	−0.9	+2.2	4.4	−1.4
+0.7	1.4	−0.4	+1.0	2.2	−0.7	+1.6	3.4	−1.0	+2.8	5.6	−1.6
+0.8	1.7	−0.5	+1.2	2.6	−0.8	+2.0	4.2	−1.2	+3.5	7.0	−2.0
+1.0	2.0	−0.6	+1.6	3.4	−1.0	+2.5	5.3	−1.6	+4.0	8.5	−2.5
+1.2	2.3	−0.7	+1.8	4.0	−1.2	+3.0	6.3	−1.8	+4.5	10.0	−3.0
+1.4	2.9	−0.9	+2.2	4.8	−1.4	+4.0	7.7	−2.2	+5.0	11.5	−3.5
+1.6	3.5	−1.0	+2.5	5.6	−1.6	+4.5	8.7	−2.5	+6.0	13.5	−4.0
+1.8	4.2	−1.2	+2.8	6.6	−1.8	+5.0	10.3	−2.8	+7.0	16.5	−4.5
+2.0	4.7	−1.2	+3.0	7.5	−2.0	+6.0	12.0	−3.0	+8.0	19	−5.0
+2.2	5.9	−1.4	+3.5	8.7	−2.2	+6.0	14.5	−3.5	+9.0	23	−6.0

Table 49 (cont'd.) Locational interference fits. (Values in thousandths of an inch.)

	Class FN3 Heavy Drive Fit			Class FN4 Shrink Fit			FN5 Heavy Shrink Fit		
Hole Tol. GR7	Maximum Interference	Shaft Tol. GR6	Hole Tol. GR7	Maximum Interference	Shaft Tol. GR6	Hole Tol. GR8	Maximum Interference	Shaft Tol. GR7
−0		+0	−0		+0	−0		+0
			+0.4	0.95	−0.25	+0.6	1.3	−0.4
			+0.5	1.2	−0.3	+0.7	1.7	−0.5
			+0.6	1.6	−0.4	+0.9	2.0	−0.6
			+0.7	1.8	−0.4	+1.0	2.3	−0.7
			+0.7	1.8	−0.4	+1.0	2.5	−0.7
			+0.8	2.1	−0.5	+1.2	3.0	−0.8
+0.8	2.1	−0.5	+0.8	2.3	−0.5	+1.2	3.3	−0.8
+1.0	2.6	−0.6	+1.0	3.1	−0.6	+1.6	4.0	−1.0
+1.0	2.8	−0.6	+1.0	3.4	−0.6	+1.6	5.0	−1.0
+1.2	3.2	−0.7	+1.2	4.2	−0.7	+1.8	6.2	−1.2
+1.2	3.7	−0.7	+1.2	4.7	−0.7	+1.8	7.2	−1.2
+1.4	4.4	−0.9	+1.4	5.9	−0.9	+2.2	8.4	−1.4

Table 50 (cont'd.) Force and shrink fits. (Values in thousandths of an inch.)

Example ⌀ 50H9/d9

Hole Size ⌀ $\frac{50.062}{50.000}$

Shaft Size ⌀ $\frac{49.920}{49.858}$

Clearance $\frac{\text{Max. } 0.204}{\text{Min. } 0.080}$

Example ⌀ 70H7/g6
(Use values midway between 60 and 80.)

Hole Size ⌀ $\frac{70.030}{70.000}$

Shaft Size ⌀ $\frac{69.990}{69.971}$

Clearance $\frac{\text{Max. } 0.059}{\text{Min. } 0.010}$

		PREFERRED HOLE BASIS CLEARANCE FITS															
		LOOSE RUNNING			FREE RUNNING			CLOSE RUNNING			SLIDING			LOCATIONAL CLEARANCE		
BASIC SIZE		Hole H11	Shaft c11	Fit	Hole H9	Shaft d9	Fit	Hole H8	Shaft f7	Fit	Hole H7	Shaft g6	Fit	Hole H7	Shaft h6	Fit
1	MAX	1.060	0.940	0.180	1.025	0.980	0.070	1.014	0.994	0.030	1.010	0.998	0.018	1.010	1.000	0.016
	MIN	1.000	0.880	0.060	1.000	0.955	0.020	1.000	0.984	0.006	1.000	0.992	0.002	1.000	0.994	0.000
1.2	MAX	1.260	1.140	0.180	1.225	1.180	0.070	1.214	1.194	0.030	1.210	1.198	0.018	1.210	1.200	0.016
	MIN	1.200	1.080	0.060	1.200	1.155	0.020	1.200	1.184	0.006	1.200	1.192	0.002	1.200	1.194	0.000
1.6	MAX	1.660	1.540	0.180	1.625	1.580	0.070	1.614	1.594	0.030	1.610	1.598	0.018	1.610	1.600	0.016
	MIN	1.600	1.480	0.060	1.600	1.555	0.020	1.600	1.584	0.006	1.600	1.592	0.002	1.600	1.594	0.000
2	MAX	2.060	1.940	0.180	2.025	1.980	0.070	2.014	1.994	0.030	2.010	1.998	0.018	2.010	2.000	0.016
	MIN	2.000	1.880	0.060	2.000	1.955	0.020	2.000	1.984	0.006	2.000	1.992	0.002	2.000	1.994	0.000
2.5	MAX	2.560	2.440	0.180	2.525	2.480	0.070	2.514	2.494	0.030	2.510	2.498	0.018	2.510	2.500	0.016
	MIN	2.500	2.380	0.060	2.500	2.455	0.020	2.500	2.484	0.006	2.500	2.492	0.002	2.500	2.494	0.000
3	MAX	3.060	2.940	0.180	3.025	2.980	0.070	3.014	2.994	0.030	3.010	2.998	0.018	3.010	3.000	0.016
	MIN	3.000	2.880	0.060	3.000	2.955	0.020	3.000	2.984	0.006	3.000	2.992	0.002	3.000	2.994	0.000
4	MAX	4.075	3.930	0.220	4.030	3.970	0.090	4.018	3.990	0.040	4.012	3.996	0.024	4.012	4.000	0.020
	MIN	4.000	3.855	0.070	4.000	3.940	0.030	4.000	3.978	0.010	4.000	3.988	0.004	4.000	3.992	0.000
5	MAX	5.075	4.930	0.220	5.030	4.970	0.090	5.018	4.990	0.040	5.012	4.996	0.024	5.012	5.000	0.020
	MIN	5.000	4.855	0.070	5.000	4.940	0.030	5.000	4.978	0.010	5.000	4.988	0.004	5.000	4.992	0.000
6	MAX	6.075	5.930	0.220	6.030	5.970	0.090	6.018	5.990	0.040	6.012	5.996	0.024	6.012	6.000	0.020
	MIN	6.000	5.855	0.070	6.000	5.940	0.030	6.000	5.978	0.010	6.000	5.988	0.004	6.000	5.992	0.000
8	MAX	8.090	7.920	0.260	8.036	7.960	0.112	8.022	7.987	0.050	8.015	7.995	0.029	8.015	8.000	0.024
	MIN	8.000	7.830	0.080	8.000	7.924	0.040	8.000	7.972	0.013	8.000	7.986	0.006	8.000	7.991	0.000
10	MAX	10.090	9.920	0.260	10.036	9.960	0.112	10.022	9.987	0.050	10.015	9.995	0.029	10.015	10.000	0.024
	MIN	10.000	9.830	0.080	10.000	9.924	0.040	10.000	9.972	0.013	10.000	9.986	0.005	10.000	9.991	0.000
12	MAX	12.110	11.905	0.315	12.043	11.950	0.136	12.027	11.984	0.061	12.018	11.994	0.035	12.018	12.000	0.029
	MIN	12.000	11.795	0.095	12.000	11.907	0.050	12.000	11.966	0.016	12.000	11.983	0.006	12.000	11.989	0.000
16	MAX	16.110	15.905	0.315	16.043	15.950	0.136	16.027	15.984	0.061	16.018	15.994	0.035	16.018	16.000	0.029
	MIN	16.000	15.795	0.095	16.000	15.907	0.050	16.000	15.966	0.016	16.000	15.983	0.006	16.000	15.989	0.000
20	MAX	20.130	19.890	0.370	20.052	19.935	0.169	20.033	19.980	0.074	20.021	19.993	0.041	20.021	20.000	0.034
	MIN	20.000	19.760	0.110	20.000	19.883	0.065	20.000	19.959	0.020	20.000	19.980	0.007	20.000	19.987	0.000
25	MAX	25.130	24.890	0.370	25.052	24.935	0.169	25.033	24.980	0.074	25.021	24.993	0.042	25.021	25.000	0.034
	MIN	25.000	24.760	0.110	25.000	24.883	0.065	25.000	24.959	0.020	25.000	24.980	0.007	25.000	24.987	0.000
30	MAX	30.130	29.890	0.370	30.052	29.935	0.169	30.033	29.980	0.074	30.021	29.993	0.041	30.021	30.000	0.034
	MIN	30.000	29.760	0.110	30.000	29.883	0.065	30.000	29.959	0.020	30.000	29.980	0.007	30.000	29.987	0.000
40	MAX	40.160	39.880	0.440	40.062	39.920	0.204	40.039	39.975	0.089	40.025	39.991	0.050	40.025	40.000	0.041
	MIN	40.000	39.720	0.120	40.000	39.858	0.080	40.000	39.950	0.025	40.000	39.975	0.009	40.000	39.984	0.000
50	MAX	50.160	49.870	0.450	50.062	49.920	0.204	50.039	49.975	0.089	50.025	49.991	0.050	50.025	50.000	0.041
	MIN	50.000	49.710	0.130	50.000	49.858	0.080	50.000	49.950	0.025	50.000	49.975	0.009	50.000	49.984	0.000
60	MAX	60.190	59.860	0.520	60.074	59.900	0.248	60.046	59.970	0.106	60.030	59.990	0.059	60.030	60.000	0.049
	MIN	60.000	59.670	0.140	60.000	59.826	0.100	60.000	59.940	0.030	60.000	59.971	0.010	60.000	59.981	0.000
80	MAX	80.190	79.850	0.530	80.074	79.900	0.248	80.046	79.970	0.106	80.030	79.990	0.059	80.030	80.000	0.049
	MIN	80.000	79.660	0.150	80.000	79.826	0.100	80.000	79.940	0.030	80.000	79.971	0.010	80.000	79.981	0.000
100	MAX	100.220	99.830	0.610	100.087	99.880	0.294	100.054	99.964	0.125	100.035	99.988	0.069	100.035	100.000	0.057
	MIN	100.000	99.610	0.170	100.000	99.793	0.120	100.000	99.929	0.036	100.000	99.966	0.012	100.000	99.978	0.000
120	MAX	120.220	119.820	0.620	120.087	119.880	0.294	120.054	119.964	0.125	120.035	119.988	0.069	120.035	120.000	0.057
	MIN	120.000	119.600	0.180	120.000	119.793	0.120	120.000	119.929	0.036	120.000	119.966	0.012	120.000	119.978	0.000
160	MAX	160.250	159.790	0.710	160.100	159.855	0.345	160.063	159.957	0.146	160.040	159.986	0.079	160.040	160.000	0.065
	MIN	160.000	159.540	0.210	160.000	159.755	0.145	160.000	159.917	0.043	160.000	159.961	0.014	160.000	159.975	0.000

Table 51 **Preferred hole basis fits.** (Dimensions in millimeters.)

Example ⌀ 10H7/n6

Hole Size ⌀ 10.015 / 10.000

Shaft Size ⌀ 10.019 / 10.010

Max. Clearance 0.015

Max. Interference 0.019

Example ⌀ 35H7/u6
(Use values midway between 30 and 40.)

Hole Size ⌀ 35.023 / 35.000

Shaft Size ⌀ 35.068 / 35.054

Min. Interference 0.031

Max. Interference 0.068

PREFERRED HOLE BASIS TRANSITION AND INTERFERENCE FITS															
BASIC SIZE	LOCATIONAL TRANSN.			LOCATIONAL TRANSN.			LOCATIONAL INTERF.			MEDIUM DRIVE			FORCE		
	Hole H7	Shaft k6	Fit	Hole H7	Shaft n6	Fit	Hole H7	Shaft p6	Fit	Hole H7	Shaft s6	Fit	Hole H7	Shaft u6	Fit
1 MAX	1.010	1.006	0.010	1.010	1.010	0.006	1.010	1.012	0.004	1.010	1.020	−0.004	1.010	1.024	−0.008
MIN	1.000	1.000	−0.006	1.000	1.004	−0.010	1.000	1.006	−0.012	1.000	1.014	−0.020	1.000	1.018	−0.024
1.2 MAX	1.210	1.206	0.010	1.210	1.210	0.006	1.210	1.212	0.004	1.210	1.220	−0.004	1.210	1.224	−0.008
MIN	1.200	1.200	−0.006	1.200	1.204	−0.010	1.200	1.206	−0.012	1.200	1.214	−0.020	1.200	1.218	−0.024
1.6 MAX	1.610	1.606	0.010	1.610	1.610	0.006	1.610	1.612	0.004	1.610	1.620	−0.004	1.610	1.624	−0.008
MIN	1.600	1.600	−0.006	1.600	1.604	−0.010	1.600	1.606	−0.012	1.600	1.614	−0.020	1.600	1.618	−0.024
2 MAX	2.010	2.006	0.010	2.010	2.010	0.006	2.010	2.012	0.004	2.010	2.020	−0.004	2.010	2.024	−0.008
MIN	2.000	2.000	−0.006	2.000	2.004	−0.010	2.000	2.006	−0.012	2.000	2.014	−0.020	2.000	2.018	−0.024
2.5 MAX	2.510	2.506	0.010	2.510	2.510	0.006	2.510	2.512	0.004	2.510	2.520	−0.004	2.510	2.524	−0.008
MIN	2.500	2.500	−0.006	2.500	2.504	−0.010	2.500	2.506	−0.012	2.500	2.514	−0.020	2.500	2.518	−0.024
3 MAX	3.010	3.006	0.010	3.010	3.010	0.006	3.010	3.012	0.004	3.010	3.020	−0.004	3.010	3.024	−0.008
MIN	3.000	3.000	−0.006	3.000	3.004	−0.010	3.000	3.006	−0.012	3.000	3.014	−0.020	3.000	3.018	−0.024
4 MAX	4.012	4.009	0.011	4.012	4.016	0.004	4.012	4.020	0.000	4.012	4.027	−0.007	4.012	4.031	−0.011
MIN	4.000	4.001	−0.009	4.000	4.008	−0.016	4.000	4.012	−0.020	4.000	4.019	−0.027	4.000	4.023	−0.031
5 MAX	5.012	5.009	0.011	5.012	5.016	0.004	5.012	5.020	0.000	5.012	5.027	−0.007	5.012	5.031	−0.011
MIN	5.000	5.001	−0.009	5.000	5.008	−0.016	5.000	5.012	−0.020	5.000	5.019	−0.027	5.000	5.023	−0.031
6 MAX	6.012	6.009	0.011	6.012	6.016	0.004	6.012	6.020	0.000	6.012	6.027	−0.007	6.012	6.031	−0.011
MIN	6.000	6.001	−0.009	6.000	6.008	−0.016	6.000	6.012	−0.020	6.000	6.019	−0.027	6.000	6.023	−0.031
8 MAX	8.015	8.010	0.014	8.015	8.019	0.005	8.015	8.024	0.000	8.015	8.032	−0.008	8.015	8.037	−0.013
MIN	8.000	8.001	−0.010	8.000	8.010	−0.019	8.000	8.015	−0.024	8.000	8.023	−0.032	8.000	8.028	−0.037
10 MAX	10.015	10.010	0.014	10.015	10.019	0.005	10.015	10.024	0.000	10.015	10.032	−0.008	10.015	10.037	−0.013
MIN	10.000	10.001	−0.010	10.000	10.010	−0.019	10.000	10.015	−0.024	10.000	10.023	−0.032	10.000	10.028	−0.037
12 MAX	12.018	12.012	0.017	12.018	12.023	0.006	12.018	12.029	0.000	12.018	12.039	−0.010	12.018	12.044	−0.015
MIN	12.000	12.001	−0.012	12.000	12.012	−0.023	12.000	12.018	−0.029	12.000	12.028	−0.039	12.000	12.033	−0.044
16 MAX	16.018	16.012	0.017	16.018	16.023	0.006	16.018	16.029	0.000	16.018	16.039	−0.010	16.018	16.044	−0.015
MIN	16.000	16.001	−0.012	16.000	16.012	−0.023	16.000	16.018	−0.029	16.000	16.028	−0.039	16.000	16.033	−0.044
20 MAX	20.021	20.015	0.019	20.021	20.028	0.006	20.021	20.035	−0.001	20.021	20.048	−0.014	20.021	20.054	−0.020
MIN	20.000	20.002	−0.015	20.000	20.015	−0.028	20.000	20.022	−0.035	20.000	20.035	−0.048	20.000	20.041	−0.054
25 MAX	25.021	25.015	0.019	25.021	25.028	0.006	25.021	25.035	−0.001	25.021	25.048	−0.014	25.021	25.061	−0.027
MIN	25.000	25.002	−0.015	25.000	25.015	−0.028	25.000	25.022	−0.035	25.000	25.035	−0.048	25.000	25.048	−0.061
30 MAX	30.021	30.015	0.019	30.021	30.028	0.006	30.021	30.035	−0.001	30.021	30.048	−0.014	30.021	30.061	−0.027
MIN	30.000	30.002	−0.015	30.000	30.015	−0.028	30.000	30.022	−0.035	30.000	30.035	−0.048	30.000	30.048	−0.061
40 MAX	40.025	40.018	0.023	40.025	40.033	0.008	40.025	40.042	−0.001	40.025	40.059	−0.018	40.025	40.076	−0.035
MIN	40.000	40.002	−0.018	40.000	40.017	−0.033	40.000	40.026	−0.042	40.000	40.043	−0.059	40.000	40.060	−0.076
50 MAX	50.025	50.018	0.023	50.025	50.033	0.008	50.025	50.042	−0.001	50.025	50.059	−0.018	50.025	50.086	−0.045
MIN	50.002	50.000	−0.018	50.000	50.017	−0.033	50.000	50.026	−0.042	50.000	50.043	−0.059	50.000	50.070	−0.086
60 MAX	60.030	60.021	0.028	60.030	60.039	0.010	60.030	60.051	−0.002	60.030	60.072	−0.023	60.030	60.106	−0.057
MIN	60.000	60.002	−0.021	60.000	60.020	−0.039	60.000	60.032	−0.051	60.000	60.053	−0.072	60.000	60.087	−0.106
80 MAX	80.030	80.021	0.028	80.030	80.039	0.010	80.030	80.051	−0.002	80.030	80.078	−0.029	80.030	80.121	−0.072
MIN	80.000	80.002	−0.021	80.000	80.020	−0.039	80.000	80.032	−0.051	80.000	80.059	−0.078	80.000	80.102	−0.121
100 MAX	100.035	100.025	0.032	100.035	100.045	0.012	100.035	100.059	−0.002	100.035	100.093	−0.036	100.035	100.146	−0.089
MIN	100.000	100.003	−0.025	100.000	100.023	−0.045	100.000	100.037	−0.059	100.000	100.071	−0.093	100.000	100.124	−0.146
120 MAX	120.035	120.025	0.032	120.035	120.045	0.012	120.035	120.059	−0.002	120.035	120.101	−0.044	120.035	120.166	−0.109
MIN	120.000	120.003	−0.025	120.000	120.023	−0.045	120.000	120.037	−0.059	120.000	120.079	−0.101	120.000	120.144	−0.166
160 MAX	160.040	160.028	0.037	160.045	160.052	0.013	160.040	160.068	−0.003	160.040	160.125	−0.060	160.040	160.215	−0.150
MIN	160.000	160.003	−0.028	160.000	160.027	−0.052	160.000	160.043	−0.068	160.000	160.000	−0.125	160.000	160.190	−0.215

Table 51 (cont'd.)   Preferred hole basis fits.

Example ⌀ 100C11/h11

Hole Size ⌀ 100.390 / 100.170

Shaft Size ⌀ 100.000 / 99.780

Clearance Max. 0.610 / Min. 0.170

Example ⌀ 11H7/h6
(Use values midway between 10 and 12.)

Hole Size ⌀ 11.016 / 11.000

Shaft Size ⌀ 11.000 / 10.990

Clearance Max. 0.026 / Min. 0.000

		PREFERRED SHAFT BASIS CLEARANCE FITS													
		LOOSE RUNNING			FREE RUNNING			CLOSE RUNNING			SLIDING			LOCATIONAL CLEARANCE		
BASIC SIZE		Hole C11	Shaft h11	Fit	Hole D9	Shaft h9	Fit	Hole F8	Shaft h7	Fit	Hole G7	Shaft h6	Fit	Hole H7	Shaft h6	Fit
1	MAX	1.120	1.000	0.180	1.045	1.000	0.070	1.020	1.000	0.030	1.012	1.000	0.018	1.010	1.000	0.016
	MIN	1.060	0.940	0.060	1.020	0.975	0.020	1.006	0.990	0.006	1.002	0.994	0.002	1.000	0.994	0.000
1.2	MAX	1.320	1.200	0.180	1.245	1.200	0.070	1.220	1.200	0.030	1.212	1.200	0.018	1.210	1.200	0.016
	MIN	1.260	1.140	0.060	1.220	1.175	0.020	1.206	1.190	0.006	1.202	1.194	0.002	1.200	1.194	0.000
1.6	MAX	1.720	1.600	0.180	1.645	1.600	0.070	1.620	1.600	0.030	1.612	1.600	0.018	1.610	1.600	0.016
	MIN	1.660	1.540	0.060	1.620	1.575	0.020	1.606	1.590	0.006	1.602	1.594	0.002	1.600	1.594	0.000
2	MAX	2.120	2.000	0.180	2.045	2.000	0.070	2.020	2.000	0.030	2.012	2.000	0.018	2.010	2.000	0.016
	MIN	2.060	1.940	0.060	2.020	1.975	0.020	2.006	1.990	0.006	2.002	1.994	0.002	2.000	1.994	0.000
2.5	MAX	2.620	2.500	0.180	2.545	2.500	0.070	2.520	2.500	0.030	2.512	2.500	0.018	2.510	2.500	0.016
	MIN	2.560	2.440	0.060	2.520	2.475	0.020	2.506	2.490	0.006	2.502	2.494	0.002	2.500	2.494	0.000
3	MAX	3.120	3.000	0.180	3.045	3.000	0.070	3.020	3.000	0.030	3.012	3.000	0.018	3.010	3.000	0.016
	MIN	3.060	2.940	0.060	3.020	2.975	0.020	3.006	2.990	0.006	3.002	2.994	0.002	3.000	2.994	0.000
4	MAX	4.145	4.000	.0220	4.060	4.000	0.090	4.028	4.000	0.040	4.016	4.000	0.024	4.012	4.000	0.020
	MIN	4.070	3.925	0.070	4.030	3.970	0.030	4.010	3.988	0.010	4.004	3.992	0.004	4.000	3.992	0.000
5	MAX	5.145	5.000	0.220	5.060	5.000	0.090	5.028	5.000	0.040	5.016	5.000	0.024	5.012	5.000	0.020
	MIN	5.070	4.925	0.070	5.030	4.970	0.030	5.010	4.988	0.010	5.004	4.992	0.004	5.000	4.992	0.000
6	MAX	6.145	6.000	0.220	6.060	6.000	0.090	6.028	6.000	0.040	6.016	6.000	0.024	6.012	6.000	0.020
	MIN	6.070	5.925	0.070	6.030	5.970	0.030	6.010	5.988	0.010	6.004	5.992	0.004	6.000	5.992	0.000
8	MAX	8.170	8.000	0.260	8.076	8.000	0.112	8.035	8.000	0.050	8.020	8.000	0.029	8.015	8.000	0.024
	MIN	8.080	7.910	0.080	8.040	7.964	0.040	8.013	7.985	0.013	8.005	7.991	0.005	8.000	7.991	0.000
10	MAX	10.170	10.000	0.260	10.076	10.000	0.112	10.035	10.000	0.050	10.020	10.000	0.029	10.015	10.000	0.024
	MIN	10.080	9.910	0.080	10.040	9.964	0.040	10.013	9.985	0.013	10.005	9.991	0.005	10.000	9.991	0.000
12	MAX	12.205	12.000	0.315	12.093	12.000	0.136	12.043	12.000	0.061	12.024	12.000	0.035	12.018	12.000	0.029
	MIN	12.095	11.890	0.095	12.050	11.957	0.050	12.016	11.982	0.016	12.006	11.989	0.006	12.000	11.989	0.000
16	MAX	16.205	16.000	0.315	16.093	16.000	0.136	16.043	16.000	0.061	16.024	16.000	0.035	16.018	16.000	0.029
	MIN	16.095	15.890	0.095	16.050	15.957	0.050	16.016	15.982	0.016	16.006	15.989	0.006	16.000	15.989	0.000
20	MAX	20.240	20.000	0.370	20.117	20.000	0.169	20.053	20.000	0.074	20.028	20.000	0.041	20.021	20.000	0.034
	MIN	20.110	19.870	0.110	20.065	19.948	0.065	20.020	19.979	0.020	20.007	19.987	0.007	20.000	19.987	0.000
25	MAX	25.240	25.000	0.370	25.117	25.000	0.169	25.053	25.000	0.074	25.028	25.000	0.041	25.021	25.000	0.034
	MIN	25.110	24.870	0.110	25.065	24.948	0.065	25.020	24.979	0.020	25.007	24.987	0.007	25.000	24.987	0.000
30	MAX	30.240	30.000	0.370	30.117	30.000	0.169	30.053	30.000	0.074	30.028	30.000	0.041	30.021	30.000	0.034
	MIN	30.110	29.870	0.110	30.065	29.948	0.065	30.020	29.979	0.020	30.007	29.987	0.007	30.000	29.987	0.000
40	MAX	40.280	40.000	0.440	40.142	40.000	0.204	40.064	40.000	0.089	40.034	40.000	0.050	40.025	40.000	0.041
	MIN	40.120	39.840	0.120	40.080	39.938	0.080	40.025	39.975	0.025	40.009	39.984	0.009	40.000	39.984	0.000
50	MAX	50.290	50.000	0.450	50.142	50.000	0.204	50.064	50.000	0.089	50.034	50.000	0.050	50.025	50.000	0.041
	MIN	50.130	49.840	0.130	50.080	49.938	0.080	50.025	49.975	0.025	50.009	49.984	0.009	50.000	49.984	0.000
60	MAX	60.330	60.000	0.520	60.174	60.000	0.248	60.076	60.000	0.106	60.040	60.000	0.059	60.030	60.000	0.049
	MIN	60.140	59.810	0.140	60.100	59.926	0.100	60.030	59.970	0.030	60.010	59.981	0.010	60.000	59.981	0.000
80	MAX	80.340	80.000	0.530	80.174	80.000	0.248	80.076	80.000	0.106	80.040	80.000	0.059	80.030	80.000	0.049
	MIN	80.150	79.810	0.150	80.100	79.926	0.100	80.030	79.970	0.030	80.010	79.981	0.010	80.000	79.981	0.000
100	MAX	100.390	100.000	0.610	100.207	100.000	0.294	100.090	100.000	0.125	100.047	100.000	0.069	100.035	100.000	0.057
	MIN	100.170	99.780	0.170	100.120	99.913	0.120	100.036	99.965	0.036	100.012	99.978	0.012	100.000	99.978	0.000
120	MAX	120.400	120.000	0.620	120.207	120.000	0.294	120.090	120.000	0.125	120.047	120.000	0.069	120.035	120.000	0.057
	MIN	120.180	119.780	0.180	120.120	119.913	0.120	120.036	119.965	0.036	120.012	119.978	0.012	120.000	119.978	0.000
160	MAX	160.460	160.000	0.710	160.245	160.000	0.345	160.106	160.000	0.146	160.054	160.000	0.079	160.040	160.000	0.065
	MIN	160.210	159.750	0.210	160.145	159.900	0.145	160.043	159.960	0.043	160.014	159.975	0.014	160.000	159.975	0.000

Table 52 **Preferred shaft basis fits.** (Dimensions in millimeters.)

Example ∅ 16N7/h6

Hole Size ∅ $\frac{15.995}{15.977}$

Shaft Size ∅ $\frac{16.000}{15.989}$

Max. Clearance 0.006

Max. Interference 0.023

Example ∅ 45U7/h6
(Use values midway between 40 and 50.)

Hole Size ∅ $\frac{44.944}{44.919}$

Shaft Size ∅ $\frac{45.000}{44.984}$

Min. Interference 0.040

Max. Interference 0.081

PREFERRED SHAFT BASIS TRANSITION AND INTERFERENCE FITS																	
BASIC SIZE		LOCATIONAL TRANSN.			LOCATIONAL TRANSN.			LOCATIONAL INTERF.			MEDIUM DRIVE			FORCE		
		Hole K7	Shaft h6	Fit	Hole N7	Shaft h6	Fit	Hole P7	Shaft h6	Fit	Hole S7	Shaft h6	Fit	Hole U7	Shaft h6	Fit
1	MAX	1.000	1.000	0.006	0.996	1.000	0.002	0.994	1.000	0.000	0.986	1.000	−0.008	0.982	1.000	−0.012
	MIN	0.990	0.994	−0.010	0.986	0.994	−0.014	0.984	0.994	−0.016	0.976	0.994	−0.024	0.972	0.994	−0.028
1.2	MAX	1.200	1.200	0.006	1.196	1.200	0.002	1.184	1.200	0.000	1.186	1.200	−0.008	1.182	1.200	−0.012
	MIN	1.190	1.194	−0.010	1.186	1.194	−0.014	1.184	1.194	−0.016	1.176	1.194	−0.024	1.172	1.194	−0.028
1.6	MAX	1.600	1.600	0.006	1.596	1.600	0.002	1.594	1.600	0.000	1.586	1.600	−0.008	1.582	1.600	−0.012
	MIN	1.590	1.594	−0.010	1.586	1.594	−0.014	1.584	1.594	−0.016	1.576	1.594	−0.024	1.572	1.594	−0.028
2	MAX	2.000	2.000	0.006	1.996	2.000	0.002	1.994	2.000	0.000	1.986	2.000	−0.008	1.982	2.000	−0.012
	MIN	1.990	1.994	−0.010	1.986	1.994	−0.014	1.984	1.994	−0.016	1.976	1.994	−0.024	1.972	1.994	−0.028
2.5	MAX	2.500	2.500	0.006	2.496	2.500	0.002	2.494	2.500	0.000	2.486	2.500	−0.008	2.482	2.500	−0.012
	MIN	2.490	2.494	−0.010	2.486	2.494	−0.014	2.484	2.494	−0.016	2.476	2.494	−0.024	2.472	2.494	−0.028
3	MAX	3.000	3.000	0.006	2.996	3.000	0.002	2.994	3.000	0.000	2.986	3.000	−0.008	2.982	3.000	−0.012
	MIN	2.990	2.994	−0.010	2.986	2.994	−0.014	2.984	2.994	−0.016	2.976	2.994	−0.024	2.972	2.994	−0.028
4	MAX	4.003	4.000	0.011	3.996	4.000	0.004	3.992	4.000	0.000	3.985	4.000	−0.007	3.981	4.000	−0.011
	MIN	3.991	3.992	−0.009	3.984	3.992	−0.016	3.980	3.992	−0.020	3.973	3.992	−0.027	3.969	3.992	−0.031
5	MAX	5.003	5.000	0.011	4.996	5.000	0.004	4.992	5.000	0.000	4.985	5.000	−0.007	4.981	5.000	−0.011
	MIN	4.991	4.992	−0.009	4.984	4.992	−0.016	4.980	4.992	−0.020	4.973	4.992	−0.027	4.969	4.992	−0.031
6	MAX	6.003	6.000	0.011	5.996	6.000	0.004	5.992	6.000	0.000	5.985	6.000	−0.007	5.981	6.000	−0.011
	MIN	5.991	5.992	−0.009	5.984	5.992	−0.016	5.980	5.992	−0.020	5.973	5.992	−0.027	5.969	5.992	−0.031
8	MAX	8.005	8.000	0.014	7.996	8.000	0.005	7.991	8.000	0.000	7.983	8.000	−0.008	7.978	8.000	−0.013
	MIN	7.990	7.991	−0.010	7.981	7.991	−0.019	7.976	7.991	−0.024	7.968	7.991	−0.032	7.963	7.991	−0.037
10	MAX	10.005	10.000	0.014	9.996	10.000	0.005	9.991	10.000	0.000	9.983	10.000	−0.008	9.978	10.000	−0.013
	MIN	9.990	9.991	−0.010	9.981	9.991	−0.019	9.976	9.991	−0.024	9.968	9.991	−0.032	9.963	9.991	−0.037
12	MAX	12.006	12.000	0.017	11.995	12.000	0.006	11.989	12.000	0.000	11.979	12.000	−0.010	11.974	12.000	−0.015
	MIN	11.988	11.989	−0.012	11.977	11.989	−0.023	11.971	11.989	−0.029	11.961	11.989	−0.039	11.956	11.989	−0.044
16	MAX	16.006	16.000	0.017	15.995	16.000	0.006	15.989	16.000	0.000	15.979	16.000	−0.010	15.974	16.000	−0.015
	MIN	15.988	15.989	−0.012	15.977	15.989	−0.023	15.971	15.989	−0.029	15.961	15.989	−0.039	15.956	15.989	−0.044
20	MAX	20.006	20.000	0.019	19.993	20.000	0.006	19.986	20.000	−0.001	19.973	20.000	−0.014	19.967	20.000	−0.020
	MIN	19.985	19.987	−0.015	19.972	19.987	−0.028	19.965	19.987	−0.035	19.952	19.987	−0.048	19.946	19.987	−0.054
25	MAX	25.006	25.000	0.019	24.993	25.000	0.006	24.986	25.000	−0.001	24.973	25.000	−0.014	24.960	25.000	−0.027
	MIN	24.985	24.987	−0.015	24.972	24.987	−0.028	24.965	24.987	−0.035	24.952	24.987	−0.048	24.939	24.987	−0.061
30	MAX	30.006	30.000	0.019	29.993	30.000	0.006	29.986	30.000	−0.001	29.973	30.000	−0.014	29.960	30.000	−0.027
	MIN	29.985	29.987	−0.015	29.972	29.987	−0.028	29.965	29.987	−0.035	29.952	29.987	−0.048	29.939	29.987	−0.061
40	MAX	40.007	40.000	0.023	39.992	40.000	0.008	39.983	40.000	−0.001	39.966	40.000	−0.018	39.949	40.000	−0.035
	MIN	39.982	39.984	−0.018	39.967	39.984	−0.033	39.958	39.984	−0.042	39.941	39.984	−0.059	39.924	39.984	−0.076
50	MAX	50.007	50.000	0.023	49.992	50.000	0.008	49.983	50.000	−0.001	49.966	50.000	−0.018	49.939	50.000	−0.045
	MIN	49.982	49.984	−0.018	49.967	49.984	−0.033	49.958	49.984	−0.042	49.941	49.984	−0.059	49.914	49.984	−0.086
60	MAX	60.009	60.000	0.028	59.991	60.000	0.010	59.979	60.000	−0.002	59.958	60.000	−0.023	59.924	60.000	−0.057
	MIN	59.979	59.981	−0.021	59.961	59.981	−0.039	59.949	59.981	−0.051	59.928	59.981	−0.072	59.894	59.981	−0.106
80	MAX	80.009	80.000	0.028	79.991	80.000	0.010	79.979	80.000	−0.002	79.952	80.000	−0.029	79.909	80.000	−0.072
	MIN	79.979	79.981	−0.021	79.961	79.981	−0.039	79.949	79.981	−0.051	79.922	79.981	−0.078	79.879	79.981	−0.121
100	MAX	100.010	100.000	0.032	99.990	100.000	0.012	99.976	100.000	−0.002	99.942	100.000	−0.036	99.889	100.000	−0.089
	MIN	99.975	99.978	−0.025	99.955	99.978	−0.045	99.941	99.978	−0.059	99.907	99.978	−0.093	99.854	99.978	−0.146
120	MAX	120.010	120.000	0.032	119.990	120.000	0.012	119.976	120.000	−0.002	119.934	120.000	−0.044	119.869	120.000	−0.109
	MIN	119.975	119.978	−0.025	119.955	119.978	−0.045	119.941	119.978	−0.059	119.899	119.978	−0.101	119.834	119.978	−0.166
160	MAX	160.012	160.000	0.037	159.988	160.000	0.013	159.972	160.000	−0.003	159.915	160.000	−0.060	159.825	160.000	−0.150
	MIN	159.972	159.975	−0.028	159.948	159.975	−0.052	159.932	159.975	−0.068	159.875	159.975	−0.125	159.785	159.975	−0.215

Table 52 (cont'd.)   Preferred shaft basis fits.

NORTH AMERICAN GAGES												EUROPEAN GAGES								
Ferrous metals, such as galvanized steel, tin plate			Nonferrous metals, such as copper, brass, aluminum						Steel and iron wire and bare copper piano wire			Galvanized steel, tin plate, copper, strip steel and steel, copper and aluminum tubes						Nonferrous		
U.S. Standard (USS)			U.S. Standard (Revised) Formerly Manufactures Standard			American Standard or Brown and Sharpe (B & S)			United States Steel Wire Gage			Birmingham (BWG)			New Birmingham (BG)			Imperial Wire Gage Imperial Standard (SWG)		
Gage	in.	mm	Gage	in.	mm	Gage	in.	mm	Gage	in.	mm	Gage	in.	mm	Gage	in.	mm	Gage	in.	mm
			3	.240	6.01	3	.229	5.83												
4	.234	5.95	4	.224	5.70	4	.204	5.19	4	.225	5.72	4	.238	6.05	4	.250	6.35	4	.232	5.89
5	.219	5.56	5	.209	5.31	5	.182	4.62	5	.207	5.26	5	.220	5.59	5	.223	5.65	5	.212	5.39
6	.203	5.16	6	.194	4.94	6	.162	4.12	6	.192	4.88	6	.203	5.16	6	.198	5.03	6	.192	4.88
7	.188	4.76	7	.179	4.55	7	.144	3.67	7	.177	4.50	7	.180	4.57	7	.176	4.48	7	.176	4.47
8	.172	4.37	8	.164	4.18	8	.129	3.26	8	.162	4.11	8	.165	4.19	8	.157	3.99	8	.160	4.06
9	.156	3.97	9	.149	3.80	9	.114	2.91	9	.148	3.77	9	.148	3.76	9	.140	3.55	9	.144	3.66
10	.141	3.57	10	.135	3.42	10	.102	2.59	10	.135	3.43	10	.134	3.40	10	.125	3.18	10	.128	3.25
11	.125	3.18	11	.120	3.04	11	.091	2.30	11	.121	3.06	11	.120	3.05	11	.111	2.83	11	.116	2.95
12	.109	2.78	12	.105	2.66	12	.081	2.05	12	.106	2.68	12	.109	2.77	12	.099	2.52	12	.104	2.64
13	.094	2.38	13	.090	2.78	13	.072	1.83	13	.092	2.32	13	.095	2.41	13	.088	2.24	13	.092	2.34
14	.078	1.98	14	.075	1.90	14	.064	1.63	14	.080	2.03	14	.083	2.11	14	.079	1.99	14	.080	2.03
15	.070	1.79	15	.067	1.71	15	.057	1.45	15	.072	1.83	15	.072	1.83	15	.070	1.78	15	.072	1.83
16	.063	1.59	16	.060	1.52	16	.051	1.29	16	.063	1.63	16	.065	1.65	16	.063	1.59	16	.064	1.63
17	.056	1.43	17	.054	1.37	17	.045	1.15	17	.054	1.37	17	.058	1.47	17	.056	1.41	17	.056	1.42
18	.050	1.27	18	.048	1.21	18	.040	1.02	18	.048	1.21	18	.049	1.25	18	.050	2.58	18	.048	1.22
19	.044	1.11	19	.042	1.06	19	.036	0.91	19	.041	1.04	19	.042	1.07	19	.044	1.19	19	.040	1.02
20	.038	0.95	20	.036	0.91	20	.032	0.81	20	.035	0.88	20	.035	0.89	20	.039	1.00	20	.036	0.91
21	.034	0.87	21	.033	0.84	21	.029	0.72	21	.032	0.81	21	.032	0.81	21	.035	0.89	21	.032	0.81
22	.031	0.79	22	.030	0.76	22	.025	0.65	22	.029	0.73	22	.028	0.71	22	.031	0.79	22	.028	0.71
23	.028	0.71	23	.027	0.68	23	.023	0.57	23	.026	0.66	23	.025	0.64	23	.028	0.71	23	.024	0.61
24	.025	0.64	24	.024	0.61	24	.020	0.51	24	.023	0.58	24	.022	0.56	24	.025	0.63	24	.022	0.56
25	.022	0.56	25	.021	0.53	25	.018	0.46	25	.020	0.52	25	.020	0.51	25	.022	0.56	25	.020	0.51
26	.019	0.48	26	.018	0.46	26	.016	0.40	26	.018	0.46	26	.018	0.46	26	.020	0.50	26	.018	0.46
27	.017	0.44	27	.016	0.42	27	.014	0.36	27	.017	0.44	27	.016	0.41	27	.017	0.44	27	.016	0.42
28	.016	0.40	28	.015	0.38	28	.013	0.32	28	.016	0.41	28	.014	0.36	28	.016	0.40	28	.015	0.38
29	.014	0.36	29	.014	0.34	29	.011	0.29	29	.015	0.38	29	.013	0.33	29	.014	0.35	29	.014	0.35
30	.013	0.32	30	.012	0.31	30	.010	0.25	30	.014	0.36	30	.012	0.31	30	.012	0.31	30	.012	0.32
31	.011	0.28	31	.011	0.27	31	.009	0.23	31	.013	0.34	31	.010	0.25	31	.011	0.28			
32	.010	0.26	32	.010	0.25	32	.008	0.20	32	.013	0.33	32	.009	0.23				32	.011	0.27
33	.009	0.24	33	.009	0.23	33	.007	0.18	33	.012	0.30	33	.008	0.20	33	.009	0.22	33	.010	0.25
34	.009	0.22	34	.008	0.21	34	.006	0.16	34	.010	0.26	34	.007	0.18	34	.008	0.20	34	.009	0.23
									35	.010	0.24	35	.005	0.13	35	.007	0.18	35	.008	0.21
36	.007	0.18	36	.007	0.17	36	.005	0.13	36	.009	0.23	36	.004	0.10	36	.006	0.16			
									37	.008	0.22							37	.007	0.17
38	.006	0.16	38	.006	0.15	38	.004	0.10	38	.008	0.20				38	.005	0.12	38	.006	0.15
									39	.008	0.19									
									40	.007	0.18				40	.004	0.10	40	.005	0.12
									41	.007	0.17							42	.004	0.10

Note: Metric standards governing gage sizes were not available at the time of publication. The sizes given in the above chart are "soft conversion" from current inch standards and are not meant to be representative of the precise metric gage sizes which may be available in the future. Conversions are given only to allow the student to compare gage sizes readily with the metric drill sizes.

Table 53 **Wire and sheet-metal gages and thicknesses.**

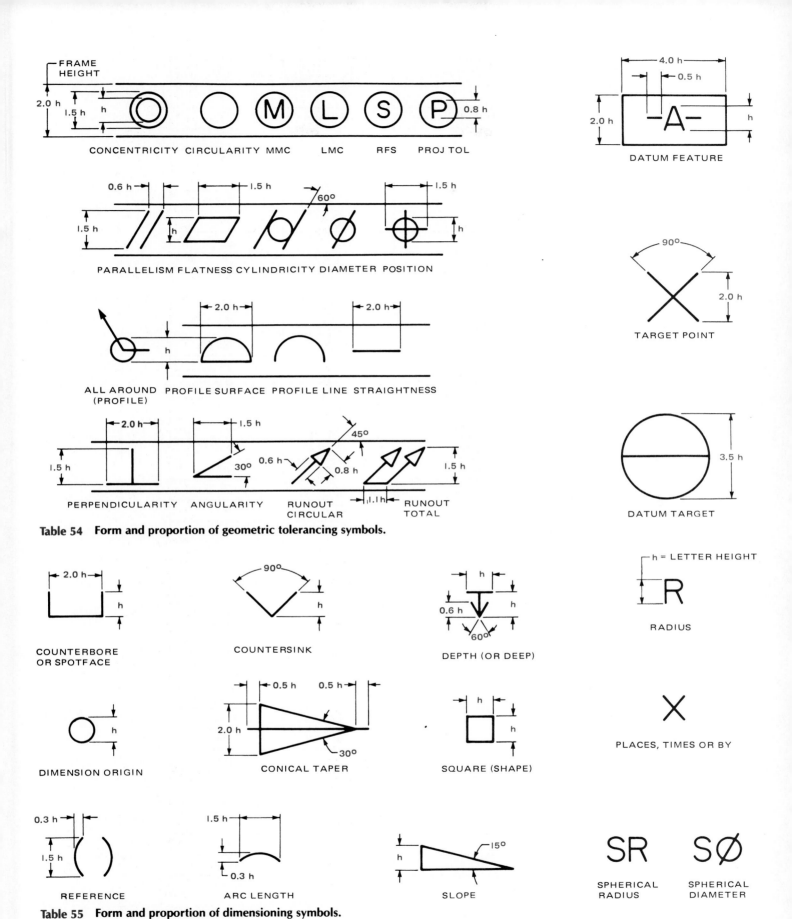

Table 54 Form and proportion of geometric tolerancing symbols.

CONCENTRICITY CIRCULARITY MMC LMC RFS PROJ TOL

PARALLELISM FLATNESS CYLINDRICITY DIAMETER POSITION

ALL AROUND PROFILE SURFACE PROFILE LINE STRAIGHTNESS
(PROFILE)

PERPENDICULARITY ANGULARITY RUNOUT RUNOUT
CIRCULAR TOTAL

DATUM FEATURE

TARGET POINT

DATUM TARGET

Table 55 Form and proportion of dimensioning symbols.

COUNTERBORE
OR SPOTFACE

COUNTERSINK

DEPTH (OR DEEP)

h = LETTER HEIGHT

RADIUS

DIMENSION ORIGIN

CONICAL TAPER

SQUARE (SHAPE)

PLACES, TIMES OR BY

REFERENCE

ARC LENGTH

SLOPE

SR
SPHERICAL
RADIUS

SØ
SPHERICAL
DIAMETER

SYMBOL FOR:	ANSI Y14.5	ISO
STRAIGHTNESS	—	—
FLATNESS	▱	▱
CIRCULARITY	○	○
CYLINDRICITY	⌭	⌭
PROFILE OF A LINE	⌒	⌒
PROFILE OF A SURFACE	⌓	⌓
ALL AROUND—PROFILE	⟲	NONE
ANGULARITY	∠	∠
PERPENDICULARITY	⊥	⊥
PARALLELISM	//	//
POSITION	⊕	⊕
CONCENTRICITY/COAXIALITY	◎	◎
SYMMETRY	NONE	═
CIRCULAR RUNOUT	*↗	↗
TOTAL RUNOUT	*↗↗	↗↗
AT MAXIMUM MATERIAL CONDITION	Ⓜ	Ⓜ
AT LEAST MATERIAL CONDITION	Ⓛ	NONE
REGARDLESS OF FEATURE SIZE	Ⓢ	NONE
PROJECTED TOLERANCE ZONE	Ⓟ	Ⓟ
DIAMETER	⌀	⌀
BASIC DIMENSION	50	50
REFERENCE DIMENSION	(50)	(50)
DATUM FEATURE	-A-	OR ▯A▮
DATUM TARGET	⌀6/A1	⌀6/A1
TARGET POINT	✕	✕
DIMENSION ORIGIN	⊕▶	NONE
FEATURE CONTROL FRAME	⊕ ⌀0.5Ⓜ A B C	⊕ ⌀0.5Ⓜ A B C
CONICAL TAPER	▷	▷
SLOPE	◁	◁
COUNTERBORE/SPOTFACE	⌴	NONE
COUNTERSINK	⌵	NONE
DEPTH/DEEP	↧	NONE
SQUARE (SHAPE)	□	□
DIMENSION NOT TO SCALE	15	15
NUMBER OF TIMES/PLACES	8X	8X
ARC LENGTH	⌒105	NONE
RADIUS	R	R
SPHERICAL RADIUS	SR	NONE
SPHERICAL DIAMETER	S⌀	NONE

*MAY BE FILLED IN

Table 56 Comparison of ANSI and ISO symbols.

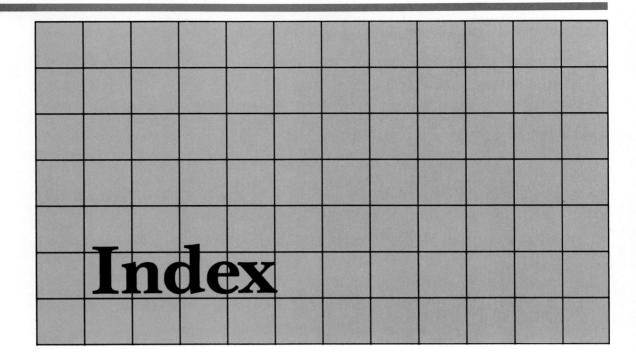

Index